Doing Physics with Scientific Notebook

Doing Physics with Scientific Notebook

A Problem-Solving Approach

Joseph Gallant

University of Massachusetts

Amherst, MA, USA

A John Wiley & Sons, Ltd., Publication

This edition first published 2012

Registered office
John Wiley & Sons Ltd, The Atrium, Southern Gate, Chichester, West Sussex, PO19 8SQ, UK

For details of our global editorial offices, for customer services and for information about how to apply for permission to reuse the copyright material in this book please see our website at www.wiley.com.

Library of Congress Cataloging-in-Publication Data

Gallant, Joseph.
Doing physics with Scientific Notebook : a problem solving approach / Joseph Gallant.
p. cm.
Includes bibliographical references and index.
ISBN 978-0-470-66597-8 (hardback) -- ISBN 978-0-470-66598-5 (paper)

1. Physics--Data processing. 2. Mathematical physics--Data processing. 3. Scientific Notebook.
4. Problem solving--Programmed instruction. I. Title.
QC52.G35 2012
530.07--dc23

2012001475

A catalogue record for this book is available from the British Library.

Print ISBN: 978-0470-665978 (Cloth) 978-0470-665985 (Paper)

Set in 10 point Times New Roman by the author with Scientific WorkPlace.

Printed and bound in Malaysia by Vivar Printing Sdn Bhd

1 2012

“It’s *all* physics!”

Donny MacNamara

Contents

Preface

Welcome to the wonderful world of physics! The study of physics is useful and important because physics is the most fundamental science. It is the framework upon which other sciences are built. As my old friend the high school physics teacher used to say "It's ***all*** physics!" Once you understand the basic principles of physics, you'll find it easier to understand other sciences. The thinking and problem-solving skills you develop here will help you in any endeavor.

This goal of this book is to teach undergraduate physics students how to use *Scientific Notebook* to solve physics problems. I've tried to choose topics that have educational value, fit within a typical physics curriculum, and show the benefits of *SNB*. Many problems come from my class notes, some from my research, while others I included because they're interesting. Some are problems I wanted to do in class but couldn't because the math was too difficult or time consuming.

Solving real-world problems usually requires more complicated mathematics than the idealized problems presented in introductory textbooks. Those "easy" problems are a good place to start. Once you can solve and understand them, we'll add some complications and let *SNB* do the math. This lets us solve interesting, more realistic problems, and this book will be a useful reference for your entire undergraduate career.

Many of you are training for careers in fields which require technical or scientific calculations and written reports. *SNB*, which you can think of as a combination math gizmo and word processor, is ideally suited for these tasks. You'll find this inexpensive software helpful and easy to use, in the classroom and beyond. However *SNB* can only assist you in solving physics problems, it cannot solve them for you.

Physics is not math. In mathematics you learn to solve equations. In physics you learn to apply equations to describe and explain the physical universe. Physicists construct equations involving physical quantities that are based on fundamental principals to describe and explain nature and we use mathematics to solve them. The bad news is that *SNB* will not solve any physics problems for anyone. The good news is that is will help you do physics by helping you with the math stuff. You do the physics, *SNB* does the math. You create the equation, *SNB* solves it. *SNB* will make the graph, but you will have to interpret it.

Undergraduate physics is taught with varying degrees of mathematical rigor, from freshmen no-calculus science-major courses to upper-level lots-of-calculus courses for physics majors. Not all physics classes use calculus so many students learn physics without it. But omitting calculus from a book about *Doing Physics with SNB* is like warping with one nacelle tied behind your impulse drive. The calculus and no-calculus problems are usually in separate sections, with the more complicated sections at the ends of chapters.

In Chapter 1, I introduce you to many *SNB* features I have used to solve physics problems. There are many more features of *SNB*, and the best way to learn them is to explore and play with *SNB*. Excellent written documentation accompanies *SNB* and it has an extensive built-in help system, all of which was written with *SNB*. If you see it in the help, you can do it with *SNB*. You can even cut-and-paste from the help into your document. The Help + Search feature is a great place to start when you need help or information.

In the subsequent chapters, I follow the basic introductory physics curriculum, extending it to include interesting problems (some with calculus but all at the undergraduate level) that show the power and usefulness of *SNB* and let your skills grow. Each section has a brief introduction to the relevant physical concepts. Since this is not a comprehensive physics textbook, I emphasize problem solving over conceptual knowledge, although both are important.

At least one example appears in almost every section. Their purpose is to enhance your understanding of the relevant physics and to provide detailed instructions on using *SNB*. You can (and should) explore the topics further with the wide selection of problems at the end of each chapter.

The two Appendices at the end of the book contain special topics in classical and modern physics that are not typically part of the traditional undergraduate physics curriculum. These are topics I find interesting, important or curiosity-inducing. They also show just how powerful *SNB* is as a problem-solving tool.

This book uses the following notation and conventions:

- The abbreviation *SNB* refers to *Scientific Notebook*. Really.
- Words written like Evaluate and Plot 2D Rectangular refer to *SNB* menu commands, buttons, and options.
- Notation like Compute + Evaluate means click on the Compute menu item and select Evaluate.
- Notation like Help + Search, Tags means click on the Help menu item and select Search. When the input box appears, type Tags.
- Words written like TAB and CTRL refer to a key on your keyboard.
- Notation like CTRL + F means hold down the CTRL key, press the F key and release the two keys simultaneously.
- Important physics words like **force** appear in bold face the first time you meet them.
- Shaded gray boxes contain *SNB* output.

This is a box of output!

I made every calculation and graph in this book with *Scientific Notebook* 5.50 (Build 2960). To write this book, I used *Scientific WorkPlace* 5.50 (Build 2960), which has all of *SNB*'s computational capabilities plus the LaTeX typesetting system. To create the final PDF file, I used Pdf995 (version 10.2).

So we're all on the same page...

For the most part, I've left *SNB*'s default settings unchanged. There is one important exception. Under the Tools menu item, on the General page of Engine Setup, you will find the Solve Options. The default setting for Principal Value only and Ignore Special Cases is unchecked, which when solving equations can lead to output from *SNB* that looks like this:

$$ax^2 + bx + c = 0, \text{ Solution is:}$$
$$\begin{cases} \mathbb{C} & \text{if} \quad a = 0 \wedge b = 0 \wedge c = 0 \\ \emptyset & \text{if} \quad c \neq 0 \wedge a = 0 \wedge b = 0 \\ \left\{-\frac{1}{b}c\right\} & \text{if} \quad b \neq 0 \wedge a = 0 \\ \left\{-\frac{1}{2a}\left(b - \sqrt{-4ac + b^2}\right), -\frac{1}{2a}\left(b + \sqrt{-4ac + b^2}\right)\right\} & \text{if} \quad a \neq 0 \end{cases}$$

This answer is correct ($\mathbb{C}$ is the set all complex numbers, $\emptyset$ is the empty set, the $\wedge$ symbol means "and") and you could cut-and-paste the parts you want. But as is, it's cumbersome and contains too much information to be helpful to most students. To simplify the output, let's check both Principal Value only and Ignore Special Cases and solve the equation again.

$$ax^2 + bx + c = 0, \text{ Solution is: } -\frac{1}{2a}\left(b - \sqrt{b^2 - 4ac}\right)$$

As you can see, *SNB* returns only the first answer and considers none of the special cases. Checking Ignore Special Cases while leaving Principal Value only unchecked produces this output.

$$ax^2 + bx + c = 0, \text{ Solution is: } -\frac{1}{2a}\left(b - \sqrt{b^2 - 4ac}\right), -\frac{1}{2a}\left(b + \sqrt{b^2 - 4ac}\right)$$

Unless otherwise stated, the Solve Options I use in this book are Ignore Special Cases checked and Principal Value only unchecked.

There are a few not-so-important exceptions as well, which are more a matter of style and preference than substance. I made the following minor adjustments to the default settings.

- Under Tools + User Setup, choose the Math page, click the Radical button, and choose the Square Root button on the left.
- Under Tools + Computation Setup, choose the 2D Plots page. Click Rectangular and set the Default Plot Interval from zero to five.
- On the same page, click Polar and set the Default Plot Interval from zero to 6.2832 (about 2π).
- On the Plot Layout page, set the Screen Display and Plot Attributes to Plot Only and choose Displayed for the Placement option.

I keep *SNB*'s default setting of five digits in its output. Beyond this, I make no attempt to do significant figures, which I leave to the instructors. To make a global change for the number of digits in your answers, go to Tools + Computation Setup, choose the General page and change the Digits Shown in Results setting.

What is science?

Science is a systematic process to obtain and explain the facts to answer questions about the physical universe. Science is something you do. The fundamental precept of science is "look first, then decide". To be scientific, you can't make up your mind and then look for corroborating facts. You must carefully and rigorously gather the facts first, and then you analyze and interpret them. This prevents you from taking a few anecdotes and interpreting them as evidence. In science, "anecdotal evidence" is an oxymoron.

Experience suggests the universe operates under cause-and-effect laws. Contrary to the views espoused by many TV detectives, coincidences are real and they do happen. Take care not to mistake the occasional coincidence for a cause-and-effect relationship. You also must be careful not to assume something is inexplicable because you can't explain it. The difference between "I don't know" and "it's not knowable" is huge. The former is a wonderful beginning; the latter is a dead end. Science minimizes these mistakes.

The three parts of science are **experiment**, **theory**, and the **facts**.

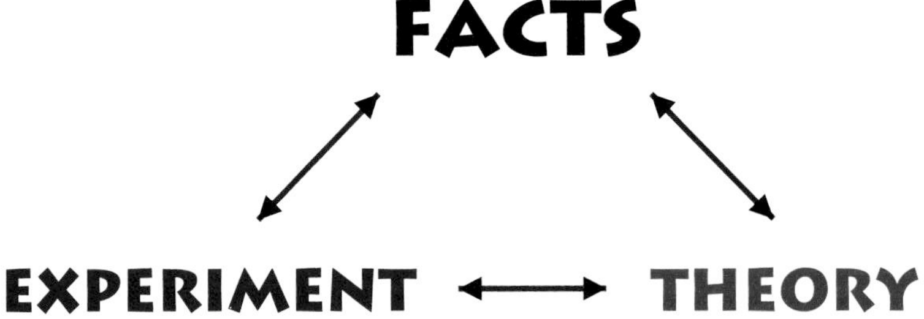

Experiments play two roles. They obtain the facts and they test the predictions of a theory. Experimenters make careful, controlled measurements to collect data. In science not all information rises to the level of data, much like not all information rises to the level of evidence in a courtroom. An experiment must be repeatable and verifiable. There is no "take my word for it" in science.

There is no truth in science either. Truth is subjective but the facts are objective. When you seek the truth, the response you get depends on who or how you ask. The facts do not depend on who or how you ask. Suppose two people decide to settle an argument with a race. When they return person A says "I won and he lost" while person B says "I finished just behind the person that won and she finished just ahead of the person that came in last". Who is telling the truth? They both are telling a version of the truth. The facts are person A ran the race in 10 seconds and person B ran the race in 11 seconds.

A scientific theory must do two things. It must explain the known facts and predict new facts that can be measured experimentally. This ensures that the theory is not simply "aimed" at only known facts. A theory is rigorous, detailed, and mathematical. A theoretical result is much more than a hypothetical idea. Don't say "I have a theory" when you only have an idea or a guess and don't say "theoretically" when you mean "hypothetically".

To the Student

Many of you are just beginning to learn physics. Perhaps you have heard that physics is a difficult course. Many students, even those who have had success in other subjects, are frightened by the prospect of taking physics for a grade. Physics is simple to *understand* but sometimes it's difficult to *learn.* Difficult, but not impossible. Ultimately you hold the key to your success and there are several ways you can help yourself.

- **Read your Textbook**
  Physics books are not novels and you might not understand everything you read the first time through. Use your text book (and this book) as a resource to look stuff up and to review. Some students prefer to read the text before the lecture, others after the lecture, some even do both. You must find the way that works best for you. Either way, you will find many answers and insights in your text.
- **Ask Questions**
  If there is something in the class or the text that is not clear, *please* ask! Do not be embarrassed. You're in a physics class and the room is full of people who don't understand. Yet.
- **Work Together**
  Students who study with other students often are more successful. One of the best ways to improve your own understanding of a concept is to explain it to someone else. But remember, the purpose of a study group is to exchange ideas not answers!
- **Seek Help**
  Some problems take more than 15 minutes to solve. Don't give up! Here's a rule of thumb: if you spend a good hour working on a problem, put it aside and visit your professor. Bring your work, as often an expert can see a small error that was preventing your success. Visit your professor, teacher, or graduate assistant with questions, comments, or just to talk physics.
- **Time Is Not Always On Your Side**
  Physics is best learned in small doses. You don't know in advance which problems will be the most time consuming, so budget your time wisely. Waiting until the last minute and trying to cram is not a good strategy. Avoid falling into the "due Monday means do Sunday" trap.
- **Pay More Than Attention**
  Physics is more a skill rather than a collection of facts. Solving physics problems is a skill, and like all skills it must be learned by doing. You cannot learn a skill by reading about it or by listening to your professor talk about it in a lecture. This is as true for riding a bicycle as it is for learning physics. How much and how well you learn physics depends mostly on the time and effort you put forth. Educators guide you, encourage you, offer you our insights. It's up to you to actually learn the physics.

In other words, as in most things in life worth doing, success in learning physics requires hard work and effort.

In most physics courses, you are required to solve word problems. There is no guaranteed recipe for successful problem solving. It is a skill that you must acquire through practice. But I can offer a few words of advice about problem solving.

- Read the problem carefully! The biggest problem with word problems is the words.
- Ask yourself two important questions: "what do I know?" and "what am I trying to find?".
- Draw a picture! Visualizing the problem will help you solve it.
- Do the math correctly. *SNB* will help you with this step.
- Check your units. Are the units right?
- Is the answer reasonable?

There is much more to do than just the math.

The problems at the end of the chapters are rarely "*SNB* problems" any more than problems in other physics books are calculator problems. After Chapter 1, there are very few *SNB* problems in this book. Most of the problems are physics problems. Remember that *SNB* is a tool and we want to ***do*** something with it.

To the Teacher

SNB has many features that teachers will find useful. You can use *SNB* to write homework assignments, exams, and solutions. You can use it to write your class notes and post them on the web so your students can access them with the Open Location command. *SNB* uses text files that are easily emailed so you and your students can exchange questions and answers.

You can create your own document types (ideal for lab reports or in-class worksheets) with the Export Document command. *SNB* exports the document as a shell file and treats it as a template. When your students click the New button, your preferred formats appears as one of their options.

We all try to be available and encourage our students to contact us. When they do, I usually use *SNB* to answer their questions. This is not the traditional approach, and the soulless minions of orthodoxy may not approve, but it seems to work.

When given a math or physics problem to solve with *SNB*, students often treat it as an *SNB* problem. Remind them *SNB* is a tool and the goal is to use that tool to do something.

Contact Information

If you have any questions, comments or suggestions about this book, please feel free to contact me directly at DPwSNB@gmail.com and I'll respond as quickly as I can.

To find the book's website, go to http://booksupport.wiley.com/ and search for this book by my name or the title. Once you're there, you'll find the *DPwSNB* e-book, the solutions manual, PowerPoint files containing all the figures in this book, and any other goodies we conjure up.

Acknowledgments

There are many people I'd like to thank for their help that made this book possible.

Barry MacKichan, President of *MacKichan Software, Inc.*, graciously responded to my initial overtures about publishing this book and turned the matter over to this staff. Patti Kearney put me in contact with the good people at Wiley. John MacKendrick and George Pearson of MacKichan Support thoroughly answered my many questions.

Many people at Wiley contributed much time and effort to this project. Christoph von Friedeburg, then Commissioning Editor for Physics, was an early and enthusiastic supporter. Jenny Cossham was the Publisher who successfully presented my proposal to the publications committee. Sarah Tilley was my Project Editor and primary contact for the last year of this project when most of the work happened. She skillfully managed every aspect of this project and did so with charm and wit. Zoë Mills, Assistant Editor and mother of the cutest pumpkin ever, oversaw the design of the book's attractive cover. Judith Egan-Shuttler, my copy editor with an extraordinary eye for detail, found many mistakes but was always gracious and gentle when pointing them out.

John Brehm was my former modern physics professor and a great teacher and author. He was also my undergraduate advisor and his simple "so where are you going to grad school?" question started this journey. His insights and suggestions improved this book immeasurably. John Dubach was my professor and the best dissertation advisor anyone could want. Barry Holstein, another former professor and quantum gator, generously gave me his time and expertise on relativity.

I'd like to thank Mike Crescimanno and John Fisher for many stimulating conversations on all thing physics, more than a few of which ended up in this book. I'm grateful to Sal Caronite for his friendship and sharing his unique perspective and keen observations on the human condition, including mine.

Over the years, many of my students were able helpers and guinea pigs. This includes my daughter Lisa, who was my student when this book was nothing more than a small handout. Our many interesting conversations and wonderful, delightful adventures are but small parts of her contribution to my life and this book.

Finally, I'd like to thank the rest of my family for encouraging and supporting me when I needed it, and for giving me the time and space to write. Words alone cannot express the gratitude and thanks I owe my mother. My wife Patti's love and support for me and this project have been unwavering. Such debts can never be paid.

Joseph Gallant .... *February 2012*

1 Introduction to SNB

The main activity of most physics classes is to teach students how to solve physics problems. Mathematics is a tool we use to solve those problems. Many of the difficulties students have in physics classes are rooted in the mathematics. They can't see the forest of physics for all the mathematical trees. *Scientific Notebook* (*SNB*) is a powerful yet easy-to-use computer algebra system that can help alleviate this problem. *SNB* is inexpensive and easy enough to be accessible to most undergraduates yet powerful enough to be useful in solving interesting physics problems.

The goal of this book is to teach students how to use *SNB* to solve physics problems. Once you have learned how (and it won't take all that long), you will use *SNB* as its name implies – as a *notebook* in which you set up a science or math problem, write and solve an equation, analyze and discuss the results. Of course a regular notebook will never help you do the math, but *SNB* will. Soon you will be able to think and write at the computer, in much the same way you use a paper and pencil now, with the power of a computer algebra system at your disposal.

Why SNB?

Scientific Notebook is powerful software that combines word processing and mathematics in standard notation with the power of symbolic computation. You enter the mathematical expressions in a form that is familiar to you and *SNB* evaluates it. This is the key to *SNB*. All the mathematics are in ***standard notation*** in a form that is ***familiar to you***. There is no arcane syntax to learn.

Consider a quick analysis of the function $y = x^2 e^{-3x} \sin 4x$. What is the area under the curve? Where is the function zero? What does the function look like? You may know how to find the answers, but you might have trouble doing the necessary mathematics. With *SNB*, one click gives the exact answer and a second click gives an approximate numerical answer.

$$\int_0^\infty x^2 e^{-3x} \sin 4x \, dx = \frac{88}{15\,625} = 5.632 \times 10^{-3}$$

With one click, *SNB* will find the first zero of the function.

$0 = x^2 e^{-3x} \sin 4x$, Solution is: 0

As you might have guessed, this function equals zero at $x = 0$.

Doing Physics with Scientific Notebook: A Problem-solving Approach, First Edition. Joseph Gallant.

You can see the other zeros with a plot of the function. It would be simple to graph this function by hand, but tedious and time consuming. To see a 2-dimensional plot of this function with *SNB*, we can again click a single button.

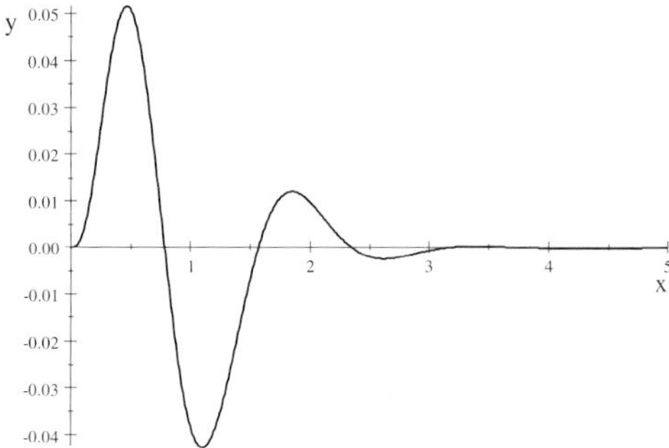

Figure 1.1 A plot of $x^2e^{-3x}\sin 4x$

Later in this chapter you'll learn how to find the other zeros.

Once we created the expressions, which was very easy to do, all it took was a few mouse clicks to answer our three questions. The entire process took about a minute. With *SNB*'s help, you will be able to spend more time thinking about physics and less time worrying about mathematics. However, keep in mind that *SNB* can only help you solve physics problems, it can not solve them for you.

This chapter presents a brief introduction to *SNB*, emphasizing features you will use in your physics class. It explains how to perform basic tasks such as entering and editing mathematics and text, solving equations and how to compute and plot mathematics. You can even use *SNB* to open and save documents available on the Internet. Keep in mind the main advantage of *SNB* over other systems. It is easy to learn and easy to use yet powerful enough to do physics. Before you start *Doing Physics with SNB*, you need to know how to use *SNB*.

The Basics

When you start *SNB*, you see a typical Windows interface containing menus, icons, and other graphics. This interface allows you to interact with the "brains" of *SNB*, the engine. The engine is the program which performs all the mathematical calculations. In version 5.5 of *SNB*, the engine is MuPAD (version 3.1). *SNB* translates your input into a form the engine can understand, sends it to the engine, translates the engine's output into a form you can understand, and shows it to you.

Since *SNB* uses a standard interface, all the editing techniques you use in other programs will work in *SNB*. If you are new to computing, all the editing techniques you learn here will be useful in other applications. The blinking vertical line on your screen is called the insertion point, and it marks the position where characters or symbols are entered when you type or click a symbol. You can change the position of the insertion point with the arrow keys, or by clicking a different screen position with your mouse. The position of the mouse is indicated by the mouse pointer, which takes the shape of an `I`-beam over text and an arrow over mathematics.

Some actions in *SNB* require you to select, or highlight, text or mathematics. When you make a selection with the mouse or the keyboard, the next action you take affects the selection. To select an individual word or mathematical object with the mouse, double-click the word or object. To make a large selection with the mouse you can either click-and-drag the pointer with the left mouse button down, or click the mouse at the start of the selection, press and hold SHIFT, move the pointer to where you want the selection to end, click the mouse and release SHIFT. For more information on selecting, look under Help + Search, Selecting Text and Mathematics.

You can access many of *SNB*'s features from various toolbars. You can display or hide any of the toolbars and you can return the toolbar display to its original setting. Also, you can dock the toolbars in the program window, let them float on the screen, or reshape them according to your preference. Use the following steps to display or hide toolbars.

1. Go to the View menu and choose Toolbars.
2. Check the box for each toolbar you want to display.
3. Choose Close. If you choose Reset, you will restore the default toolbar display.

The Standard Toolbar contains most of the commands you will need to manage files and to edit and manipulate text and mathematics in your *SNB* documents. Many of these are probably familiar to you. The Open (CTRL + O) command opens an existing file and the Save command saves the active file and keeps it open. You can Cut (CTRL + X), Copy (CTRL + C) and Paste (CTRL + V) text, mathematics, and graphics.

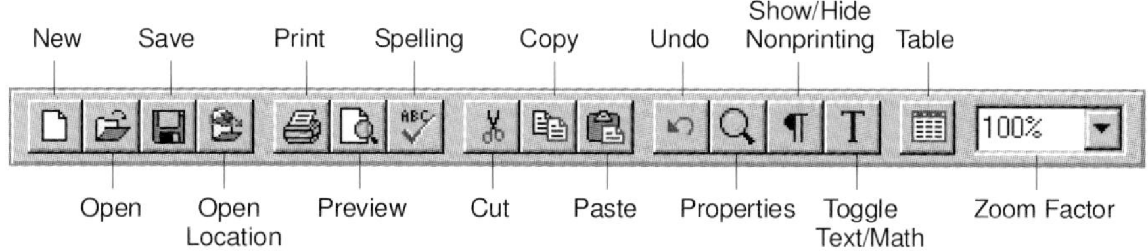

The *SNB* interface is not what-you-see-is-what-you-get, so use the Preview button to see what the printed document will look like before you Print (CTRL + P) it. The Zoom Factor only affects the on-screen appearance of your document and has no effect on the printed version.

As anyone who has ever graded papers will tell you, it is a good idea to check the Spelling in your document before printing. With the Spelling tool you can check the spelling in a selection, from the insertion point to the end of the document, or in the entire document. You can even check the spelling of a single word by selecting it and clicking the Spelling button. A spell check does not check mathematics or words embedded in mathematics.

The Standard Toolbar also includes some *SNB* commands, including the important Math/Text toggle button. With the New command you can create a new file by selecting the type of document from a list of shells provided with *SNB*. Each shell is a template for a different type of *SNB* document. You can create your own shells by using File + Export Document to place any *SNB* document as a shell file in one of the Shells folders. Once there, your new file will appear in the shell list displayed when you start a new document. If you have a required format for lab reports, you could create a shell file organized in that format. When you need to write a lab report, click New and choose that shell. You can even create new shell folders to organize your shells. For more information on creating shells, look in Help + Shells, Creating a Document Shell.

The Open Location command allows you to open an existing *SNB* file that is posted on the web as long as you know its URL. Look in the Preface to this book for any information on a website.

By changing the Properties of any text or mathematical object, you can alter the behavior of mathematical objects and the appearance of your document. Select the item you want to adjust and click the Properties button. A context-sensitive dialog box will appear that allows you to change the properties of the item. If you don't select anything, *SNB* chooses the item to the left of the insertion point. Any changes you make only affect that item.

The Compute Toolbar contains many commands you will use to carry out mathematical calculations. These are the commands you'll use most often to solve physics problems.

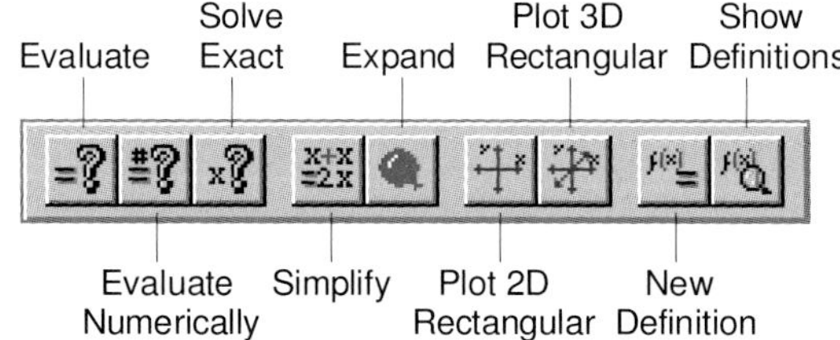

This chapter devotes significant time and space to these important commands and your success using *SNB* depends on you doing the same.

The Stop Toolbar contains a single button

that you can use to stop two operations, linking to the Internet and performing computations. You can also stop these operations by pressing CTRL + BREAK. The Stop operation is not available from a menu.

Before you carry out any calculations, you need to create mathematical expressions using the mathematical objects on the Math Templates and Math Objects toolbars.

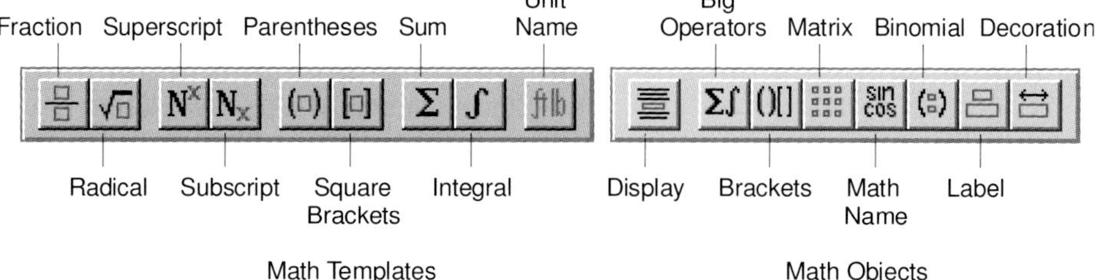

Notice the Table button is not here (it is on the Standard Toolbar). Both tables and matrices are two-dimensional arrays of boxes called cells. Each cell of a table can hold mathematics, text, or graphics. But a table is not a mathematical object, so you can't perform mathematical computations on a table as a whole as you can on a matrix. A good rule of thumb in *SNB*: matrices are for numbers and tables are for words.

The Symbol Cache contains 18 commonly used mathematical symbols

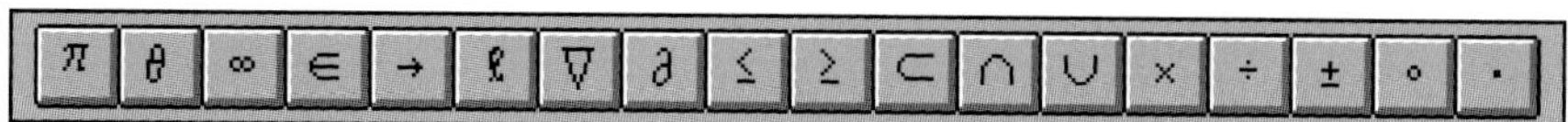

including two reserved symbols (π and ∞) and the times sign ($\times$) used in multiplication and scientific notation. You will also find these symbols and more in the Symbol Panels.

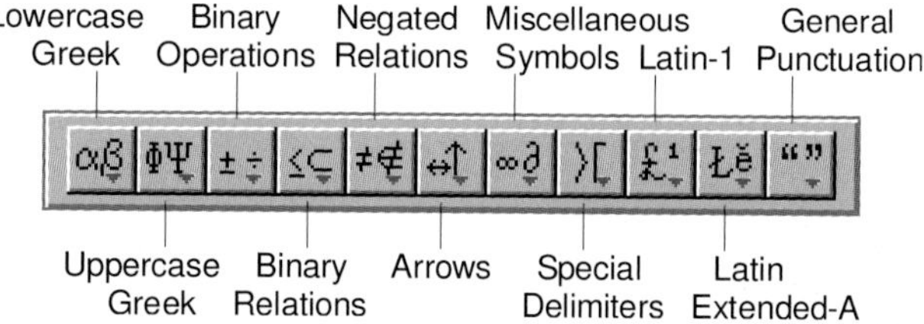

Each button opens a popup panel of symbols which you can customize to remain open all the time or dock in a different location. For a detailed look at the symbols on each panel, look under Help + Search, Symbol Panels.

The buttons on the Editing Toolbar allow you to alter the appearance of the text in your document. The first four buttons apply frequently used Text Tags: Normal, **Bold**, *Italics*, and ***Emphasized***. To change the appearance of your text, select the text and click one of these four buttons.

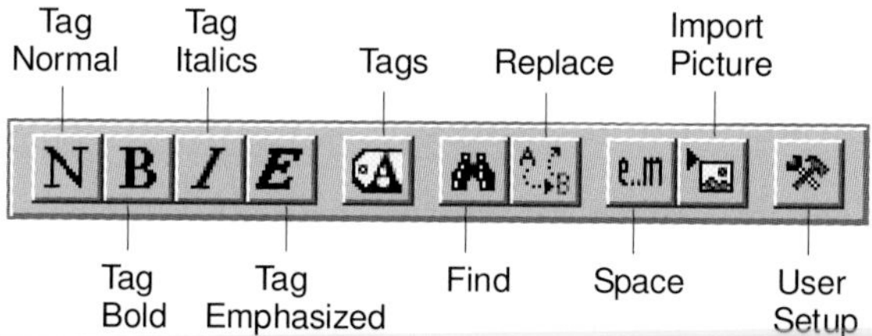

The Find (CTRL + Q) and Replace (CTRL + W) commands let you search for and replace text or mathematics in your document. You can search for all occurrences of any combination of mathematics and text, including those with a specific Tag. You can also access the Find and Replace commands from the Edit menu.

With User Setup you can customize many of *SNB*'s default values. From the User Setup dialog box, you can choose which shell *SNB* uses as the start-up document, set the properties of mathematical objects and operations, the properties of new graphics, tables and matrices, and many other general program properties. Be *very* careful when you alter any settings with User Setup. The changes you make with it are global and affect every document you open. Use Compute + Settings to make local changes that affect the current document only.

The Tag Toolbar consists of three popup lists that contain all the item tags, section and body tags, and text tags available for the current shell. With these tags, you can organize your document and alter its appearance.

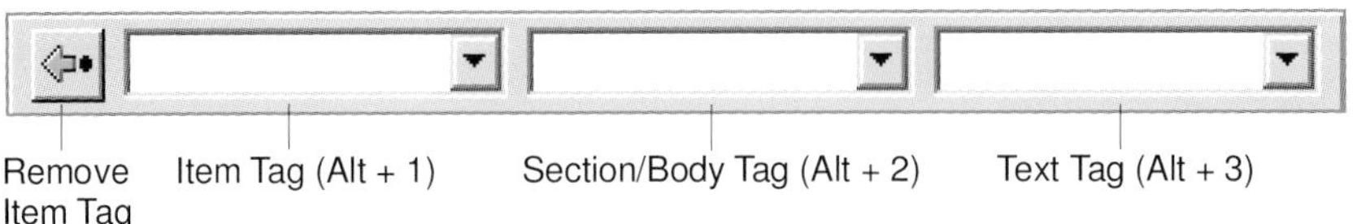

As we saw earlier, the Text Tags alter the appearance of text. Besides the four on the Editing Toolbar, you can find more Text Tags in the right-hand popup list of the Tag Toolbar. When you click the Text Tag popup box (or press ALT + 3), a list of all available text tags pop up.

The middle popup list contains Section/Body Tags. You can use the various headings, centered text, and quotations to organize your document. You can apply Item Tags to create various kinds of lists. With the Numbered List Item tag you can create a list of items that are automatically numbered sequentially. With the Bullet List Item tag you can create a list of items that are preceded by a bullet. All the numbered and bulleted lists in this book were created with Item Tags. The Description List Item tag allows you to create a customized text label for each item on your list.

The Fragment Toolbar offers an easy way to save and access frequently used expressions or equations. A fragment contains information (text, mathematics or both) that has been saved in a separate file for later recall. You can import a previously saved fragment into the current document, or you can save information in the current document as a new fragment. A fragment saved in one document is available to all documents. The Fragment Toolbar consists of the Save Fragment button and the fragment popup box.

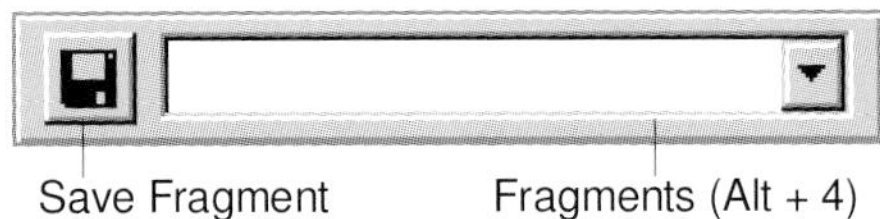

When you click the fragment popup box (or press ALT + 4), a list of fragments that you can insert in your document pops up. *SNB* comes with many predefined fragments, including an extensive list of physical constants.

It is very easy to import a fragment into your document.

1. Place the insertion point where you want the fragment to appear.

2. Click the fragment popup box (or press ALT + 4).

3. Click on the fragment you want to import.

You can also use File + Import Fragment... menu item. Just select the fragment you want from the Import Fragment dialog box and choose OK. When you import a fragment, its contents are pasted into your document at the insertion point.

It is also easy to create your own fragments.

1. Select any text or mathematics from any *SNB* document.
2. Click the Save Fragment button on the Fragment Toolbar. The Save Fragment dialog box will open.
3. Type a file name for your fragment.
4. Click Save.

Your fragment will immediately be added to the popup list of available fragments for your future reference. If you want to save your fragment with the other constants, open the Constants subdirectory in the Save Fragment dialog box before you do step 3.

Figure 1.2 shows a typical screen for *SNB*. The Symbol Cache is docked on the left, the Editing Toolbar is docked on the right, the Tag and Fragment toolbars are docked on the bottom, and some excellent reading appears to be on screen.

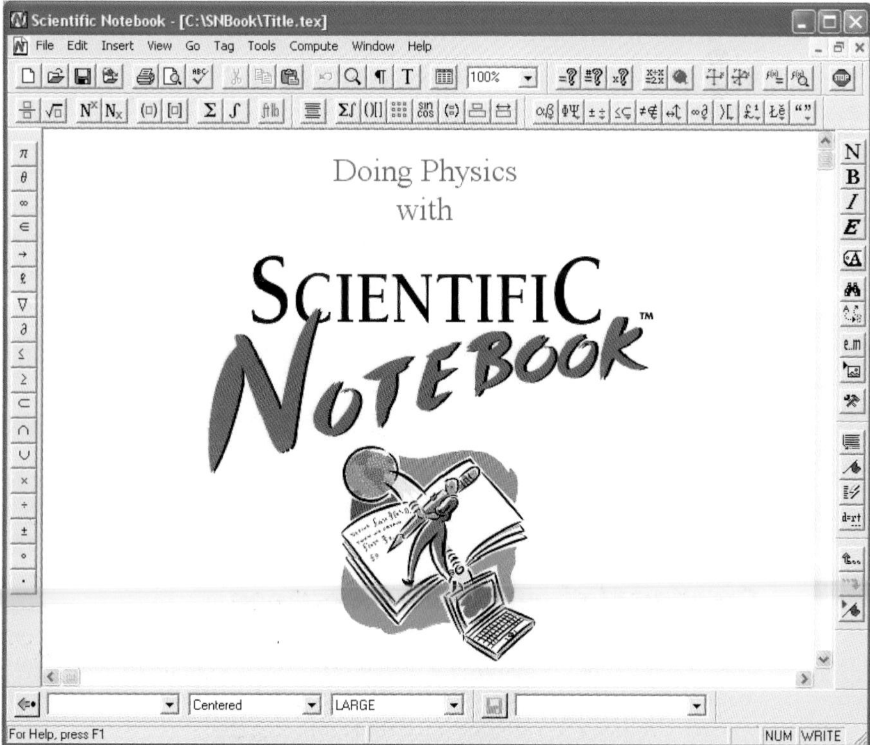

Figure 1.2 A typical screen for *SNB* 5.5

Now that we have access to many of *SNB*'s features, we are ready to start using them.

Physics *à la mode*: Math M or Text T

Since *SNB* is more than a word processor, it needs a way to distinguish between plain text which the engine ignores, and mathematical objects which are the engine's input. To make this distinction, *SNB* uses two modes of input, Text mode or Math mode. When you enter information, you do so in one of the two modes.

In Text mode, you input characters that *SNB* treats like any word processor would. Such text can be formatted in various ways using Tags. In Math mode, *SNB* treats the characters as mathematical objects that can be passed along to the engine as input. The Math/Text button on the Standard Toolbar indicates whether you are entering text or mathematics.

When the button looks like T you are in Text mode entering text.

When the button looks like M you are in Math mode entering mathematics.

Right away you'll notice that text and mathematics appear differently on the screen. In Text mode, the characters appear black and upright while in Math mode they are red and italicized. Because mathematics spacing is automatic, the spacebar moves the insertion point to the right in Math mode but does not insert spaces.

Note The Math/Text button tells you the mode at the position of the insertion point.

There are four ways to change from one mode to the other.

- Click the Math/Text button on the Standard Toolbar
- Use the first item of the Insert menu
- Use the INSERT key on your keyboard
- Press CTRL + T for Text or CTRL + M for Math.

Creating Mathematical Expressions

Since there is no programming syntax in *SNB*, it is important that you learn to create mathematical expressions. If the mathematical expression you create is not correct, then you are not likely to generate a useful result.

When you create a mathematical object, *SNB* puts you into Math mode automatically. For example, when you click the Fraction button, you are automatically in Math mode and the insertion point is in the numerator of the fraction. When you click the Radical button, you are automatically in Math mode and the insertion point is inside the square root symbol. When you click the expanding Parentheses button, you are automatically in Math mode and the insertion point is between the two parentheses.

Example 1.1 *An old friend*
Create a mathematical expression for the quadratic formula.
Solution. First, make sure you are in Math mode.

1. Use your keyboard to enter $x =$
2. Click the Fraction button (or enter CTRL + F).
3. Use your keyboard to enter $-b$
4. Click on the $\pm$ symbol on the Symbol Cache.
5. Click the Radical button (or enter CTRL + R).
6. Use your keyboard to enter b
7. Click the Superscript button (or enter CTRL + UPARROW) and type 2
8. Press the SPACEBAR to move the insertion point out of the superscript.
9. Use your keyboard to enter $-4ac$
10. Press the SPACEBAR to move the insertion point out of the radical.
11. Press the TAB key to move the insertion point to the denominator.
12. Use your keyboard to enter $2a$

Your final expression should look familiar.

$$x = \frac{-b \pm \sqrt{b^2 - 4ac}}{2a}$$

At first, all those steps seem like a lot to remember. But if you think about it, those are exactly the same steps you would use if you were writing that formula with a pencil. This is not the only way to create this expression. You can also get the plus/minus "±" symbol from the Binary Operations panel of the Symbol Panels and you can move the insertion point with mouse clicks or the arrow keys. ◄

Hint You may find the [Fraction], [Radical], [Superscript], [± ÷] buttons useful.

Example 1.2 *A new friend*
Create a mathematical expression for the Law of Cosines.
Solution. First, make sure you are in Math mode.

1. Use your keyboard to enter c
2. Click the Superscript button (or enter CTRL + UPARROW) and type 2
3. Press the SPACEBAR to move the insertion point out of the superscript.
4. Use your keyboard to enter $= a$

5. Click the Superscript button (or enter CTRL + UPARROW) and type 2
6. Press the SPACEBAR to move the insertion point out of the superscript.
7. Use your keyboard to enter $+b$
8. Click the Superscript button (or enter CTRL + UPARROW) and type 2
9. Press the SPACEBAR to move the insertion point out of the superscript.
10. Use your keyboard to enter $-2ab$
11. Use your keyboard to enter cos, which automatically turns into *SNB*'s cos function.
12. Click on the θ button on the Symbol Cache.

Your final expression should look this.

$$c^2 = a^2 + b^2 - 2ab\cos\theta$$

The Law of Cosines gives the relationship among the sides and angles in any triangle. The angle θ is the angle between sides of length a and b, and c is the length of the third side. The Pythagorean Theorem is a special case of the Law of Cosines where the angle $\theta = 90^\circ$ and c is the hypotenuse. ◄

SNB has many Keyboard Shortcuts that allow you to enter many mathematical objects quickly. The following table lists some of the most useful ones.

To enter		**Press**
Fraction	$\frac{}{}$	CTRL + F
Radical	$\sqrt{}$	CTRL + R
Superscript		CTRL + UPARROW
Subscript		CTRL + DOWNARROW
Integral	$\int$	CTRL + I
Summation	$\sum$	CTRL + 7
Expanding Parentheses	( )	CTRL + (
Expanding Square Brackets	[ ]	CTRL + [
Expanding Angle Brackets	⟨ ⟩	CTRL + SHIFT + ,
Expanding Braces	{ }	CTRL + SHIFT + [
Expanding Absolute Value	| |	CTRL + \

Table 1.1

The occasional "+ SHIFT" is there because braces are the uppercase of square brackets and the less-than symbol "<" is the uppercase of a comma (take a peek at your keyboard). For a complete list of keyboard shortcuts, look under Help + Search, Keyboard Shortcuts.

Evaluate and Evaluate Numerically

You can create mathematical expressions with any word processor, many of which have impressive equation editors. In *SNB*, these expressions are active mathematical objects that you can evaluate. To evaluate an expression, place the insertion point in or immediately to the right of it and choose Evaluate or Evaluate Numerically.

The results of your evaluation depend on the numbers in your expression. *SNB* represents integers, rational and irrational numbers such as $\sqrt{2}$, π, and e exactly and Evaluate uses the exact values. When you Evaluate an expression, *SNB* returns the result of the computation as an exact or symbolic answer whenever it can. If you Evaluate the following sum *SNB* returns the exact answer.

$$\tfrac{1}{3}\sqrt{2} + \tfrac{2}{7}\sqrt{2} = \tfrac{13}{21}\sqrt{2}$$

If you write a number in a fraction in decimal form but leave the $\sqrt{2}$ in exact form and use Evaluate, *SNB* leaves the exact value intact.

$$\tfrac{1}{3}\sqrt{2} + \tfrac{2.0}{7}\sqrt{2} = 0.619\,05\sqrt{2}$$

Symbolic real numbers such as $\sqrt{2}$ and π will retain symbolic form unless Evaluated Numerically. But if you write the numbers in the square roots in decimal form and use Evaluate, *SNB* returns the approximate numerical value of the sum.

$$\tfrac{1}{3}\sqrt{2.0} + \tfrac{2}{7}\sqrt{2.0} = 0.875\,47$$

You can force a numerical result to any evaluation if you write the numbers in the expression in decimal notation or you use Evaluate Numerically. If you Evaluate Numerically the original sum, *SNB* returns the same numerical answer.

$$\tfrac{1}{3}\sqrt{2} + \tfrac{2}{7}\sqrt{2} = 0.875\,47$$

Evaluate returns an exact answer whenever possible while Evaluate Numerically always returns an approximate numerical result.

Example 1.3 *A useful difference*
Examine the differences between Evaluate and Evaluate Numerically.
Solution. Place the insertion point anywhere in each expression and first click Evaluate then Evaluate Numerically.

$$\frac{1}{3} \times \frac{2}{5} = \frac{2}{15} = 0.133\,33$$

$$\cos\frac{\pi}{4} = \tfrac{1}{2}\sqrt{2} = 0.707\,11$$

$$\textstyle\sum_{n=1}^{10} \dfrac{1}{2^n} = \dfrac{1023}{1024} = 0.999\,02$$

Evaluate Numerically returns numerical approximations to the accuracy set in Engine Setup + Digits Used in Computations and Computation Setup + Digits Shown in Results. In the following example, both are set to 25.

$\pi = 3.141\,592\,653\,589\,793\,238\,462\,643$

$e = 2.718\,281\,828\,459\,045\,235\,360\,287$

$\sqrt{2} = 1.414\,213\,562\,373\,095\,048\,801\,689$

$\frac{8}{7} = 1.142\,857\,142\,857\,142\,857\,142\,857$

Notice that the numerical approximations are broken into 3-digit blocks to make them more readable. The symbolic numbers on the left are exact, while the numbers on the right are merely numerical approximations. So while $\left(\sqrt{2}\right)^2 = 2$ exactly, the approximate result is off in the 24th decimal place.

$1.414\,213\,562\,373\,095\,048\,801\,689^2 = 2.000\,000\,000\,000\,000\,000\,000\,001$

◄

Note You can change the number of digits shown for the output in the current document only. Go to Compute + Settings and select the General page. Click on the Set Document Values radio button and change the Digits Shown in Results value. We'll use the default setting of 5 for the rest of this book.

Example 1.4 *A numerical example*
What are the approximate numerical values for the constants π, e, $\sqrt{2}$, and i^2?
Solution. Place the insertion point to the right of each expression and click the Evaluate Numerically button.

$\pi = 3.141\,6$

$e = 2.718\,3$

$\sqrt{2} = 1.414\,2$

$i^2 = -1.0$

◄

An important calculation in physics is the **percent deviation**. In many experiments, you may have to compare two numbers because you measured a quantity two different ways or you want to compare an experimental result with a theoretical prediction. The percent deviation is a numerical way to quantify the agreement between two numbers. If you're asked to "compare a with b", then the percent deviation between these two numbers is

$$pd = 100\,\frac{a-b}{b}\;. \tag{1.1}$$

When a is less then b, the percent deviation is negative, and when a is greater than b the percent deviation is positive. If $a = b$, then the percent deviation is zero.

Example 1.5 *A circle is just a square without corners*
Compare the area of an 8×8 square with that of a circle with diameter 9.
Solution. The area of the square is the length of a side squared and the area of a circle is π times the radius squared. Evaluate Numerically the following expression which gives the percent deviation between these two areas.

$$100\frac{8^2 - \pi\left(4\frac{1}{2}\right)^2}{\pi\left(4\frac{1}{2}\right)^2} = 0.601\,64$$

The two areas are very close, and the square contains approximately 0.6% more area than the circle. ◀

We can use the percent deviation and the Evaluate Numerically command to check the accuracy of our 5-digit approximation of π.

$$100\frac{3.141\,6 - \pi}{\pi} = 2.338\,4 \times 10^{-4}$$

The approximation to π is less than one quarter of a thousandth of a percent larger than the exact value.

Scientific Notation

Sometimes you will have to deal with numbers that are very large or very small. For example, a light year is the distance light travels in one year, which is about 6 trillion miles. The Bohr radius is the radius of the ground state hydrogen orbit, which is about 2 billionths of an inch.

One way to help you use and understand such extreme numbers is to use **scientific notation**. You can write any number as the product of a number between one and ten and a power of ten. For example, Ted Williams hit $521 = 5.21 \times 10^2$ major league home runs and the fine structure constant is approximately $0.00730 = 7.30 \times 10^{-3}$. Scientific notation also eliminates any ambiguity in the significant digits of a number, which are reflected in the number of digits in the number between one and ten.

Use the following steps to write a number in scientific notation.

1. Enter the number between 1 and 10 in math mode.
2. Choose the times symbol $\times$ from the Symbol Cache toolbar or from the Binary Operations symbol panel. The times symbol is *not* the letter x.
3. Enter the number 10.
4. Click the Superscript button on the Math Templates toolbar, and enter the power in the input box.

Example 1.6 *One really big number*
Define Avogadro's number, enter it in scientific notation and write it in words.
Solution. Avogadro's number tells us the number of molecules in a mole of stuff. It is just another "bunch" number. There are 12 donuts in a dozen, 500 sheets of paper in a ream, and an Avogadro's number of molecules in a mole. Follow the four steps above to enter 6.0221×10^{23}, which is approximately 602 billion trillion. ◄

SNB provides you with a convenient keyboard shortcut that simplifies this process.

1. Enter the number between 1 and 10 in math mode.
2. Type ttt while still in math mode. This automatically turns into $\times 10$. The Superscript input box is there, but you must check View + Input Boxes to see it.
3. Place the insertion point in the superscript input box and enter the power.

Think of the "ttt" as meaning "*t*imes *t*en *t*o-the".

You can get *SNB* to convert your numbers into scientific notation automatically. *SNB* returns the result of a numerical computation in scientific notation if the number of digits in the result exceeds the setting for Threshold for Scientific Notation.

$$123450 = 1.2345 \times 10^5$$

$$0.012345 = 1.2345 \times 10^{-2}$$

If the threshold is set to 1, then *SNB* will return any result larger or equal to 10 (or less than 0.1) in scientific notation. With this setting, you can Evaluate Numerically any number and *SNB* will convert it into scientific notation.

Note To change the scientific notation threshold in your current document only, click on Compute + Settings and select the General page of the Document Computation Settings dialog box. Click on the Set Document Values radio button and change the Threshold for Scientific Notation value (the default value is 5).

Substitution and Endpoint Evaluation

You can substitute particular values or other expressions into any expression in *SNB*. Use the following steps to Substitute a number or new expression for a variable:

1. Select the expression with the mouse or SHIFT + ARROW.
2. Enclose the expression in expanding square brackets.
3. Create a Subscript to the right of the brackets.
4. List the values in the subscript input box separated by commas.
5. Click Evaluate or Evaluate Numerically.

If you want to Substitute into an expression that you have not yet created, you can create the expanding square brackets first and then create the expression inside the brackets.

When there is only one variable in the expression, you need only to include its value in the subscript input box without assigning it to a variable.

$$\left[5+20t-4.9t^2\right]_2 = 25.4$$

But if there is more than one variable, you must tell *SNB* which variable gets which value. You do this with one equation for each variable in the subscript, separated by commas.

$$\left[x_0+vt-4.9t^2\right]_{x_0=5,v=20,t=2} = 25.4$$

You can Substitute a particular value for one variable into an expression.

$$\left[x_0+vt-4.9t^2\right]_{t=0} = x_0$$

You can also Substitute other expressions with variables into your expression.

$$\left[x_0+vt+at^2\right]_{x_0=5,v=2t,a=-4.9/t} = 2t^2-4.9t+5$$

Be careful when you're substituting both variables and numerical values. *SNB* does the substitutions in the same order you list them in the subscript, so the order matters.

$$\left[x_0+vt+at^2\right]_{t=2,x_0=5,v=2t,a=-4.9/t} = 4t-\frac{19.6}{t}+5$$

$$\left[x_0+vt+at^2\right]_{x_0=5,v=2t,a=-4.9/t,t=2} = 3.2$$

Make sure the numerical values are to the right of the variables.

Example 1.7 *An old friend revisited*
Evaluate the two solutions to the quadratic equation when $a=-1$, $b=2$, and $c=3$.
Solution. The quadratic formula gives the two general solutions to a quadratic equation. Use the result from Example 1.1 and follow the above steps to create the following two expressions. Then click Evaluate (or Evaluate Numerically).

$$x=\left[\frac{-b+\sqrt{b^2-4ac}}{2a}\right]_{a=-1,b=2,c=3} = -1$$

$$x=\left[\frac{-b-\sqrt{b^2-4ac}}{2a}\right]_{a=-1,b=2,c=3} = 3$$

Notice that *SNB* interprets the plus/minus sign "±" as a plus sign only.

$$x=\left[\frac{-b\pm\sqrt{b^2-4ac}}{2a}\right]_{a=-1,b=2,c=3} = -1$$

◀

You can also use Substitution to compute the difference between the results of an expression evaluated at two different points. This is called Evaluating at Endpoints. Use the following steps to perform Evaluate at Endpoints on an expression:

1. Select the expression with the mouse or SHIFT+ARROW.
2. Click [□] or press CTRL + [to enclose the expression in expanding square brackets.
3. Click N_x, choose Insert+Subscript, or press CTRL + DOWNARROW.
4. List the values in the subscript input box separated by commas.
5. Press TAB to create a superscript box.
6. Enter another assignment for the variable in the superscript input box.
7. Click Evaluate or Evaluate Numerically.

In physics, we often talk about the change in some quantity. When we use Evaluate at Endpoints, we are calculating the change in the expression inside the brackets.

$$\Delta x = [x]_{x_0}^{x_f} = x_f - x_0$$

The change in x equals its final value (x_f) minus its initial value (x_0).

Example 1.8 *Batter up!*

When the effects of air resistance are considered, the height in meters of a thrown baseball is given by the expression

$$y = 2 - 43t + 185\left(1 - e^{-0.23t}\right)$$

where t is in seconds. If the ball is in the air for 0.45 seconds, what is the ball's change in height during the first half of its trip? What is the ball's change in height during the rest of its trip?

Solution. To find the ball's change in height during the first $0.45/2 = 0.225$ seconds, Evaluate the given expression between the endpoints of 0 and 0.225.

$$\Delta y_1 = \left[2 - 43t + 185\left(1 - e^{-0.23t}\right)\right]_0^{0.225} = -0.344\,75$$

The ball's change in height is negative, so it dropped 0.34475 meters. To find the ball's change in height during the second half of its trip, Evaluate the given expression between the endpoints of 0.225 and 0.45.

$$\Delta y_2 = \left[2 - 43t + 185\left(1 - e^{-0.23t}\right)\right]_{0.225}^{0.450} = -0.815\,31$$

During the second half of its trip, the ball drops another 0.81531 meters, so it dropped a total of 1.1606 meters (about 3.81 feet). ◀

Evaluating at Endpoints is also useful when you want to calculate the slope of a straight line. The slope of the line passing through two points (x_1, y_1) and (x_2, y_2) is the change in the y-coordinates divided by the change in the x-coordinates.

$$slope = \frac{y_2 - y_1}{x_2 - x_1} \tag{1.2}$$

In *SNB* this is the ratio of two quantities each Evaluated at Endpoints.

$$slope = \frac{[y]_{y_1}^{y_2}}{[x]_{x_1}^{x_2}} = \frac{1}{x_2 - x_1}(y_2 - y_1)$$

Example 1.9 *Hit the slope*
Calculate the slope of the line that passes through the points $(1, 3)$ and $(2, 11)$.
Solution. To calculate the slope of the line passing through these two points, create and Evaluate at Endpoints the following expression.

$$slope = \frac{[y]_3^{11}}{[x]_1^2} = 8$$

Notice that this is *not* the same as the ratio evaluated at the endpoints.

$$slope \neq \left[\frac{y}{x}\right]_{x=1,y=3}^{x=2,y=11} = \frac{5}{2}$$ ◀

Example 1.10 *Give me a sine*
Calculate the average value of the sine function between 0 and π.
Solution. The average value of the function $y = \sin x$ between $x = a$ and $x = b$ is

$$y_{\text{ave}} = -\frac{\cos b - \cos a}{b - a}$$

since $-\cos x$ is the antiderivative of $\sin x$. Create the following expression, apply Evaluate at Endpoints it, and then apply Evaluate Numerically to the result.

$$y_{\text{ave}} = -\frac{[\cos x]_0^{\pi}}{[x]_0^{\pi}} = \frac{2}{\pi} = 0.636\,62$$ ◀

Solving Equations

While there is a lot more to solving physics problems than doing math, the ability to correctly solve equations is an important part of the process. *SNB* can help you by solving the equations. You will use physics to assemble an equation and then use *SNB* to solve it. *SNB* provides four options for solving equations, Exact, Numeric, Integer, and Recursion, which you will find under Compute + Solve. You can also use the Solve Exact button on the Compute Toolbar.

Note Unless otherwise noted, the settings for the Solve Options are Ignore Special Cases (ISC) checked and Principal Value only (PVO) unchecked.

Solve Exact

Solve Exact is the most general of the four solving options. You can use it to solve equations with polynomials, logarithmic and exponential functions, and trigonometric functions. If your equation has an algebraic solution, there is a good chance Solve Exact will find it.

Once you have created an equation, you can solve it by placing the insertion point anywhere inside the equation and choosing Solve Exact. If the equation only has one variable, *SNB* will attempt to solve it immediately. Otherwise, *SNB* will prompt you with the Solution Variable(s) window. Enter the appropriate variable names in the Variable(s) to Solve for box (separated by commas) and then click OK.

Example 1.11 *Let's start simple*
Use Solve Exact to solve the simple equation $x^2 - 9 = 0$.
Solution. Enter the equation in math mode, place the insertion point anywhere inside the equation and choose Solve Exact.

$x^2 - 9 = 0$, Solution is: $-3, 3$

As you probably expected, with Principal Value Only unchecked, *SNB* returns the two solutions $x = 3$ and $x = -3$. ◀

Example 1.12 *An old friend revisited again*
Use Solve Exact to verify that the quadratic formula gives the two solutions to the quadratic equation $ax^2 + bx + c = 0$.
Solution. In math mode, create an expression for the quadratic equation, place the insertion point anywhere in the equation and click Solve Exact. Type x into the Variable(s) to Solve for box.

$ax^2 + bx + c = 0$, Solution is: $-\frac{1}{2a}\left(b - \sqrt{b^2 - 4ac}\right), -\frac{1}{2a}\left(b + \sqrt{b^2 - 4ac}\right)$

SNB returns the two results of the quadratic formula, although not in their usual form. ◀

There are some rules about variable names that *SNB* considers acceptable. A variable or function name must be either a single character or a custom math name, both with or without a subscript. The symbols π, e, and i are reserved for mathematical constants, although as the following example shows you can use them with a subscript.

Example 1.13 *An i for an i*
Use Solve Exact to solve the simple equation $10i_1 - 2 = 0$.
Solution. Enter the equation in math mode, place the insertion point anywhere inside the equation and choose Solve Exact.

$10i_1 - 2 = 0$, Solution is: $\frac{1}{5}$

Solve Exact cannot solve this equation when the reserved symbol i is the variable. ◀

If you use two or more subscripts on a variable, they must all be letters or all be numbers. *SNB* does not like "mixed" subscripts. The variable name v_{123} is acceptable as is v_{ab}, but v_{1x} will not work. When you use Solve Exact on a variable with more than one letter in a subscript, *SNB* will prompt you with the Solution Variable(s) box.

You can always use the uppercase letters I, J, K, and Y without subscripts as variable names. If you want to use them to refer to Bessel functions (in the traditional $I_v(z)$ notation) check the Use I, J, K, and Y with Subscripts check box under Bessel Function Notation on the General page of the Computation Setup dialog box. If this box is unchecked, then you can use I, J, K, and Y as variable names with subscripts as well.

Example 1.14 *An I for an I*
Solve the not-so simple equation $0 = I_x^3 + (2 - \pi) I_x^2 - (3 + 2\pi) I_x + 3\pi$.
Solution. Enter the equation in math mode, place the insertion point anywhere inside the equation and choose Solve Exact.

$0 = I_x^3 + (2 - \pi) I_x^2 - (3 + 2\pi) I_x + 3\pi$, Solution is: $1, -3, \pi$

This cubic equation for I_x has three solutions. ◀

Solve Exact can also handle more advanced equations containing trigonometric, logarithmic and exponential functions. Equations involving these functions can have repeating solutions, and that can complicate *SNB*'s output. To alleviate this problem, both PVO and ISC are checked for the rest of this section.

Solve Exact also works on equations with trigonometric functions. The default unit for the argument of trigonometric functions is the radian. To force *SNB* to use degrees, place the **red** degree symbol after the argument of the functions. The **red** degree symbol is the "o" symbol in a Superscript. You will find the "o" symbol on the Symbol Cache.

Example 1.15 *The arc of the cotangent*
Solve the equation $\cot\theta = 1/\sqrt{3}$ exactly for θ in degrees.
Solution. Create the equation with θ° as the argument of the cotangent function. Place the insertion point anywhere in the equation, and click the Solve Exact button.

$\cot\theta^\circ = \frac{1}{\sqrt{3}}$, Solution is: 60

Solving this equation is equivalent to using Evaluate on the inverse trigonometric function arccot, which gives the angle θ in radians.

$$\operatorname{arccot}\frac{1}{\sqrt{3}} = \frac{1}{3}\pi$$

When *SNB* performs an operation on trigonometric functions, it automatically converts to radians. If you want your answer in degrees, just use Evaluate to multiply by $180/\pi$.

$$\frac{180}{\pi}\cot^{-1}\frac{1}{\sqrt{3}} = 60$$

As you can see, *SNB* allows two ways to write inverse trigonometric functions. ◀

Example 1.16 *Survey says!*
Find the distance to an object that is 10 feet tall whose top is $15°$ above the horizontal.
Solution. Basic trigonometry tells us that the tangent of the angle is the object's height divided by the distance to the object. Create an equation for this condition, place the insertion point anywhere in the equation, and click the Solve Exact button. Then use Evaluate Numerically on the exact answer.

$$\tan 15° = \frac{10}{x}, \text{ Solution is: } -\frac{1}{\frac{1}{10}\sqrt{3}-\frac{1}{5}} = 37.321$$

The object is about 37.321 feet away.

The logarithmic and exponential functions are inverses of one another, so each undoes the other. If we Evaluate the exponential of the logarithm of a variable we get the variable back.

$$e^{\ln x} = x$$

The exact solution of an equation with a variable in an exponential will contain a natural logarithm. ◀

Example 1.17 *I'm a lumberjack and I'm OK*
Solve the equation $y = 2e^{x/3}$ exactly for x.
Solution. Place the insertion point anywhere in the equation, click the Solve Exact button, enter x in the Solution Variable(s) box and click OK.

$$y = 2e^{x/3}, \text{ Solution is: } 3\ln y - 3\ln 2$$

The solution is $x = 3\ln y - 3\ln 2$, where the function $\ln x$ is the natural logarithm, which is logarithm base-e. The constant e is a naturally occurring constant which has the approximate numerical value of $e = 2.7183$. ◀

SNB has two logarithm functions, the natural log $\ln x$ and the more flexible $\log x$. You can use logarithms with different bases by putting a subscript on the log function. Two common bases are base-2 and base-10, which you write as $\log_2 x$ and $\log_{10} x$ in *SNB*. There is a simple connection between logarithms of any base and natural logarithms.

$$\log_b x = \frac{\ln x}{\ln b} \tag{1.3}$$

We can use Evaluate Numerically to verify this for base-10.

$$\log_{10} x = 0.434\,29 \ln x$$

$$\frac{1}{\ln 10} = 0.434\,29$$

SNB interprets $\log x$ with no subscript as the natural logarithm unless you change the default setting on the General page of the Tools + Computation Setup dialog. Checking the Base for log function box tells *SNB* to interpret $\log x$ as the base-10 logarithm. Leaving the box unchecked tells *SNB* to interpret $\log x$ as the natural logarithm. Logarithms with explicit subscripts are unaffected.

Example 1.18 *We're radioactive*
How many half-lives have elapsed when two-thirds of a radioactive sample has decayed?
Solution. The half-life is a property of the radioactive material and equals the amount of time it takes for half the sample to decay. When x half-lives have elapsed, the fraction of the sample that has not yet decayed equals 2^{-x}. Create an equation for this condition, place the insertion point anywhere in the equation, and click the Solve Exact button. Then apply Evaluate Numerically to the exact answer.

$\frac{1}{3} = \frac{1}{2^x}$, Solution is: $\log_2 3 = 1.585\,0$

After an elapsed time of $1.585\,0$ times the half-life, one-third of the radioactive sample remains. *SNB* can solve this equation for an unspecified remaining fraction.

$F = \frac{1}{2^x}$, Solution is: $-\log_2 F$

After an elapsed time of $-\log_2 F$ times the half-life, a fraction F of the radioactive sample remains. The fraction F is less than 1 so the answer is a positive number. ◄

Solve Numeric

Some equations do not have exact algebraic solutions, so you must solve them numerically. To solve these transcendental equations, *SNB* provides the Solve Numeric command, a particularly useful command especially when you want to specify a search interval for the solution. Unlike Solve Exact, you cannot apply the Solve Numeric command to an equation containing units.

Important For the remainder of this book, whenever computing choices are specified, the preceding Compute is implied. For example, when you see Solve + Numeric, perform Compute + Solve + Numeric.

The following example involves a type of transcendental equation that arises in the determination of the ground-state energy of the square-well potential in one-dimensional Quantum Mechanics (see [2], page 258).

Example 1.19 *A transcendental experience*

Solve the equation $\arctan\sqrt{\frac{2-x}{x}} = \frac{\pi}{2}\sqrt{x}$ numerically.

Solution. Create an expression for the equation, place the insertion point anywhere in it and choose Solve + Numeric.

$\arctan\sqrt{\frac{2-x}{x}} = \frac{\pi}{2}\sqrt{x}$, Solution is: $\{[x = 0.463]\}$

As the intersection point of the two curves in Figure 1.3 shows, this answer is correct.

Can you tell which curve is which?

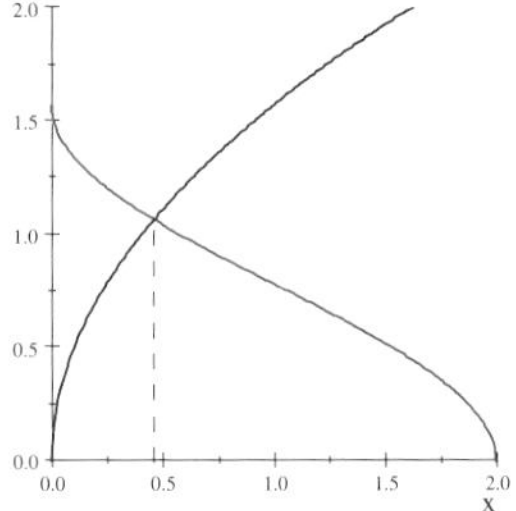

Figure 1.3 Which is which?

◀

When there is more than one numerical solution to an equation, you may have to specify the range of the variable where you want *SNB* to look for a solution. You can do this easily in *SNB* by putting the equation and the range in a matrix. A **matrix** is a 2-dimensional rectangular array of numbers or mathematical expressions. *SNB* uses matrices to pass along separate but related input to the engine, in this case the equation to be solved and the range of the variable to find the solution.

Use the following steps to create a matrix.

1. Click the [matrix] button on the Math Objects toolbar, or choose Insert + Matrix.
2. In the Matrix dialog box, set the number of rows and columns of the matrix.
3. Select one of the optional built-in delimiters to enclose the matrix.
4. Choose OK. The program places the insertion point in the input box in the top left cell of the matrix.
5. Enter the contents of the top left cell.
6. Press TAB to move to the next cell.
7. Press the SPACEBAR or the RIGHT ARROW key to leave the matrix.

There are six choices for the optional built-in delimiters. The four choices of round, square, curly brackets or no brackets are aesthetic and do not affect the mathematical properties of the matrix. But *SNB* will interpret the single vertical bars as the determinant and the double vertical bars as the norm of the matrix. Unless you want to perform those matrix operations, you should avoid those delimiters.

Shortcut To create a matrix with the same properties as your last matrix, enter CTRL + S then press M. To create a 2×2 matrix, enter CTRL + S then press SHIFT + M.

Now that you can create a matrix, you're ready to find a numerical solution within a specified range of the variable.

1. Create a 1-column, 2-row matrix.
2. Place your equation in the first row.

3. Enter your choice of the variable interval in the second row

4. Leave the insertion point anywhere in the matrix and click Solve + Numeric.

Use the membership symbol $\in$ to indicate that the variable lies in that interval. You can put the interval in parentheses or curly brackets. For example, you can indicate your choice of interval as $x \in (1, 4)$ or as $x \in \{1, 4\}$.

Note You can find the membership symbol $\in$ in either the Binary Relations panel of the Symbols Panels or the Symbol Cache. The membership symbol $\in$ is *not* the same as the lowercase Greek letter epsilon ϵ.

Example 1.20 *Not that one, that one!*
Find the "other" numerical solution to the equation $x + \sin x = -x^2 + 9x - 8$.
Solution. If you place the insertion point anywhere in the equation and choose Solve + Numeric, *SNB* returns a correct solution at $x = 1.3497$.

$x + \sin x = -x^2 + 9x - 8$, Solution is: $\{[x = 1.349\,7]\}$

This plot of the two curves on each side of the equation shows us there is a second solution near $x = 7$.
To force *SNB* to find this solution, let's have it look between $x = 6$ and $x = 8$.
Create a 1-column, 2-row matrix and put the equation in the first row.
Place the expression $x \in (6, 8)$ for the search interval in the second row.

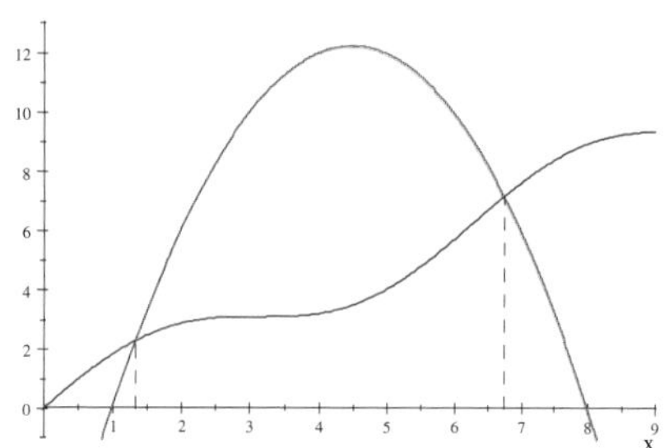

Figure 1.4 Two solutions

To find the solution, place the insertion point anywhere in the matrix and select Solve + Numeric.

$$\begin{bmatrix} x + \sin x = -x^2 + 9x - 8 \\ x \in (6, 8) \end{bmatrix}, \text{Solution is: } \{[x = 6.748]\}$$

Let's look again at our original example from the beginning of the chapter. *SNB* found the first zero at $x = 0$. We can now find the second zero, which is the first non-zero zero.

$$\begin{bmatrix} 0 = x^2 e^{-3x} \sin 4x \\ x \in \{0.25, 1.25\} \end{bmatrix}, \text{Solution is: } \{[x = 0.785\,40]\}$$

A look back at the graph in Figure 1.1 suggests this answer is correct. Let's Substitute this answer back into the expression.

$$\left[x^2 e^{-3x} \sin 4x\right]_{0.785\,40} = -4.295\,1 \times 10^{-7}$$

The answer is only approximately correct, good to the sixth decimal point. ◀

Systems of Equations

SNB is very helpful if the solution to your physics problem requires solving more than one equation. All you have to do is place each equation in a row of a 1-column matrix and click the Solve Exact button or choose Solve + Numeric. Remember, you can apply Solve Exact to equations with units but Solve Numeric cannot handle equations with units.

The solution to a typical electric circuit produces a set of simultaneous equations that you must solve for the electric current through each resistor. The following example contains three equations which *SNB* solves with the click of a button.

Example 1.21 *It's not the volts, it's the amps*
Solve a set of three simultaneous equations for a typical electric circuit problem.
Solution. Create a 1-column matrix, 3-row matrix and enter one equation in each row. Place the insertion point anywhere in the matrix and click the Solve Exact button.

$$\begin{bmatrix} 20\,\Omega\, i_1 + 10\,\mathrm{V} = 10\,\mathrm{V} + 10\,\Omega\, i_2 \\ 10\,\Omega\, i_2 + 5\,\Omega\, i_3 = 30\,\mathrm{V} \\ i_1 + i_2 = i_3 \end{bmatrix}, \text{ Solution is: } \left[i_1 = \tfrac{6}{7}\,\mathrm{A}, i_2 = \tfrac{12}{7}\,\mathrm{A}, i_3 = \tfrac{18}{7}\,\mathrm{A}\right]$$

The solution gives the electric current (in Amperes) flowing through each of the three resistors. ◀

Sometimes you need to tell *SNB* where to look for numerical solutions to a system of equations. To find a numerical solution within a specified range for more than one variable, you must include a specified range for each variable.

Example 1.22 *Watch your P's and Q's*
Find the values of a and x for which the parabola x^2 and the quartic $1 - ax^4$ both equal $\sin x$.
Solution. Create a 1-column matrix, 4-row matrix. Place the equation $\sin x = x^2$ in the first row and $1 - ax^4 = \sin x$ in the second. Since positive values of a less than one give real solutions, enter the condition $a \in \{0, 1\}$ in the third row. The sine function never exceeds one, so the first equation tells us that x must be less than one. Enter the condition $x \in \{0, 1\}$ in the last row. Place the insertion point anywhere in the matrix and choose Solve + Numeric.

$$\begin{bmatrix} \sin x = x^2 \\ 1 - ax^4 = \sin x \\ a \in \{0, 1\} \\ x \in \{0, 1\} \end{bmatrix}, \text{ Solution is: } \{[a = 0.391\,58, x = 0.876\,73]\}$$

Let's use Substitute (with Evaluate) to check these answers.

$$1 - \left[\frac{1 - ax^4}{\sin x}\right]_{a=0.391\,58, x=0.876\,73} = 1.168\,9 \times 10^{-5}$$

The approximate numerical solution is good to about 1 part per hundred-thousand. ◀

The Compute Menu

The Compute Toolbar contains some of *SNB*'s most important and useful commands, including Evaluate, Evaluate Numerically, and Solve Exact. It also contains Simplify, Expand, New Definitions, and Show Definitions. All of these choices (and much more) can be found in the Compute menu.

When you click the Compute menu item at the top of the screen, you see a drop-down menu that contains many more computing commands. Like those on the Compute Toolbar, these commands send your input to the engine and return its output to you. In this section, we will explore more of these commands.

Simplify and Expand

When you Evaluate an expression, the result you get from *SNB* may not be in the form you want. The Simplify and Expand commands can help you fix that. When applied to decimal numbers, the Evaluate and Simplify commands usually produce the same result, but Simplify is often more effective with symbolic expressions and expressions involving radicals or exponential notation for roots.

Example 1.23 *Let's get to the root of the problem*
Apply Evaluate and Simplify to the cube root of $4913/256$ in both radical and exponential notation.
Solution. Create expressions for $\sqrt[3]{\frac{4913}{256}}$ and $\left(\frac{4913}{256}\right)^{1/3}$. Apply both commands to the expressions with the exponential notation.

Evaluate: $\left(\dfrac{4913}{256}\right)^{1/3} = \dfrac{17}{256}256^{\frac{2}{3}}$

Simplify: $\left(\dfrac{4913}{256}\right)^{1/3} = \dfrac{17}{8}\sqrt[3]{2}$

Now apply both commands to the expressions with the radicals.

Evaluate: $\sqrt[3]{\dfrac{4913}{256}} = \dfrac{17}{256}256^{\frac{2}{3}}$

Simplify: $\sqrt[3]{\dfrac{4913}{256}} = \dfrac{17}{8}\sqrt[3]{2}$

In both cases, Simplify gave a simpler results than Evaluate. When applied to floating-point numbers, the two commands return the same result. ◀

You can use Simplify and Expand to convert between fractions and mixed numbers. Expand converts a fraction into a mixed number.

$$\frac{296}{167} = 1\tfrac{129}{167}$$

Use Simplify (or Evaluate) to convert a mixed number into a fraction.

$$1\tfrac{129}{167} = \frac{296}{167}$$

The fraction $296/167$ is an excellent approximation to $\sqrt{\pi}$.

You can use Simplify or Expand to manipulate expressions with exponents.

Simplify: $a^x a^y a^{-z} = a^{x+y-z}$

Expand: $a^x a^y a^{-z} = \frac{1}{a^z} a^x a^y$

These two results show the behavior of exponents.

You can use Expand to generate multi-angle trigonometric expressions.

$$\sin(\theta + \phi) = \cos\theta \sin\phi + \cos\phi \sin\theta$$

$$\cos 3\theta = \cos^3\theta - 3\cos\theta \sin^2\theta$$

You can use Simplify to reduce them too.

$$\cos\theta \sin\phi + \cos\phi \sin\theta = \sin(\theta + \phi)$$

$$\cos^3\theta - 3\cos\theta \sin^2\theta = \cos 3\theta$$

Applying Expand to a product of polynomials has the effect of what is often called "multiplying it out".

$$(x+2)(x-1)(2y+3)(y-1) = 2x^2y^2 + x^2y - 3x^2 + 2xy^2 + xy - 3x - 4y^2 - 2y + 6$$

$$(x+a)^7 = a^7 + 7a^6x + 21a^5x^2 + 35a^4x^3 + 35a^3x^4 + 21a^2x^5 + 7ax^6 + x^7$$

The result for $(x+a)^7$ is an example of a binomial expansion, and that name can remind you of which command to use.

When applied to fractions, mixed numbers, exponential and trigonometric expressions, Expand and Simplify undo each other. When applied to polynomials, Expand and Factor undo each other.

Factor

The ability to factor polynomials and integers is a useful algebraic tool. *SNB* provides the Factor command which can handle polynomials and integers. With the Factor command, you can either

- factor an integer into a product of powers of prime numbers, or
- factor a polynomial.

The Factor command is not listed under Polynomials in the Compute menu because it also factors integers.

When applied to an integer, Factor returns all the prime factors of that integer.

$$1956 = 2^2 3 \times 163$$

$$1983 = 3 \times 661$$

$$1987 = 1987$$

Oops! Apparently 1987 is a prime number. When you try to Factor a prime number, which doesn't have any integer factors besides itself and one, *SNB* just returns the number itself. You can use Simplify or Evaluate to return the results of Factor to the integer.

$$2^2 3 \times 163 = 1956$$

You can use Factor on polynomials with integer or rational coefficients to find the roots of a polynomial. Factor does not handle polynomials with decimal coefficients. If you have a polynomial with decimal coefficients, use Rewrite + Rational to convert it into a polynomial with rational coefficients and then apply Factor to the resulting polynomial.

Example 1.24 *An easy polynomial example*
Factor the quadratic polynomial $x^2 - 2x - 3$.
Solution. Place the insertion point anywhere in the polynomial and click Factor.

$$x^2 - 2x - 3 = (x + 1)(x - 3)$$

We see that the roots of this polynomial are -1 and $+3$, which agree with the results of Example 1.7. ◄

We can use Factor on complicated polynomials to find the roots.

Example 1.25 *An ugly polynomial example*
Factor the quadratic polynomial $2x^2y^2 + x^2y - 3x^2 + 2xy^2 + xy - 3x - 4y^2 - 2y + 6$.
Solution. Place the insertion point anywhere in the polynomial and click Factor.

$$2x^2y^2 + x^2y - 3x^2 + 2xy^2 + xy - 3x - 4y^2 - 2y + 6 = (2y + 3)(y - 1)(x + 2)(x - 1)$$

We see that the roots of this polynomial are $x = -2$, $x = 1$, $y = -\frac{3}{2}$, and $y = 1$. ◄

We can use Factor on mathematical expressions that appear in the solution of a physics problem.

Example 1.26 *A physics example*
Factor the expression $\frac{1}{2}mv_0^2 + mgh_0 - \mu\, mgd \cos\theta$, which gives the final energy of an object that slid down a ramp under the influence of gravity and friction.
Solution. Place the insertion point anywhere in the expression and click Factor.

$$\frac{1}{2}mv_0^2 + mgh_0 - \mu\, mgd \cos\theta = -\frac{1}{2}m\left(-2gh_0 - v_0^2 + 2dg\mu\cos\theta\right)$$

The common "m" term is factored out of this expression. ◄

Rewrite and Combine

Simplify, Expand and Factor are general commands which offer no further options. You apply them to part or all of your expression and they return a result. The Rewrite and Combine commands provide more options and they sometimes give better results.

The Rewrite command lets you write your expression in terms of other mathematical functions. You Rewrite what-you-have into what-you-choose from the Rewrite options. For example, if you want to express $\sin 2\theta$ in terms of the tangent function, choose Rewrite + Tan.

$$\sin 2\theta = 2\frac{\tan\theta}{\tan^2\theta + 1}$$

You can explore the relationship between hyperbolic and exponential functions with Rewrite + Exponential.

$$\sinh x = \tfrac{1}{2}e^x - \tfrac{1}{2}e^{-x}$$

$$\cosh x = \tfrac{1}{2}e^x + \tfrac{1}{2}e^{-x}$$

There is a similar relationship between the inverse hyperbolic function and logarithms that you can see with Rewrite + Logarithm.

$$\sinh^{-1} x = \operatorname{arcsinh} x = \ln\left(x + \sqrt{x^2+1}\right)$$

$$\cosh^{-1} x = \operatorname{arccosh} x = \ln\left(x + \sqrt{x^2-1}\right)$$

The following example looks at the relationship between trigonometric and exponential functions.

Example 1.27 *DeMoivre's Theorem*
Use the Rewrite command to verify DeMoivre's theorem.
Solution. DeMoivre's theorem says that if n is a positive integer, then

$$(\cos x + i\sin x)^n = \cos nx + i\sin nx$$

To verify this, first use Rewrite + Exponential on $(\cos x + i\sin x)^n$ and then Expand the result.

$$(\cos x + i\sin x)^n = e^{n\ln\left(e^{ix}\right)} = e^{n(ix)}$$

Now use Rewrite + Sin and Cos.

$$e^{n(ix)} = \cos nx + i\sin nx$$

DeMoivre's theorem is useful in deriving multi-angle trigonometric formulas and extracting the roots of complex numbers. ◀

The Factor command only works on polynomials with rational coefficients. If you have a polynomial with decimal coefficients, you can use Rewrite to change the coefficients to rational numbers.

Example 1.28 *Author!*
Factor the polynomial $x^2 + 0.8\,x - 3.84$.
Solution. First use Rewrite + Rational to express the polynomial with rational coefficients, and then Factor the result.

$$x^2 + 0.8\,x - 3.84 = x^2 + \tfrac{4}{5}x - \tfrac{96}{25} = \tfrac{1}{25}(5x + 12)(5x - 8)$$ ◀

Where Rewrite lets you write your expressions in terms of other functions, Combine works on similar functions. With the Combine command, you can combine Exponentials, Logs, Powers, and Trig Functions. For example, you can use Combine + Trig Functions to combine $\sin x$ and $\cos x$.

$$\sin\theta\cos\theta = \tfrac{1}{2}\sin 2\theta$$

The Combine + Powers command produces the same result as Simplify when it is applied to numbers other than e raised to a power.

Combine + Powers: $a^x a^y a^{-z} = a^{x+y-z}$
Simplify: $a^x a^y a^{-z} = a^{x+y-z}$

You must use the Combine + Exponentials command when e is raised to a power because Simplify does not work.

Combine + Exponentials: $e^x e^y e^{-z} = e^{x+y-z}$
Simplify: $e^x e^y e^{-z} = e^x e^y e^{-z}$

Check Equality

You can use *SNB* to verify equalities and inequalities with the Check Equality command. This command works on numerical and symbolic expressions. When you use Check Equality, *SNB* returns one of three possible responses: true, false, or undecidable. The last means that the test is inconclusive and the equality may be true or false. Use the following steps to use Check Equality to verify an equality or inequality.

1. Create an expression for your equality or inequality.
2. Place the insertion point anywhere in the equation.
3. Choose Compute + Check Equality.

Example 1.29 *Just checking*
Verify the two answers from Example 1.15 are equal.
Solution. Create an expression equating the two answers. Place the insertion point anywhere in the equation and choose Check Equality.

$\frac{1}{3}\pi = 60^\circ$ is true

The two answers are equal. ◀

Even with its diverse collection of commands, *SNB* does not always present the results from the engine in the form you want. You may still have to edit the engine output (or any expression) the "old-fashioned" way. With *SNB*, you can Cut, Copy and Paste mathematical expressions and change them by-hand, but this introduces the possibility of human error. Use Check Equality to make sure you did your algebra correctly.

Example 1.30 *Equality for all!*

Edit by-hand one of the solutions to the quadratic equation returned by Solve Exact into a more standard form, and verify that your expression equals *SNB*'s result.

Solution. As we saw in Example 1.12, *SNB* returns the correct answers, but not in a standard form. To edit the "positive" solution by-hand, take the minus sign in front and move it to create "$-b$". Then replace the minus sign before the radical with a plus sign. Place the insertion point anywhere in the expression and choose Check Equality.

$-\frac{1}{2a}\left(b-\sqrt{b^2-4ac}\right)=\frac{1}{2a}\left(-b+\sqrt{b^2-4ac}\right)$ is true

When editing an expression by-hand in *SNB*, it is a good idea to Copy it, set the copy equal to the original expression, and work on the copy. After you've made a few changes, use Check Equality, Save your work, and repeat the process. That way, you'll have a record of your work just as you would if you were using paper and pencil. ◀

As a simple example of an inconclusive test, consider the apparently obvious equation $x = \ln e^x$. Exponentiation and taking the natural logarithm of a number are inverse operations, so mathematically they "undo" each other. You might expect Check Equality to verify that the equation is true.

$x = \ln e^x$ is undecidable

Since x can be real or complex, the right-hand side may be a multivalued function so this equation may or may not be true. Later we'll see how to tell *SNB* that x is real.

Here is an example of an equality test that yields a false result.

$x = \frac{1}{2}\ln e^x$ is false

There is no value of x, real or complex, that satisfies this equation.

Example 1.31 *A special case of Euler's formula*

Is the equation $e^{i\pi}+1=0$ correct?

Solution. Place the insertion point anywhere in the equation and choose Check Equality.

$e^{i\pi}+1=0$ is true

This equation is quite interesting because it contains five important fundamental mathematical constants. It also has the added attraction of being true. ◀

Polynomials

Many of *SNB*'s commands, such as Evaluate, Simplify, Factor, Expand and Combine + Powers, work on polynomials as well as other kinds of expressions. The Polynomials commands provide more options that are only applicable to polynomials. When you use these commands on a polynomial with more than one variable, specify your polynomial variable in the Need Polynomial Variable dialog box that appears.

The Polynomials + Sort command returns the polynomial with the terms in order of decreasing degree, so the largest power of the polynomial variable is the first term. It also collects the coefficients of terms in the polynomial

$$x + x^4 + 3x^2 - 2x^3 + bx^2 - ax = x^4 - 2x^3 + x^2\,(b+3) - x\,(a-1)$$

The Polynomials + Collect command collects all the coefficients of terms of the polynomial, but it does not necessarily sort the terms by degree.

$$x + x^4 + 3x^2 - 2x^3 + bx^2 - ax = x^4 - 2x^3 + (b+3)\,x^2 + (1-a)\,x$$

In both of these computations, x was the polynomial variable.

The roots of a polynomial are the solutions to the equation $polynomial = 0$. The Polynomials + Roots command returns all the roots of a degree-n polynomial in a 1-column, n-row matrix. For polynomials up to degree-4, *SNB* finds exact symbolic roots for polynomials with rational coefficients and approximate numerical roots for polynomials with decimal coefficients. For polynomials of degree-5 and higher, *SNB* always finds the roots numerically.

To find all the roots of a polynomial, use the following steps.

1. Create an expression for the polynomial.
2. Leave the insertion point in the expression.
3. Choose Polynomials + Roots from the Compute menu.

As the following example shows, the roots of a polynomial are not restricted to the real numbers.

Example 1.32 *Is this real?*
Find all the roots to the polynomial $x^4 - 1$.
Solution. Create an expression for the polynomial, leave the insertion point anywhere in it and choose Polynomials + Roots. Notice the expression is not an equation but just the polynomial.

$$x^4 - 1\text{, roots: } \begin{matrix} -1 \\ 1 \\ -i \\ i \end{matrix}$$

This degree-4 polynomial has four roots, two real ($x = \pm 1$) and two complex ($x = \pm i$).
◀

Power Series

Many functions can be written as an infinite sum of the product of constants and powers of the variable.

$$\begin{aligned} f(x) &= \sum_{n=0}^{\infty} a_n (x-a)^n \qquad (1.4) \\ &= a_0 + a_1 (x-a) + a_2 (x-a)^2 + a_3 (x-a)^3 + \cdots \end{aligned}$$

This is called expanding the function in a power series about $x = a$. Here are the power series expansions about $x = 0$ for the sine, cosine, and exponential functions.

$$\sin x = x - \tfrac{1}{6}x^3 + \tfrac{1}{120}x^5 - \tfrac{1}{5040}x^7 + \cdots \qquad (1.5a)$$

$$\cos x = 1 - \tfrac{1}{2}x^2 + \tfrac{1}{24}x^4 - \tfrac{1}{720}x^6 + \cdots \qquad (1.5b)$$

$$e^x = 1 + x + \tfrac{1}{2}x^2 + \tfrac{1}{6}x^3 + \tfrac{1}{24}x^4 + \cdots \qquad (1.5c)$$

As long as we include all the terms in the sum, these expansions are exact. Of course, since there are an infinite number of terms we can never include them all.

Why would we do this? Some physics problems cannot be solved without approximations and these expansions provide a method to approximate functions. When we truncate the sum, we replace the function with a polynomial. This introduces some error, and the trick is to keep the error as small as possible. As long as the expansion variable is small (usually as compared to 1) we can ignore some of the higher-order terms. The number of terms we keep depends on the situation. If x is small enough that we can ignore any terms with x^2 or higher, then we can replace $\sin x \approx x$, $\cos x \approx 1$, and $e^x \approx 1 + x$.

Use the following steps to expand a function in a power series with *SNB*.

1. Create the expression for the function and place the insertion point in it.
2. Choose Power Series from the Compute menu. The Series Expansion of f(x) dialog box will appear (see Figure 1.5).
3. Specify the desired Number of Terms.
4. Enter the expansion variable in the Expand in Powers of window and choose OK.

Enter the expansion variable in the Expand in Powers of window to produce an expansion about $x = 0$. You can expand about a particular non-zero point, say $x = 1$, by entering the expression $x - 1$ in the window. You can even expand about general points like $x = a$ by entering $x - a$.

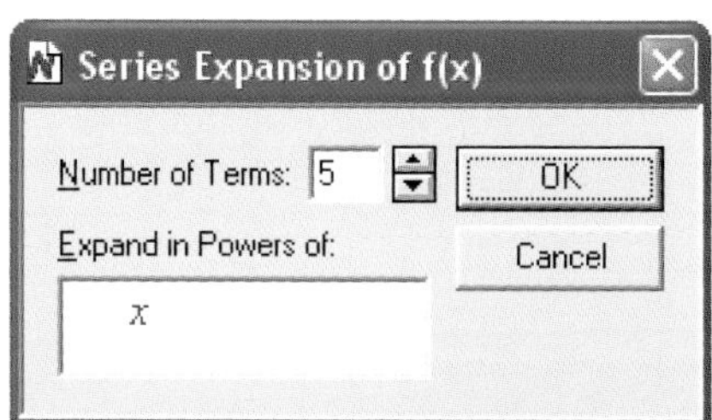

Figure 1.5 Series Expansion dialog

A series expanded about $x = 0$ is called a Maclaurin series and a series expanded about a non-zero point is called a Taylor series.

Example 1.33 *Is this what they mean by term limits?*
Find the first 4 non-zero terms in the power series expansion of the function $\tan x$.
Solution. Create an expression for $\tan x$, leave the insertion point anywhere in it and choose Power Series. Expand in Powers of x, and pick 4 as your Number of Terms.

$$\tan x = x + \tfrac{1}{3}x^3 + O\left(x^5\right)$$

This may look like it only has two terms, but *SNB* starts with the first term and counts all the terms that follow, even those with zero coefficients. For this power series, *SNB*'s 4 terms includes the x^2 and x^4 terms even though their coefficients are zero. To get the first 4 non-zero terms, you have to pick 7 as your Number of Terms.

$$\tan x = x + \tfrac{1}{3}x^3 + \tfrac{2}{15}x^5 + \tfrac{17}{315}x^7 + O\left(x^9\right)$$

The $O\left(x^9\right)$ means "order x^9" and tells you the power of x in the next non-zero term in the expansion. ◀

Consider the power series expansion of the binomial $(1+x)^n$.

$$(1+x)^n = 1 + nx + \tfrac{1}{2}n(n-1)x^2 + \tfrac{1}{6}n(n-1)(n-2)x^3 + \cdots \tag{1.6}$$

When n is an integer, this is just an nth-order polynomial with $n+1$ terms. When n is not an integer, we can use the power series expansion to replace the binomial with a polynomial. As long as x is much smaller than 1, we can approximate the expression $(1+x)^n$ with the much simpler linear expression $1+nx$.

Example 1.34 *When is close close enough?*
Compare the exact value and the first-order approximation for $(1+x)^{5/2}$ when $x = \frac{1}{10}$ and $x = \frac{1}{100}$.
Solution. Let's use Power Series to produce a 2-term expansion of the binomial.

$$(1+x)^{5/2} = 1 + \tfrac{5}{2}x + O\left(x^2\right)$$

Substitute (with Evaluate Numerically) both values of x into the two expressions.

$$\left[(1+x)^{5/2}\right]_{x=1/10} = 1.269\,1$$

$$\left[1 + \tfrac{5}{2}x\right]_{x=1/10} = 1.25$$

The percent deviation between these two numbers is about $1\frac{1}{2}\%$.

$$\left[(1+x)^{5/2}\right]_{x=1/100} = 1.025\,2$$

$$\left[1 + \tfrac{5}{2}x\right]_{x=1/100} = 1.025$$

The percent deviation between these two numbers is less than 0.02%. The approximation gets better as the value of x gets smaller. ◀

It is easy to expand a function in a power series with *SNB*. It is up to you to decide whether the approximation is appropriate and how many terms you need in your expansion to keep the answer mathematically accurate so that it is physically meaningful.

Example 1.35 *Now you know I know you know the answer...*
Use a power series expansion to estimate algebraically the first positive solution to the equation $\cos x = 0$.
Solution. You can use Solve Exact and Evaluate Numerically to remind yourself of the exact answer and its approximate numerical value.

$\cos x = 0$, Solution is: $\frac{1}{2}\pi = 1.5708$

For your first estimate, Expand in Powers of x, and pick 3 as your Number of Terms.

$$\cos x = 1 - \tfrac{1}{2}x^2 + O\left(x^4\right)$$

Delete the $O\left(x^4\right)$, set the expansion equal to zero and use Solve Exact to find x.

$0 = 1 - \frac{1}{2}x^2$, Solution is: $\sqrt{2}, -\sqrt{2}$

The result $x = \sqrt{2} \approx 1.4142$ is almost 10% smaller than the exact answer. To get a better result, try an expansion with 5 as your Number of Terms.

$$\cos x = 1 - \tfrac{1}{2}x^2 + \tfrac{1}{24}x^4 + O\left(x^6\right)$$

Delete the $O\left(x^6\right)$, set the expansion equal to zero and use Solve Numeric to find x between $x = 1$ and $x = 2$.

$\begin{bmatrix} 0 = 1 - \frac{1}{2}x^2 + \frac{1}{24}x^4 \\ x \in (1, 2) \end{bmatrix}$, Solution is: $[x = 1.5925]$

This result is only about 1.38% larger than the exact answer. The approximation gets better as the number of terms we include in the expansion gets larger. ◀

Power series expansions also allow you to extract information from analytic solutions.

Example 1.36 *What am I doing hangin' round?*
Show that the shape of a cable hanging under its own weight is approximately parabolic when the cable's weight is much smaller than the tension.
Solution. The shape of a cable hanging under its own weight is given by $y = A\cosh\frac{x}{A}$ where A is the ratio of the tension in the cable to the cable's weight. When the cable's weight is much smaller than the tension, the ratio x/A is small. Create an expression for the shape of the cable, leave the insertion point anywhere in the expression and choose Power Series. Set the Number of Terms to 3 and Expand in Powers of x.

$$y = A\cosh\frac{x}{A} = A + \tfrac{1}{2A}x^2 + O\left(x^4\right)$$

The cable's shape is approximately parabolic since $y \approx A + \frac{1}{2A}x^2$ for large A. ◀

Definitions

It is standard mathematical notation to represent a function as $f(x)$. If you Evaluate the expression $f(x)$, *SNB* interprets it as meaning $f \times x$, the product of two variables.

$$f(x) = fx$$

There is a way to define a function in *SNB* so that the expression $f(x)$ works like a function. To demonstrate, let's use the following steps to define a function f whose value at x is $x^3 + \sin^2 x$.

1. Create the equation $f(x) = x^3 + \sin^2 x$.
2. Place the insertion point anywhere in the equation.
3. Click the New Definition button on the Compute toolbar, or choose New Definition from the Definitions submenu of the Compute menu.

To define a different function, just replace the right-hand side of the equation in Step 1.

Now the symbol f represents the defined function and it behaves like a function. When you Evaluate the expression $f(x)$, you get the function you defined.

$$f(x) = x^3 + \sin^2 x$$

You can Evaluate it at particular values of x.

$$f(\pi) = \pi^3 \qquad f\left(\frac{\pi}{2}\right) = \frac{1}{8}\pi^3 + 1$$

You can Evaluate the function's derivative using standard calculus notation.

$$\frac{d}{dx} f(x) = \sin 2x + 3x^2 \qquad f'(x) = 2\cos x \sin x + 3x^2$$

You can use Evaluate to calculate the indefinite integral of the function.

$$\int f(x)\, dx = \frac{1}{2}x - \frac{1}{4}\sin 2x + \frac{1}{4}x^4$$

You can use Substitute (with Evaluate) to evaluate the derivative of the function at a particular value.

$$\left[\frac{d f(x)}{dx}\right]_{x=\pi} = 3\pi^2$$

You can use Evaluate at Endpoints (using Simplify) to calculate the change in the function.

$$\left[\ f(x)\ \right]_{x=\pi/2}^{x=\pi} = \frac{7}{8}\pi^3 - 1$$

Example 1.37 *Fibonacci Numbers*
Define a function to calculate the nth Fibonacci Number and use it to calculate the 10th, 25th, and 100th Fibonacci Numbers.

Solution. A Fibonacci Number is a member of a sequence where each number equals the sum of the two previous numbers. The first eleven are 0, 1, 1, 2, 3, 5, 8, 13, 21, 34, and 55. One way to calculate the nth Fibonacci Number uses a binomial coefficient.

$$F_n = \sum_{m=1}^{n} i^{m-1} (1 - 2i)^{n-m} \binom{2n-m}{m-1}$$

Create an expression for the above function, using the Sum button on the Math Templates toolbar and the Insert + Binomial menu item. The upper and lower limits on the Sum go in a Superscript and Subscript, respectively. Leave the insertion point anywhere in the expression and click the New Definition button.

$$F(n) = \sum_{m=1}^{n} i^{m-1} (1 - 2i)^{n-m} \binom{2n-m}{m-1}$$

To find the 10th, 25th, and 100th Fibonacci Numbers, Evaluate the function for $n = 10$, $n = 25$ and $n = 100$.

$$F(10) = 55$$

$$F(25) = 75\,025$$

$$F(100) = 354\,224\,848\,179\,261\,915\,075$$

You can use Evaluate Numerically to see just how big the 100th Fibonacci Number is.

$$F(100) = 3.542\,2 \times 10^{20}$$

◀

Once you Define a function, the symbol acts like a function. There is an issue here. Once you have defined a function, *SNB* will interpret every occurrence of that symbol as the defined function. Don't use that symbol for anything else until you have removed the definition. The New Definition command is case sensitive, so *SNB* treats our two functions f and F as different mathematical objects.

There are two ways to remove a New Definition that you created.

- Select the defining equation or select the name of the defined expression or function and choose Undefine from the Definitions submenu, or
- Choose Clear Definitions from the Definitions submenu. This will cancel **all** definitions displayed under Show Definitions that were created with New Definition.

To look at the complete list of currently defined variables and functions, click [icon] or choose Show Definitions from the Definitions submenu to open the Definitions and Mappings window. This will show all the definitions active in your document, listing the defined variables and functions in the order in which you made the definitions.

Other Good Stuff

There is often more than one way to get *SNB* to do what you want. Other times it is difficult to get *SNB* to do exactly what you want. When deciding which *SNB* command to use, there is a certain amount of trial and error. To get your results in the form you want, you may have to experiment with different combinations of commands. You may need to apply commands in a different order, or edit their output by-hand, to get the exact result you want. Or you can apply *SNB* commands to part of an expression and replace that part of the expression with the result of the computation.

Computing In-place

Using Computing in-place, you can replace part or all of an expression with the result of a computation on that part. *SNB* will replace the selected expression with the output of the command. When you hold the CTRL key while applying a command, *SNB* replaces the selected input expression with the output result. If all of the expression is selected, *SNB* will take the entire expression as input and replace the entire expression with the result. If part of the expression is selected, then the command is applied only to the selected part and only the selected part is replaced.

Here is how to execute a command in-place.

1. Use the mouse or the SHIFT and ARROW keys to select all or part of an expression.
2. Press and hold the CTRL key while applying a command to the expression.

To replace an expression by its value, you can apply Evaluate in-place by holding down the CTRL key and clicking the Evaluate button. The keyboard shortcut CTRL + SHIFT + E will also Evaluate the selection in-place.

Example 1.38 *Factor In-place*
Use Computing in-place to factor the expression $a^2 + 3b - 5a + ab - 15 + b^2$.
Solution. If you Factor or Simplify the entire expression, *SNB* returns it unchanged. Instead, create an expression equating the polynomial to itself. Select these four terms $3b - 5a + ab - 15$, hold down the CTRL key and choose Compute + Factor.

$$a^2 + 3b - 5a + ab - 15 + b^2 = a^2 + (b - 5)(a + 3) + b^2$$

The four terms are replaced by $(b - 5)(a + 3)$. ◄

Making Assumptions About Variables

Sometimes you have to place restrictions on a variable. You may want it to be real or positive, or both. *SNB* has four built-in functions that let you do this. The assume function lets you apply restrictions to a variable. You can use the additionally function to place additional restrictions. The about function returns information about the current restrictions and unassume removes all restrictions.

The six allowable assumptions you can place on a variable are real, complex, integer, positive, negative, and nonzero. When you type these or any of *SNB*'s built-in function names in Math mode, they automatically turn gray. *SNB* treats them as a single mathematical object.

Use the following steps to make an assumption about a particular variable.

1. In Math mode type assume
2. Click the expanding parentheses button (□) or enter parentheses from the keyboard.
3. Type the variable name followed by a comma and one of the allowable assumptions.
4. Leave the insertion point anywhere in the expression and choose Evaluate.

Let's look at the four solutions to the quartic equation $x^4 = 1$.

$x^4 = 1$, Solution is: $-i, -1, i, 1$

There are four solutions but only two are real and the other two are imaginary. Now we restrict the variable x to be real only.

$\text{assume}\,(x, \text{real}) = \mathbb{R}$

When we use Solve Exact on the equation, the only solutions *SNB* returns are the two that are real.

$x^4 = 1$, Solution is: $-1, 1$

If we further restrict ourselves to only the positive solutions, then *SNB* returns the only solution that is both real and positive.

$\text{additionally}\,(x, \text{positive}) = (0, \infty)$

$x^4 = 1$, Solution is: 1

You can see what assumptions you've made by using the about command.

$\text{about}\,(x) = (0, \infty)$

This tells you the value of x ranges from zero to infinity. To remove the assumptions on x, use the unassume command.

$\text{unassume}\,(x)$

To make sure there are no active assumptions about the variable x, use the about command again.

$\text{about}\,(x) = x$

This is *SNB*'s way of telling you there are no active assumptions about the variable x. You can also use the "greater than" and "less than" signs to make assumptions about a variable.

assume $(x \geq 0) = [0, \infty)$ additionally $(x < 1) = [0, 1)$

In this case, *SNB* assumes the variable x has values from $x = 0$ to $x < 1$. You can also make global assumptions about variables so that the assumption affects all variables. When you Evaluate the command

assume (positive) $= (0, \infty)$

you are telling *SNB* to treat all variables as positive. To remove this assumption, Evaluate the unassume command with no argument.

unassume ()

Inexplicably, the unassume command does not produce any output, but you can check the status of the global default with the about command.

about () = Global

This is *SNB*'s way of telling you that no special global assumptions have been made about variables.

Example 1.39 *I sleep all night and I work all day*
Solve the equation $x = 3 \ln \frac{y}{2}$ exactly for y.
Solution. Let's first try without any assumptions. Create the equation, place the insertion point anywhere in the equation, click the Solve Exact button and enter y in the Solution Variable(s) box.

$x = 3 \ln \frac{y}{2}$, No solution found.

SNB can't find the solution without more information. Evaluate the following expression and use the assume command to tell *SNB* that the variable x is a real number.

assume $(x, \text{real}) = \mathbb{R}$

Then place the insertion point anywhere in the equation, click the Solve Exact button and enter y in the Solution Variable(s) box.

$x = 3 \ln \frac{y}{2}$, Solution is: $2e^{\frac{1}{3}x}$

Our solution is $y = 2e^{x/3}$. Now that *SNB* thinks x is real, we can use Check Equality to verify $x = \ln e^{x}$.

$x = \ln e^{x}$ is true

◄

Limits

You hated them in math class, but limits are useful and important and *SNB* makes them easy to handle. You can access *SNB*'s limit operator by typing *lim* in Math mode (it automatically turns to the gray lim operator) or by clicking the Math Name button on the Math Objects toolbar.

Use the following steps to calculate a limit of a mathematical expression with *SNB*.

1. Type *lim* while in Math mode or click [sin cos] and choose lim from the list.
2. Click [Nx] or CTRL + DOWNARROW to create a Subscript.
3. Enter the limit condition in the Subscript.
4. Press the SPACEBAR to move the insertion point out of the Subscript, then enter the mathematical expression.
5. Click [=?] or choose Evaluate (or Evaluate Numerically).

The limit condition uses standard mathematical notation with the right arrow symbol ($\rightarrow$), which you can find on the Symbol Cache or the Arrows panel on the Symbol Panels. If your limit condition is "as x goes to infinity", then enter $x \rightarrow \infty$. For the limit condition "as θ approaches zero from above" enter $\theta \rightarrow 0^+$, where the plus sign is in a Superscript. For a limit "from below", replace the plus sign with a minus sign.

Several important mathematical constants are defined in terms of limits. Evaluate (and then Evaluate Numerically) the following expression to see one such constant: e, the base of the natural logarithm

$$\lim_{n \rightarrow \infty} \left(1 + \frac{1}{n}\right)^n = e = 2.718\,3$$

This is what "naturally occurring constant" means. Another is the Euler-Mascheroni constant, the most famous mathematical constant after π and e.

$$\lim_{n \rightarrow \infty} \left(\sum_{m=1}^{n} \frac{1}{m} - \ln n \right) = \text{gamma} = 0.577\,22$$

The Euler-Mascheroni constant appears, among other places, in a product formula for the gamma function. It is also one of *SNB*'s built-in constants. For the complete list of these constants, look under Help + Search, Constants (MuPAD constants).

Sometimes you need to take a Limit instead of using Evaluate because a straightforward evaluation produces a zero in the denominator.

$$\lim_{\theta \rightarrow 0} \frac{\sin a\theta}{b\theta} = \frac{a}{b}$$

You cannot Evaluate this ratio at $\theta = 0$ because the denominator equals zero there.

Example 1.40 *Newton's Nose-cone Problem*
Find the reduced drag coefficient for a hemispherical nose cone with a radius equal to its height.

Solution. The reduced drag coefficient for an elliptical nose cone is

$$C = \frac{1 - \frac{H^2}{R^2}\left(1 - \ln \frac{H^2}{R^2}\right)}{\left(1 - \frac{H^2}{R^2}\right)^2}$$

where H is the nose-cone's height and R its radius. When the height equals the radius ($H = R$) the denominator of this expression is zero, so you cannot simply Evaluate the expression. Instead take the limit as $H \to R$.

$$\lim_{H \to R} \frac{1 - \frac{H^2}{R^2}\left(1 - \ln \frac{H^2}{R^2}\right)}{\left(1 - \frac{H^2}{R^2}\right)^2} = \frac{1}{2}$$

The reduced drag coefficient of this hemispherical nose cone is $\frac{1}{2}$, so it experiences half as much air resistance as a flat circular surface with the same radius. See Appendix A for more information on the air resistance of nose cones. ◀

The limit is the fundamental concept used in defining a derivative in calculus.

Example 1.41 *Calculating a derivative the old-fashioned way*
Use the definition of a derivative to calculate the derivative of $\sin kx$ at $x = a$.

Solution. Use *SNB*'s built-in help (Help + Search, Derivatives) or a math textbook to find the expression for the definition of a derivative.

$$f'(a) = \lim_{h \to 0} \frac{f(a+h) - f(a)}{h}$$

Use Evaluate at Endpoints on the function between $x = a$ and $x = a+h$, and Expand the result.

$$[\sin kx]_{x=a}^{x=a+h} = \sin k(a+h) - \sin ak = \cos ak \sin hk - \sin ak + \cos hk \sin ak$$

Now divide by h and Evaluate the limit as h goes to zero.

$$f'(a) = \lim_{h \to 0} \left(\frac{\cos ka \sin kh - \sin ka + \cos kh \sin ka}{h} \right) = k \cos ak$$

You can also Evaluate a less transparent but more efficient one-fell-swoop expression.

$$f'(a) = \lim_{h \to 0} \left(\frac{1}{h} [\sin kx]_{x=a}^{x=a+h} \right) = k \cos ak$$

Either way, the derivative of $\sin kx$ at $x = a$ equals $k \cos ka$. ◀

Note To put the limit condition directly underneath, select the lim operator, click the Properties button and change the Operator Limit Placement to Above/Below.

A Few Words About Calculus ∫

Calculus provides us with a collection of powerful problem-solving tools, many of which you can use easily in *SNB*. The two most common, the derivative and the integral, do not have their own menu item, but you can create and Evaluate expressions for them. When in doubt, create expressions that look just like those in your math or physics book.

A derivative in *SNB* uses a Fraction and standard mathematical notation. Use the following steps to take the derivative of an expression using $\frac{d}{dx}$.

1. Place the insertion point where you want your derivative.
2. Click on the Fraction button, or choose Insert + Fraction.
3. Type a d in the numerator.
4. Press DOWN ARROW, TAB or click the denominator to move the insertion point to the denominator.
5. Type dx in the denominator.
6. Press RIGHT ARROW or the SPACEBAR to leave the fraction.
7. Click the Parentheses button and enter the mathematical expression inside them.
8. Place the insertion point anywhere in the expression and click Evaluate.

To take the second derivative, replace the d in the numerator with a d^2 and the dx in the denominator with a dx^2. To take a derivative with respect to another variable, replace each x with that variable.

Example 1.42 *Calculating a derivative a new fangled way*
Calculate the first and second derivatives with respect to time of the polynomial $4t + 4t^2 - 9t^3$.
Solution. Use the steps above to create and Evaluate the following expression for the first derivative of the polynomial.

$$\frac{d}{dt}\left(4t + 4t^2 - 9t^3\right) = -27t^2 + 8t + 4$$

Repeat the process, making the appropriate changes to create an expression for the second derivative, and Evaluate it.

$$\frac{d^2}{dt^2}\left(4t + 4t^2 - 9t^3\right) = 8 - 54t$$

Placing the polynomial in parentheses tells *SNB* you want to take the derivative of the entire polynomial. Without the parentheses, *SNB* will only take the derivative of the first term. ◀

An integral in *SNB* looks exactly like it does in your textbooks. Just like in math class, you put the expression you want to integrate between an integral sign and the dx. Use the following steps to integrate an expression.

1. Place the insertion point where you want your integral.
2. Click the Integral button to enter an integral sign. You could also click the Big Operators button or choose Insert + Operator, and click $\int$ on the choice of operators.
3. If the integral is definite, place the lower limit in a Subscript of the integral sign and the upper limit in a Superscript.
4. Enter the expression you want to integrate to the right of the integral sign.
5. Enter dx to the right of the expression.
6. Place the insertion point anywhere in the integral and click Evaluate.

To integrate with respect to another variable, replace the x in step 5 with that variable.

Example 1.43 *Area* $\frac{3}{2}\sqrt{\pi} + 2^4 3 + 0.341$
Calculate the area between the function $e^{-x^2}\left(1+x^2\right)$ and the entire x-axis.
Solution. The area between a function and the x-axis is given by the integral of the function from negative infinity to positive infinity. Use the above steps to create and Evaluate the following definite integral.

$$\int_{-\infty}^{\infty} e^{-x^2}\left(1+x^2\right) dx = \frac{3}{2}\sqrt{\pi}$$

◀

To calculate the arc length of a path along a curve, you need to take derivatives and calculate a definite integral.

Example 1.44 *Spinning Wheel got to go 'round*
Calculate the arc length of the path followed by a point on the edge of a rolling wheel for one complete revolution.
Solution. The path traced out by a point on the circumference of a rolling circle is a cycloid. The parametric equations for a cycloid created by a wheel of radius R are $x = R(t - \sin t)$ and $y = R(1 - \cos t)$. The **arc length** of the path travelled between times t_1 and t_2 described by parametric equations is

$$L = \int_{t_1}^{t_2} \sqrt{\left(\frac{dx}{dt}\right)^2 + \left(\frac{dy}{dt}\right)^2}\, dt$$

First, tell *SNB* that the radius of the wheel is a positive number.

$$\text{assume}\,(R, \text{positive}) = (0, \infty)$$

Create an expression for the integrand and Simplify it.

$$\sqrt{\left(\frac{d}{dt}\left(R\left(t-\sin t\right)\right)\right)^2+\left(\frac{d}{dt}\left(R\left(1-\cos t\right)\right)\right)^2}=\sqrt{2}R\sqrt{1-\cos t}$$

Use the result to create an expression for a definite integral from $t_1 = 0$ to $t_2 = 2\pi$, and Evaluate it.

$$L=\sqrt{2}R\int_0^{2\pi}\sqrt{1-\cos t}\,dt=8R$$

The arc length is 8 times the wheel's radius. ◀

Occasionally, *SNB* has trouble doing some rather simple definite integrals, particularly when one of the limits is zero. For example, *SNB* can easily do this indefinite integral.

$$\int\frac{x}{k^2+x^2}dx=\tfrac{1}{2}\ln\left(k^2+x^2\right)$$

But it cannot do the corresponding definite integral.

$$\int_a^b\frac{x}{k^2+x^2}dx=\int_a^b\frac{x}{k^2+x^2}\,dx$$

Sometimes, you can get around this by telling *SNB* all the variables are positive.

$$\text{assume}\left(\text{positive}\right)=\left(0,\infty\right)$$

$$\int_a^b\frac{x}{k^2+x^2}dx=\tfrac{1}{2}\ln\left(b^2+k^2\right)-\tfrac{1}{2}\ln\left(a^2+k^2\right)$$

If that does not work, you can try using Evaluate at Endpoints. Place the indefinite integral inside the Square Brackets, and use Evaluate at Endpoints between the two limits of the definite integral you want to calculate.

$$\left[\int\frac{x}{k^2+x^2}dx\right]_{x=a}^{x=b}=\tfrac{1}{2}\ln\left(b^2+k^2\right)-\tfrac{1}{2}\ln\left(a^2+k^2\right)$$

This **definite integral fudge** lets you work around a minor difficulty with *SNB*.

The following example shows you still need to think when you use *SNB* to solve a physics problem.

Example 1.45 *The ambiguity has put on weight*
Find the electric potential along the z-axis inside and outside a thin spherical shell, with radius R and uniform surface charge σ, centered at the origin. ([11], page 85)
Solution. Here is the electric potential of this shell at any point on the z-axis.

$$V\left(z\right)=\tfrac{1}{2}\tfrac{\sigma R^2}{\epsilon_0}\int_0^{\pi}\frac{\sin\theta\,d\theta}{\sqrt{R^2+z^2-2Rz\cos\theta}}$$

The value of the integral depends on whether the point is inside ($z < R$) or outside ($z > R$) the shell. To see this, tell *SNB* the variables are positive and then Evaluate the integral.

$$\text{assume}\,(\text{positive}) = (0, \infty)$$

$$\int_0^\pi \frac{\sin\theta\, d\theta}{\sqrt{R^2 + z^2 - 2Rz\cos\theta}} = \frac{1}{Rz}\left(R + z - R\,\text{signum}\,(R - z) + z\,\text{signum}\,(R - z)\right)$$

The answer includes *SNB*'s built-in sign function signum. You could edit the answer by-hand, replacing the signum functions with the appropriate values ($+1$ for inside, -1 for outside) or you could try the definite integral fudge.

$$\left[\int \frac{\sin\theta\, d\theta}{\sqrt{R^2 + z^2 - 2Rz\cos\theta}}\right]_{\theta=0}^{\theta=\pi} = \frac{1}{Rz}\sqrt{(R+z)^2} - \frac{1}{Rz}\sqrt{(R-z)^2}$$

This doesn't fix the problem. There is a mathematical ambiguity in the second term, depending on whether the point is inside or outside the shell.

$$\sqrt{(R-z)^2} = \begin{cases} R - z & (z < R) \quad \text{Inside} \\ z - R & (z > R) \quad \text{Outside} \end{cases}$$

You can avoid this ambiguity by using *SNB*'s assume function. We've already told *SNB* the variables are positive, so tell *SNB* the point is inside the shell.

$$\text{assume}\,(z < R) = (-\infty, R)$$

You can check that the range of z is between 0 and R with the about function.

$$\text{about}\,(z) = (0, R)$$

Now when you Evaluate the integral, *SNB* returns the result for inside the shell.

$$\int_0^\pi \frac{\sin\theta\, d\theta}{\sqrt{R^2 + z^2 - 2Rz\cos\theta}} = \frac{2}{R}$$

Repeat the process, but this time tell *SNB* the point is outside the shell.

$$\text{assume}\,(z > R) = (-\infty, z)$$

$$\text{about}\,(z) = (R, \infty)$$

$$\int_0^\pi \frac{\sin\theta\, d\theta}{\sqrt{R^2 + z^2 - 2Rz\cos\theta}} = \frac{2}{z}$$

So the electric potential of our spherical shell is

$$V(z) = \begin{cases} \frac{\sigma}{\epsilon_0} R & (z < R) \quad \text{Inside} \\ \frac{\sigma}{\epsilon_0} \frac{R^2}{z} & (z > R) \quad \text{Outside} \end{cases}$$

The potential is constant inside the shell, falls off as $1/z$ outside, and it is continuous at the boundary $z = R$. ◀

Units ftlb

The answer to a physics question is rarely "2" but is often "2 seconds" or "2 hours". Units are important. *SNB* comes with a complete set of built-in units from both the American and Metric Systems. Each system has three fundamental units from which other units are derived.

	Metric		**American**	
Physical quantity	*Name*	*Symbol*	*Name*	*Symbol*
Length	meter	m	foot	ft
Mass	kilogram	kg	slug	slug
Time	second	s	second	s

Table 1.2

To access *SNB*'s built-in units, you can either click the Unit Name ftlb button on the Math Templates bar or choose Insert + Unit Name from the menu bar so that the Unit Name Box appears. It displays the physical quantities on the left and available units for the selected quantity on the right. Figure 1.6 shows the units available in *SNB* for electric Current. Once you open the Unit Name Box, *SNB* allows you to keep it open continuously while you work, which is a convenient time-saving feature when you're creating expressions with units.

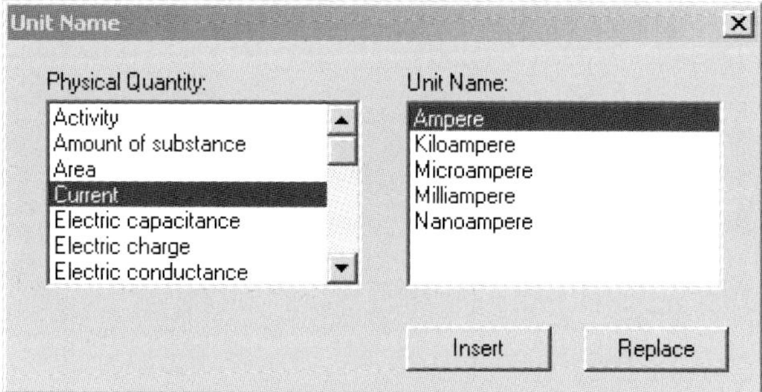

Figure 1.6 The Unit Name dialog box

Use the following steps to enter a unit using the Unit Name dialog box.

1. Place the insertion point at the position where you want the unit.
2. Choose Insert + Unit Name, or click the Unit Name button.
3. Select a category from the Physical Quantity list.
4. Select a name from the Unit Name list.
5. Choose Insert or double-click the name you selected.

The unit name will appear at the position of the insertion point. Although units appear on your screen as green characters, units are in Math mode and are active mathematical objects.

You can also enter units from the keyboard and *SNB* will automatically recognize them. The following table lists some commonly used units and their keyboard shortcuts.

Unit Name	Unit Symbol	Shortcut	Unit Name	Unit Symbol	Shortcut
kilogram	kg	*ukg*	kilometer	km	*ukme*
meter	m	*ume*	mile	mi	*umi*
second	s	*use*	foot	ft	*uft*
Newton	N	*uN*	hour	h	*uhr*
Joule	J	*uJ*	degree	°	*udeg*
centimeter	cm	*ucm*	radian	rad	*urad*

Table 1.3

To enter a unit from the keyboard, place the insertion point where you want the unit, enter Math mode, and type the shortcut. These shortcuts are case sensitive, so type them exactly as shown. For a complete list of units and their keyboard shortcuts, look under Help + Search, Units of Measure.

Converting Units

You will often need to convert the units of some physical quantity. A car's speed may be given in miles-per-hour, but its stopping distance should be calculated in feet or meters (at least for any car I'm willing to drive). There are several methods for converting units available to you with *SNB*.

The Standard Method

In the **Standard Method**, you use *SNB* to convert units in the same way you would use your calculator. You multiply the original quantity by a conversion factor that is equal to one!

Suppose we want to convert the typical (at least in principal) highway speed of 55 miles per hour to some other unit. Obviously, we can multiply the speed by 1. The trick is to multiply by the right "1".

The internal conversion factor used by *SNB* to convert between miles and meters is $1\,\mathrm{mi} = 1609.344\,\mathrm{m}$. To verify this, set up the ratio and Evaluate Numerically:

$$\frac{1.609344\,\mathrm{km}}{1\,\mathrm{mi}} = 1.0$$

The units are important! Obviously $1.609344 \div 1$ does not equal 1, but $1.609344\,\mathrm{km} \div 1\,\mathrm{mi}$ does. For our purposes, the $1\,\mathrm{mi} = 1609.3\,\mathrm{m}$ value will suffice.

Note You will find an extensive list of conversion factors under Help + Search, Conversion Factors.

Example 1.46 *I can't drive...*
Convert the typical highway speed limit from miles-per-hour to kilometers-per-hour.
Solution. To convert the $55\,\mathrm{mi/\,h}$ to $\mathrm{km/\,h}$, you multiply it by the miles-to-kilometers conversion factor. Place the insertion point anywhere in the expression and click Evaluate.

$$55.0\frac{\mathrm{mi}}{\mathrm{h}} = 55.0\frac{\mathrm{mi}}{\mathrm{h}} \times \frac{1.6093\,\mathrm{km}}{1\,\mathrm{mi}} = \frac{88.512}{\mathrm{h}}\,\mathrm{km}$$

Perhaps the more metrically inclined among us should sing "I can't drive... 88.5" in the key of $\mathrm{km/\,h}$. ◀

Example 1.47 *I still can't drive...*
Convert the typical highway speed limit from miles-per-hour to meters-per-second.
Solution. You have already converted to $\mathrm{km/\,h}$, so let's start there. To convert $88.512\,\mathrm{km/\,h}$ to $\mathrm{m/\,s}$, you must multiply it by two conversion factors. One factor converts the kilometers to meters and the other converts "per hour" to "per second". Use $1\,\mathrm{km} = 1000\,\mathrm{m}$ and $1\,\mathrm{h} = 3600\,\mathrm{s}$ to create the following expression, place the insertion point anywhere in the expression and Evaluate it.

$$88.512\frac{\mathrm{km}}{\mathrm{h}} = 88.512\frac{\mathrm{km}}{\mathrm{h}} \times \frac{1000\,\mathrm{m}}{1\,\mathrm{km}} \times \frac{1\,\mathrm{h}}{3600\,\mathrm{s}} = 24.587\frac{\mathrm{m}}{\mathrm{s}}$$

It seems that $55\,\mathrm{mi/\,h}$ is about $24.587\ \mathrm{m/\,s}$. ◀

The Solve Method

To use the Standard Method of converting units, you must know the appropriate conversion factors. Some of them are fairly obscure (such as the hectare to square-meter conversion factor) and difficult to remember. The **Solve Method** of converting units avoids this problem. To convert with the Solve Method, write the conversion equation in the following form:

Quantity in original Units = a Variable times new Unit

Then use Solve Exact in the usual way by placing the insertion point anywhere in the equation and clicking x?.

Example 1.48 *I really can't drive...*
Convert the typical highway speed limit from miles-per-hour to meters-per-second using the Solve Method.
Solution. Create an expression equating $55\,\mathrm{mi/\,h}$ to $v\,\mathrm{m/\,s}$, put the insertion point anywhere in the expression and click Solve Exact.

$$55\frac{\mathrm{mi}}{\mathrm{h}} = v\frac{\mathrm{m}}{\mathrm{s}}, \text{ Solution is: } 24.587$$

If you use the Standard Method with the $1\,\mathrm{mi} = 1609\,\mathrm{m}$ conversion factor value found in many textbooks, your result will be slightly different because *SNB* uses its more precise internal conversion factor. ◀

In physics, we usually measure angles in radians. *SNB*'s built-in trigonometric and inverse trigonometric functions all default to radians. You may be more familiar with degrees, and you may want to convert between the two. There are two forms of the degree unit in *SNB*.

The **green** degree unit is listed under Plane Angle in the Unit Name Box, or you can create it with the keyboard shortcut *udeg*. The **red** degree symbol is the small circle symbol in a Superscript immediately after a symbol or number. You'll find the circle on the Symbol Cache or the Binary Operations panel.

The two degree units behave differently. When you Evaluate an expression or use Solve Exact on an equation with the **green** degree unit, *SNB* gives an approximate numerical result. With the **red** degree symbol, *SNB* gives exact symbolic results.

Example 1.49 *Converting from degrees to radians*
Convert 30° to radians using both degree units.
Solution. Enter the following equations in Math mode, leave the insertion point in each one and click the Solve Exact button.

Red: $30^\circ = \theta$, Solution is: $\frac{1}{6}\pi$

Green: $30\,{}^\circ = \theta\,\mathrm{rad}$, Solution is: $0.523\,60$

You can see the solutions $\frac{1}{6}\pi = 0.523\,60$ are the same with Evaluate Numerically. ◄

Example 1.50 *Converting from radians to degrees*
Convert $\pi/6\,\mathrm{rad}$ to degrees using both degree units.
Solution. Enter the following equations in Math mode, leave the insertion point in each one and click the Solve Exact button.

Red: $\frac{1}{6}\pi = \theta^\circ$, Solution is: 30

Green: $\frac{1}{6}\pi\,\mathrm{rad} = \theta\,{}^\circ$, Solution is: 30.0

If you prefer the Standard Method, you can try either degree unit with the conversion factor $180\,{}^\circ = \pi\,\mathrm{rad}$ and Evaluate the following expressions.

Red: $\frac{\pi}{6} \times \frac{180^\circ}{\pi} = \frac{1}{6}\pi$

Green: $\frac{\pi}{6}\,\mathrm{rad} \times \frac{180\,{}^\circ}{\pi\,\mathrm{rad}} = 30\,{}^\circ$

Because the **red** degree symbol produces exact symbolic results, Evaluate returns $180^\circ/\pi$ as 1. Use the **green** degree unit to convert with the Standard Method. ◄

You can place $180^\circ/\pi$ in front of an inverse trigonometric function and Evaluate the expression. *SNB* will return the angle in degrees with the **green** degree unit.

Red: $\frac{180^\circ}{\pi}\csc^{-1} 2 = \frac{1}{6}\pi$ **Green**: $\frac{180\,{}^\circ}{\pi}\csc^{-1} 2 = 30\,{}^\circ$

So 30° is the angle whose cosecant equals 2.

You can also force Solve Exact to return a solution in degrees by using $\pi/180$ in the argument of the trigonometric function.

$\tan \frac{\pi\theta}{180} = 1$, Solution is: 45 $\qquad$ $\tan \theta^\circ = 1$, Solution is: 45

This also works with Solve Numeric.

$\begin{bmatrix} \cos \frac{\pi\theta}{180} = -\frac{\sqrt{3}}{2} \\ \theta \in (180, 270) \end{bmatrix}$, Solution is: $\{[\theta = 210.0]\}$

$\begin{bmatrix} \cos \theta^\circ = -\frac{\sqrt{3}}{2} \\ \theta \in (180, 270) \end{bmatrix}$, Solution is: $\{[\theta = 210.0]\}$

SNB treats the **red** degree symbol and $\frac{\pi}{180}$ the same way.

The Default Method

The Evaluate Numerically command defaults to the metric system's fundamental units. This **Default Method** works even if you mix units from more than one system, which is not a good idea usually, and provides an easy fool-proof method for converting ***to*** the metric system. For example, let's apply Evaluate Numerically directly to our highway speed.

$$55\frac{\text{mi}}{\text{h}} = 24.587\frac{\text{m}}{\text{s}}$$

Since the metric unit for time is the second, if we apply Evaluate Numerically to one year we find out how many seconds are in one year.

$$1\,\text{y} = 3.155\,7 \times 10^7\,\text{s}$$

One year is approximately thirty-one million, five hundred and fifty-seven thousand seconds. The Default Method lets us compare quantities in different units easily. Let's see how 100 feet-per-minute compares with half a millimeter-per-millisecond.

$$\frac{100\,\text{ft}/\,\text{min}}{0.5\,\text{mm}/\,\text{ms}} = 1.016$$

It is about 1.6% larger. The metric unit for temperature is the kelvin, which *SNB* denotes in the Unit Name box as K. The result from the Solve Method is in a slightly different form than the result from the Default Method.

Solve Method: $70\,^\circ\text{F} = T\ \text{K}$, Solution is: 294.26

Default Method: $70\,^\circ\text{F} = 294.26\,\text{tmpK}$

SNB has two equivalent ways to denote a kelvin. Remember, any time you use the Evaluate Numerically command on an expression with units, the result will always be in the metric system's fundamental units.

User-Defined Units

Even though *SNB* has an impressive collection of predefined units, there is always room for more. You can use Insert + Math Name to create your own used-defined unit names. These names will appear gray onscreen, the same as any Math Name used for a function. Then you can use the Define command to relate your new unit to existing units. To use your new unit again, click Insert + Math Name and choose your unit from the alphabetical list.

There are two steps to define a new unit.

1. Create a Math Name

   Place the insertion point where you want the new unit and click Insert + Math Name. Type the name of your new unit in the Name box, click Function or Variable for the Name Type and click OK.

2. Define your new Unit

   Create an equation that defines your new unit in terms of other *SNB* units. With the insertion point in the equation, click the New Definition button.

The addition of your new name to the Math Name list is global but the defining equation is local. If you want to use your new unit in another document, you will have to repeat the "Define your new Unit" step. To save your new unit definition for future use in your current document, be sure the Always Restore and Always Save options are selected on the Definition Options page of the Computation Setup dialog box.

Example 1.51 *To boldLY gAU...*
Two important units of distance in Astronomy are the light-year and the astronomical unit. How many astronomical units are there in one light-year?
Solution. One light-year, defined as the distance light travels in one year, is 9.4605×10^{15} m. One AU, defined as the average distance between the Earth and the Sun, is 1.4960×10^{11} m. Neither is on *SNB*'s list of built-in units, so create a Math Name for each and use the following two equations to Define each unit.

$$\mathrm{AU} = 1.4960 \times 10^{11}\,\mathrm{m}$$

$$\mathrm{LY} = 9.4605 \times 10^{15}\,\mathrm{m}$$

Now you can use Solve Exact to convert.

$1\,\mathrm{LY} = x\,\mathrm{AU}$, Solution is: 63239.

Light travels $63\,239$ AU in one year. As with any unit, you can convert your newly defined unit to its fundamental SI components with Evaluate Numerically.

Let's use the Solve Method to see how many miles are in a light-year.

$1\,\mathrm{LY} = x\,\mathrm{mi}$, Solution is: $5.878\,5 \times 10^{12}$

There are almost 5.9 trillion miles in one light-year. ◀

Plotting

The capability to create plots is a strength of *SNB*. You can plot expressions, data, functions, and numerical solutions to ordinary differential equations with the click of a button. You can easily add new items to an existing plot just by dragging and dropping them onto the plot. You can also adjust the appearance of each item in your plot individually.

To see how easy it is to make a plot with *SNB*, let's reproduce Figure 1.1. Create the expression $x^2e^{-3x}\sin 4x$, leave the insertion point anywhere in it and click the Plot 2D Rectangular button. If you set the Default Plot Intervals from 0 to 5, your plot should look like Figure 1.1; otherwise you need to adjust the appearance of your plot.

To adjust your plot's appearance, click on the plot to select it and click the Properties button on the Standard Toolbar. When you do so, the Plot Properties dialog box opens. The Plot Properties dialog has five pages that you can use to alter your plot's Layout, Labeling, Items Plotted, Axes, and View.

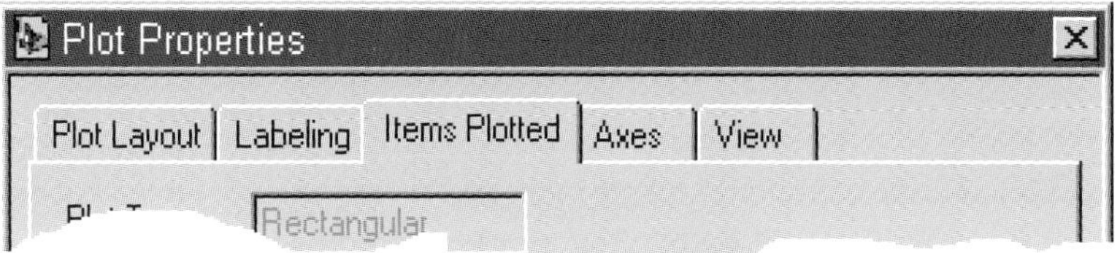

Figure 1.7 The Five Pages

From the Plot Layout page, you can choose whether your plot appears In-line or is Displayed centered on a separate line. Most of the plots in this chapter are Displayed. You can change the Size of your plot, and adjust its Screen Display Attributes and Print Attributes. From the Labeling page you can enter a label in any combination of Text and Math. If you move your plot, the label moves with it.

The Items Plotted page gives you access to each item in the plot. Each item has its own Item Number, so you can adjust each item separately. When you choose an item, it appears in the Expressions and Relations box where you can edit it directly. You can use Delete Item to remove the current item from the plot. Click Add Item and type or Paste a new expression directly into the Expressions and Relations box.

For each item, you can adjust the interval to be plotted. Click the Variables and Intervals button and change the Plot Intervals. You can also change the number of points *SNB* uses to draw the plot by adjusting the number of Points Sampled. Using more points gives your plot better resolution, but it also takes more time to plot.

The Items Plotted page also lets you adjust the details of the curve drawn for each item. The Plot Style can be either a Line or a series of Points. The Line Style can be Solid, Dash, Dot, DotDash or DotDotDash. The options for Point Marker are Dot, Circle, Cross, Box or Diamond. The Plot Color is the color of the Line or Point Markers. You can use one of the twenty named colors or click the Edit Color button and choose from the additional unnamed colors. The three choices for Line Thickness are Thin, Medium, and Thick.

The Axes page allows you to control the details of the coordinate axes of your plot. There are four choices for Axes Type, including None. Normal axes intersect in the middle of the plot, Framed axes intersect in a left corner, and Boxed axes form a rectangular box around the plot. For Axis Scaling you can choose a Linear plot, Log (a semi-log) plot, or a Log-Log plot. Check the Equal Scaling Along Each Axis check box when plotting circles or trajectories. You can change the number of Tick Marks along either axis, and you can add an x-axis label or a y-axis label, but they can be in Text only (not in Math mode).

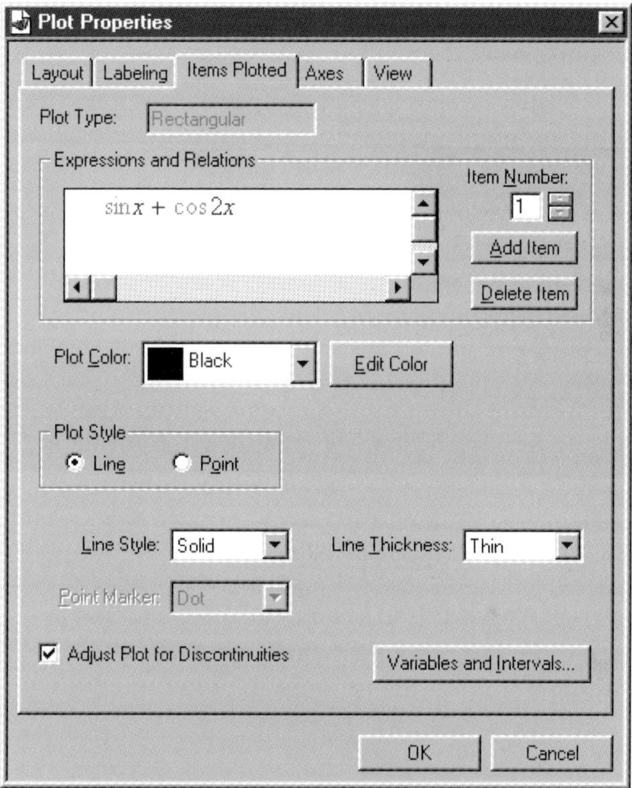

Figure 1.8 The Items Plotted page.

From the View page you can set the View Intervals for a 2-dimensional plot. The Plot Intervals on the Items Plotted page set the range of points that *SNB* evaluates when making the plot and the View Intervals determine the coordinates that are visible. When you click the Generate Snapshop button, *SNB* generates a graphic file of your plot in WMF format, gives it a random name and stores it in the same folder as your *SNB* document. You can rename the file and use it in many other applications.

Every graph in this book was created with *SNB* and its various plotting capabilities including the Generate Snapshop button.

Plot 2D Rectangular

With the Plot 2D Rectangular button (found on the Compute Toolbar) you can make rectangular x-y plots of expressions, data, functions, and numeric solutions to ordinary differential equations. These are useful in lab reports, homework problems, and research results. Oh, and physics books too.

It's easy to make a 2-dimensional rectangular plot of an expression with one variable.

1. Create the expression in your document where you want your plot.

2. Place the insertion point anywhere in the expression, and click the Plot 2D Rectangular button or choose Plot 2D + Rectangular.

If your expression is an equation, it must be in the form "y = one-variable expression" and you must put the insertion point in or just to the right of the "one-variable expression" to plot it with Plot 2D Rectangular.

The following example revisits the graph in Figure 1.1 of the expression $x^2e^{-3x}\sin 4x$.

Example 1.52 *The envelope please...*
Make a 2-dimensional plot of $y = x^2e^{-3x}\sin 4x$ for values of x running from 0 to 4, and add the expressions $\pm x^2e^{-3x}$ to your plot.
Solution. Create the expression $x^2e^{-3x}\sin 4x$, leave the insertion point anywhere inside it and click the Plot 2D Rectangular button. Select the plot and open the Plot Properties dialog box. Choose the Items Plotted page and click the Variable and Intervals button. Set the Interval from 0 to 4. Close the Plot Properties dialog box.

To add the x^2e^{-3x} part of your expression to the plot, select it the with the mouse, and drag it onto the plot. Create the expression $-x^2e^{-3x}$, select it with the mouse and drag it onto the plot. Open the Plot Properties dialog box and change the Plot Color of the last two items to Light Red and their Line Style to Dash.

Here is the resulting plot shown with its Placement set to Displayed.

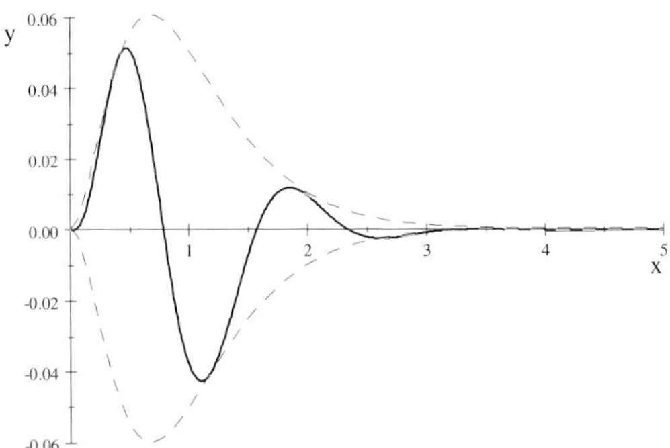

Figure 1.9 The envelope please

As you can see, the two expressions $\pm x^2e^{-3x}$ form an envelope inside of which the function $y = x^2e^{-3x}\sin 4x$ oscillates. ◀

When plotting trigonometric functions, *SNB* uses radians as the default unit. The following example shows you how to force *SNB* to plot trigonometric expressions in degrees rather than radians.

Example 1.53 *Sines and Cosines*
Plot the two trigonometric functions $\sin\theta$ and $\frac{3}{5}\cos 2\theta$ in degrees from $\theta = -180°$ to $\theta = +180°$.

Solution. You can force *SNB* to plot the functions in degrees by putting the **red** degree symbol after the argument of the functions. Create the expression $\sin\theta°$ with the **red** degree unit, place the insertion point anywhere in it and click the Plot 2D Rectangular button. Open the Plot Properties dialog. On the Items Plotted page, click the Variable and Intervals button and set the Interval from -180 to 180. Change the PlotColor to LightBlue and the Line Thickness to Medium. Click OK.

To plot the cosine function, create the expression $\frac{3}{5}\cos 2\theta°$, select it with the mouse and drag it onto the plot. Change its PlotColor to LightRed and the Line Thickness to Medium. Click OK.

Notice the effects of the constants in the expression $\frac{3}{5}\cos 2\theta°$.

The overall factor $\frac{3}{5}$ reduces the size of the wiggle and the 2 multiplying the θ increases the frequency of the wiggle.

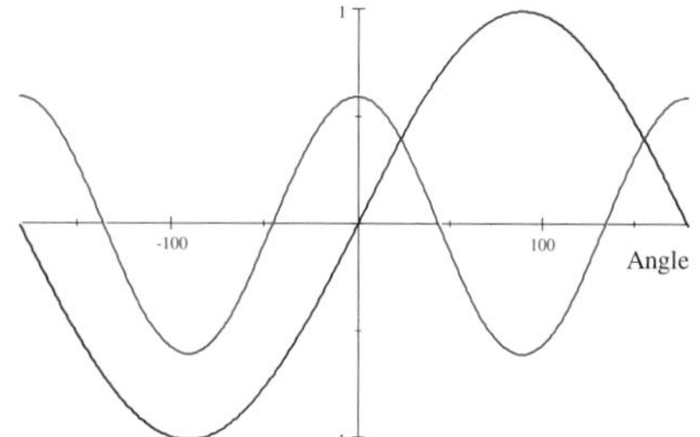

Figure 1.10 Two trigonometric functions

◀

Other 2-Dimensional Plots

SNB offers many other choices for 2-dimensional plots, including Polar plots, Implicit plots, and Parametric plots. There is no button for these options, but you will find them on the Compute menu under Plot + 2D.

In a 2-dimensional rectangular plot, each point is specified by its x and y coordinates. In a polar plot, a point is specified by its distance r from the origin and the angle ϕ the line connecting the point with the origin makes with the x-axis. The rectangular and polar coordinates are related by the usual trigonometric functions.

$$x = r\cos\phi \tag{1.7a}$$

$$y = r\sin\phi \tag{1.7b}$$

When r is constant, these equations describe a circle. When r varies with ϕ, these equations describe many interesting shapes. In its documentation, *SNB* uses θ for the angle in the x-y plane, but we'll use the notation found in most physics books.

To make a 2-dimensional polar plot of an expression with a single variable, create the expression in your document where you want your plot. Place the insertion point anywhere in the expression and choose Plot 2D + Polar. Do not click the Plot 2D Rectangular button unless you want a rectangular plot of your expression instead of a polar plot.

Example 1.54 *A wiggly-piggly orbit*
Make a polar plot of the expression $r = 1 + \frac{1}{4}\sin 5\phi$, and include a reference unit circle.
Solution. Place the insertion point anywhere in the expression $1 + \frac{1}{4}\sin 5\phi$ and choose Plot 2D + Polar. Open the Plot Properties dialog. On the Items Plotted page, click the Variable and Intervals button and set the Interval from 0 to 6.2832. Change the PlotColor to LightRed and the Line Thickness to Medium. On the Axes page, click the Equal Scaling Along Each Axis box. Create the expression 1 and add it to the plot. Change its LineStyle to Dash.

Figure 1.11 shows the result: a lovely 5-point star-shaped periodic orbit that oscillates about the unit circle. See Appendix A for the force that produces such an orbit.

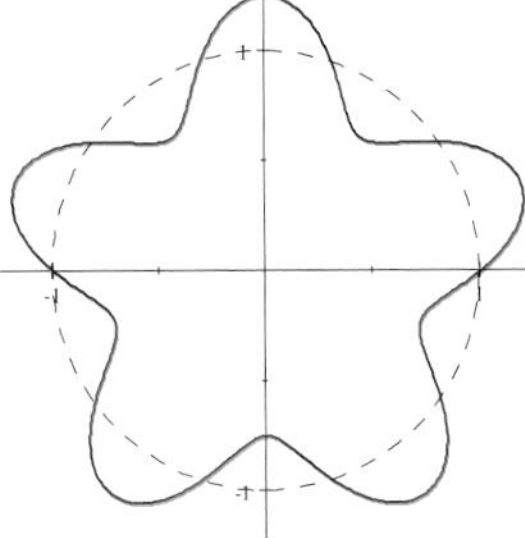

Figure 1.11 Wiggly-piggly

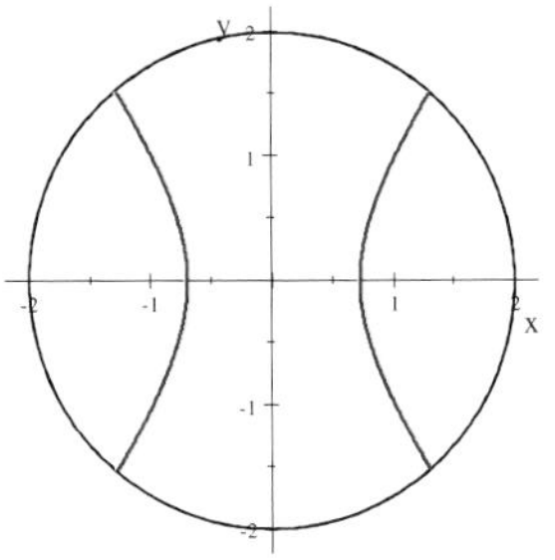

Figure 1.12 Circle the hyperbola ◀

You need an equation relating variables in the form "y = 1-variable expression" to make a rectangular plot with Plot 2D Rectangular. Sometimes it is inconvenient or impossible to create such an equation. In such cases, you can make an Implicit plot.

Use these steps to make a 2-dimensional Implicit plot of an equation with two variables.

1. Create an ***equation*** in your document where you want your plot.

2. Place the insertion point anywhere in the equation and choose Plot 2D + Implicit.

If you click the Plot 2D Rectangular button, *SNB* will attempt to create a 2-dimensional rectangular plot of whichever side of the equation you left the insertion point.

Example 1.55 *A two-seam fastball*
Make an implicit plot of the circle $x^2 + y^2 = 4$ and the hyperbola $x^2 - y^2/2 = 1/2$ such that the circle encloses the hyperbola.
Solution. We need to use Solve + Numeric on the two equations simultaneously to calculate the hyperbola's Plot Intervals.

$$\begin{bmatrix} x^2 + y^2 = 4 \\ x^2 - \frac{1}{2}y^2 = \frac{1}{2} \end{bmatrix}, \text{ Solution is: } \{[x = 1.2910, y = 1.5275]\}$$

Place the insertion point anywhere in the first equation and choose Plot 2D + Implicit. Select the hyperbola equation and add it to the plot. Open the Plot Properties dialog box and change the hyperbola's Line Thickness to Thick. Change the hyperbola's Plot Intervals, letting x run from -1.291 to $+1.291$ and y from -1.5275 to $+1.5275$.

Figure 1.12 shows the result: a plot that resembles the stitches on a baseball. ◀

To make a Rectangular or Implicit plot, you need one equation that relates x and y directly. In physics, we sometimes have two separate equations relating the x and y coordinates to some third parameter, which is often time. These defining equations are called parametric equations and you can make a 2-dimensional plot of them with *SNB*'s Parametric plot option.

Use the following steps to make a 2-dimensional Parametric plot.

1. Create an expression in your document which encloses the two one-parameter expressions in parentheses, separated by a comma, in the form $(x(t), y(t))$. Make sure there is only one variable in this expression (the parameter) and no equal signs.

2. Place the insertion point anywhere in the expression and choose Plot 2D + Parametric.

You can also use the Plot 2D Rectangular button to make a Parametric plot, and you can use a 1-column, 2-row or a 1-row, 2-column matrix to enclose the expressions.

Example 1.56 *A cycloid*
Make a parametric plot of the cycloid created by one revolution of a wheel with radius $R = 2$.
Solution. For a cycloid created by a wheel of radius $R = 2$, the equation for the horizontal distance is $x = 2(t - \sin t)$ and $y = 2(1 - \cos t)$ is for the height. Use these parametric equations to create the following expression.

$$(2(t - \sin t), 2(1 - \cos t))$$

Place the insertion point anywhere in the expression and choose Plot 2D + Parametric. Open the Plot Properties dialog box and change the cycloid's Line Thickness to Medium and PlotColor to LightBlue. Change the cycloid's Plot Intervals, letting the parameter t run from 0 to $+6.283$ (which is about 2π). On the Axes page, check the Equal Scaling Along Each Axis checkbox.

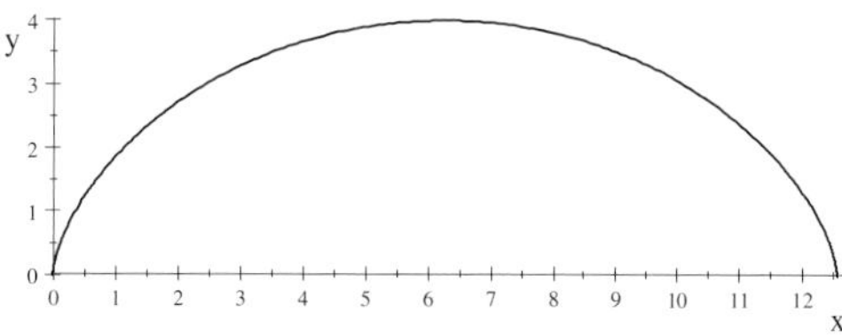

Figure 1.13 A cycloid

That's what a cycloid looks like! ◀

To make a parametric polar plot, create an expression in the form $(r(t), \theta(t))$, place the insertion point anywhere in it, and choose Plot 2D + Parametric. Open the Plot Properties dialog box. Look at the top of the Items Plotted page and change the Plot Type to Polar. To make this change, you must use Plot 2D + Parametric to make the plot, and not the Plot 2D Rectangular button.

Plot 3D Rectangular

With the Plot 3D Rectangular button (found on the Compute Toolbar) you can make rectangular x-y-z plots of expressions, data, functions, and numeric solutions to ordinary differential equations. Like their 2-d counterparts, these are useful in lab reports, homework problems, research results, and other noteworthy documents.

Here's how to make a 3-d rectangular plot of an expression with two variables.

1. Create the expression in your document where you want your plot.

2. Place the insertion point anywhere in the expression, and click the Plot 3D Rectangular button or choose Plot 3D + Rectangular.

If your expression is an equation, it must be in the form "z = two-variable expression" and you must put the insertion point in or just to the right of the "two-variable expression" to plot it with Plot 3D Rectangular.

The Plot Orientation Tool 

SNB allows you to look at your 3-dimensional plot from any direction by changing the plot's orientation. Double-click a 3-d plot and the Plot Orientation Tool appears in the upper right-hand corner of the plot frame and eight gray handles, which do not resize the plot, appear along the outside of the frame. Once the tool is activated, you can change the plot's orientation by left-clicking on it, holding the button down, and moving the mouse. As you do, a 3-dimensional rectangular box indicates the plot's new orientation. When you release the mouse button, *SNB* redraws the plot in its new orientation.

You can also change the orientation of your plot on the View page of the Plot Properties dialog, where you can set the Tilt and Turn. The Tilt is the polar angle which sets the orientation of the positive z-axis. The Tilt can have integer values from -180 to $+180$. The Turn is the azimuthal angle which sets the orientation of the x-y plane relative to the z-axis, and can have integer values from -360 to $+360$. The default orientation has the Turn and Tilt both set at 45. When you use the Plot Orientation Tool, vertical motions of the mouse change the Tilt and horizontal motions change the Turn.

When the Tilt is zero, the positive z-axis points out of your computer screen toward you. From this orientation, increasing the Tilt aims the z-axis toward the top of the screen. Table 1.4a gives the direction of the $+\,z$-axis for various Tilt values.

Tilt	Direction of + z-axis
0	Toward you
+ 90	Up
$\pm$ 180	Away from you
– 90	Down

Table 1.4a

Turn	Direction of + x-axis
0	Down
90	Left
180	Up
270	Right

Table 1.4b

Table 1.4b gives the direction of the positive x-axis for various values of the Turn when the Tilt is zero so the $+\,z$-axis points toward you. From this orientation, an increase in the Turn moves the $+\,x$-axis clockwise and a decrease moves it counterclockwise.

Table 1.5 may help you pick the best orientation for your 3-dimensional plot. A Turn of $+270$ is the same as a Turn of -90.

To see the	from the	Turn	Tilt
x-y plane	+ z-axis	− 90	0
x-z plane	− y-axis	− 90	+ 90
y-z plane	+ x-axis	0	+ 90

Table 1.5

When you plot an expression in the form "z = two-variable expression" with Plot 3D Rectangular, the result is a 2-dimensional surface.

Example 1.57 *A surface that features many surface features*
Make a 3-dimensional plot of the surface $z = 2\sin x \cos 2y$.
Solution. Create the expression $2\sin x \cos 2y$, leave the insertion point anywhere in it, and click the Plot 3D Rectangular button. Open the Plot Properties dialog. On the Items Plotted page, use the Variables and Intervals button to set the Plot Intervals for both x and y from 0 to 6.28 (about 2π). On the Axes Page, choose Framed as the Axes Type. Set the Turn to -50 and the Tilt $+60$ on the View Page.

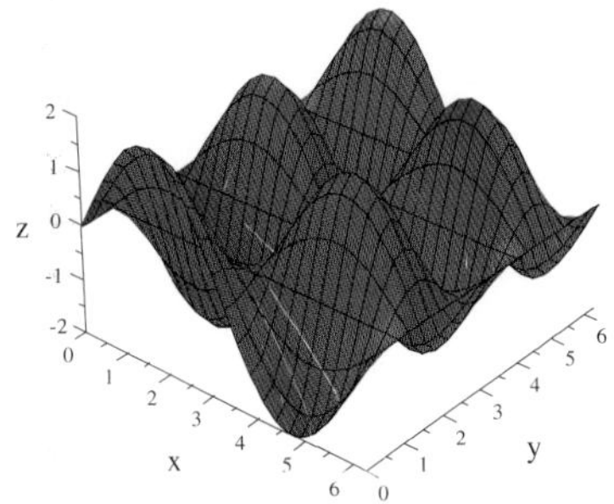

Figure 1.14 The surface $z = 2\sin x \cos 2y$

After you make this or any 3-d plot, experiment with the Plot Orientation Tool and look at it from different perspectives. ◄

A **trajectory** is the path followed by an object moving through space. To plot a 3-dimensional trajectory in *SNB*, you need three 1-parameter expressions, one each for the three coordinates (x, y, z). Although *SNB* doesn't offer a "parametric" option on the Plot 3D menu, this is essentially a 3-dimensional parametric plot.

Here's how to make a plot of a 3-d trajectory (what *SNB* calls a "curve in space").

1. Create an expression in your document which encloses three one-parameter expressions in parentheses, separated by commas in the form $(x(t), y(t), z(t))$.

2. Place the insertion point in the expression and choose Plot 3D + Rectangular.

You can also use the Plot 3D Rectangular button to make a 3-dimensional plot of a trajectory. Make sure there is only one variable in the expression (the parameter) and no equal signs. You can also use a 1-column, 3-row or a 1-row, 3-column matrix to enclose the parametric expressions.

Example 1.58 *A routine fly ball*
Make a 3-dimensional plot of the trajectory of a typical major league fly ball hit into a significant cross wind.
Solution. The trajectory of a typical major league fly ball under the influence of gravity, air resistance, and a significant cross wind is described in meters by the following parametric equations.

$$\begin{aligned} x &= 147\left(1 - e^{-0.211t}\right) \\ y &= 4t \\ z &= 1 - 46.4t + 367\left(1 - e^{-0.211t}\right) \end{aligned}$$

Create the expression $\left(147\left(1 - e^{-0.211t}\right), 4t, 1 - 46.4t + 367\left(1 - e^{-0.211t}\right)\right)$, leave the insertion point anywhere in it, and click the Plot 3D Rectangular button. Open the Plot Properties dialog. On the Items Plotted page, use the Variables and Intervals button to set the Plot Interval for t from 0 to 5.4. On the View page, set the Turn to -82 and the Tilt to 85. On the Axes page, Label the x, y, and z axes "Range", "Yaw", and "Height" respectively, choose Framed for the Axes Type, and check the Equal Scaling Along Each Axis checkbox.

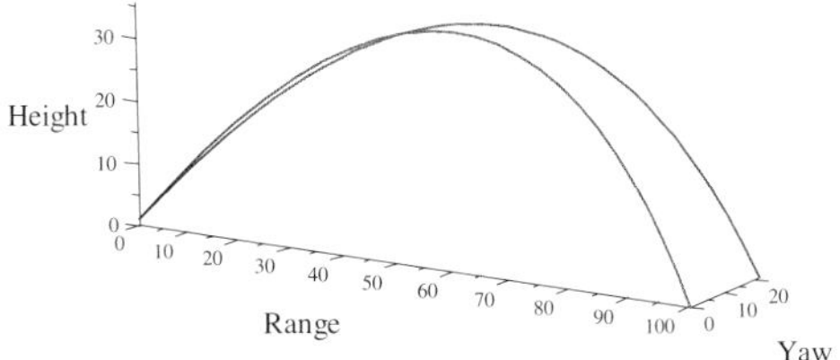

Figure 1.15 I got it!

Copy the expression for the ball's trajectory and change the y-component to zero for the ball's trajectory without the cross wind, and add it to the plot. We'll explain these expressions for the ball's trajectory in the chapter on Projectile Motion. ◀

Cylindrical and Spherical Plots

SNB offers many other choices for 3-dimensional plots, including Cylindrical plots, Spherical plots, and Tube plots. None of these choices has a button, but you will find them on the Compute menu under Plot + 3D.

In a 3-d rectangular plot, each point is specified by its (x, y, z) coordinates. In a cylindrical plot, a point is specified by (r, ϕ, z). The distance r and the angle ϕ are the same coordinates used in 2-d polar plots, and z is the point's distance above the x-y plane. Rectangular and cylindrical coordinates are related by trigonometric functions.

$$x = r\cos\phi \tag{1.8a}$$
$$y = r\sin\phi \tag{1.8b}$$
$$z = z \tag{1.8c}$$

When r is constant, these equations describe a cylinder. When r is a function of ϕ and z, these equations describe many interesting 3-dimensional shapes.

As the following example shows, when r is a function of z only, the resulting shape has a circular cross section.

Example 1.59 *La Tour d'Eiffel circulaire*?
Use Plot 3D + Cylindrical to plot the shape of the Eiffel Tower as if the Tower had a circular cross section.
Solution. The shape of the Eiffel Tower can be described as

$$r(z) = -\,z\,w(z) + \sqrt{r_0^2 + z^2 w^2(z)}$$

where the height-dependent wind profile is

$$w(z) = 0.690 - 1.53 \times 10^{-3}\,z + 3.96 \times 10^{-5}\,z^2 - 9.22 \times 10^{-8}\,z^3$$

and $r_0 = 62.5$ is the Tower's radius in meters at the bottom ($z = 0$). [23]

Create a New Definition for the function $w(z)$ as described above. Create the expression $-\,z\,w(z) + \sqrt{62.5^2 + z^2 w^2(z)}$, place the insertion point in it and choose Plot 3D + Cylindrical. Open the Plot Properties dialog. On the Items Plotted page, click the Variables and Intervals button and set the range of z from 0 to 300.

Choose Z (hue) as the Directional Shading, so different colors denote changes in height.
On the Axes page, check the Equal Scaling Along Each Axis box and set the Axes Type to Framed.
Chose Custom Tick Marks and set the number on the x, y, and z axes to 3, 3, and 4 respectively.
Go to the View page and set the Turn to -45 and the Tilt to 83.
Voilà! See Appendix A for more on the shape of the Eiffel Tower.

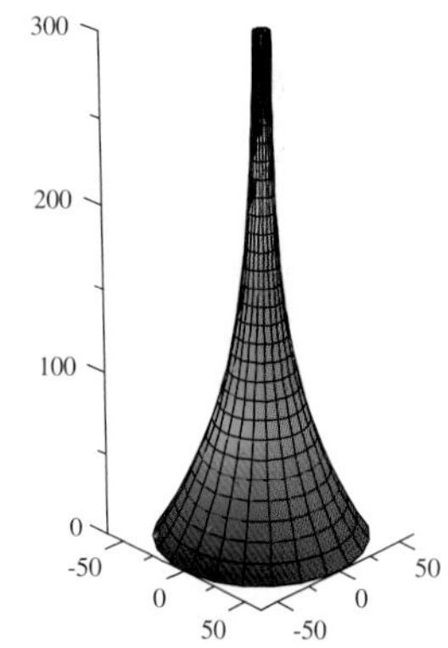

Figure 1.16 The Eiffel Tower

◀

In a 3-d spherical plot, each point is specified by its (r, θ, ϕ) coordinates. The radial coordinate r is the distance from the point to the origin. The azimuthal angle ϕ is measured relative to the positive x-axis and the polar angle θ is measured relative to the positive z-axis. Spherical and rectangular coordinates are related.

$$x = r\cos\phi\sin\theta \quad (1.9a)$$

$$y = r\sin\phi\sin\theta \quad (1.9b)$$

$$z = r\cos\theta \quad (1.9c)$$

When r is constant, these equations describe a sphere. When r is a function of θ and ϕ, these equations describe many interesting 3-dimensional shapes.

Most physicists use ϕ for the azimuthal angle and θ for the polar angle, but *SNB* does not. You can fix this potential problem after you create your spherical plot. When you open the Plot Properties dialog and click the Variables and Intervals button, you'll see the Switch Variables button. Clicking it switches the definitions of θ and ϕ. Be sure you only click it once!

Example 1.60 *Lately it o-Kerrs to me...*
Make a 3-dimensional spherical plot of the ergosphere and event horizon around an extreme Kerr black hole.
Solution. The size of the event horizon around a rotating black hole is given by

$$R_{\text{EH}} = M + \sqrt{M^2 - a^2}$$

where M is the mass and a is related to the rotational angular momentum of the black hole. An extreme Kerr black hole has the maximum value for the rotational angular momentum, $a = M$.

The ergosphere is the region between the event horizon and the static limit

$$R_{\text{SL}} = M + \sqrt{M^2 - a^2 \cos^2 \theta}$$

where θ is the polar angle. The rotation is in the azimuthal ϕ-direction.

Let's use Substitute to get expressions for the static limit and event horizon of a rapidly spinning black hole with mass $M = 1$.

$$R_{\text{EH}} = \left[M + \sqrt{M^2 - a^2}\right]_{a=M, M=1} = 1$$

$$R_{\text{SL}} = \left[M + \sqrt{M^2 - a^2 \cos^2 \theta}\right]_{a=M, M=1} = \sqrt{1 - \cos^2 \theta} + 1$$

Put the insertion point in the expression for R_{SL} and choose Plot 3D + Spherical. Open the Plot Properties dialog. Click the Variables and Intervals button on the Items Plotted page and click the Switch Variables button so *SNB* interprets θ as the polar angle. Change the Surface Style to Wire Frame. Click the Add Item button and enter 1 in the Expressions and Relations window.

For Item 2, change the Surface Style to Color Patch, the Surface Mesh to None, and the Directional Shading to Z (grayscale). Choose LightGray for the Base Color and Gray for the Secondary Color.

On the Axes page, check the Equal Scaling Along Each Axis checkbox, and set the Axes Type to Framed. On the View page, set the Turn to -70 and the Tilt to 70.

The black hole's rotation drags the neighboring space-time so much that it is physically impossible for anything within the ergosphere to be at rest. Everything in the ergosphere must move in the direction of the black hole's rotation. The spherical event horizon defines the region of no return.

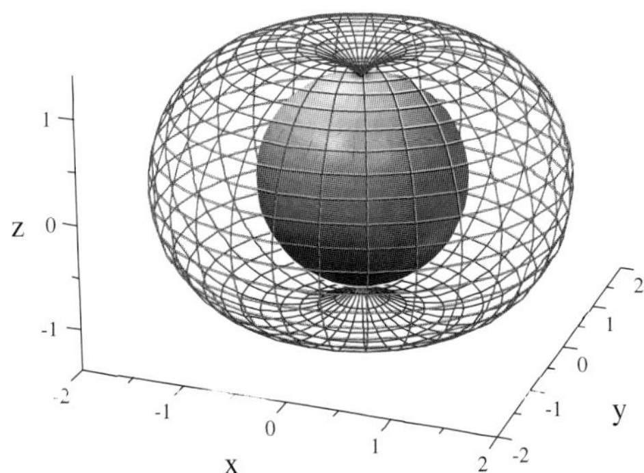

Figure 1.17 A Kerr black hole

Traveling through the ergosphere and moving beyond the event horizon would certainly be a long, strange trip. ◀

Plotting Data

One of the most important parts of science is the description of experimental data. Experimental data must be analyzed and rendered in a form that permits comparison with theoretical predictions. One way of doing this is to find a mathematical expression that best describes the data, and plotting the fit and the data on the same graph.

To make a 2-dimensional plot of numerical data, put the data in a 2-column matrix with as many rows as you have data points. The left column holds the numbers for the x-axis and the right column holds the numbers for the y-axis.

Your 2-column, n-row data matrix should look something like this: $\begin{bmatrix} x_1 & y_1 \\ x_2 & y_2 \\ x_3 & y_3 \\ \vdots & \vdots \\ x_n & y_n \end{bmatrix}$

Once you have the data in the matrix, place the insertion point anywhere in the matrix and click the Plot 2D Rectangular button. If the columns have labels (a variable name in the first row of each column) select just the numerical data before you click the Plot 2D Rectangular button. You can also plot the points as a set of ordered pairs $\{(x_1, y_1), (x_2, y_2), (x_3, y_3), \ldots, (x_n, y_n)\}$, but the matrix form is easier to read.

Use the following steps to make a 2-dimensional plot of n data points.

1. Create a 2-column, n-row matrix containing the data points.
2. Place the insertion point in the matrix and click the Plot 2D + Rectangular button.
3. Select the plot and click the Properties button.
4. Go to the Items Plotted page and choose Point as the Plot Style.

To make a 3-dimensional plot of some data, make a 3-column matrix with as many rows as you have data points. The left column holds the numbers for the x-axis, the middle column holds the numbers for the y-axis, and the right column holds the numbers for the z-axis. Once you have the data in the matrix, place the insertion point anywhere in the matrix and click the Plot 3D Rectangular button.

Fitting a Curve to Data

The process of finding a mathematical expression that best describes data is called "fitting a curve to the data". *SNB* has several curve-fitting options, all of which can be found on the Compute + Statistics menu. All *SNB*'s curve fitting options use the least-square fitting technique and they can all handle units.

To fit a curve to data in *SNB*, the data must be in a column matrix with one column for each variable. If the data are presented as a collection of ordered pairs in the form

$$(x_1, y_1), (x_2, y_2), (x_3, y_3), (x_4, y_4)$$

you can Reshape them into a two-column matrix. First remove the parentheses, then use the Reshape command from the Matrices submenu.

Place the insertion point in the data and choose Matrices + Reshape. Select 2 as the Number of Columns.

$$x_1, y_1, x_2, y_2, x_3, y_3, x_4, y_4, \begin{bmatrix} x_1 & y_1 \\ x_2 & y_2 \\ x_3 & y_3 \\ x_4 & y_4 \end{bmatrix}$$

Using a built-in delimiter (a square bracket here) for the matrix is not necessary, but it makes the data easier to read when printed. *SNB* does not print any lines within a matrix.

Occasionally, data are presented as two separate lists in the form x_1, x_2, x_3, x_4 and y_1, y_2, y_3, y_4. In this case, you must first Reshape each list into a 1-column matrix (by selecting 1 as the number of columns) and then Concatenate the two 1-column matrices into one 2-column matrix. Place the two 1-column matrices side-by-side, leave the insertion point in the data and choose Matrices + Concatenate.

$$\begin{bmatrix} x_1 \\ x_2 \\ x_3 \\ x_4 \end{bmatrix} \begin{bmatrix} y_1 \\ y_2 \\ y_3 \\ y_4 \end{bmatrix}, \text{ concatenate: } \begin{bmatrix} x_1 & y_1 \\ x_2 & y_2 \\ x_3 & y_3 \\ x_4 & y_4 \end{bmatrix}$$

If your data are in a 2-row, multiple-column matrix, click Matrices + Transpose to transform the data into the desired 2-column, multiple-row matrix.

$$\begin{bmatrix} x_1 & x_2 & x_3 & x_4 \\ y_1 & y_2 & y_3 & y_4 \end{bmatrix}, \text{ transpose: } \begin{bmatrix} x_1 & y_1 \\ x_2 & y_2 \\ x_3 & y_3 \\ x_4 & y_4 \end{bmatrix}$$

Once your data are in the correct form for *SNB*, you can begin your fit. When you choose Fit Curve to Data from the Statistics submenu, this dialog box appears.

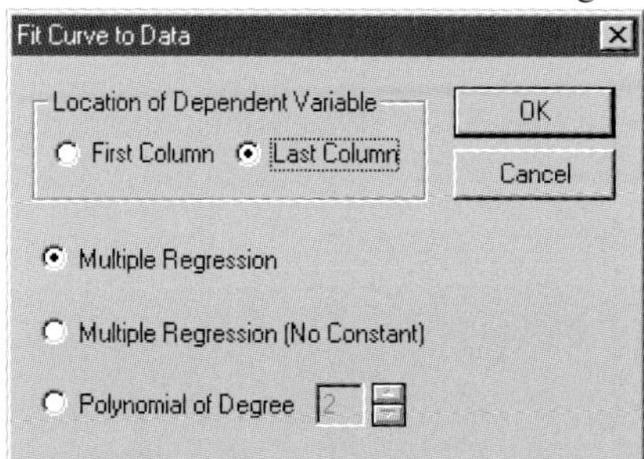

Figure 1.18 The Fit Curve to Data dialog box

There are three fit options – Multiple Regression, Multiple Regression (No Constant), and Polynomial of Degree, plus a choice for the "Location of Dependent Variable". Usually we take x as the independent and y as the dependent variable, so that the value of y depends on the value of x. The figure depicts the default situation with the dependent variable in the right column.

Multiple regression is a method of determining a linear relationship between some result and several factors. More than one independent variable may be used to predict the result, but the relationship among the variables is always linear. You can have more than one independent variable, so the columns of the matrix must be labelled to use either Multiple Regression option. The resulting equation relates the dependent variable (in the first or last column) to a linear combination of the other variables, plus a constant (unless you chose the Multiple Regression (no constant) option). You can even do a multiple regression on columns filled with variables instead of numbers.

Labels are not required to do a fit with Polynomial of Degree, and *SNB* defaults to x and y if there are no labels. This option *requires* a two-column matrix of numbers, one independent variable and one dependent variable. The resulting equation relates the dependent variable to a polynomial of the chosen degree in the independent variable.

Example 1.61 *Position as a function of time*
Use the following height data from a free-fall experiment to calculate the height as a function of time for an object thrown straight up.

$$\begin{bmatrix} t & 0.0 & 0.20 & 0.40 & 0.60 & 0.80 & 1.00 & 1.20 & 1.40 \\ h & 1.03 & 3.46 & 5.47 & 7.14 & 8.43 & 9.23 & 9.69 & 9.74 \end{bmatrix}$$

The heights are given in meters and the times are in seconds.
Solution. As we will see in Chapter 2, the height versus time graph for an object in free fall is a parabola. Transpose the data into a 2-column matrix, place the insertion point anywhere in the matrix, and click Statistics + Fit Curve to Data. Select a Polynomial of Degree 2.

Polynomial fit: $h = -4.971\,7t^2 + 13.192t + 1.019\,6$

It is useful (and often required) to include a plot of the data and the fit in a lab report, so let's make such a plot. Place the insertion point in the right-hand-side of the polynomial fit and click the Plot 2D Rectangular button. Select the data (but *not* the column labels) with your mouse and drag it onto the plot. Open the Plot Properties dialog box. On the Items Plotted page, change the fit's Line Thickness to Medium and PlotColor to LightBlue. Change the Plot Intervals so time runs from 0 to 1.5. For the data, change the Pointer Marker to Circle. On the Axes page, change the x-axis label to "Time (s)" and the y-axis label to "Height (m)". On the View page, change the View Intervals to 0 to 1.5 for the Time axis and 0 to 10.5 for the Height axis.

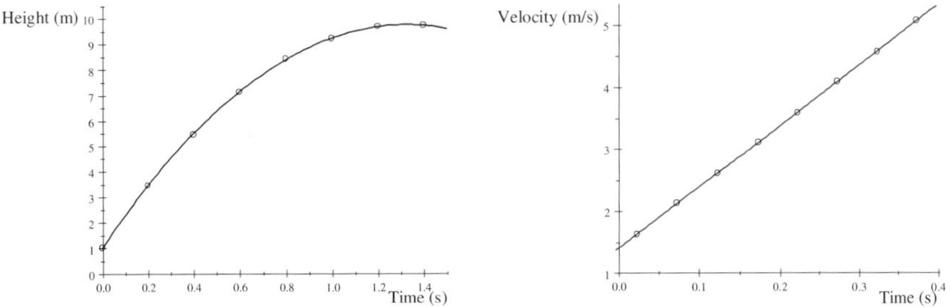

Figure 1.19 Height vs Time **Figure 1.20** Velocity vs Time

Figure 1.19 shows a graph of these data and the quadratic fit. ◀

Example 1.62 *The acceleration due to gravity*
Use the following velocity data from a free-fall experiment to calculate g, the acceleration due to gravity near the Earth's surface.

$$\begin{bmatrix} t & 0.025 & 0.075 & 0.125 & 0.175 & 0.225 & 0.275 & 0.325 & 0.375 \\ v & 1.64 & 2.14 & 2.62 & 3.12 & 3.60 & 4.10 & 4.58 & 5.08 \end{bmatrix}$$

The velocities are given in meters per second and the times are in seconds.

Solution. We will also see in Chapter 2 that the velocity versus time graph for an object in free fall is a straight line whose slope is the acceleration due to gravity and intercept is the initial velocity. Repeat the process from the previous example on these data, but this time select a Polynomial of Degree 1.

Polynomial fit: $v = 9.809\,5t + 1.398\,1$

According to our data, the initial velocity is $1.3981\,\mathrm{m/s}$ and the acceleration due to gravity is $9.8095\,\mathrm{m/s^2}$. Figure 1.20 shows a graph of these data and the linear fit. ◀

The following example uses *SNB*'s logarithmic plot capability and also reveals your author's inner Trekkie.

Example 1.63 *Ahead, warp factor 1*
Faster-than-light travel is possible in the Star Trek universe, and the warp factor (w) describes the speed as a number (s) times the speed of light. The following data represent the warp factor of a starship and the corresponding speed. [12]

$$\begin{bmatrix} w & 0 & 1 & 2 & 3 & 4 & 5 & 6 & 7 & 8 & 9 & 9.2 & 9.6 \\ s & 0 & 1 & 10 & 39 & 102 & 214 & 392 & 656 & 1024 & 1516 & 1649 & 1909 \end{bmatrix}$$

Find an expression that gives the speed as a function of the warp factor.

Solution. With the insertion point in the data, select Matrices + Transpose to convert the data into a 2-column matrix. Then place the insertion point anywhere in the data and select Statistics + Fit Curve to Data, and select Polynomial of Degree 4. The resulting fourth-degree polynomial is

Polynomial fit: $s = 0.117\,35w^4 + 0.800\,97w^3 + 2.742\,5w^2 - 5.622\,1w + 1.094\,9$

The actual data used in The Original Series were generated with a simple power-law expression $s = w^n$. A log-log plot of such a function is a straight line whose slope is the exponent n so we can use Simplify to calculate the exponent from the data with a logarithmic slope.

$$n = \frac{\ln 1024 - \ln 1}{\ln 8 - \ln 1} = \frac{10}{3}$$

This gives us the actual exponent used by the show's staff.

Follow the steps from the previous examples and make a plot of this polynomial, $w^{10/3}$, and the data. When you select and drag the data, omit the labels and the $(0, 0)$ point. On the Axes page of the Plot Properties dialog box, change the Axis Scaling to Log-Log and on the View Page, start the Speed axis at 1.

The straight line through the point $(1, 1)$ is a graph of $w^{10/3}$ and the other curve is the polynomial fit. The fit's behavior for small-w shows it is not accurate for warp factors less than two, but it seems good for the larger warp factors. Let's check the fit for $w > 2$ by using it to calculate the exponent.

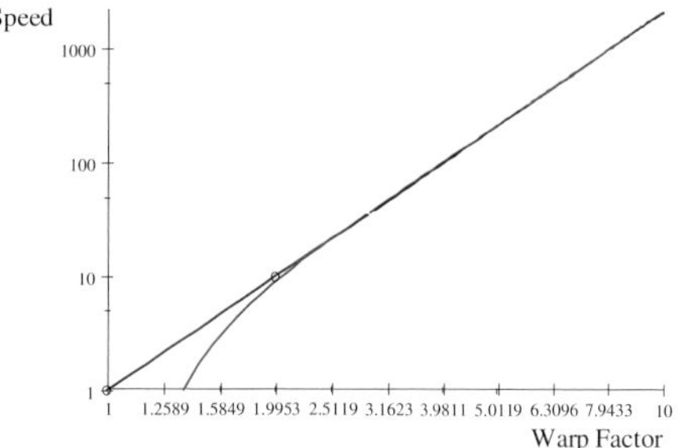

Figure 1.21 Ahead warp factor 7.9433

Use the polynomial fit to Define a function $s(w)$ for the speed and apply Evaluate Numerically to calculate the logarithmic slope.

$$n = \frac{\ln(s(10)) - \ln(s(4))}{\ln 10 - \ln 4} = 3.3296$$

The percent deviation between this result and the actual exponent is small.

$$100\left(\frac{3.3296 - 10/3}{10/3}\right) = -0.112$$

The two values differ by only about one-tenth of a percent. ◄

Differential Equations

Differential equations are important mathematical tools for describing and explaining the physical world. Differential equations often arise in the solution to physics problems. Application of the most important dynamic rule of classical physics, Newton's Second Law, produces a differential equation for all but the simplest problems. Describing the physical world is often done in the language of differential equations.

The "order" of a differential equation is the highest derivative in the equation. If the highest derivative in the equation is a second derivative, then that equation is a second-order differential equation. Most of the differential equations you will encounter in physics are either first or second order. A differential equation is "linear" if the unknown variable only appears to the first power. Otherwise the equation is "nonlinear". An Ordinary Differential Equation (ODE) contains ordinary derivatives with only one independent variable. The solution to the ODE is any mathematical function that satisfies the equation.

Use the following steps to solve a first-order differential equation with *SNB*.

1. Create a 1-column, 2-row matrix.
2. Place the differential equation in standard mathematical notation in the first row.
3. Place the initial condition for the unknown variable in the second row.
4. Choose a method of solution from the Solve ODE submenu of the Compute menu.

The initial condition for the unknown variable is an equation in the form of $y(0) =$ either a numeric or symbolic quantity. For example, you might have $y(0) = 5$ or $y(0) = y_0$.

Use the following steps to solve a second-order differential equation with *SNB*.

1. Create a 1-column, 3-row matrix.
2. Place the differential equation in standard mathematical notation in the first row.
3. Place the initial condition for the unknown variable in the second row.
4. Place the initial condition for the derivative of the unknown in the third row.
5. Choose a method of solution from the Solve ODE submenu of the Compute menu.

The initial condition for the derivative of the unknown variable is an equation in the form of $y'(0) =$ either a numeric or symbolic quantity. For example, you might have $y'(0) = 5$ or $y'(0) = v_0$.

Note To create the expressions y' or y'', put the insertion point just to the right of the character y and, while still in Math mode, type an apostrophe (the key just to the left of ENTER) or two. *SNB* will put the apostrophes in a Superscript automatically.

Don't be intimidated by the complicated nature of differential equations. They are simply mathematical rules for how an unknown variable changes. *SNB* offers several methods you can use to solve differential equations exactly, approximately, and numerically.

Solve ODE Exact and Laplace

SNB offers two methods (Exact and Laplace) that return exact solutions to linear differential equations. Both methods allow you to use either standard mathematical notation in your ODE to indicate derivatives.

To indicate a	you can use	To indicate a	you can use
first derivative	$\frac{dy}{dx}$ or y'	second derivative	$\frac{d^2y}{dx^2}$ or y''

If you use only the prime notation, the ODE Independent Variable dialog box will pop up. Enter your choice of independent variable (which is often x or t) in the Independent Variable window.

As its name suggests, the Laplace method uses the Laplace transform to solve linear ordinary differential equations. If you do not specify the appropriate initial conditions, the Laplace method will return the solution in terms of the generic initial conditions $y(0)$ and $y'(0)$.

The Exact method is more general since it works for some nonlinear differential equations as well. If you do not specify the appropriate initial conditions, the Exact option will return the solution with any arbitrary constants represented as C_1, C_2, and so on.

In the following example, we use Solve ODE + Exact to solve a first-order differential equation for the velocity of an object experiencing air resistance.

Example 1.64 *An ode to an ODE*
Find the velocity as a function of time (with arbitrary initial velocity) for an object that experiences a linear air resistance after being thrown straight up.

Solution. The equation of motion for this scenario is

$$\frac{dv}{dt} = -kv - g \tag{1.10}$$

where g is the acceleration due to gravity and k is a positive constant. To solve the equation for the object's velocity v, use the above steps to create the appropriate matrix. Set the initial velocity equal to v_0 and choose Solve ODE + Exact. The use Simplify in-place on the solution.

$$\begin{bmatrix} \frac{dv}{dt} = -kv - g \\ v(0) = v_0 \end{bmatrix}, \text{ Exact solution is: } \left\{ \frac{1}{k}\left(ge^{-kt} - g + kv_0 e^{-kt}\right) \right\}$$

With a little editing by-hand you can show that this equals $v = \left(\frac{g}{k} + v_0\right) e^{-kt} - \frac{g}{k}$. ◄

In the next example, we use Solve ODE + Laplace to solve a second-order differential equation for the trajectory of a simple pendulum.

Example 1.65 *A simple pendulum*
Calculate the angular position as a function of time for a simple pendulum released from rest at an arbitrary initial angle.

Solution. The equation of motion for a simple pendulum is

$$\frac{d^2\theta}{dt^2} = -\frac{g}{l}\theta \tag{1.11}$$

where g is the acceleration due to gravity, l is the length of the pendulum, and the angle θ is measured from the vertical. Create a 1-column, 3-row matrix, place the equation in the first row and the initial conditions in the other two rows. Place the insertion point anywhere in the matrix and choose Solve ODE + Laplace from the Compute menu.

$$\begin{bmatrix} \frac{d^2\theta}{dt^2} = -\frac{g}{l}\theta \\ \theta(0) = \theta_0 \\ \theta'(0) = 0 \end{bmatrix}, \text{ Laplace solution is: } \left\{\theta_0 \cos t\sqrt{\tfrac{g}{l}}\right\}$$

The pendulum's trajectory is $\theta\left(t\right) = \theta_0 \cos\sqrt{\frac{g}{l}}t$. For small angles, the simple pendulum undergoes simple harmonic motion, so its period is constant and does not depend on the initial angle. You could also use the more general Solve ODE + Exact to solve this problem.

$$\begin{bmatrix} \frac{d^2\theta}{dt^2} = -\frac{g}{l}\theta \\ \theta(0) = \theta_0 \\ \theta'(0) = 0 \end{bmatrix}, \text{ Exact solution is: } \left\{\theta_0 \cos \frac{\sqrt{g}}{\sqrt{l}}t\right\}$$

The two solutions are essentially the same. ◄

Solve ODE Numeric

Many of the differential equations that describe interesting physical situations do not have analytic solutions and we have to solve them numerically. The purpose of solving an ODE numerically is to find an approximation to the function that satisfies the ODE for the given initial conditions. You should solve an ODE numerically when finding the analytical solution is impossible or infeasible.

Most of the differential equations you will encounter in physics can be solved numerically. *SNB* treats these numeric solutions as functions that you can Evaluate at various points or Plot. It is important to note that *SNB* does not save numeric solutions to ODEs when you close your document.

Unlike the two exact methods for solving differential equations, you must use the prime y'' notation for derivatives when you use Solve ODE Numeric. *SNB* does not support the use of dots as shorthand for time derivatives. You can create expressions like $\ddot{r}$ and $\dot{\theta}$ by selecting a variable in Math mode, clicking the Properties button and choosing the double-dots from the Character Properties dialog box. However the dots are just ornamental and carry no mathematical meaning.

Example 1.66 *The full pendulum*
Calculate the trajectory for a full pendulum (length 1 meter) released from rest at an initial angle of 90° and compare the trajectory to a simple pendulum with the same length released from the same angle. Which one completes two cycles first?
Solution. The equation of motion for a full pendulum is

$$\frac{d^2\theta}{dt^2} = -\frac{g}{l}\sin\theta \tag{1.12}$$

where the angle θ is measured from the vertical. There is no exact solution to this equation, so you have to solve it numerically. Create a 1-column, 3-row matrix and place the equation in the first row and the initial conditions in the other rows. Place the insertion point anywhere in the matrix and choose Solve ODE + Numeric from the Compute menu.

$$\begin{bmatrix} \theta'' = -9.8067\sin\theta \\ \theta(0) = 1.5708 \\ \theta'(0) = 0 \end{bmatrix}, \text{ Functions defined: } \theta$$

Leave the insertion point to the right of the θ and click the Plot 2D Rectangular button. Add the simple pendulum result $\frac{\pi}{2}\cos\sqrt{9.8067}t$ to the plot by click-and-drag. Open the Plot Properties dialog. Click on the Axes page and change the x-axis label to "Time (s)" and the y-axis label to "Angle".

On the Items Plotted page for Item 1, click the Variable and Intervals button and set the Interval from 0 to 5. Change the PlotColor to Red and the Line Thickness to Medium. For Item 2, click the Variable and Intervals button and set the Interval from 0 to 4.5. Change the Line Style to Dash.

The following figure shows the angle as a function of time for the simple (dashed line) and full (solid) pendula with the same initial conditions.

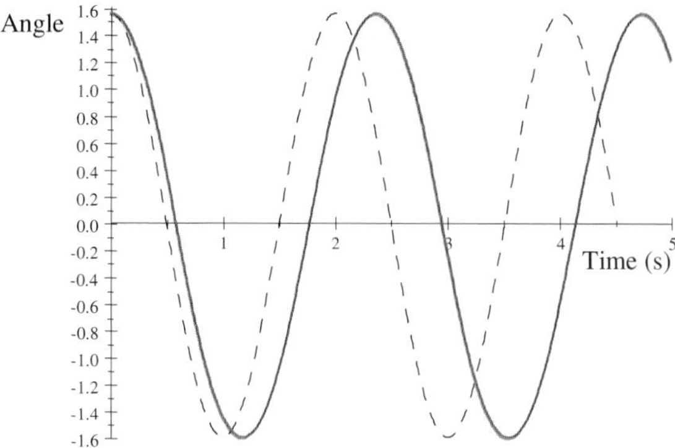

Figure 1.22 Dueling pendula

Let's use Evaluate Numerically to calculate the period of the simple pendulum.

$$2\pi\sqrt{\frac{l}{g}} = \frac{2\pi}{\sqrt{9.8067}} = 2.006\,4$$

We can use Evaluate Numerically, the graph, and a little trial and error to estimate the period of the full pendulum.

$$\frac{\theta\,(T)}{\theta_0} = \frac{\theta\,(2.3691)}{\pi/2} = 1.0$$

The period of the simple pendulum is 2.0064 s, which is less than the 2.3691 s it takes the full pendulum to complete its first cycle. The simple pendulum completes the two cycles sooner. Of course, 90° is not a small angle and the simple pendulum result is not appropriate, as the plot shows. ◀

Some problems require you to solve coupled differential equations. When differential equations are coupled, the unknown functions appear in more than one equation so the equations must be solved together simultaneously. Each unknown function requires its own initial conditions.

One such problem is the **Swinging Atwood's Machine** (SAM). [33] A SAM is an Atwood's Machine that allows one of the masses to swing in a vertical plane. The equations of motion for the SAM (in a coordinate system where $\phi = 0$ is along the $+x$-axis) are

$$(1+\mu)\frac{d^2r}{dt^2} = r\left(\frac{d\phi}{dt}\right)^2 - g\,(\sin\phi + \mu) \tag{1.13a}$$

$$r\frac{d^2\phi}{dt^2} = -2\frac{dr}{dt}\frac{d\phi}{dt} - g\cos\phi \tag{1.13b}$$

where μ is the ratio of the hanging mass to the swinging mass. These are two coupled 2nd-order differential equations for the radial coordinate r and the angular coordinate ϕ as functions of time.

Example 1.67 *SAM I am*
Calculate and plot the loop-the-loop periodic orbit for the Swinging Atwood's Machine that starts with a radius of 1 meter and is released from a horizontal position from rest.
Solution. The loop-the-loop periodic orbit starts from rest at $r_0 = 1$ meter and $\phi_0 = 0$. The mass ratio is $\mu = 2.812$ and the period of the orbit is 3.1841 seconds. Place the two equations and four initial conditions in a 1-column, 6-row matrix, leave the insertion point anywhere in the matrix and choose Compute + Solve ODE + Numeric.

$$\begin{bmatrix} (1+2.812)\,r'' = r\,(\phi')^2 - 9.8067\,(\sin\phi + 2.812) \\ r\phi'' + 2r'\phi' = -9.8067\cos\phi \\ r(0) = 1 \\ \phi(0) = 0 \\ r'(0) = 0 \\ \phi'(0) = 0 \end{bmatrix}, \text{ Functions defined: } r, \phi$$

To plot the radial coordinate as a function of time $r(t)$, highlight the defined function r and click the Plot 2D button. Open the Plot Properties dialog. On the Items Plotted page, click the Variable and Intervals button and set the Interval from 0 to 3.2. Change the PlotColor to LightRed and the Line Thickness to Medium. Click on the Axes page and change the x-axis label to "Time" and the y-axis label to "Radial". On the View Page, start the Radial axis at 0.45.

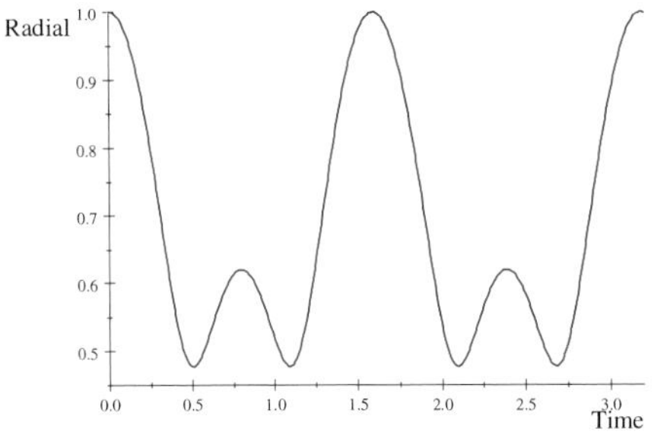

Figure 1.23a SAM's $r(t)$

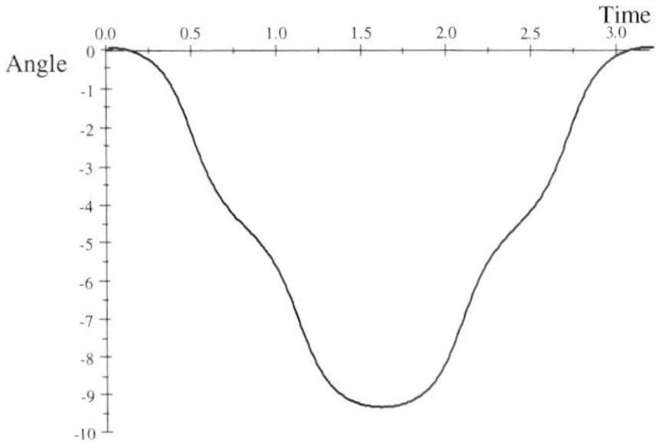

Figure 1.23b SAM's $\phi(t)$

To plot the angular coordinate as a function of time $\phi(t)$, highlight the defined function ϕ and click the Plot 2D button. Open the Plot Properties dialog. On the Items Plotted page, click the Variable and Intervals button and set the Interval from 0 to 3.2. Change the PlotColor to LightBlue and the Line Thickness to Medium. Click on the Axes page and change the x-axis label to "Time" and the y-axis label to "Angle". On the View Page, let the Angle axis run from -10 to 0.

Figures 1.23 show the SAM's position coordinates as functions of time, but it's more interesting and fun to plot the trajectory $r(\phi)$. Our numerical solutions depend on time, so we'll have to plot the trajectory parametrically. We could use the polar form $(r(t), \phi(t))$ but let's use the following parametric expression.

$$(r(t)\cos(\phi(t)), r(t)\sin(\phi(t)))$$

Place the insertion point in the expression and choose Plot 2D + Parametric.

Open the Plot Properties dialog. On the Items Plotted page, click the Variable and Intervals button and set the Interval from 0 to 1.6. Change the PlotColor to Purple and the Line Thickness to Medium. On the Axes page, select Equal Scaling Along Each Axis. Click OK.

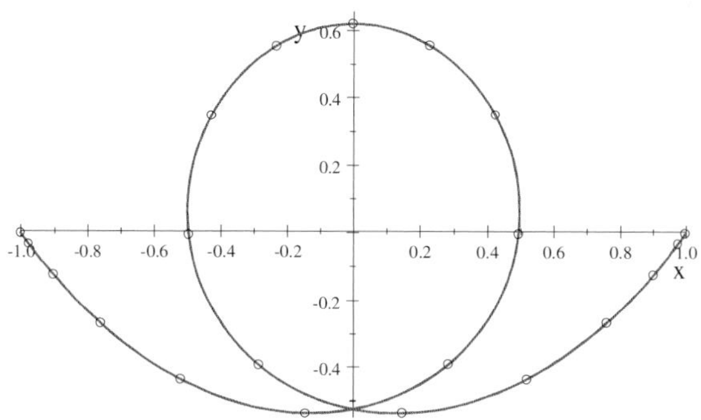

Figure 1.24 SAM's trajectory $r(\phi)$

To include the 21 points (one point every $\frac{1}{40}$ of the period), use the defined functions r, ϕ to fill a matrix with the swinging mass's x and y coordinates. Create a 2-column, 21-row matrix, put the $x = r\cos\phi$ coordinates in the left column, the $y = r\sin\phi$ coordinates in the right, and click Evaluate.

In the interest of space, here is an abridged version of the matrix.

$$\begin{bmatrix} r(0)\cos(\phi(0)) & r(0)\sin(\phi(0)) \\ r(0.39766)\cos(\phi(0.39766)) & r(0.39766)\sin(\phi(0.39766)) \\ r(0.79533)\cos(\phi(0.79533)) & r(0.79533)\sin(\phi(0.79533)) \\ r(1.1930)\cos(\phi(1.1930)) & r(1.1930)\sin(\phi(1.1930)) \\ r(1.5907)\cos(\phi(1.5907)) & r(1.5907)\sin(\phi(1.5907)) \end{bmatrix} = \begin{bmatrix} 1.0 & 0 \\ 0.14909 & -0.53273 \\ -1.8959\times 10^{-3} & 0.62064 \\ -0.14173 & -0.53297 \\ -1.0 & -2.7155\times 10^{-5} \end{bmatrix}$$

Once you've filled the matrix, select it and click-and-drag it to the plot. Open the Plot Properties dialog. Go to Item 2 on the Items Plotted page and change the Plot Style to Points, the Point Marker to Circle, and the PlotColor to Black.

As you can see from the plots, the swinging mass swings clockwise through $1\frac{1}{2}$ orbits from $\phi = 0$ to $\phi = -3\pi$ and then swings back to $\phi = 0$ once every 3.1841 seconds. During this time the radial coordinate goes through two complete cycles. The time interval between the points is constant, so the separation between points shows the swinging mass's speed. This would make a very interesting ride at an amusement park. ◀

Example 1.68 *More SAM I am?*
Where is the swinging mass when the elapsed time is one-quarter and one-half a period?
Solution. *SNB*'s numerical solutions to ODEs are defined functions that you can Evaluate or Evaluate Numerically. Let's Evaluate the results of the previous example to verify the given period $T = 3.1841$ s.

$$r\,(3.1841) = 1.0$$

$$\phi\,(3.1841) = -7.517 \times 10^{-5}$$

At $t = T$ both the radial and angular coordinates approximately equal their initial values, which tells us the numerical solution is accurate.

Let's use Evaluate Numerically to calculate the swinging mass's position at $t = \frac{1}{4}T$.

$$r\left(\tfrac{1}{4} \times 3.1841\right) = 0.620\,64$$

$$\frac{1}{\pi}\phi\left(\tfrac{1}{4} \times 3.1841\right) = -1.500\,1$$

After one-quarter of a period, the swinging mass is at its highest point $0.620\,64\,\mathrm{m}$ from the origin. It is crossing the $+y$-axis at $\phi = -\frac{3}{2}\pi$ moving from left to right.

Let's use Evaluate Numerically to calculate the swinging mass's position at $t = \frac{1}{2}T$.

$$r\left(\tfrac{1}{2} \times 3.1841\right) = 1.0$$

$$\frac{1}{\pi}\phi\left(\tfrac{1}{2} \times 3.1841\right) = -3.000\,0$$

After half a period, the swinging mass is 1 meter from the origin and has rotated to an angle of $-\,3\pi$ radians. It is stopped on the $-\,x$-axis and will now start to swing back in the counterclockwise direction.

You can use Evaluate Numerically on this expression

$$\overline{r} = \frac{1}{T/2}\int_0^{T/2} r\,(t)\,dt$$

to calculate the time-average of the radial coordinate over half the period.

$$\overline{r} = \frac{1}{1.592\,1}\int_0^{1.592\,1} r\,(t)\,dt = 0.686\,55$$

A look at the trajectory in Figure 1.24 shows this is a reasonable answer. ◀

This chapter provides only a brief introduction to *SNB*'s capabilities. There are many other features of *SNB*, some of which we'll meet along the way and others you'll discover yourself. You can find much useful information and many examples in *SNB*'s help. I encourage you to explore it, play with it, learn it.

Problems

1. Set up your *SNB* screen so it looks the same as the one depicted in Figure 1.2. Which toolbars are visible?

2. Create an expression for the quadratic formula which gives the two solutions to $\alpha x^2 + \beta c + \gamma = 0$.

3. Create an expression for the integration-by-parts formula. Consult *SNB*'s help or a textbook as needed.

4. Create an expression for any trigonometric identity (look inside the front cover of your physics text).

5. Create a fragment containing the acceleration due to gravity on the Earth's surface and place it in the Constants folder.

6. Use the Evaluate command to compute the following expressions.

   a. $27 + 33 - 16$

   b. $|-11.3|$

   c. $\left(3x^2 + 3x\right) + \left(8x^2 + 7\right)$

   d. $\int_0^a x^{98} dx$

7. Compute these expressions with Evaluate Numerically.

   a. $\frac{8}{9}$

   b. $\sqrt{2}$

   c. $\displaystyle\int_0^1 e^{x^2} dx$

   d. The factorial of the number of inches of your height (just the number without the inch unit). Consult *SNB*'s help if you are not sure what "factorial" means.

8. Compute these expressions with Evaluate and Evaluate Numerically, noting the different results.

   a. $\frac{5}{8} \times \frac{1}{7}$

   b. $(x + 3) + (x - y)$

   c. $\sum_{n=1}^{100} \dfrac{1}{2n}$

   d. Square your age and then find the factorial of that number.

9. Prove that a log-log plot of $y = ax^n$ is a straight line whose slope is n and crosses the $x = 1$ line at $y = a$. Pick reasonable values for a and n and make a log-log plot.

10. Calculate the percent deviation between $\pi \times 10^7$ seconds and one year.

11. Calculate the percent deviation between the irrational number π and the ratio $\frac{22}{7}$.

12. Calculate the percent deviation between the irrational number π and the ratio $\frac{1080}{343}$.

13. Calculate the percent deviation between the irrational number $\sqrt{\pi}$ and the ratio $\frac{296}{167}$.

14. Calculate the percent deviation between the irrational number $\sqrt{2}$ and the ratio $\frac{99}{70}$.

15. A typical college class lasts 55 minutes. Calculate the percent deviation between one class period and a micro-century. (This problem originates from Enrico Fermi.)

16. Calculate the percent deviation between the area under the curve of $x^2e^{-3x}\sin 4x$ from $x = 0$ to $x = 3$ and the total area under the positive x-axis.

17. Use Solve Exact to solve the following equations.
    a. $\frac{1}{x} + \frac{1}{y} = 1$ (Solve for x)
    b. $\frac{1}{x} + \frac{1}{y} = 1$ (Solve for y)
    c. $x^2 - 5x + 4 = 0$
    d. $2x + y = 5$ and $3x - 7y = 2$

18. Use Solve Numeric to solve the following equations.
    a. $16 - 7y = 10y - 4$
    b. $x^5 - 5x^4 + 3x + 4 = 0$
    c. $5\left(e^x - 1\right) = xe^x$
    d. $\sin x = \cos x$ between $x = 9$ and $x = 12$

19. Find the third and fourth zeroes of the function plotted in Figure 1.1.

20. The two solutions to the equation $ax^2 + bx + c = 0$ are given by the quadratic formula. There is an analogous result for the cubic equation $ax^3 + bx + c = 0$. Use Solve Exact on the cubic equation for unspecified a, b, and c. Delete the two possibly complex solutions and Evaluate the third for $a = 1$, $b = 2$, and $c = 3$.

21. Simplify these expressions:
    a. $\sqrt[3]{8} + 3$
    b. $\sin^2 x + \cos^2 x$
    c. $\int_1^a \frac{1}{t}dt$

22. Apply Rewrite + Logarithm to these expressions before and after you tell *SNB* that x is a positive number.
    a. $\log_{10} e^x$
    b. $\log_b e^x$
    c. $\log_b 10^x$

23. Use Rewrite + Logarithm to verify the relation in Eq. (1.3) between $\log_b x$ and the natural logarithm.

24. Factor these expressions. Would you like to verify your results without *SNB*?

    a. $5x^5 + 5x^4 - 10x^3 - 10x^2 + 5x + 5$

    b. $\frac{1}{2}x^2 + 3x - \frac{20}{9}$

    c. $x^6 - y^6$

    d. The product of the month, day, and year of your birth (for example December 8, 1956 $\Longrightarrow 12 \times 8 \times 1956$).

25. Find all the prime factors for the 50th, 83rd, and 100th Fibonacci numbers.

26. Expand the following expressions:

    a. $\left(3x^2 + 3x\right)^3$

    b. $(x + y)^9$

    c. $\sin(x + y)$

    d. $\left(3x^2 + 3x\right)\left(8x^2 + 7\right)$

27. Collect and Sort the terms in these polynomial expressions.

    a. $5t^2 + 2t - 16t^5 + t^3 - 2t^2 + 9$

    b. $x^2 + y + 5 - 3x^3y + 5x^2 + 4y^3 + 13 + 2x^4$ (Use x as the variable.)

    c. $3x - 7x^2 + 8x - 3 + x^5$

28. Find the roots of the following polynomials.

    a. $x^3 - 2x - 2x^2 + 4$

    b. $x^3 - \frac{13}{5}ix^2 - 8x^2 + \frac{29}{5}ix + \frac{81}{5}x + 6i - \frac{18}{5}$

    c. $x^5 - 3x^4 - 23x^3 + 51x^2 + 94x - 1$

29. Expand the function $\ln x$ in a power series to order-x^2 about the point $x = 1$. Evaluate your expansion at $x = 1.1$ and compare the result to the exact value.

30. Show that the result from Example 1.64 reduces to $v = v_0 - gt$ in the absence of air resistance (where $k = 0$).

31. Use Solve Exact to find the exact solution to the equation $0 = 1 - \frac{1}{2}x^2 + \frac{1}{24}x^4$ that corresponds to the approximate solution $x \approx 1.5925$. (Hint: the equation is quadratic in x^2.)

32. Verify that the relation $e^{-\pi/2} = i^i$ is correct.

33. Use a power series expansion and Solve Numeric to find a numerical solution to the equation $\cos x = 0$ that is within 0.25% of the exact solution.

34. Use Solve Numeric to verify the time of flight for the fly ball in Example 1.58.

35. Find the first three non-zero terms in the power series expansion of $e^{-at}(1+\sin bt)$. What is the lowest-order power of t when $a=b$? What is the lowest-order power of t when $a \neq b$?

36. Define a function for the expression $\frac{1}{\sqrt{5}}\left(\left(\frac{1+\sqrt{5}}{2}\right)^n - \left(\frac{2}{1+\sqrt{5}}\right)^n \cos n\pi\right)$ and verify that it is an expression for the nth Fibonacci Number.

37. Find the two exact solutions to the equation $\phi^2 - \phi - 1 = 0$. Show that the positive solution equals $\phi = 2\cos\frac{\pi}{5}$. This solution is called the Golden Ratio and gives the ratio of the length of a diagonal to the length of a side of a regular pentagon.

38. Show that the nth Fibonacci Number can be written as

$$F_n = \frac{1}{\sqrt{5}}\left(\phi^n - \frac{\cos n\pi}{\phi^n}\right)$$

where ϕ is the Golden Ratio.

39. The coefficient of the x^n term in a power series expansion of the function $f(x)$ about the point $x=a$ is

$$a_n = \frac{1}{n!}\left.\frac{d^n f(x)}{dx^n}\right|_{x=a}$$

Use this expression to verify the first three terms (including those with zero coefficients) in the power series expansions (about $x=0$) of

a. $\sin x$ b. $\cos x$ c. e^x d. $(1+x)^n$.

40. Use the definition of a derivative to calculate the derivative of e^{kx} and $\ln kx$ at $x=a$.

41. Use the definition of a derivative to calculate the derivative of the following functions at $x=a$.

a. $\tan kx$ b. $\sinh kx$ c. $\cos kx$

42. Use the definition of a derivative to calculate the derivative of the following functions at $x=a$.

a. $\sin x^2$ and $\sin^2 x$ b. $\tan x^2$ and $\tan^2 x$ c. $\sinh x^2$ and $\sinh^2 x$

43. Use the definition of a derivative to calculate the derivative of the following inverse functions at $x=a$.

a. $\arcsin x^2$ b. $\arctan x^2$ c. $\sinh^{-1} x^2$

44. Consider the curve described by this parametric expression.

$$(x,y) = \left(0.47\left(\frac{t}{1.82} - \sin\frac{t}{1.82}\right), 0.47\left(1-\cos\frac{t}{1.82}\right)\right)$$

What is the arc length of one cycle of this curve?

45. There is another way to calculate the arc length of a path besides parametrically. If you know the path $x(y)$ then the arc length is

$$L = \int_a^b \sqrt{1 + \left(\frac{dx}{dy}\right)^2}\, dy$$

where a and b are the minimum and maximum heights respectively. For a cycloid, the path is

$$x = R\cos^{-1}\left(1 - \frac{y}{R}\right) - \sqrt{2Ry - y^2}$$

where R is the wheel's radius. Use these two expressions to calculate the arc length of the path followed by a point on the edge of a rolling wheel for one revolution.

46. One way to calculate the magnetic field of a spinning spherical shell, rotating at angular velocity ω with radius R and uniform surface charge σ, involves the integral

$$\int_0^\pi \frac{\cos\theta \sin\theta}{\sqrt{R^2 + r^2 - 2Rr\cos\theta}}\, d\theta$$

where r is the distance from the origin to the point where you're calculating the field. Evaluate this integral inside and outside the shell.

47. Evaluate the indefinite integral $\int x^2 e^{-3x} \sin 4x\, dx$ from earlier in the chapter. Feel free to verify the result without *SNB*.

48. Evaluate the indefinite integral $\int \frac{dx}{x^2 + a^2}$ before and after telling *SNB* the variables are positive. Sometimes it matters...

49. Evaluate the indefinite integral $\int \frac{dx}{x^2 - a^2}$ before and after telling *SNB* the variables are positive. Sometimes it doesn't...

50. Here are a few more definite integrals for you to explore:

   a. $\int \tan ax\, dx$ and $\int \arctan \frac{x}{a}\, dx$

   b. $\int \ln ax\, dx$ and $\int e^{ax}\, dx$

   c. $\int x \ln ax\, dx$ and $\int \sqrt{x^2 + a^2}\, dx$

51. Evaluate the indefinite integral $\int \frac{dx}{a^2 - x^2}$. Look up the answer in a table of integrals and compare. Even *SNB* doesn't get the right answer every time.

52. Evaluate the indefinite integral $\int \frac{dx}{\sqrt{a^2 - x^2}}$. This one is right.

53. Use Evaluate and some editing by-hand to verify this relation.

$$\int \frac{dx}{ax^2 + bx + c} = \sqrt{\frac{1}{b^2 - 4ac}} \ln \frac{b + 2ax - \sqrt{b^2 - 4ac}}{b + 2ax + \sqrt{b^2 - 4ac}}$$

54. Convert $60\,\mathrm{mi}/\,\mathrm{h}$ into kilometers per hour, miles per minute, and feet per second.

55. Use the Solve Method to determine how many light-years are in one mile.

56. How many years do you have to live to have lived a billion seconds?

57. Calculate your height in inches and convert it from inches to centimeters, from inches to meters, and from centimeters to meters.

58. Calculate your *exact* age, as of 10:00 AM on the morning of September 1 of this year. If you do not know what time you were born, use 2:00 PM. Give your answer in seconds, days, and years.

59. Starting with *SNB*'s built-in unit for the year $(1\,\mathrm{y})$, calculate how many days there are in one year. The answer is **not** 365.

60. Perform the following conversions:

 a. $100\,\mathrm{in}^2$ to square meters,

 b. $1234\,\mathrm{kg}/\,\mathrm{m}^3$ to grams per cubic centimeter, and

 c. $14.7\,\mathrm{lb}/\,\mathrm{in}^2$ to Newtons per square meter.

61. The speed of light in a vacuum is $2.9979\times 10^8\,\mathrm{m}/\,\mathrm{s}$. Use the Standard Method to convert this speed into

 a. miles per second,

 b. miles per hour,

 c. astronomical units per year, and

 d. light years per year.

62. Use your name and height to create a unit of length. For example, one Lisa might be 5.375 feet. Then convert the following distances to your unit.

 a. the height of an official basketball hoop

 b. the distance from home plate to second base on an official baseball diamond

 c. the height of the Eiffel Tower

63. Find a better fit for the warp factor function for $w < 3$.

64. Approximately how fast is warp factor 7.389?

65. What warp factor corresponds to $v = 47c$?

66. Make a plot of $x \sin x$ versus x. Add two curves to this plot: the straight line x and the sine curve $\sin x$.

67. Calculate the integral of $x \sin x$ and plot the result. Look at the result and make an educated guess as to how *SNB* did this integral.

68. Calculate the derivative of $x \sin x$ and plot the result. To familiarize yourself with the extensive help available in *SNB*, consult the built-in help on how to perform a derivative.

69. Plot the three expressions from Example 1.22, letting x range from 0 to 1. Notice where the three graphs intersect.

70. Make a 2-dimensional rectangular plot of the expression $x \sin \frac{1}{x}$.

    a. Let x run from -1 to $+1$.

    b. Add the Label $y = x \sin \frac{1}{x}$.

    c. Adjust the Line Thickness to "medium".

71. Make a 2-dimensional rectangular plot of the expression $e^{-x} \sin 5x$.

    a. Let x run from 0 to $+4$.

    b. Add two more items to this plot, e^{-x} and $-e^{-x}$.

    c. Plot the e^{-x} with a red, dotted line and the $-e^{-x}$ with a blue, dashed line.

    d. Add an appropriate label.

72. Make a 2-dimensional rectangular plot of the expressions $\frac{3}{7} \sin x$ and $\frac{2}{5} \cos 2x$, where x is in degrees.

    a. Let x run from -180 to $+180$.

    b. Adjust the $\sin x$ curve so that its line is red and of medium thickness.

    c. Add an appropriate label.

73. Make a 2-d plot of $\cos x$ for the complete cycle from $x = 0$ to $x = 2\pi$. Do a 9-term power series expansion of the function. Add the first 2 terms to the plot for $x < 1.5$. Add the first 3 terms for $x < 2.5$. Add the first 4 terms for $x < 3.5$. Add the first 5 terms for $x < 4.5$. What does this tell you about power series?

74. Make a 2-d plot of $\sin x$ for the complete cycle from $x = 0$ to $x = 2\pi$. Do a 9-term power series expansion of the function. Add the first term to the plot for $x < 2$. Add the first 2 terms for $x < 3$. Add the first 3 terms for $x < 4$. Add the first 4 terms for $x < 5$. What does this tell you about power series?

75. Use Eqs. (1.7) to make a parametric plot of the wiggly-piggly orbit $r = 1 + \frac{1}{3} \sin 7\phi$. What effect does the "7" have on the orbit?

76. Make a 2-dimensional polar plot of the following expressions. For each plot, choose equal scaling along each axis and use a sufficiently large number of sampled points.

    a. $r = \sin^5 \frac{\phi}{12}$ (for $0 < \phi < 24\pi$)

    b. $r = e^{\sin \phi} - 2 \cos 4\phi$ (for $0 < \phi < 2\pi$)

    c. $r = e^{\sin \phi} - 2 \cos 4\phi + \sin^5 \frac{\phi}{12}$ (for $0 < \phi < 24\pi$)

77. Plot the following set of data as points with the cross symbol.

$$(-1, 3.24), (-0.5, 1.69), (0, 0.64), (0.5, 0.09), (1, 0.04), (1.5, 0.49), (2, 1.44).$$

a. Find the best fit by a polynomial to the set of points. What "power" should the polynomial be?

b. Add the best fit polynomial to your plot with a red Line Color.

c. Change the domain (values of x) so that x runs from -1.2 to $+2.5$.

78. One way to measure the velocity-dependence of air resistance is to drop coffee filters and measure their terminal velocity. If the force of air resistance is bv^n, then the terminal velocity is $v = (mg/b)^{1/n}$. A crack team of experimenters recently performed such an experiment and collected the following data.

$$\begin{bmatrix} v & 1.45 & 1.77 & 1.89 & 2.18 & 2.31 & 2.72 \\ m & 2 & 3 & 4 & 5 & 6 & 7 \end{bmatrix}$$

The terminal velocities are given in meters per second and the masses are in grams.

a. Show the terminal velocity expression is equivalent to $\ln m = n \ln v + \ln (b/g)$.

b. Create a matrix containing data in the form $\begin{bmatrix} \ln v & \ln m \end{bmatrix}$.

c. Use a linear fit to find the exponent n and the drag coefficient b (with units).

d. Plot the data and the fit.

79. Calculate the trajectory for a simple pendulum that starts at $\theta_0 = 0$ with an arbitrary initial speed v_0.

80. Calculate the trajectory for a simple pendulum that starts at with an arbitrary initial angle θ_0 and an arbitrary initial speed v_0.

81. When you use Solve ODE + Numeric to find the numerical solution for the position $x(t)$, you can approximate the velocity $v(t)$ accurately with

$$v(t) \approx \frac{x(t + \varepsilon/2) - x(t - \varepsilon/2)}{\varepsilon}$$

as long as ε is a small number.

a. Solve the equation $x'' = 3x(x - 2)$, with $x(0) = 1$ and $x'(0) = 0$ numerically.

b. Plot the result from $t = 0$ to $t = 5.64$.

c. Add the numerical approximation to the velocity to the plot. Use $\varepsilon = 10^{-6}$.

Here are more SAM periodic orbits with the same initial conditions as Example 1.67.

82. Plot the trajectory for $\mu = 1.665$, which has a period of 0.9129 seconds. Don't worry, be happy.

83. Plot the trajectory for $\mu = 1.1185$, which has a period of 13.11 seconds. See if you agree this one looks like a caduceus.

84. Plot the trajectory for $\mu = 2.394$, which has a period of 4.074 seconds. This one looks like an SEG (super-evil grin).

2 One-Dimensional Kinematics

We start *Doing Physics with SNB* in the branch of physics known as Classical Mechanics. Our study of Classical Mechanics begins with one-dimensional kinematics, the description of motion in a straight line. This includes horizontal motion to the right or left and vertical motion straight up or down. This description will tell us where an object is, where it is going, and how much time it took to get there.

Constant Acceleration

Studying motion with a constant acceleration is a good place to start. We can describe this motion completely without using calculus. Let's start with some definitions and important distinctions, and then we'll solve some one-dimensional problems.

Displacement and Position

To describe an object's motion, we need to establish a coordinate system so we can specify location. For 1-dimensional motion the coordinate system is just the x-axis with the origin at $x = 0$ and positive values to the right. The object's **position**, specified by its x-coordinate, tells us how far from the origin it is and in which direction.

The **displacement** is the object's change in position.

$$\Delta x = x - x_0 \tag{2.1}$$

This difference between the object's final and initial position tells us how far from its original position it ends. A positive displacement means the object ends to the right of x_0 and a negative displacement means the object ends to the left of x_0. For a round trip, the initial and final positions are equal and the displacement is zero. The SI unit for both position and displacement is the meter (m).

Note The use of the upper case Greek letter delta ("Δ") to mean "change in" is standard mathematical notation, but it has no special significance in *SNB*. In a calculation, *SNB* interprets the expression Δf as $\Delta \times f$, the product of two variables.

Even though they have the same unit, displacement and distance are different. Suppose you start 2 meters from the origin, walk 9 meters to the right, turn around and walk 5 meters back toward the origin. Your initial position was $x_0 = 2\,\mathrm{m}$ and your final position is $x = 6\,\mathrm{m}$, so your displacement is $\Delta x = 6\,\mathrm{m} - 2\,\mathrm{m} = +4\,\mathrm{m}$. Your final position is $4\,\mathrm{m}$ to the right of your initial position, but you traveled $9\,\mathrm{m} + 5\,\mathrm{m} = 14\,\mathrm{m}$. Your displacement is $+4\,\mathrm{m}$ but your total distance traveled is $14\,\mathrm{m}$.

Doing Physics with Scientific Notebook: A Problem-solving Approach, First Edition. Joseph Gallant.

To further illustrate the distinction, consider this round-trip. You start at x_0, drive a distance D and then return to x_0. You traveled a distance of $2D$; your odometer reading increased by $2D$. But your displacement is zero because you returned to your original position. Distance tells you how far the object travels regardless of any changes in direction. Distance is a running total while displacement is a net change.

Example 2.1 *Hey wait a minute, this guy can run!*
In game 4 of the 2004 ALCS, pinch runner Dave Roberts stood on first base. Three times he took a 9 ft lead but returned to first base when the pitcher tried to pick him off. He then ran on the first pitch and stole second. What was his displacement during the steal? What total distance did he travel to get to second base?
Solution. Let's have the x-axis connect the two bases with first base at the origin. During the steal, Roberts moved from an initial position of $x_0 = 9\,\mathrm{ft}$ to a final position of $x = 90\,\mathrm{ft}$. Create and Evaluate an expression for the displacement.

$$\Delta x = 90\,\mathrm{ft} - 9\,\mathrm{ft} = 81\,\mathrm{ft}$$

His displacement was $\Delta x = +81\,\mathrm{ft}$. Before the steal, he moved toward or away from first base 7 times so let's create and Evaluate this expression for his total distance.

$$D = 7\,(9\,\mathrm{ft}) + 81\,\mathrm{ft} = 144\,\mathrm{ft}$$

So began the biggest comeback, and the biggest choke, in professional sports history. ◀

Velocity and Acceleration

As an object moves, its position changes during an elapsed time. The object's **average velocity** is the change in its position divided by the elapsed time.

$$v_{ave} = \frac{\Delta x}{\Delta t} \tag{2.2}$$

Since the change in position is the displacement, the average velocity is the rate of change of the displacement calculated over a finite time interval. The elapsed time is $\Delta t = t - t_0$ and the initial time is usually zero. When the acceleration is constant, the average velocity is also the average of the initial and final velocities.

$$v_{ave} = \tfrac{1}{2}\left(v_0 + v\right) \tag{2.3}$$

The SI unit for velocity is the meter per second ($\mathrm{m/s}$). The American unit for velocity is the foot per second ($\mathrm{ft/s}$), although the mile per hour ($\mathrm{mi/h}$) is also used.

Example 2.2 *Walk him and pitch to the giraffe*
After game 5 of the 1999 ALDS, some delirious Red Sox fans drove all the way home from Cleveland to Boston. How long did it take at an average velocity of $62.5\,\mathrm{mi/h}$?
Solution. Boston is 650 miles east of Cleveland. Use Eq. (2.2) to create and Evaluate an expression relating the elapsed time to the displacement and average velocity.

$$\Delta t = \frac{\Delta x}{v_{ave}} = \frac{+650\,\mathrm{mi}}{62.5\,\mathrm{mi/h}} = 10.4\,\mathrm{h}$$

The happy trip took $10.4\,\mathrm{h}$. Walking Nomar twice seemed like such a good idea. ◀

The **average speed** is the total distance traveled divided by the elapsed time.

$$\text{Average Speed} = \frac{\text{Distance}}{\text{Elapsed Time}}$$

There is an important difference between the average speed and average velocity. The average speed is always positive and conveys no information about direction. The average velocity contains information about direction and can be positive, negative or zero. It tells us how fast the object moved and the direction of the motion.

The **instantaneous velocity** tells us how fast the object is moving right now, at this instant. As the time interval in Eq. (2.2) gets smaller, the average velocity gives a better approximation to the instantaneous velocity. The instantaneous velocity is the average velocity calculated over a vanishingly small time interval. Of course, "vanishingly small" brings us dangerously close to calculus. However, as you will soon see, we can derive equations for the instantaneous velocity without using calculus.

Acceleration

As an object moves, its velocity can change. The object's **average acceleration** is the rate of change in its velocity

$$a_{ave} = \frac{\Delta v}{\Delta t} \tag{2.4}$$

where $\Delta v = v - v_0$ is the change in velocity and Δt is the finite time interval. For constant acceleration, the average and instantaneous acceleration are always the same. The SI unit for acceleration is the meter per second per second ($\mathrm{m/s^2}$). The American unit for acceleration is the foot per second per second ($\mathrm{ft/s^2}$), although the mile per hour per second ($\mathrm{mi/h/s}$) is also used.

Example 2.3 *A test drive*

What is the constant acceleration of a car that goes from zero to sixty in $4\frac{1}{2}$ seconds?

Solution. The phrase "zero to sixty" refers to miles per hour, so you need to convert 60.0 miles per hour to meters per second. The following expression converts the final velocity with the Solve Method.

$60.0\frac{\mathrm{mi}}{\mathrm{h}} = v\frac{\mathrm{m}}{\mathrm{s}}$, Solution is: 26.822

You can use Eq. (2.4) and Evaluate to calculate the acceleration.

$$a = \frac{26.822\,\mathrm{m/s} - 0\,\mathrm{m/s}}{4.5\,\mathrm{s}} = 5.960\,4\frac{\mathrm{m}}{\mathrm{s}^2}$$

Since this problem calls for an answer in meters and seconds, you can also use Evaluate Numerically directly.

$$a = \frac{60\,\mathrm{mi/h} - 0\,\mathrm{mi/h}}{4\frac{1}{2}\,\mathrm{s}} = 5.960\,5\frac{\mathrm{m}}{\mathrm{s}^2}$$

The answers are slightly different because $60\frac{\mathrm{mi}}{\mathrm{h}}$ is only approximately $26.822\frac{\mathrm{m}}{\mathrm{s}}$. ◀

Another convenient unit of acceleration is g, the acceleration due to gravity near the Earth's surface. In SI units, this acceleration is approximately $9.8067\,\mathrm{m/s^2}$. The following expression converts the acceleration into units of g with the Standard Method.

$$a = 5.960\,4\frac{\mathrm{m}}{\mathrm{s}^2} \times \frac{g}{9.8067\frac{\mathrm{m}}{\mathrm{s}^2}} = 0.607\,79g$$

The car's acceleration is about 60.8% of a g.

Note The accepted value for the acceleration produced by gravity at the Earth's surface is $g = 9.80665\,\mathrm{m/s^2}$. [37]

Example 2.4 *A more-fun test drive*
How much time does it take a car to go from "zero to sixty" if its acceleration is $1g$?
Solution. Use the definition of acceleration in Eq. (2.4) and Solve Exact to find the time t.

$$a = \frac{v - v_0}{t - 0}, \text{ Solution is: } \frac{1}{a}(v - v_0)$$

Use this result to create an expression for the time, and Evaluate it.

$$t = \frac{26.822\,\mathrm{m/s} - 0\,\mathrm{m/s}}{9.806\,7\,\mathrm{m/s^2}} = 2.735\,1\,\mathrm{s}$$

You can use Eq. (2.4) and Solve Exact to do this in one step.

$$9.806\,7\frac{\mathrm{m}}{\mathrm{s}^2} = \frac{26.822\,\mathrm{m/s} - 0\,\mathrm{m/s}}{t}, \text{ Solution is: } 2.735\,1\,\mathrm{s}$$

It takes the car $2.7351\,\mathrm{s}$ to reach the speed of $60\frac{\mathrm{mi}}{\mathrm{h}}$ starting from rest with an acceleration of $1g$. ◄

Equations of Motion

From the definitions of displacement, average velocity, and acceleration, we can derive four basic equations which describe the position and velocity of objects moving with a constant acceleration.

Position as a function of Time: $x = x_0 + v_{ave}\,t$
This equation relates the object's final position to its average velocity. It follows directly from the definition of average velocity, so use Solve Exact and Eq. (2.2) to find x.

$$v_{ave} = \frac{x - x_0}{t - 0}, \text{ Solution is: } x_0 + tv_{ave}$$

This result is valid whether or not the acceleration is constant, but the arithmetic averaging expressed in Eq. (2.3) is only valid if the acceleration is constant!

Velocity as a function of Time: $v = v_0 + at$

This equation relates the object's instantaneous velocity to its acceleration. It follows directly from the definition of acceleration, so use Solve Exact and Eq. (2.4) to find v.

$$a = \frac{v - v_0}{t - 0}, \text{ Solution is: } v_0 + at$$

The graph of this equation is a straight line whose slope is the acceleration a and whose intercept is the initial velocity v_0. The area under the velocity curve is the object's displacement.

Position as a function of Time: $x = x_0 + v_0 t + \frac{1}{2}at^2$

This equation relates the objects final position to the initial velocity and acceleration. First, Substitute the velocity as a function of time into the definition of average velocity.

$$v_{ave} = [(v_0 + v)/2]_{v=v_0+at} = v_0 + \tfrac{1}{2}at$$

Now Substitute (with Expand) the following expression for x.

$$[x_0 + v_{ave}\, t]_{v_{ave}=v_0+\frac{1}{2}at} = \tfrac{1}{2}at^2 + v_0 t + x_0$$

The graph of this equation is a parabola whose intercept is the initial position x_0.

Velocity as a function of Position: $v^2 = v_0^2 + 2a(x - x_0)$

This equation is useful and important because it does not depend on time. It's a combination of the two previous equations. Use the velocity as a function of time to Substitute the time into the position versus time equation.

$$x = \left[x_0 + v_0 t + \tfrac{1}{2}at^2\right]_{t=(v-v_0)/a} = x_0 + \tfrac{1}{2a}(v - v_0)^2 + \tfrac{1}{a}v_0(v - v_0)$$

Now use Solve Exact to find the velocity v.

$$x = x_0 + \tfrac{1}{a}v_0(v - v_0) + \tfrac{1}{2a}(v - v_0)^2,$$
$$\text{Solution is: } -\sqrt{v_0^2 - 2ax_0 + 2ax}, \sqrt{v_0^2 - 2ax_0 + 2ax}$$

In problems where the time is not explicitly given, use this result to relate the velocity directly to the position, or you can calculate the time and use the other three equations.

A little editing by-hand rearranges these results into a more traditional form.

$$x = x_0 + v_0 t + \tfrac{1}{2}at^2 \quad (2.5a)$$
$$x = x_0 + v_{ave}\, t \quad (2.5b)$$
$$v = v_0 + at \quad (2.5c)$$
$$v^2 = v_0^2 + 2a(x - x_0) \quad (2.5d)$$

These four equations of motion describe the 1-dimensional motion of an object moving with a constant acceleration. They relate the object's position and velocity to its constant acceleration and the elapsed time.

Example 2.5 *A test drive again*
How far does the car in Example 2.3 travel while it is accelerating?
Solution. There are several ways to solve this problem. You can use Eq. (2.5d) with $v_0 = 0$ and $x_0 = 0$ and Solve Exact to get an expression for x to find the distance traveled.

$$v^2 = 2ax, \text{ Solution is: } \frac{1}{2a}v^2$$

Then Evaluate the result.

$$x = \frac{1}{2\left(5.9604\dfrac{\mathrm{m}}{\mathrm{s}^2}\right)}\left(26.822\frac{\mathrm{m}}{\mathrm{s}}\right)^2 = 60.350\,\mathrm{m}$$

Since you know the time, you can also Evaluate Eq. (2.5a) with $x_0 = 0$ and $v_0 = 0$.

$$x = 0 + 0 + \tfrac{1}{2}\left(5.9604\frac{\mathrm{m}}{\mathrm{s}^2}\right)(4.5\,\mathrm{s})^2 = 60.349\,\mathrm{m}$$

You can even Evaluate Eq. (2.5b), taking the average of the initial and final velocities.

$$x = 0 + \tfrac{1}{2}\left(0 + 26.822\frac{\mathrm{m}}{\mathrm{s}}\right)(4.5\,\mathrm{s}) = 60.350\,\mathrm{m}$$

Again the answers are slightly different because $60\,\mathrm{mi/\,h}$ is only approximately $26.822\,\mathrm{m/\,s}$. All three results are $60.35\,\mathrm{m}$ to four significant figures. ◂

Signs of the Times

Displacement, velocity, and acceleration can all be positive, negative, or zero. The sign of each tells you something about the motion. A positive displacement means the object's position is to the right of its starting position. A negative velocity means the object is moving to the left. A zero acceleration means the object's velocity is not changing.

Table 2.1 summarizes the sign information for displacement and velocity for horizontal motion.

Displacement	**Velocity**	**Interpretation**
Positive	Positive	To the right, moving right
Positive	Negative	To the right, moving left
Negative	Positive	To the left, moving right
Negative	Negative	To the left, moving left

Table 2.1

If the displacement and velocity have the same sign, the object is moving away from the origin. If they have opposite signs, the object is moving toward the origin.

Table 2.2 summarizes the sign information for velocity and acceleration for horizontal motion.

Velocity	Acceleration	Interpretation
Positive	Positive	Moving right, speed increasing
Positive	Negative	Moving right, speed decreasing
Negative	Positive	Moving left, speed decreasing
Negative	Negative	Moving left, speed increasing

Table 2.2

If the velocity and the acceleration have the same sign, the object's speed is increasing. If they have opposite signs, the object is slowing down.

For vertical motion, just replace the word "right" with "up" and replace the word "left" with "down" in both tables.

Free Fall

One famous and important example of constant acceleration motion is the motion of an object under the influence of gravity near the Earth's surface. This vertical acceleration is $a = -g$ where $g = 9.8067\,\mathrm{m/\,s^2}$ is the acceleration's magnitude and the minus sign indicates gravity acts down. Because the acceleration due to gravity is approximately constant, all objects fall at the same constant rate.

When gravity is the only force acting on an object, we say it is in free fall. As long as the object is not too light or moving too fast, air resistance is negligible and gravity determines its motion. An object is in free fall as soon as it is released, whether it is dropped or thrown, whether it is going up or down.

We can apply Eqs. (2.5) to this problem. To distinguish horizontal motion from vertical, let's rewrite the equations of motion letting y denote vertical position (height). When you change $x \to y$ and let $a = -g$, you have the following four equations.

$$y = y_0 + v_0 t - \tfrac{1}{2}\,gt^2 \qquad (2.6a)$$

$$y = y_0 + v_{ave}\,t \qquad (2.6b)$$

$$v = v_0 - g\,t \qquad (2.6c)$$

$$v^2 = v_0^2 - 2g\,(y - y_0) \qquad (2.6d)$$

These four equations describe the motion of an object that is dropped or thrown straight up or down under the influence of gravity with negligible air resistance. They relate the object's height and vertical velocity to the acceleration due to gravity and the time, with up as positive and down as negative.

Example 2.6 *What goes up...*
A ball is thrown straight up with an initial speed of $22\,\mathrm{m/\,s}$. How high does it travel?
Solution. At its maximum height, the ball stops going up so its vertical velocity is zero. Use Eq. (2.6d) with $y_0 = 0$ and $v = 0$, and Solve Exact to find the height y.

$0 = v_0^2 - 2gy$, Solution is: $\dfrac{1}{2g}v_0^2$

Now Substitute the appropriate values into the resulting expression.

$$y_{max} = \left[\frac{v_0^2}{2g}\right]_{v_0=22\,\mathrm{m/s},g=9.8067\,\mathrm{m/s^2}} = 24.677\,\mathrm{m}$$

You can use Eq. (2.6d) and Solve Exact to do this in one step.

$$0 = \left(22\frac{\mathrm{m}}{\mathrm{s}}\right)^2 - 2\left(9.8067\frac{\mathrm{m}}{\mathrm{s}^2}\right) y, \text{ Solution is: } 24.677\,\mathrm{m}$$

The ball reaches a maximum height of $24.677\,\mathrm{m}$ above its release point. ◀

Example 2.7 *...must come down.*
How much time does the ball spend in the air?
Solution. We can use Eq. (2.6c) to find the time it takes the ball to reach its maximum height.

$$v = v_0 - g\,t, \text{ Solution is: } -\frac{1}{g}(v - v_0)$$

Create an expression for the time with the appropriate values and Evaluate it.

$$t = -\frac{1}{9.8067\,\mathrm{m/s^2}}\left(0\frac{\mathrm{m}}{\mathrm{s}} - 22\frac{\mathrm{m}}{\mathrm{s}}\right) = 2.2434\,\mathrm{s}$$

This is the time it takes the ball to go up, so it takes our ball $2 \times 2.2434\,\mathrm{s} = 4.4868\,\mathrm{s}$ to attain a height of $24.677\,\mathrm{m}$ and fall back to its initial height. This equal-time symmetry only exists when the initial and final heights are equal. ◀

Example 2.8 *Way down!*
If the ball in Example 2.6 is thrown with the same initial velocity from an initial height of $30\,\mathrm{m}$, how much time does it spend in the air before it hits the ground? How fast was it moving when it hit the ground?
Solution. The ball's height as a function of time is given by Eq. (2.6a) where the final height is $y = 0\,\mathrm{m}$. Create an expression with the appropriate values and Solve it for the time.

$$0 = 30\,\mathrm{m} + \left(22\frac{\mathrm{m}}{\mathrm{s}}\right)t - \tfrac{1}{2}\left(9.8067\frac{\mathrm{m}}{\mathrm{s}^2}\right)t^2, \text{ Solution is: } 5.5827\,\mathrm{s}, -1.0959\,\mathrm{s}$$

Discard the negative solution (it actually tells you how much earlier the ball would have to be thrown from the ground to follow the same trajectory and land at the same time) so the ball hits the ground $5.5827\,\mathrm{s}$ after it was thrown.

The ball's velocity as a function of time is given by Eq. (2.6d). Create an expression with the appropriate values and Evaluate it.

$$\left(22\frac{\mathrm{m}}{\mathrm{s}}\right) - \left(9.8067\frac{\mathrm{m}}{\mathrm{s}^2}\right)(5.5827\,\mathrm{s}) = -32.748\frac{\mathrm{m}}{\mathrm{s}}$$

Since the velocity is negative, the ball is moving down at a speed of $32.748\,\mathrm{m/s}$. ◀

Varying Acceleration

If the acceleration is not constant, then we have to use the techniques of calculus to solve the equations of motion and describe the motion. While the mathematics are more complicated, this lets us solve more realistic problems by including the effects of air resistance on falling objects or the dependence of gravity on height.

Displacement, Velocity, and Acceleration

Let's start with the same three basic kinematic definitions, now using calculus.

Displacement

During an infinitesimal time interval dt, an object's change of position is dx. The displacement is the net change in position.

$$\Delta x \equiv x - x_0 = \int_{x_0}^{x} dx$$

Velocity

The instantaneous velocity is the rate of change of the position. This is what your speedometer reads, the velocity at this instant. To find the instantaneous velocity, take the derivative of the position with respect to time.

$$v = \frac{dx}{dt} \tag{2.7}$$

If you know the position as a function of time, then you differentiate to find the velocity as a function of time.

Example 2.9 *Engage!*

As the starship *Enterprise* goes into warp from rest, its position is given by

$$x = \frac{d_w}{2 - \dfrac{t^2}{t_w^2}}$$

where d_w is the ship's displacement while it is going to warp. How fast is the *Enterprise* moving when it enters warp at time t_w?

Solution. To find the ship's velocity as a function of time, Evaluate the following expression for the derivative of the position with respect to time.

$$v = \frac{d}{dt}\left(\frac{d_w}{2 - \frac{t^2}{t_w^2}}\right) = 2td_w \frac{t_w^2}{\left(2t_w^2 - t^2\right)^2}$$

Use Substitute to find the ship's velocity as it enters warp at $t = t_w$.

$$v_w = \left[2td_w \frac{t_w^2}{\left(2t_w^2 - t^2\right)^2}\right]_{t=t_w} = 2\frac{d_w}{t_w}$$

The ship's final velocity is twice the average given by Eq. (2.2). The velocity has the same sign as the displacement, so the *Enterprise* is moving away from the origin. ◄

Acceleration

The instantaneous acceleration of an object is the rate of change of the velocity.

$$a = \frac{dv}{dt} \tag{2.8}$$

Using the definition of velocity,

$$a = \frac{d}{dt}\frac{dx}{dt} = \frac{d^2x}{dt^2} \tag{2.9}$$

we see that the acceleration is the first derivative of the velocity and the second derivative of the position, both with respect to time. The average acceleration is still given by Eq. (2.4).

Example 2.10 *Steady as she goes*

What is the starship *Enterprise*'s acceleration as she goes into warp?

Solution. To find the ship's acceleration as a function of time, Evaluate the following expression for the second derivative of the position.

$$a = \frac{d^2}{dt^2}\left(\frac{d_w}{2 - \frac{t^2}{t_w^2}}\right) = 2d_w \frac{t_w^2}{\left(2t_w^2 - t^2\right)^3}\left(3t^2 + 2t_w^2\right)$$

To find the acceleration at $t = t_w$, Substitute the value for the time into the expression for the acceleration.

$$a_w = \left[2d_w \frac{t_w^2}{\left(2t_w^2 - t^2\right)^3}\left(3t^2 + 2t_w^2\right)\right]_{t=t_w} = 10\frac{d_w}{t_w^2}$$

The ship's final acceleration is 5 times the average given by Eq. (2.4). The acceleration and velocity have the same sign, so the ship's velocity is increasing as it enters warp. ◀

Here is a time-saving *SNB* tip: Unless you need the explicit expressions for the velocity and acceleration as functions of time, you can use Evaluate to take the derivative and Substitute the time-value in one step.

$$v_w = \left[\frac{d}{dt}\left(\frac{d_w}{2 - \frac{t^2}{t_w^2}}\right)\right]_{t=t_w} = 2\frac{d_w}{t_w}$$

$$a_w = \left[\frac{d^2}{dt^2}\left(\frac{d_w}{2 - \frac{t^2}{t_w^2}}\right)\right]_{t=t_w} = 10\frac{d_w}{t_w^2}$$

SNB often allows such a one-fell-swoop approach.

Equations of Motion

We can derive equations to describe the position and velocity of objects moving with a varying acceleration from the definitions of displacement, velocity, and acceleration. Unlike the equations of motion for constant acceleration, which have algebraic solutions, solving these equations requires calculus.

Position as a function of Time

We can derive an equation for a moving object's position as a function of time from the definition of velocity by solving Eq. (2.7) for the position.

$$x = x_0 + \int_0^t v\,dt \tag{2.10}$$

If you know the velocity as a function of time, then you integrate to find the position as a function of time.

Example 2.11 *Resistance is futile*
Find the position as a function of time (with arbitrary initial position) for an object whose velocity is $v = v_0\, e^{-kt}$, where v_0 is the initial velocity.
Solution. Tell *SNB* that the variable k in this problem is positive.

$$\text{assume}\,(k > 0) = (0, \infty)$$

Use Eq. (2.10) to calculate the position as a function of time. Create and Evaluate an expression for the integral of the velocity.

$$v_0 \int_0^t e^{-kt} dt = -\frac{1}{k} v_0 \left(e^{-kt} - 1\right)$$

The final result for the position is $x = x_0 + \frac{v_0}{k}\left(1 - e^{-kt}\right)$. These expressions for the position and velocity describe an object experiencing a resistive acceleration that is proportional to its velocity. ◀

If the velocity is a function of position, we can still use the definition of velocity to find the position as a function of time. Solving Eq. (2.7) for the time gives us

$$t = \int_{x_0}^{x} \frac{dx}{v} \tag{2.11}$$

the amount of time it takes the object to go from x_0 to x. To find the position as a function of time, do the integral on the right-hand side of Eq. (2.11) and solve the resulting equation for the position.

Example 2.12 *Spring forward*
Find the position as a function of time for an object that starts at the origin whose velocity is $v = v_0\sqrt{1 - \frac{x^2}{A^2}}$ where A is a positive constant and v_0 is the velocity at $x = 0$.
Solution. Use assume (positive) to tell *SNB* the variables in this problem are positive. Create an expression for the integral in Eq. (2.11) and Evaluate it.

$$\frac{1}{v_0}\int_0^x \frac{dx}{\sqrt{1 - \frac{x^2}{A^2}}} = \frac{A}{v_0} \arcsin \frac{1}{A} x$$

Now use this result to create an expression for Eq. (2.11) and Solve the resulting equation for the position.

$$t = \frac{A}{v_0} \arcsin \frac{1}{A} x, \text{ Solution is: } A \sin \frac{1}{A} t v_0$$

The position $x = A \sin \frac{v_0}{A} t$ describes the oscillatory motion of an object that is attached to a spring and starts from the origin with a velocity of v_0. ◀

Velocity as a function of Time

We've already seen how to use Eq. (2.7) to find an object's velocity once we know its position. There is also a way to find its velocity if we know its acceleration. From the definition of acceleration in Eq. (2.8), we can derive an equation for the velocity of a moving object as a function of time.

$$v = v_0 + \int_0^t a\, dt \tag{2.12}$$

You integrate a time-dependent acceleration to find the velocity as a function of time.

Example 2.13 *Not exactly the same trip*
Suppose the car in Example 2.3 accelerated for $4\frac{1}{2}$ seconds with the varying acceleration $a = 28\, t\, e^{-t}$ (where t is in seconds and a is in $\mathrm{m/s^2}$). How fast is the car moving at the end? How far did it go?
Solution. Let's use Eq. (2.12) to create and Evaluate the following expression for the car's velocity as a function of time.

$$v = 0 + \int_0^t 28t\, e^{-t}\, dt = 28 - 28te^{-t} - 28e^{-t}$$

To find the final velocity at $t = 4.5\,\mathrm{s}$, Substitute the value for the time.

$$v_f = \left[28 - 28te^{-t} - 28e^{-t}\right]_{t=4.5} = 26.289$$

The car's final velocity ($26.289\,\mathrm{m/s}$) is slightly less than the result from Example 2.3. Use Eq. (2.10) to Evaluate this expression for the car's position as a function of time.

$$x = 0 + 28 \int_0^t \left(1 - te^{-t} - e^{-t}\right) dt = 28t + 56e^{-t} + 28te^{-t} - 56$$

To find the car's final position at $t = 4.5\,\mathrm{s}$, Substitute the value for the time.

$$x_f = \left[28t + 56e^{-t} + 28te^{-t} - 56\right]_{t=4.5} = 72.022$$

This car traveled almost 20% further than the car in Example 2.3. Let's use Eq. (2.2) to Evaluate the car's average velocity.

$$v_{ave} = \frac{72.022\,\mathrm{m}}{4.50\,\mathrm{s}} = 16.005\frac{\mathrm{m}}{\mathrm{s}}$$

The average velocity is more than half of the car's final velocity, and it is *not* the arithmetic average of the initial and final velocities. ◀

If the acceleration is a function of velocity, we can still use the definition of acceleration to find the velocity as a function of time. Solving Eq. (2.8) for the time gives us

$$t = \int_{v_0}^{v} \frac{dv}{a} \tag{2.13}$$

the amount of time it takes for the object's velocity to change from v_0 to v. To find the velocity as a function of time, do the integral on the right-hand side of Eq. (2.13) and solve the resulting equation for the velocity.

Example 2.14 *Not so futile after all*
Find the velocity as a function of time when the acceleration is $a = -k\,v$.
Solution. Tell *SNB* that the variables in this problem are positive.

$$\text{assume}\,(\text{positive}) = (0, \infty)$$

Create an expression for the right-hand side of Eq. (2.13), and Evaluate it.

$$-\frac{1}{k}\int_{v_0}^{v} \frac{dv}{v} = -\frac{1}{k}\left(\ln v - \ln v_0\right)$$

Then use Solve Exact on the resulting equation to find the velocity, and Simplify the result.

$$t = -\frac{1}{k}\left(\ln v - \ln v_0\right), \text{ Solution is: } e^{\ln v_0 - kt} = v_0 e^{-kt}$$

As claimed in Example 2.11, the velocity as a function of time for an object experiencing a resistive acceleration that is proportional to its velocity is $v = v_0 e^{-kt}$. ◄

Velocity as a function of Position

Using the chain rule, we can relate the acceleration to the velocity as a function of position

$$a = \frac{dv}{dt} = \frac{dv}{dx}\frac{dx}{dt} = v\,\frac{dv}{dx} \tag{2.14}$$

so that $v\,dv = a\,dx$. For an acceleration that depends on position, we can integrate both sides of this equation.

$$v^2 - v_0^2 = 2\int_{x_0}^{x} a\,dx \tag{2.15}$$

If you know the acceleration as a function of position, you can use Eq. (2.15) to find the velocity.

Example 2.15 *Simple harmonic motion*
Calculate the time it takes an object released from rest at $x_0 = +A$ to reach the origin when the acceleration is $a = -\omega^2 x$.
Solution. Use Eq. (2.15) to create and Expand an expression for the object's velocity as a function of position.

$$v^2 = 2\left(-\omega^2\right)\int_{A}^{x} x dx = A^2\omega^2 - x^2\omega^2$$

The object will move to the left so take the negative root for the velocity.

$$v = -\omega\sqrt{A^2 - x^2}$$

Tell *SNB* that the variables in this problem are positive. Use Eq. (2.11) to create and Evaluate an expression for the time.

$$t = \int_A^0 \frac{dx}{\left(-\omega\sqrt{A^2 - x^2}\right)} = \frac{1}{2}\frac{\pi}{\omega}$$

You may recognize this as $1/4$ of the period of a simple harmonic oscillator. Notice the answer does **not** depend on the initial position A. For this acceleration, the time it takes the object to reach to origin is independent of its starting point. ◂

For an acceleration that depends on velocity, we can derive an equation for the velocity by solving Eq. (2.14) for the position.

$$x = x_0 + \int_{v_0}^{v} \frac{v\,dv}{a} \tag{2.16}$$

If you know the acceleration as a function of velocity, you can use Eq. (2.16) to find the velocity. After you do the integral in Eq. (2.16), you need to solve the resulting equation for the velocity.

Example 2.16 *Resistance is even more futile*
Find the velocity as a function of position for an object starting from initial position x_0 with initial velocity v_0 when the acceleration is $a = -\kappa v^2$.
Solution. Tell *SNB* that the variables in this problem are positive.

$$\text{assume}\,(\text{positive}) = (0, \infty)$$

Now create an expression for the integral in Eq. (2.16), and Evaluate it.

$$-\frac{1}{\kappa}\int_{v_0}^{v} \frac{v\,dv}{v^2} = -\frac{1}{\kappa}(\ln v - \ln v_0)$$

Then Solve the resulting equation for the velocity, and Simplify the result.

$$x = x_0 - \frac{1}{\kappa}(\ln v - \ln v_0), \text{ Solution is: } \exp(\ln v_0 - \kappa(x - x_0)) = v_0 e^{\kappa x_0 - x\kappa}$$

The velocity as a function of position for an object experiencing a resistive acceleration that is proportional to the square of its velocity is $v = v_0 e^{-\kappa(x - x_0)}$. ◂

Gravity and Air Resistance

In our previous discussion of falling objects, we ignored the effects of air resistance and altitude and treated the acceleration due to gravity as a constant. This approximation is valid for heavy or slow-moving objects falling near the Earth's surface. Otherwise, this approximation is not valid and the acceleration is not constant.

Resisting Air Resistance is Futile

Near the Earth's surface, air resistance has a significant effect on light or fast-moving objects. Baseballs wouldn't curve, party balloons wouldn't slow down so quickly, and parachutes wouldn't work without the effects of air resistance.

Let's look at the 1-d motion of objects under the influence of gravity and air resistance, with up as the positive direction. One model for air resistance uses a velocity-dependent acceleration that depends quadratically on the velocity. On the way up, gravity and air resistance are both slowing the object down so they have the same sign.

$$a_{\uparrow} = -g - \kappa v^2 \tag{2.17a}$$

On the way down, gravity is speeding up the object but air resistance is still slowing it down so they have opposite signs.

$$a_{\downarrow} = -g + \kappa v^2 \tag{2.17b}$$

The air resistance parameter κ depends on the object's mass and aerodynamic properties.

An object thrown straight up starts with velocity v_0 and stops at its maximum height. Let's use Eqs. (2.13) and (2.17a) and Evaluate the integral for the time the object spends moving upward.

$$-\int_{v_0}^{0} \frac{dv}{\kappa v^2 + g} = \frac{1}{\sqrt{g}\sqrt{\kappa}} \arctan \frac{1}{\sqrt{g}} \sqrt{\kappa} v_0$$

A little editing by-hand tells us the time the object spends moving up.

$$t_{\uparrow} = \frac{1}{\sqrt{\kappa g}} \arctan \sqrt{\frac{\kappa}{g}} v_0 \tag{2.18a}$$

Use Eqs. (2.16) and (2.17a) and the definite integral fudge to Evaluate the integral for the maximum height.

$$\left[-\int \frac{v dv}{\kappa v^2 + g} \right]_{v=v_0}^{v=0} = \frac{1}{2\kappa} \left(\ln \left(\kappa v_0^2 + g \right) - \ln \kappa \right) - \frac{1}{2\kappa} \left(\ln g - \ln \kappa \right)$$

Then edit the result by-hand, using Expand in-place on the argument of the natural logarithm.

$$y_{max} = y_0 + \frac{1}{2\kappa} \ln \left(1 + \frac{\kappa}{g} v_0^2 \right) \tag{2.18b}$$

This relates the initial velocity to the maximum height.

Let's revisit two previous examples where we threw a ball straight up. This time we'll take air resistance into account.

Example 2.17 *What goes up must come down, eventually*
A baseball is thrown straight up with an initial velocity of 22 $\mathrm{m/s}$. Taking air resistance into account, how high does the ball travel? How much time does it take for the ball to reach its maximum height?
Solution. For a slow-moving baseball, a typical value for κ is $0.00945/\mathrm{m}$. The maximum height is given by Eq. (2.18b) so let's use it to create an expression.

Then Substitute (with Evaluate) the appropriate numerical values.

$$y_{max} = \left[\frac{1}{2\kappa}\ln\left(\frac{1}{g}\kappa v_0^2+1\right)\right]_{\kappa=0.00945/\,\mathrm{m},v_0=22\,\mathrm{m/\,s},g=9.806\,7\,\mathrm{m/\,s^2}} = 20.254\,\mathrm{m}$$

The ball reaches a maximum height of $20.254\,\mathrm{m}$ above its release point. This is approximately 17.9% lower than the result from Example 2.6 which ignored air resistance.

The time to reach the maximum height is given by Eq. (2.18a) so create an expression and Substitute the appropriate values using Simplify.

$$t_\uparrow = \left[\frac{1}{\sqrt{\kappa g}}\arctan\sqrt{\frac{\kappa}{g}}v_0\right]_{\kappa=0.00945/\,\mathrm{m},v_0=22\,\mathrm{m/\,s},g=9.806\,7\,\mathrm{m/\,s^2}} = 1.968\,2\,\mathrm{s}$$

The baseball takes approximately 12.3% less time than the result of Example 2.7 which ignored air resistance. The effect of air resistance reduces both the maximum height and the time required to reach that height by a significant amount. ◀

A dropped object starts with velocity $v_0 = 0$ and reaches its final height y_f moving at velocity v_f. Use Eqs. (2.13) and (2.17b) and the definite integral fudge to Evaluate the integral for the time the object spends moving downward.

$$\left[\int\frac{dv}{\kappa v^2-g}\right]_{v=0}^{v=v_f} = \frac{1}{2\sqrt{g}\sqrt{\kappa}}\left(\ln\left(\sqrt{\kappa}v_f-\sqrt{g}\right)-\ln\left(\sqrt{\kappa}v_f+\sqrt{g}\right)\right)$$
$$-\frac{1}{2\sqrt{g}\sqrt{\kappa}}\left(\ln\left(-\sqrt{g}\right)-\ln\sqrt{g}\right)$$

A little editing by-hand tells us the time the object spends moving down.

$$t_\downarrow = \tfrac{1}{2}\sqrt{\frac{1}{g\kappa}}\ln\frac{1-\sqrt{\frac{\kappa}{g}}v_f}{1+\sqrt{\frac{\kappa}{g}}v_f} \tag{2.19a}$$

Use Eqs. (2.16) and (2.17b) and the definite integral fudge to Evaluate the integral for the final height.

$$\left[\int\frac{vdv}{\kappa v^2-g}\right]_{v=0}^{v=v_f} = \frac{1}{2\kappa}\left(\ln\kappa-\ln(-g)\right)-\frac{1}{2\kappa}\left(\ln\kappa-\ln\left(\kappa v_f^2-g\right)\right)$$

Edit the result by-hand, using Expand in-place on the argument of the natural logarithm.

$$y_f = y_0 + \frac{1}{2\kappa}\ln\left(1-\frac{\kappa}{g}v_f^2\right) \tag{2.19b}$$

This relates the final velocity to the final height (which is often zero).

Example 2.18 *Coffee, tea or...*
Four embedded coffee filters, dropped from a height of $2\,\mathrm{m}$, hit the floor $1.17\,\mathrm{s}$ later. Find the air resistance parameter and the landing speed of the filters.
Solution. Use Eqs. (2.19) to create equations involving the two unknowns. Since both equations contain both unknowns, let's put them in a 2-row, 1-column matrix with the appropriate numerical values, and use Solve Numeric.

$$\begin{bmatrix} 1.17 = \frac{1}{2}\sqrt{\frac{1}{9.806\,7\kappa}}\ln\left(\frac{1-\sqrt{\frac{\kappa}{9.806\,7}}v_f}{1+\sqrt{\frac{\kappa}{9.806\,7}}v_f}\right) \\ 0 = 2+\frac{1}{2\kappa}\ln\left(1-\frac{\kappa v_f^2}{9.806\,7}\right) \end{bmatrix},$$

Solution is: $\{[\kappa = 2.617\,1, v_f = -1.935\,7]\}$

The filters have an air resistance parameter of $\kappa = 2.617\,1\,\mathrm{m}^{-1}$ and they land with a velocity of $v = -1.935\,7\,\mathrm{m/s}$. The large value of κ tells us coffee filters are very susceptible to the effects of air resistance. The negative velocity tells us the filters were moving down. ◀

Long-Distance Free Fall

An object falling from a large distance experiences a position-dependent acceleration that depends on its distance from the Earth's center.

$$a = -g\frac{R^2}{r^2} \tag{2.20}$$

As usual, g is the acceleration due to gravity on the surface and R is the Earth's radius. The change from y to r for the vertical position reminds you to measure the radial distance from the Earth's center and not the surface.

Let's look at the long-distance free fall of an object dropped from rest. We can use Eq. (2.15) with $v_0 = 0$ to create an expression for the object's velocity as a function of position. First tell *SNB* to assume the variables are positive.

$$\text{assume}\,(\text{positive}) = (0, \infty)$$

Then Evaluate the integral in the expression for the velocity.

$$v^2 = -2gR^2\int_{r_0}^{r}\frac{dr}{r^2} = -2R^2\frac{g}{rr_0}(r - r_0)$$

A little editing by-hand produces the velocity as a function of vertical position

$$v = -\sqrt{\frac{2gR^2}{r_0}}\sqrt{\frac{r_0 - r}{r}} \tag{2.21}$$

where the minus sign indicates downward motion.

Now we can use Eq. (2.11) to calculate the fall-in time.

$$t = -\sqrt{\frac{r_0}{2gR^2}} \int_{r_0}^{R} \sqrt{\frac{r}{r_0 - r}}\, dr \tag{2.22}$$

This integral is not in a form *SNB* can do, so we need to adjust it. Create the following expression multiplying numerator and denominator by $\sqrt{r}$. Then use Simplify in-place on the numerator and Expand in-place on the term inside the radical in the denominator.

$$\sqrt{\frac{r}{r_0 - r}} = \frac{\sqrt{r}\sqrt{r}}{\sqrt{r\,(r_0 - r)}} = \frac{r}{\sqrt{rr_0 - r^2}}$$

The integral is now in a form *SNB* can Evaluate.

$$-\int_{r_0}^{R} \frac{r}{\sqrt{rr_0 - r^2}} dr = \sqrt{R}\sqrt{r_0 - R} + \tfrac{1}{2} r_0 \arcsin \frac{1}{r_0} (r_0 - 2R) + \tfrac{1}{4}\pi r_0$$

A little editing by-hand produces a complicated equation relating the fall-in time to the initial position.

$$t = \sqrt{\frac{r_0}{2gR^2}} \left(\sqrt{R\,(r_0 - R)} + \tfrac{1}{2} r_0 \arcsin\left(1 - \frac{2R}{r_0}\right) + \tfrac{1}{4}\pi r_0 \right) \tag{2.23}$$

The arcsine term prevents an exact solution for the initial position, but we can solve the equation numerically or in the large-r_0 limit. Either way, we'll need values for the Earth's radius and surface gravity.

$$gR^2 = \left(9.8067 \frac{\text{m}}{\text{s}^2}\right) \left(6.3781 \times 10^6\ \text{m}\right)^2 = 3.989\,4 \times 10^{14} \frac{\text{m}^3}{\text{s}^2}$$

That's a big number so let's work in mega-meters and days (leaving off the units since we'll be using Solve Numeric).

$$gR^2 = 3.989\,4 \times 10^{14} \frac{\text{m}^3}{\text{s}^2} \times \left(\frac{1}{10^6\ \text{m}}\right)^3 \times \left(\frac{86400\ \text{s}}{1}\right)^2 = 2.978\,1 \times 10^6$$

As an example, let's calculate the time for the Moon to free-fall from rest to the Earth. The initial position is the average Earth-Moon distance.

$$r_0 = 384400\ \text{km} = 3.844 \times 10^8\ \text{m} \times \left(\frac{1}{10^6\ \text{m}}\right) = 384.4$$

Use Eq. (2.23) with the appropriate numerical values and apply Evaluate Numerically.

$$t = \sqrt{\frac{384.4}{2(2.978\,1 \times 10^6)}} \left(\sqrt{6.371\,(378.03)} + \tfrac{1}{2} 384.4 \arcsin\left(1 - \tfrac{2(6.371)}{384.4}\right) + \tfrac{\pi}{4} 384.4\right)$$
$$= 4.846\,4$$

It takes the free-falling Moon just under 5 days to reach the Earth.

Let's square Eq. (2.23), impose the large initial position condition (where $r_0 \gg R$) by-hand and Simplify the resulting expression.

$$t^2 \approx \left(\sqrt{\frac{r_0}{2gR^2}}\left(\tfrac{1}{2}r_0 \arcsin(1) + \tfrac{1}{4}\pi r_0\right)\right)^2 = \tfrac{1}{8}\frac{\pi^2}{R^2 g} r_0^3$$

This is a special case of a famous result that we'll see again in Chapter 9. It's Kepler's 3rd Law for half an orbit with an initial position that's twice the semi-major axis.

Let's use it with Evaluate Numerically to calculate the fall-in time.

$$t = \sqrt{\tfrac{1}{8}\frac{\pi^2}{2.9781 \times 10^6}(384.4)^3} = 4.8508$$

The large-r_0 approximation is within 0.01 percent of our numerical solution to the complete equation.

If you know the fall-in time, you can solve Eq. (2.23) numerically for the initial height. In his epic poem "Paradise Lost", Milton wrote that once cast out of heaven, the devil fell for nine days before landing in hell. In Milton's universe, hell is a separate place from Earth. But if we suppose it's at the Earth's center, we can use Milton's time to calculate the distance from heaven to Earth.

Use Eq. (2.23) with Milton's $t = 9\,\mathrm{d}$ and apply Solve Numeric to find the initial height.

$$9 = \sqrt{\frac{r_0}{2(2.9781 \times 10^6)}}\left(\sqrt{6.371(r_0 - 6.371)} + \tfrac{1}{2}r_0 \arcsin\left(1 - 2\tfrac{6.371}{r_0}\right) + \tfrac{1}{4}\pi r_0\right),$$

Solution is: $\{[r_0 = 580.6]\}$

Let's compare 580.6 mega-meters with the average Earth-Moon distance.

$$r_0 = 580.6 \times 10^6\,\mathrm{m} \times \frac{1}{3.844 \times 10^8\,\mathrm{m}} = 1.5104$$

A 9-day fall from rest to Earth starts from a height of $5.806 \times 10^8\,\mathrm{m}$, which is about 1.51 times the Earth-Moon distance (and 30 times larger than Milton's value of $3R$).

Some might consider this a delicate topic that seems to combine science and religion. Please don't. This is not an attempt to validate or refute any philosophy, theology or ideology. It is simply a typical intellectual exercise a physicist (or a physics student) might carry out when confronted with such information. That's it.

Here is my message for students: we applied some basic physics to a situation described in a work of literature. Though motivated by a literary work that had nothing to do with physics, the analysis produced an interesting, useful result. That is often the case when we combine our natural curiosity and physics. You were born with a wonderful brain; feel free to use it.

Problems

1. A baseball is dropped from a height of $30\,\mathrm{m}$. Ignoring air resistance, how long does it take the ball to hit the ground? How fast is it going when it hits the ground?

2. A baseball is thrown down with an initial speed of $22\,\mathrm{m/\,s}$ from an initial height of $30\,\mathrm{m}$. How much time does it spend in the air before it hits the ground? How fast was it moving when it hit the ground? Ignore air resistance.

3. Some major league pitchers routinely throw a fastball with an initial horizontal velocity of $95\,\mathrm{mi/\,h}$. How long does it take the ball to reach home plate $60\,\mathrm{ft}$, $6\,\mathrm{in}$ away? Ignore air resistance.

4. The length of the barrel of a simple blowgun is $1.2\,\mathrm{m}$. Upon leaving the barrel, a dart has a speed of $18\,\mathrm{m/\,s}$. Assuming that the dart's acceleration is constant, calculate

   a. the acceleration of the dart, and

   b. the time it takes the dart to travel the length of the barrel.

5. Suppose your car can go from zero to $60\,\mathrm{mi/\,h}$ in $3.3\,\mathrm{s}$. Find the magnitude of your car's acceleration, and give your answer in *both* $\mathrm{m/\,s^2}$ and units of g.

6. Create a single table that contains the information in Tables 2.1 and 2.2 about the signs of displacement, velocity, and acceleration. Include all possible combination of signs and don't forget to label the columns.

7. The position of an object moving in a straight line is given by $x = t^3 - 5.0t^2 + 5.25t$, where x is in meters and t in seconds.

   a. What is the object's displacement between $t = 0\,\mathrm{s}$ and $t = 2\,\mathrm{s}$?

   b. What is the object's average velocity between $t = 0\,\mathrm{s}$ and $t = 2\,\mathrm{s}$?

   c. What is the average velocity for the time interval from $t = 0\,\mathrm{s}$ to $t = 4\,\mathrm{s}$?

   d. Graph x versus t for $0 \leq t \leq 4\,\mathrm{s}$.

8. Calculate the velocity as a function of time for the object in the previous problem. Where is the velocity zero? Add the expression for the velocity to the position plot.

9. The position of particle moving along the x-axis is given in centimeters by $x = 12.5 + 1.50t^3$, where t is in seconds. Consider the time interval $t = -2.00\,\mathrm{s}$ to $t = +2.00\,\mathrm{s}$ and calculate

   a. the velocity as a function of time,

   b. the average velocity, and

   c. the instantaneous velocity at $t = -2.00\,\mathrm{s}$, $0\,\mathrm{s}$, and $+2.00\,\mathrm{s}$.

   d. Graph x versus t and v versus t on the same plot.

10. Suppose a particle's velocity is given by $v = 6t - 12$ (where t is in seconds and v is in meters per second), and its initial position is $4\,\mathrm{m}$.

    a. What is the particle's initial velocity?

    b. What is its velocity at $t = 1\,\mathrm{s}$?

    c. Where is it at $t = 1\,\mathrm{s}$?

    d. Does the particle ever stop? If so, where?

    e. What is the particle's acceleration?

11. Show that the acceleration of the starship *Enterprise* from Example 2.10 is equal to

$$a = 2\frac{d_w}{t_w^2}\left(2 + 3\frac{t^2}{t_w^2}\right)\left(2 - \frac{t^2}{t_w^2}\right)^{-3}$$

12. Show that the acceleration of the starship *Enterprise* from the previous problem is approximately constant for large values of t_w. What is that constant? What is the lowest-order correction to that constant?

13. Find the velocity as a function of position for the velocity-dependent acceleration $a = -\kappa v^3$.

14. For the time-dependent acceleration $a = \alpha_0 te^{-t}$, find the value of the constant α_0 so that the average acceleration over the first 5 seconds of motion from rest is $1g$. What are the units of α_0?

15. Make a graph of the time-dependent acceleration $a = 28te^{-t}$ and the constant acceleration $a = 5.9604$. Use the graph to explain how the car in Example 2.13 had a smaller final velocity but traveled further than the car in Example 2.3.

16. Calculate the average acceleration for the car in Example 2.13 using the following equation.

$$a_{ave} = \tfrac{1}{\Delta t}\int a\,dt$$

    How does your result compare to the value using Eq. (2.4)? How does your result compare to that of Example 2.3?

17. Calculate the velocity and acceleration as a function of time for the time-dependent position

$$x = \frac{\frac{1}{2}a_0\,t^2}{1 + \dfrac{t^2}{T^2}}$$

    where a_0 and T are constants. Show that for small values of t, these expressions reduce to the usual constant-acceleration results. Find the lowest-order time-dependent correction to those results.

18. A baseball is dropped from a height of $30\,\mathrm{m}$. Including the effects of air resistance, how long does it take the ball to hit the ground? How fast is it going when it hits the ground?

19. In the episode "Air Plane Hour", the MythBusters dropped a dummy from $4000\,\text{ft}$ and the dummy hit the ground $31\,\text{s}$ later. Assume the dummy starts from rest and experiences a quadratic air resistance.

    a. Find the dummy's air-resistance parameter and hit-the-ground speed.

    b. From what altitude must the dummy be dropped to have a fall time of $90\,\text{s}$?

20. How high would the MythBuster's dummy go if it was launched straight up with its landing speed from the previous problem? How high would it go if you ignore air resistance?

21. How much time does it take the baseball in Example 2.17 to come down? How fast is it moving when it returns to its initial height? Compare your answers to the no-air-resistance results.

22. A baseball is thrown down with an initial speed of $22\,\text{m/s}$ from an initial height of $30\,\text{m}$. How much time does it spend in the air before it hits the ground? How fast was it moving when it hit the ground? Do not ignore air resistance.

23. Some major league pitchers routinely throw a fastball with an initial horizontal velocity of $95\,\text{mi/h}$. How long does it take the ball to reach home plate, $60\,\text{ft}\ 6\,\text{in}$ away? Do not ignore air resistance.

24. Show that an object thrown straight up with initial velocity v_0 has velocity

$$v_f = -\frac{v_0}{\sqrt{1 + \dfrac{\kappa}{g} v_0^2}}$$

when it returns to its initial height. Assume a quadratic air resistance.

25. In the episode "Exploding Water Heater", the MythBusters launched a water heater straight up to a maximum height of $500\,\text{ft}$ with an initial speed of about $300\,\text{mi/h}$. How long was the heater in the air? What was its landing speed?

26. The Sun's radius is 109 times that of Earth and its "surface" gravity is $28g$. How long would it take a free-falling Earth to fall into the Sun?

27. Do a little research, find the appropriate data for Venus and Mars, and calculate their fall-in times.

28. Let's verify our expression for the fall-in time reduces to the correct result when the initial height is only slightly larger than the Earth's radius. Tell *SNB* to assume the variables are positive.

    a. Substitute $r_0 = R + y_0$ into the right side of Eq. (2.23).

    b. Do a 2-term Power Series expansion in terms of y_0.

    c. Delete the $O\left(y_0^{3/2}\right)$ and Simplify what's left.

    d. Use Solve Exact to find the initial height y_0. Look familiar?

3 Vectors

In the previous chapter, we described 1-dimensional motion. We specified an object's position along one axis as either positive or negative. All information about the direction was carried by the sign.

The universe is not 1-dimensional. There are more complicated motions that require more complicated descriptions. Fortunately, there are mathematical tools called vectors that will allow us to extend our study of motion to 2 and 3 dimensions. Vectors turn a difficult 2-d problem into two easy 1-d problems. That is a good trade.

Many physical quantities can be described by a single number and a unit. Room temperature is approximately $21\,^{\circ}\mathrm{C}$ and a typical class lasts about $3600\,\mathrm{s}$. These numbers are called **scalars.** Temperature and time are examples of scalars. Scalars obey the usual rules of arithmetic. You can punch them into your calculator or use *SNB*. Scalars are old friends, the numbers you've been using for a long time. We just gave them a new name.

Sometimes scalars are inadequate. Consider the following simple example. Suppose you walk $5\,\mathrm{m}$, stop, then walk another $12\,\mathrm{m}$. How far from your original location are you? It depends on the direction! Where you end up depends on both how far and in which direction you walk.

If you only walk in one direction, you end up $5\,\mathrm{m} + 12\,\mathrm{m} = 17\,\mathrm{m}$ away from your initial position. If you walk $5\,\mathrm{m}$ to the left and then turn around and walk $12\,\mathrm{m}$ to the right, you end up $12\,\mathrm{m} - 5\,\mathrm{m} = 7\,\mathrm{m}$ to the right of your initial position. If you walk forward $5\,\mathrm{m}$, turn right and walk another $12\,\mathrm{m}$, you end up $13\,\mathrm{m}$ away from your initial position.

Such quantities are called **vectors**. Vectors are mathematical quantities that have magnitude and direction. The vector's magnitude tells us how much of the physical quantity we have. As Figure 3.1 shows, we can represent these quantities with arrows. When we use an arrow to represent a vector, the arrow's "length" tells us the vector's magnitude and the arrow points in the vector's direction.

Many physical quantities are vectors. As our simple example showed, displacement is a vector. Velocity and acceleration are vectors, as are momentum and force. In this chapter, you will learn now to use *SNB* to do vector addition, subtraction, scalar multiplication, dot products, and cross products.

We'll indicate a vector with an arrow over a bold-faced symbol (such as $\overrightarrow{\mathbf{r}}$). This is one example of a Decoration in *SNB*. Decorations are bars, arrows, wide accents, and braces placed above or below an expression, or small boxes placed around an expression.

Doing Physics with Scientific Notebook: A Problem-solving Approach, First Edition. Joseph Gallant.

SNB provides a template for these decorations which you can access with the Decoration button on the Math Objects toolbar or from the Insert + Decoration menu item.

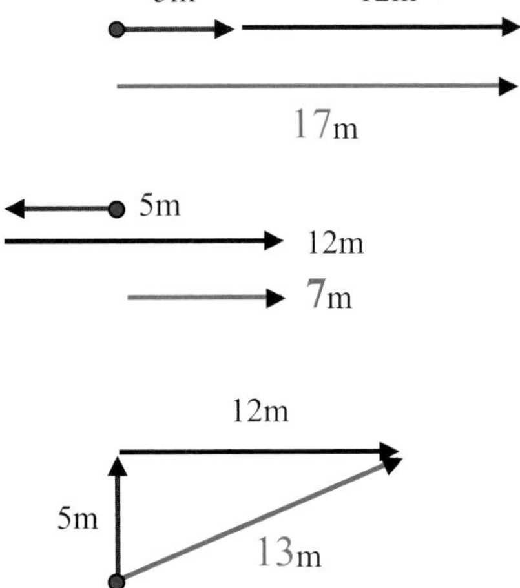

Figure 3.1 Vectors are your friends

When you click the Decoration button, a panel of templates appears (Figure 3.2). To choose a Decoration, click one of the 14 choices and then click OK. If you selected an expression before opening the panel, *SNB* will apply the decoration to that expression. Otherwise, the decoration will appear with the insertion point in an empty input box.

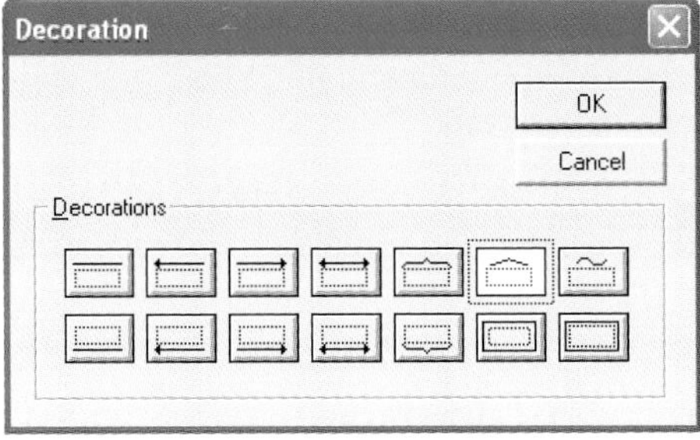

Figure 3.2 Decorations

With the exception of the underline (first column, bottom row), applying a Decoration converts the selected text to Math mode. Even though most of the Decorations leave you in Math mode, they really are just decorations. They contain no mathematical meaning to *SNB*'s engine.

Use the following steps to create an expression for the name of a vector.

1. Enter the letter in either Text or Math mode.
2. Select it and make it Bold faced.
3. While it is still selected, click the Decoration button and choose "right arrow above" (top row, third from the left).
4. Click OK.

Components of a Vector

Suppose you are standing somewhere in the x-y plane. There are two ways to specify your position. We could use your (x, y) coordinates or we could state the length and direction of the vector $\vec{\mathbf{r}}$ that points to your position.

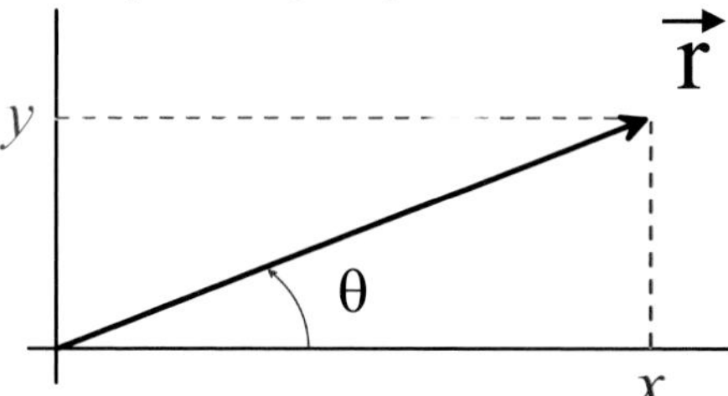

Figure 3.3 A position vector $\vec{\mathbf{r}}$

The coordinates form a right triangle with the vector, so they are related to the vector's length (r) and direction (θ) by the usual trigonometric relations.

$$x = r\cos\theta \tag{3.1a}$$

$$y = r\sin\theta \tag{3.1b}$$

The length of the position vector forms the hypotenuse of the right triangle.

Noteworthy Notation Note A boldfaced symbol with an arrow over it ($\vec{\mathbf{r}}$) indicates a vector. The same symbol in Math mode without the arrow (r) indicates the magnitude of that vector. The vector's direction is indicated by the angle (θ) it makes with the positive x-axis.

Now consider a generic 2-dimensional vector $\vec{\mathbf{A}}$ which connects the origin to a point with coordinates (A_x, A_y). These coordinates are called the **components** of the vector.

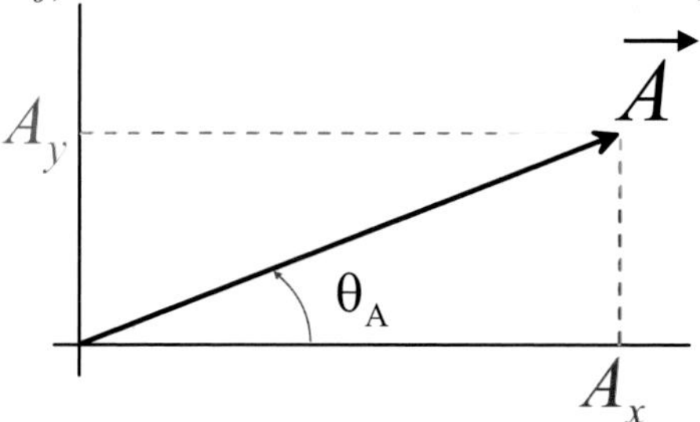

Figure 3.4 A generic vector $\vec{\mathbf{A}}$

As Figure 3.4 shows, the relationships between the vector's components and its magnitude (A) and direction (θ_A) are the same as those between your position vector and coordinates.

$$A_x = A\cos\theta_A \tag{3.2a}$$

$$A_y = A\sin\theta_A \tag{3.2b}$$

The quantity A_x is called the x-component of the vector $\vec{\mathbf{A}}$, and the quantity A_y is called the y-component. Once you know a vector's magnitude and direction, use Eqs. (3.2) to calculate its components. This is called resolving a vector into its components.

Example 3.1 *Isn't that the Beta Quadrant?*
Find the components of a vector whose magnitude is 12 and direction is 125°.
Solution. Use Eqs. (3.2) with either the **red** or **green** degree unit to create an expression for the x-component of the vector and apply Evaluate Numerically to it.

$$A_x = 12\cos 125° = -6.8829$$

Repeat the process for the y-component of the vector

$$A_y = 12\sin 125° = 9.8298$$

The x-component is $A_x = -6.8829$ and the y-component is $A_y = 9.8298$. This vector is in the second quadrant. ◀

Magnitude and Direction

In *SNB*, a vector is a matrix with either one row or one column. Each cell of the matrix holds one component of the vector. For a 2-dimensional vector, you can use either a 2-row, 1-column or a 2-column, 1-row matrix.

$$\vec{\mathbf{A}} = \begin{Bmatrix} A_x \\ A_y \end{Bmatrix} \quad \text{or} \quad \vec{\mathbf{A}} = \begin{Bmatrix} A_x & A_y \end{Bmatrix}$$

To distinguish our vectors from other matrices and binomial coefficients, we'll enclose them with Braces.

Use the following steps to create a vector.

1. Click on the Matrix button on the Math Objects toolbar, or choose Matrix from the Insert menu.
2. Use the Matrix dialog box to set the Dimensions of your vector.
3. Select one of the Built-in Delimiters to enclose the matrix with Braces.
4. Choose OK.

SNB places the insertion point in the top left cell of the vector, where you can type in the x-component of your vector. Use TAB or the ARROW KEYS to move to the next cell, and type in the y-component of your vector. If your vector has a third cell for a z-component, repeat the last step.

Example 3.2 *The Beta Quadrant it is*
Create a 2-dimensional vector using the components from the previous example.
Solution. Follow the above steps and create a 1-column, 2-row matrix with Braces for the Built-in Delimiters. For the x-component, enter -6.8829 in the first cell. Enter 9.8298 in the second cell for the y-component. Your vector should look like this:

$$\vec{\mathbf{A}} = \begin{Bmatrix} -6.8829 \\ 9.8298 \end{Bmatrix} \text{ or } \vec{\mathbf{A}} = \begin{Bmatrix} -6.8829 & 9.8298 \end{Bmatrix}$$

You can use Matrices + Transpose to change a column vector into a row vector (or a row vector into a column vector). ◀

When you know a vector's magnitude and direction, you can use Eqs. (3.2) to calculate its components. Next we see how to calculate a vector's magnitude and direction if we know its components.

The **magnitude** of a 2-dimensional vector is given by

$$A = \left\| \vec{\mathbf{A}} \right\| = \sqrt{A_x^2 + A_y^2} \tag{3.3}$$

where the double vertical lines are *SNB*'s notation for the magnitude of the vector. *SNB* calls them the "norm bracket", using the mathematical name for magnitude.

The double vertical lines are one example of the *SNB*'s many Brackets. *SNB* provides templates (Figure 3.5) for these brackets which you can access with the Brackets button on the Math Objects toolbar or from the Insert + Brackets menu item.

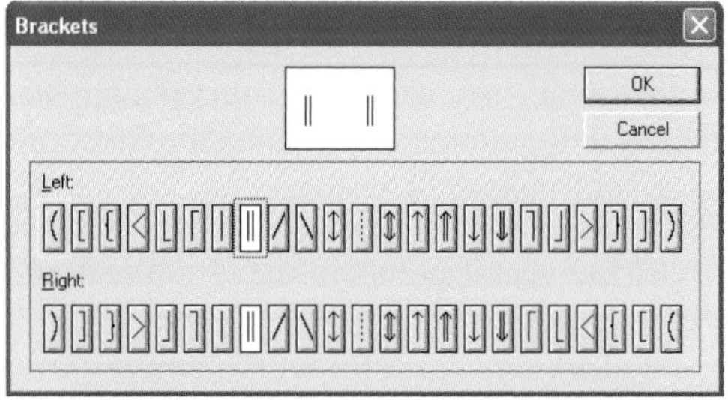

Figure 3.5 Bracketology *á la SNB*

Most of these expanding brackets are for grouping, and they can be used as left or right members of matched or unmatched bracket pairs. You can use the single vertical lines to calculate the absolute value of a number or the determinant of a matrix. You can use the double vertical lines to calculate the magnitude of a vector.

Use these steps to calculate a vector's magnitude.

1. Create the vector with symbolic or numeric components, and select it.
2. Click the Brackets button ()[] and choose the double vertical lines for both the left and right.
3. Leave the insertion point anywhere in the expression and click Evaluate, Evaluate Numerically, or Simplify.

You can also do step 2 first and then create the vector inside the brackets.

Example 3.3 *Just checking*

Verify that the magnitude of the vector you created in the previous example is correct.

Solution. Use the above steps to enclose the vector $\{-6.8829 \quad 9.8298\}$ in double vertical line Brackets, and calculate the magnitude using Evaluate.

$$A = \left\| \begin{Bmatrix} -6.8829 \\ 9.8298 \end{Bmatrix} \right\| = 12.000$$

The magnitude of the vector is correct, agreeing with the value used in Example 3.1. ◀

In trigonometry, the tangent of the angle is the ratio of the opposite side to the adjacent side. In physics, the tangent of the direction is the ratio of the y-component to the x-component. The **direction** of a vector relative to the $+x$-axis is given by the inverse tangent function.

There are two ways to do this in *SNB*. You can use the arctan function, and *SNB* also supports the $\tan^{-1}$ notation, where the "−1" is in a Superscript after the tan function.

$$\theta_A = \tan^{-1}\frac{A_y}{A_x} \qquad (3.4a)$$

$$\theta_A = \arctan\frac{A_y}{A_x} \qquad (3.4b)$$

SNB always returns the direction in radians. To convert it to degrees, you can use the Solve Method (with the **green** degree unit) or multiply it by $\frac{180}{\pi}$.

Note To create the tan and arctan functions, you can type *tan* or *arctan* in Math mode and *SNB* will automatically create the functions, or you can use the Math Name button on the Math Objects toolbar.

When $\theta = 0°$ the vector points in the $+x$-direction. A vector with direction between $0°$ and $90°$ is in the first quadrant. A vector with direction between $90°$ and $180°$ is in the second quadrant. A vector with direction between $180°$ and $270°$ is in the third quadrant. A vector with direction between $0°$ and $-90°$ is in the fourth quadrant.

Here are the angles and corresponding directions that form the quadrant boundaries.

When the angle equals...	...your vector points in the
$0°$	$+x$-direction.
$90°$	$+y$-direction.
$\pm 180°$	$-x$-direction.
$270°$ or $-90°$	$-y$-direction.

Table 3.1

Example 3.4 *North and east*

You walk $5\,\mathrm{m}$ due north, then make a right turn and walk $12\,\mathrm{m}$ due east. What is the magnitude and direction of your final position?

Solution. Let east be the x-direction and north be the y-direction. The vector describing your final position is $\vec{\mathbf{r}} = \{12\,\mathrm{m} \quad 5\,\mathrm{m}\}$. Create and Simplify the following expression for the magnitude of the vector.

$$r = \|\{12\,\mathrm{m} \quad 5\,\mathrm{m}\}\| = 13\,\mathrm{m}$$

The direction of the vector is $\tan^{-1}\frac{5}{12} = 0.39479\,\mathrm{rad}$, which you can convert to degrees directly with Evaluate Numerically.

$$\theta = \frac{180}{\pi}\tan^{-1}\frac{5}{12} = 22.620$$

You ended up $13\,\mathrm{m}$ from your starting point in a direction $22.620°$ north of east. ◀

The following example shows that you can't always take *SNB*'s results at face value.

Example 3.5 *Not south and east*

You walk 5 m due north, then make a left turn and walk 12 m due west. What is the magnitude and direction of your final position?

Solution. Let east be the x-direction and north be the y-direction. The vector describing your final position is $\vec{\mathbf{r}} = \{-12\,\text{m} \quad 5\,\text{m}\}$. Create and Simplify the following expression for the magnitude of the vector.

$$r = \left\|\{-12\,\text{m} \quad 5\,\text{m}\}\right\| = 13\,\text{m}$$

According to *SNB*, here is the vector's direction in degrees.

$$\theta = \frac{180}{\pi}\tan^{-1}\frac{5}{-12} = -22.620$$

This incorrectly places you in the 4th quadrant, east and south of your initial position. *SNB* (and most likely your calculator) always returns the results of inverse trigonometric functions in the 1st or 4th quadrants. It cannot distinguish $\tan^{-1}\frac{5}{-12}$ from $\tan^{-1}\frac{-5}{12}$. When the correct answer is in the second or third quadrants, you must add $180\,°$ (using Evaluate and the **green** degree unit) to *SNB*'s result.

$$\theta = -22.620\,° + 180\,°$$

This correctly places you in the 2nd quadrant, north and west of your initial position. ◄

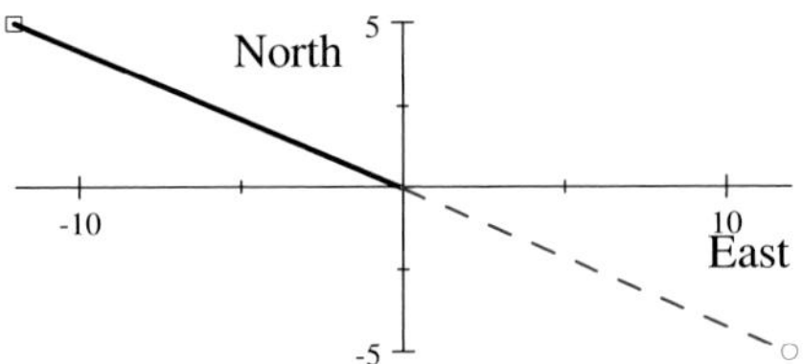

Figure 3.6 North and west

Adding Vectors

We have two equivalent ways to specify a vector, either by its components or its magnitude and direction. A vector is defined by its magnitude and direction, but we need the components to add and plot vectors.

We already know how to add vectors in three special cases (see Figure 3.1):

1. Colinear, same direction: add the magnitudes
2. Colinear, opposite direction: subtract the magnitudes
3. Perpendicular: use the Pythagorean theorem

Most vectors are not one of these special cases. To add those generic vectors, we could use the Law of Cosines, but that only works for two vectors at a time.

Instead, we want a general method for adding several vectors, given their magnitudes and arbitrary directions.

$$\vec{R} = \vec{A} + \vec{B} + \vec{C} + \cdots \tag{3.5}$$

The vector $\vec{R}$ is called the "resultant" or the "vector sum" of the other vectors.

We also want to subtract vectors. To subtract a vector, you just add the negative of the vector.

$$\vec{C} - \vec{B} = \vec{C} + \left(-\vec{B}\right)$$

The vector $-\vec{B}$ has the same magnitude but opposite direction as the vector $\vec{B}$.

To reverse the direction of a vector, you negate every component, which adds 180° to the direction. When $\vec{A} = -\vec{B}$, the two vectors have the same magnitude $A = B$, opposite direction $\theta_A = \theta_B + 180°$, and opposite components $A_x = -B_x$ and $A_y = -B_y$. If a vector is zero then every component is zero and its magnitude is zero.

The Component Method

The **Component Method** is a general, algebraic way to add vectors. It is based on a simple idea. The x-components of all vectors point in the same direction, so they are colinear and we can just add them. The same statement applies to the y-components.

Here are the components of the vector sum $\vec{R}$ in Eq. (3.5).

$$R_x = A_x + B_x + C_x + \cdots \tag{3.6a}$$

$$R_y = A_y + B_y + C_y + \cdots \tag{3.6b}$$

The x-component of the sum (R_x) is the sum of the x-components. The y-component of the sum (R_y) is the sum of the y-components. No matter how many 2-dimensional vectors you add, you end up with only two scalars (R_x and R_y).

In fact, every quantity in Eqs. (3.6) is a scalar! They can be positive or negative, but since they are scalars they obey the rules of simple arithmetic. To add two 2-dimensional vectors, all you have to do is add two sets of scalars.

Here is a simple way to remember the Component Method:

The component of the sum equals the sum of the components!

Once you have the components of the vector sum, you can calculate its magnitude and direction in the usual way.

$$R = \sqrt{R_x^2 + R_y^2} \tag{3.7a}$$

$$\theta = \tan^{-1} \frac{R_y}{R_x} \tag{3.7b}$$

Here is the procedure for adding vectors with the Component Method:

1. Resolve all of the individual vectors into their components using Eqs. (3.2).
2. Add the corresponding components with Eqs. (3.6).
3. Calculate the magnitude and direction of the vector sum with Eqs. (3.7).

Example 3.6 *Tug-of-war*
In a strange game of 2-d tug-of-war, Team A applies a force of 200 N at 0 ° and Team B applies a force of 100 N at 125 °. What force must Team C apply to balance the other two teams?
Solution. Team C must balance the net force applied by Teams A and B, which is the vector sum of the two forces $\overrightarrow{\boldsymbol{C}} = \overrightarrow{\boldsymbol{A}} + \overrightarrow{\boldsymbol{B}}$. First, use Evaluate Numerically to resolve vectors $\overrightarrow{\boldsymbol{A}}$ and $\overrightarrow{\boldsymbol{B}}$ into their x-components.

$$A_x = (200 \cos 0\,°)\ \mathrm{N} = 200\,\mathrm{N}$$

$$B_x = (100 \cos 125\,°)\ \mathrm{N} = -57.358\,\mathrm{N}$$

Do the same for the y-components.

$$A_y = (200 \sin 0\,°)\ \mathrm{N} = 0\,\mathrm{N}$$

$$B_y = (100 \sin 125\,°)\ \mathrm{N} = 81.915\,\mathrm{N}$$

Now use Evaluate to add the corresponding components.

$$C_x = (200\,\mathrm{N}) + (-\,57.358\,\mathrm{N}) = 142.64\,\mathrm{N}$$

$$C_y = (0\,\mathrm{N}) + (81.915\,\mathrm{N}) = 81.915\,\mathrm{N}$$

Finally, use Evaluate Numerically to calculate the magnitude and direction (in degrees) of the net force vector.

$$C = \sqrt{(142.64\,\mathrm{N})^2 + (81.915\,\mathrm{N})^2} = 164.49\,\mathrm{N}$$

$$\theta_C = \frac{180}{\pi} \tan^{-1} \frac{81.915\,\mathrm{N}}{142.64\,\mathrm{N}} = 29.868$$

To counter this force, Team C must pull in the *opposite* direction and apply a force of 164.49 N in a direction $29.868\,° + 180\,° = 209.868\,°$. ◀

Note To avoid Evaluate Numerically reducing the N to $\mathrm{kg\,m/\,s^2}$, select everything but the N and do the Evaluate Numerically in-place.

The SNB Method

It is important that you understand how the Component Method works so that you can use it to add vectors. Once you do, it is easier to let *SNB* do most of the work. With the ***SNB* Method**, you can use the New Definition button and the Evaluate Numerically command to add vectors easily.

To define a vector, create an equation for the vector, put the insertion point anywhere in the equation and click the New Definition button. You cannot use any Decorations with New Definition, so just use the bold faced symbol. Define the vector in terms of the magnitude and direction, using the **red** degree symbol.

For example, let's define a position vector $\vec{\mathbf{r}}$ that has magnitude $12.0\,\mathrm{m}$ and direction $55°$. Remember, you can't use any Decorations with New Definition.

$$\mathbf{r} = \{12.0\,\mathrm{m}\cos 55° \quad 12.0\,\mathrm{m}\sin 55°\}$$

Let's define a second position vector $\vec{\mathbf{s}}$ that has magnitude $10.0\,\mathrm{m}$ and direction $20°$.

$$\mathbf{s} = \{10.0\,\mathrm{m}\cos 20° \quad 10.0\,\mathrm{m}\sin 20°\}$$

To add the vectors, create an expression for the sum of the two vectors, and apply Evaluate Numerically to it. Here it is OK to use the vector Decorations.

$$\vec{\mathbf{r}} + \vec{\mathbf{s}} = \{16.280\,\mathrm{m} \quad 13.25\,\mathrm{m}\}$$

To calculate the magnitude of the vector sum, create an expression with the sum of the vectors inside the double vertical line Brackets and Simplify it.

$$\left\|\vec{\mathbf{r}} + \vec{\mathbf{s}}\right\| = 20.99\,\mathrm{m}$$

Use the components of the vector sum to calculate the direction in degrees with Evaluate Numerically.

$$\theta = \frac{180}{\pi}\tan^{-1}\frac{13.25\,\mathrm{m}}{16.280\,\mathrm{m}} = 39.142$$

The vector sum $\vec{\mathbf{r}} + \vec{\mathbf{s}}$ has a magnitude of $20.99\,\mathrm{m}$ and a direction of $39.142°$.

That's all there is to the *SNB* Method: one New Definition for each vector you want to add, and three simple expressions to Simplify or Evaluate Numerically. Once you've defined a vector in this way, you can even use Evaluate Numerically to find the components.

$$\vec{\mathbf{r}} = \{6.8829\,\mathrm{m} \quad 9.8298\,\mathrm{m}\}$$

$$\vec{\mathbf{s}} = \{9.3969\,\mathrm{m} \quad 3.4202\,\mathrm{m}\}$$

Example 3.7 *Tug-of-war redux*

Use the *SNB* Method to find the net force exerted by teams A and B in our strange game of tug-of-war.

Solution. Define vectors for the forces using the magnitudes and directions with the **red** degree symbol.

$$\mathbf{A} = \{200\cos 0° \quad 200\sin 0°\}$$

$$\mathbf{B} = \{100\cos 125° \quad 100\sin 125°\}$$

Now use Evaluate Numerically to add the vectors.

$$\vec{\mathbf{A}} + \vec{\mathbf{B}} = \begin{Bmatrix} 142.64 & 81.915 \end{Bmatrix}$$

Next, calculate the magnitude of the vector sum with Evaluate Numerically.

$$\left\| \vec{\mathbf{A}} + \vec{\mathbf{B}} \right\| = 164.49$$

Calculate the direction of the vector sum in degrees with Evaluate Numerically.

$$\theta_{AB} = \frac{180}{\pi} \tan^{-1} \frac{81.915}{142.64} = 29.868$$

The net force exerted by teams A and B is 164.49 N in a direction of $29.868°$. ◀

The Graphing Method

The **Graphing Method** provides another way to add vectors. The Component and *SNB* Methods are more accurate, easier and faster than the Graphing Method, but the Graphing Method provides a visual representation that is useful. With the Graphing Method, you plot the vectors "tail-to-head", print the plot, and measure the magnitude and direction of the vector sum. When the vectors are arranged "tail-to-head", the vector sum is the vector that connects the tail of the first vector to the head of the last.

We can plot individual 2-dimensional vectors and their components as matrices with Plot 2D Rectangular. A vector is an arrow that connects two points, so we can plot a vector as two points connected by a line (*SNB* will ***not*** put arrow heads on the vectors). We do this with a 2×2 matrix, with the coordinates of the "tail" in the first row and the coordinates of the "head" in the second row.

Use the following steps to plot a 2-dimensional vector $\vec{\mathbf{A}}$ and its components.

1. Create the matrix $\begin{bmatrix} 0 & 0 \\ A_x & A_y \end{bmatrix}$ with all numeric or defined components.
2. Leave the insertion point anywhere in the matrix and click the Plot 2D Rectangular button.
3. To plot the projections of the components along the axes, create the matrix $\begin{bmatrix} A_x & 0 \\ A_x & A_y \\ 0 & A_y \end{bmatrix}$ with all numeric or defined components, and add it to the plot.
4. Open the Plot Properties dialog. Change the Line Thickness of Item Number 1 to Thick and the Line Style of Item Number 2 to Dash.

Example 3.8 *A 2-d vector and its components*

Plot the 2-dimensional vector $\{3 \quad 2\}$ and its components.

Solution.

Use the above steps to create and plot this matrix for the vector.

$$\begin{bmatrix} 0 & 0 \\ 3 & 2 \end{bmatrix}$$

Then create this matrix for the component projections

$$\begin{bmatrix} 3 & 0 \\ 3 & 2 \\ 0 & 2 \end{bmatrix}$$

and add it to the plot.

Make the appropriate adjustments to the plot's appearance.

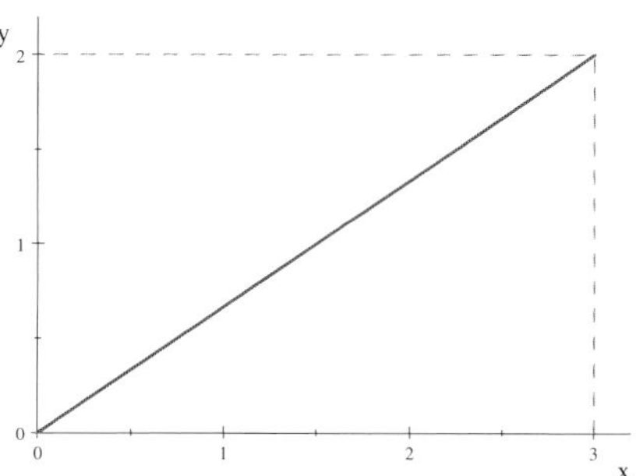

Figure 3.7 A 2-d vector

◀

We can also plot individual 3-dimensional vectors and their components as matrices with Plot 3D Rectangular. We do this with a 2-row, 3-column matrix, with the coordinates of the "tail" in the first row and the coordinates of the "head" in the second row.

Use the following steps to plot a 3-dimensional vector $\vec{\boldsymbol{B}}$ and its components.

1. Create the matrix $\begin{bmatrix} 0 & 0 & 0 \\ B_x & B_y & B_z \end{bmatrix}$ with all numeric or defined components.

2. Leave the insertion point anywhere in the matrix and click the Plot 3D Rectangular button.

3. To plot the projections of the components in the x-y plane, create the matrix $\begin{bmatrix} B_x & 0 & 0 \\ B_x & B_y & 0 \\ 0 & B_y & 0 \end{bmatrix}$ with all numeric or defined components, and add it to the plot.

4. To extend the projections to the z-axis, create the matrix $\begin{bmatrix} B_x & B_y & 0 \\ B_x & B_y & B_z \\ 0 & 0 & B_z \end{bmatrix}$ with all numeric or defined components, and add it to the plot.

5. Open the Plot Properties dialog. Change the Line Thickness of Item Number 1 to Thick and the Line Style of Items 2 and 3 to Dash.

Example 3.9 *A 3-d vector and its components*

Plot the 3-dimensional vector $\{3 \quad 8 \quad 6\}$ and its components.

Solution. Use the above steps to create and plot the matrix $\begin{bmatrix} 0 & 0 & 0 \\ 3 & 8 & 6 \end{bmatrix}$.

To plot the projections in the x-y plane, add this matrix to the plot:

$$\begin{bmatrix} 3 & 0 & 0 \\ 3 & 8 & 0 \\ 0 & 8 & 0 \end{bmatrix}$$

To plot the extension to the z-axis, add this matrix to the plot:

$$\begin{bmatrix} 3 & 8 & 0 \\ 3 & 8 & 6 \\ 0 & 0 & 6 \end{bmatrix}$$

Make the appropriate adjustments to the plot's appearance.

Figure 3.8 shows the result with the Tilt set to 85 and the Turn set to 25.

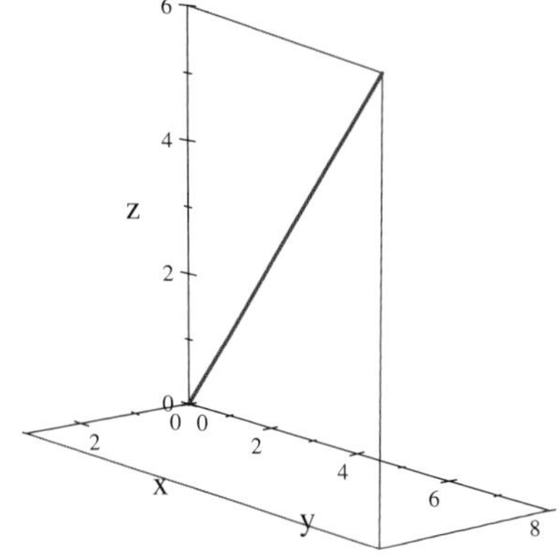

Figure 3.8 A 3-d vector

◀

Now that we can plot vectors, we are ready to add vectors graphically with *SNB*. When drawn "tail-to-head", the coordinates of each vector's "tail" are the same as those of the previous vector's "head". Since these coordinates add as scalars according to Eqs. (3.6), the coordinates of each vector equals its coordinates plus the sum of the coordinates of the previous vectors.

Use the following steps to add three vectors graphically "tail-to-head".

1. Create the "tail-to-head" matrix $\begin{bmatrix} 0 & 0 \\ A_x & A_y \\ A_x + B_x & A_y + B_y \\ A_x + B_x + C_x & A_y + B_y + C_y \end{bmatrix}$ with all numeric or defined components. The number of rows in the "tail-to-head" matrix is always one more than the number of vectors you're adding.

2. Leave the insertion point anywhere in the matrix and click the Plot 2D Rectangular button.

3. For the vector sum, add the matrix $\begin{bmatrix} 0 & 0 \\ A_x + B_x + C_x & A_y + B_y + C_y \end{bmatrix}$ to the plot.

4. Open the Plot Properties dialog. Change the Line Thickness of Item Number 2 to Thick.

This method also works with 3-dimensional vectors. Create 3-column "tail-to-head" and vector sum matrices with the z-components in the third column, and click the Plot 3D Rectangular button. It takes two angles to specify the direction of a 3-dimensional vector, which makes the Graphing Method less useful for measuring the sum of 3-d vectors but still a good way to visualize them.

Before you Print your plot to measure it, make it as large as possible and use the Preview button to check your plot's appearance. Also, be sure to check Equal Scaling Along Each Axis on the Axes tab of the Plot Properties dialog box.

Example 3.10 *Position vectors*

Use the Graphing Method to add the position vectors $\overrightarrow{\mathbf{r}} = \{6.882\,9\,\mathrm{m} \quad 9.829\,8\,\mathrm{m}\}$ and $\overrightarrow{\mathbf{s}} = \{9.396\,9\,\mathrm{m} \quad 3.420\,2\,\mathrm{m}\}$ that we met earlier.

Solution. *SNB*'s plotter does not handle units, so don't include them in your matrices. Create and plot the "tail-to-head" matrix $\begin{bmatrix} 0.0 & 0.0 \\ 6.882\,9 & 9.829\,8 \\ 16.280 & 13.25 \end{bmatrix}$.

For the vector sum, add this matrix to the plot:

$$\begin{bmatrix} 0.0 & 0.0 \\ 16.280 & 13.25 \end{bmatrix}$$

Let's include the components of the sum by adding this matrix to the plot:

$$\begin{bmatrix} 16.280 & 0 \\ 16.280 & 13.25 \\ 0 & 13.25 \end{bmatrix}$$

Make the appropriate adjustments to the plot's appearance.

It's a good idea to put the units in both the x-axis Label and y-axis Label.

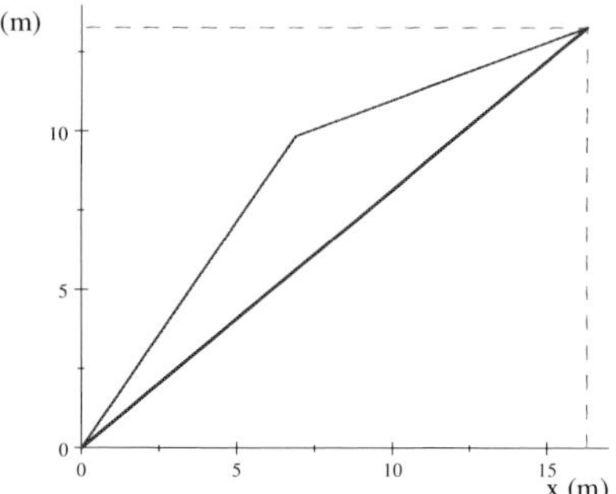

Figure 3.9 Adding position vectors

◀

In Example 3.6, we used the Component Method to find the net force exerted by teams A and B. Let's use the Graphing Method on that same problem.

Example 3.11 *Tug-of-war yet again*

Make a plot of the force vectors from that strange game of tug-of-war in Example 3.6.

Solution. Using the components we labeled A_x, A_y, B_x, and B_y, create and plot the "tail-to-head" matrix $\begin{bmatrix} 0 & 0 \\ 200 & 0 \\ 142.64 & 81.915 \end{bmatrix}$. To plot the vector sum, add the the matrix $\begin{bmatrix} 0 & 0 \\ 142.64 & 81.915 \end{bmatrix}$ to the plot. Make the appropriate adjustments to the plot's appearance. Team C pulls in the opposite direction of the vector sum. We can include that force by adding the matrix $\begin{bmatrix} 0 & 0 \\ -142.64 & -81.915 \end{bmatrix}$ to the plot. Here are the results.

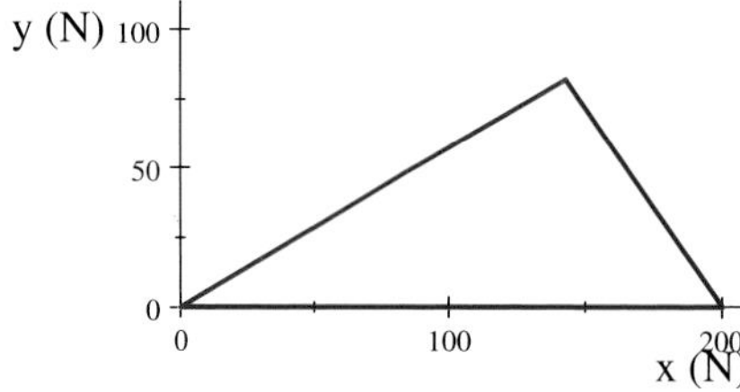

Figure 3.10 Adding force vectors

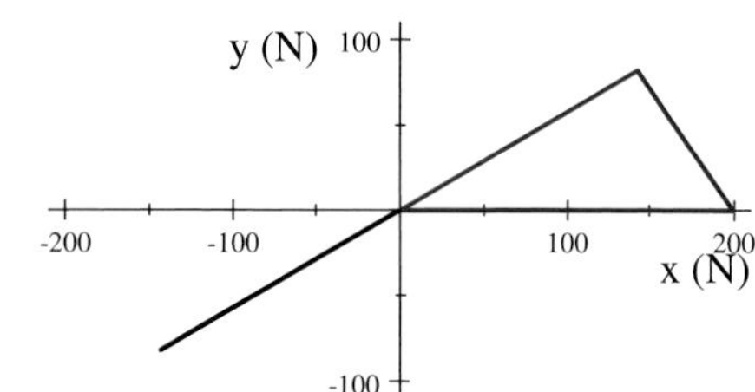

Figure 3.11 Finding equilibrium

◀

Unit Vectors

A **unit vector** is a dimensionless vector whose magnitude equals one that is used to specify direction. The unit vector $\widehat{\mathbf{x}}$ points in the x-direction, and the unit vector $\widehat{\mathbf{y}}$ points in the y-direction. The expression $\widehat{\mathbf{x}}$ is pronounced "x-hat".

Use the following steps to create an expression for the name of a unit vector.

1. Enter the letter in Text or Math mode.
2. Select it and make it Bold faced.
3. While it is still selected, click the Decoration button and choose the "hat". Figure 3.2 shows the "hat" template chosen (top row, second from right).

There are several different ways to denote unit vectors. Table 3.2 lists four of the most common sets of unit vectors.

x-direction	y-direction	z-direction
$\widehat{\mathbf{x}}$	$\widehat{\mathbf{y}}$	$\widehat{\mathbf{z}}$
$\widehat{\mathbf{i}}$	$\widehat{\mathbf{j}}$	$\widehat{\mathbf{k}}$
$\widehat{\mathbf{e}}_1$	$\widehat{\mathbf{e}}_2$	$\widehat{\mathbf{e}}_3$
$\widehat{\mathbf{u}}_x$	$\widehat{\mathbf{u}}_y$	$\widehat{\mathbf{u}}_z$

Table 3.2

The first two sets are the most common, but they do not lend themselves to use in *SNB*. We want to write vectors, in unit-vector notation, that act like vectors. We can use New Definition to define unit vectors (with no "hat" decoration). The letters x, y, and z occur frequently in physics expressions, and the letters i and e are *SNB* reserved symbols. To avoid any definition conflicts, let's use the last set (the "**u**" stands for "unit vector").

To define 2-dimensional unit vectors, create the following expressions for each, leave the insertion point in each expression and click the New Definition button.

$\mathbf{u}_x = \{1 \quad 0\}$ $\qquad$ $\mathbf{u}_y = \{0 \quad 1\}$

When you click the New Definition button, the Interpret Subscript dialog box will appear. Click the Part of the name radio button, then click OK. Now you can define a vector in terms of the unit vectors.

$\boldsymbol{A} = \mathbf{12}\,\mathbf{u}_x + 5\,\mathbf{u}_y$

Once you've defined your vector, you can Evaluate it and *SNB* returns the vector in matrix form.

$\overrightarrow{\boldsymbol{A}} = \{12 \quad 5\}$

You can use Simplify or Evaluate to calculate its magnitude.

$\left\|\overrightarrow{\boldsymbol{A}}\right\| = 13$

Example 3.12 *May the net force be with you*
Write the net force vector from the *Tug-of-war* example in unit-vector notation.
Solution. The x-component of the net force is $142.64\,\mathrm{N}$ and the y-component is $81.915\,\mathrm{N}$. In unit-vector notation, the net force looks like this:

$$\vec{\mathbf{C}} = 142.64\,\mathrm{N}\,\mathbf{u}_x + 81.915\,\mathrm{N}\,\mathbf{u}_y$$

You can still use Simplify to calculate its magnitude.

$$\|142.64\,\mathrm{N}\,\mathbf{u}_x + 81.915\,\mathrm{N}\,\mathbf{u}_y\| = 164.49\,\mathrm{N}$$ ◀

You can create a unit vector in the direction of any particular vector by dividing that vector by its magnitude. The unit vector pointing in the direction of vector $\vec{\mathbf{A}}$ is

$$\hat{\mathbf{u}}_{\mathbf{A}} = \frac{\vec{\mathbf{A}}}{\|\vec{\mathbf{A}}\|} \tag{3.8}$$

The unit vector $\hat{\mathbf{u}}_{\mathbf{A}}$ is parallel to $\vec{\mathbf{A}}$, it is dimensionless, and has magnitude 1.

Example 3.13 *One step at a time*
Find the unit vector pointing in the direction of the position vector $\{12\,\mathrm{m} \quad 5\,\mathrm{m}\}$.
Solution. Create an expression for the vector (in either matrix or unit-vector notation) divided by its magnitude, Simplify it, and apply Evaluate Numerically to the result.

$$\hat{\mathbf{u}}_{\mathbf{A}} = \frac{\mathbf{12}\,\mathrm{m}\,\mathbf{u}_x + 5\,\mathrm{m}\,\mathbf{u}_y}{\|\mathbf{12}\,\mathrm{m}\,\mathbf{u}_x + 5\,\mathrm{m}\,\mathbf{u}_y\|} = \left\{\frac{12}{13} \quad \frac{5}{13}\right\} = \{0.923\,08 \quad 0.384\,62\}$$

$$\hat{\mathbf{u}}_{\mathbf{A}} = \frac{\{12\,\mathrm{m} \quad 5\,\mathrm{m}\}}{\|\{12\,\mathrm{m} \quad 5\,\mathrm{m}\}\|} = \left\{\frac{12}{13} \quad \frac{5}{13}\right\} = \{0.923\,08 \quad 0.384\,62\}$$

We can verify that this is a unit vector by calculating its magnitude with Evaluate.

$$\|\hat{\mathbf{u}}_{\mathbf{A}}\| = \|\{0.923\,08 \quad 0.384\,62\}\| = 1.0$$ ◀

Multiplying Vectors

Multiplication involving vectors shows up in many physics problems. You can multiply a vector by a scalar to produce a vector, and you can multiply two vectors to produce a scalar, a vector, or a second-rank tensor.

In our discussion of vector multiplication, we will use 3-dimensional vectors. As in the 2-d case, once we define the 3-dimensional unit vectors, we can write and manipulate vectors in unit-vector notation.

$$\mathbf{u}_x = \{1 \quad 0 \quad 0\} \qquad \mathbf{u}_y = \{0 \quad 1 \quad 0\} \qquad \mathbf{u}_z = \{0 \quad 0 \quad 1\}$$

When you click the New Definition button, the Interpret Subscript dialog box will appear. Click the Part of the name radio button, then click OK.

We can write a generic 3-d vector in unit-vector or matrix notation.

$$\vec{\mathbf{A}} = A_x\,\mathbf{u}_x + A_y\,\mathbf{u}_y + A_z\,\mathbf{u}_z = \{A_x \quad A_y \quad A_z\}$$

When you multiply a vector by a scalar, each component of the vector is multiplied by the scalar.

$$n\,\vec{\mathbf{A}} = n\{A_x \quad A_y \quad A_z\} = \{nA_x \quad nA_y \quad nA_z\}$$

So to reverse a vector, you just multiply it by -1.

$$-\vec{\mathbf{A}} = -1\{A_x \quad A_y \quad A_z\} = \{-A_x \quad -A_y \quad -A_z\}$$

Dot Product

The **dot product** combines two vectors to produce a scalar.

$$\vec{\mathbf{A}} \cdot \vec{\boldsymbol{B}} = A_x B_x + A_y B_y + A_z B_z \tag{3.9}$$

You will find the "dot" symbol on the Symbol Cache or on the Binary Operations panel of the Symbol Panels toolbar. Geometrically, the dot product is

$$\vec{\mathbf{A}} \cdot \vec{\boldsymbol{B}} = \left\|\vec{\mathbf{A}}\right\| \left\|\vec{\mathbf{B}}\right\| \cos\phi \tag{3.10}$$

where ϕ is the angle between $\vec{\mathbf{A}}$ and $\vec{\boldsymbol{B}}$. The dot product between two perpendicular vectors is zero, since $\cos 90° = 0$. The magnitude of a vector is the square root of the dot product of the vector with itself.

With *SNB*, you can calculate the dot product of two vectors with Evaluate, Evaluate Numerically, or Simplify. *SNB* doesn't assume the components of a generic vector are real unless you tell it to.

$$\{A_x \quad A_y \quad A_z\} \cdot \{B_x \quad B_y \quad B_z\} = A_x B_x^* + A_y B_y^* + A_z B_z^*$$

$$\text{assume}\,(\text{real}) = \mathbb{R}$$

$$\{A_x \quad A_y \quad A_z\} \cdot \{B_x \quad B_y \quad B_z\} = A_x B_x + A_y B_y + A_z B_z$$

An asterisk in a Superscript indicates the complex conjugate.

Example 3.14 *Get the dot ready*

Calculate the dot product of the vectors $\vec{\mathbf{A}} = \{3 \quad 8 \quad 10\}$ and $\vec{\boldsymbol{B}} = \{7 \quad -3 \quad 8\}$.

Solution. Create and Evaluate an expression for the dot product of the two vectors.

$$\vec{\mathbf{A}} \cdot \vec{\boldsymbol{B}} = \{3 \quad 8 \quad 10\} \cdot \{7 \quad -3 \quad 8\} = 77$$

You can also use unit-vector notation, but be sure to enclose each vector in parentheses.

$$\vec{\mathbf{A}} \cdot \vec{\boldsymbol{B}} = (3\,\mathbf{u}_x + 8\,\mathbf{u}_y + 10\,\mathbf{u}_z) \cdot (7\,\mathbf{u}_x - 3\,\mathbf{u}_y + 8\,\mathbf{u}_z) = 77$$

◀

Example 3.15 *What's the angle?*
Find the angle between the same two 3-dimensional vectors.
Solution. According to Eq. (3.10), we can use the inverse cosine function to find the angle between the two vectors.

$$\phi = \cos^{-1} \frac{\vec{\mathbf{A}} \cdot \vec{\mathbf{B}}}{\left\|\vec{\mathbf{A}}\right\| \left\|\vec{\mathbf{B}}\right\|}$$

Let's use the actual vectors to create an expression for the angle in degrees, and apply Evaluate Numerically to it.

$$\phi = \frac{180}{\pi} \cos^{-1} \left(\frac{\{3 \quad 8 \quad 10\} \cdot \{7 \quad -3 \quad 8\}}{\|\{3 \quad 8 \quad 10\}\| \, \|\{7 \quad -3 \quad 8\}\|} \right) = 57.994$$

The angle between the two vectors is $\phi = 57.994°$. Figure 3.12 shows a plot of the two vectors from this example, with their three component projections. ◄

The component of a vector in a particular direction is the dot product of the vector and the unit vector in that direction.

$$\{A_x \quad A_y \quad A_z\} \cdot \mathbf{u}_y = A_y$$
$$\{A_x \quad A_y \quad A_z\} \cdot \mathbf{u}_z = A_z$$

More generally, the component of vector $\vec{\boldsymbol{B}}$ in the direction of vector $\vec{\mathbf{A}}$ is

$$\vec{\boldsymbol{B}} \cdot \hat{\mathbf{u}}_{\mathbf{A}} = \frac{\vec{\boldsymbol{B}} \cdot \vec{\mathbf{A}}}{\left\|\vec{\mathbf{A}}\right\|} = \left\|\vec{\boldsymbol{B}}\right\| \cos\phi \tag{3.11}$$

This **projection** of $\vec{\boldsymbol{B}}$ onto $\vec{\mathbf{A}}$ tells you the component of vector $\vec{\boldsymbol{B}}$ along the direction of vector $\vec{\mathbf{A}}$. You can think of it as the overlap between the two vectors.

Cross Product

The **cross product** combines two vectors to produce a new vector.

$$\vec{\mathbf{A}} \times \vec{\boldsymbol{B}} = \begin{Bmatrix} A_y B_z - A_z B_y \\ -A_x B_z + A_z B_x \\ A_x B_y - A_y B_x \end{Bmatrix} \tag{3.12}$$

You will find the "cross" symbol on the Symbol Cache or on the Binary Operations panel of the Symbol Panels toolbar. This is the same "times" symbol we used in scientific notation.

You can easily find the cross product of two vectors with Evaluate, Evaluate Numerically, Simplify or Expand.

$$\vec{\mathbf{A}} \times \vec{\boldsymbol{B}} = \begin{Bmatrix} A_x \\ A_y \\ A_z \end{Bmatrix} \times \begin{Bmatrix} B_x \\ B_y \\ B_z \end{Bmatrix} = \begin{Bmatrix} A_y B_z - A_z B_y \\ A_z B_x - A_x B_z \\ A_x B_y - A_y B_x \end{Bmatrix}$$

Example 3.16 *Cross referenced*
Find the cross product of the same vectors $\vec{\boldsymbol{A}} = \{3 \quad 8 \quad 10\}$ and $\vec{\boldsymbol{B}} = \{7 \quad -3 \quad 8\}$.

Solution. We want the vector $\vec{\boldsymbol{A}} \times \vec{\boldsymbol{B}}$, so create and Evaluate the following expression for the cross product of the two vectors.

$$\vec{\boldsymbol{A}} \times \vec{\boldsymbol{B}} = \{3 \quad 8 \quad 10\} \times \{7 \quad -3 \quad 8\} = \{94 \quad 46 \quad -65\}$$

The cross product is $\vec{\boldsymbol{A}} \times \vec{\boldsymbol{B}} = \{94 \quad 46 \quad -65\}$. ◀

The cross product is not defined for 2-dimensional vectors. To understand the 3-dimensional nature of the cross product, let's look at the cross product of two vectors in the x-y plane.

$$\begin{Bmatrix} A_x \\ A_y \\ 0 \end{Bmatrix} \times \begin{Bmatrix} B_x \\ B_y \\ 0 \end{Bmatrix} = \begin{Bmatrix} 0 \\ 0 \\ A_x B_y - A_y B_x \end{Bmatrix}$$

The cross product of two vectors in the x-y plane is a third vector that is parallel to the z-axis and therefore perpendicular to both vectors. This is true for all cross products: any two vectors $\vec{\boldsymbol{A}}$ and $\vec{\boldsymbol{B}}$ define a plane and the vector $\vec{\boldsymbol{A}} \times \vec{\boldsymbol{B}}$ is always perpendicular to that plane. The unit vector perpendicular to the plane defined by the two vectors is

$$\widehat{n} = \frac{\vec{\boldsymbol{A}} \times \vec{\boldsymbol{B}}}{\left\| \vec{\boldsymbol{A}} \times \vec{\boldsymbol{B}} \right\|} \tag{3.13}$$

and is called the **unit normal vector**.

Example 3.17 *Perfectly normal behavior*
Find the unit vector normal to the plane defined by those same two 3-dimensional vectors $\vec{\boldsymbol{A}} = \{3 \quad 8 \quad 10\}$ and $\vec{\boldsymbol{B}} = \{7 \quad -3 \quad 8\}$.

Solution. We want the unit vector pointing in the direction of the cross product $\vec{\boldsymbol{A}} \times \vec{\boldsymbol{B}}$. Create and Evaluate Numerically the following expression for the cross product of the two vectors divided by its magnitude.

$$\widehat{n} = \frac{\{94 \quad 46 \quad -65\}}{\left\|\{94 \quad 46 \quad -65\}\right\|} = \{0.763\,02 \quad 0.373\,39 \quad -0.527\,62\}$$

Let's verify that this is a unit vector by calculating its magnitude. Create the following expression, leave the insertion point anywhere in it and click Evaluate.

$$\|\widehat{n}\| = \left\|\{0.763\,02 \quad 0.373\,39 \quad -0.527\,62\}\right\| = 1.0$$

Let's verify this unit vector is perpendicular to vector $\vec{\boldsymbol{B}}$.

$$\widehat{n} \cdot \vec{\boldsymbol{B}} = \{0.763\,02 \quad 0.373\,39 \quad -0.527\,62\} \cdot \{7 \quad -3 \quad 8\} = 0.000\,01$$

The dot product is approximately zero within a small rounding error. ◀

Geometrically, the magnitude of the cross product is

$$\left\|\vec{A} \times \vec{B}\right\| = \left\|\vec{A}\right\| \left\|\vec{B}\right\| \sin\phi \tag{3.14}$$

where ϕ is the angle between $\vec{A}$ and $\vec{B}$ and it is equal to the area of the parallelogram defined by $\vec{A}$ and $\vec{B}$. The cross product between two parallel vectors is zero. When the two vectors are perpendicular, the magnitude of the cross product is the product of the magnitudes of the two vectors, and the parallelogram is a rectangle.

Example 3.18 *Land shark*

Find the area of the parallelogram defined by our old friends $\vec{A} = \{3 \quad 8 \quad 10\}$ and $\vec{B} = \{7 \quad -3 \quad 8\}$.

Solution. The area of the parallelogram is the magnitude of the cross product of the two vectors. We previously calculated the cross product $\vec{A} \times \vec{B} = \{94 \quad 46 \quad -65\}$. Create and Evaluate Numerically an expression for the magnitude of this vector.

$$\left\|\vec{A} \times \vec{B}\right\| = \|\{94 \quad 46 \quad -65\}\| = 123.19$$

The area of the parallelogram is 123.19 (see Figure 3.13). If the two vectors were perpendicular, they would define a rectangle. The area of that rectangle, which is the product of the two magnitudes,

$$\|\{3 \quad 8 \quad 10\}\| \, \|\{7 \quad -3 \quad 8\}\| = 145.28$$

is the maximum area these two vectors can define.

Figure 3.13 shows the parallelogram defined by the two vectors from the last few examples. The two vectors $\vec{A}$ and $\vec{B}$ emanate from the origin. ◄

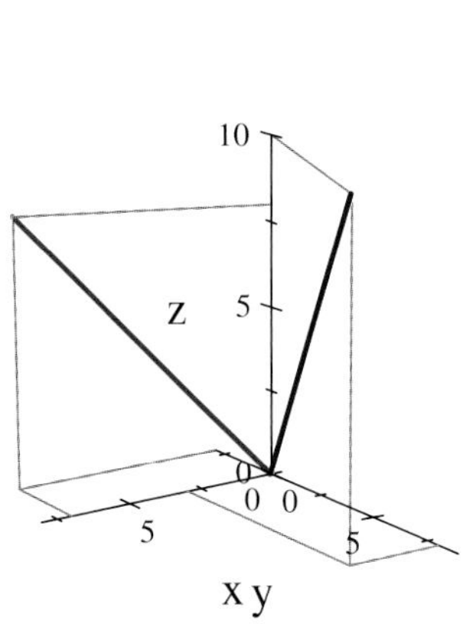

Figure 3.12 Two 3-d vectors...

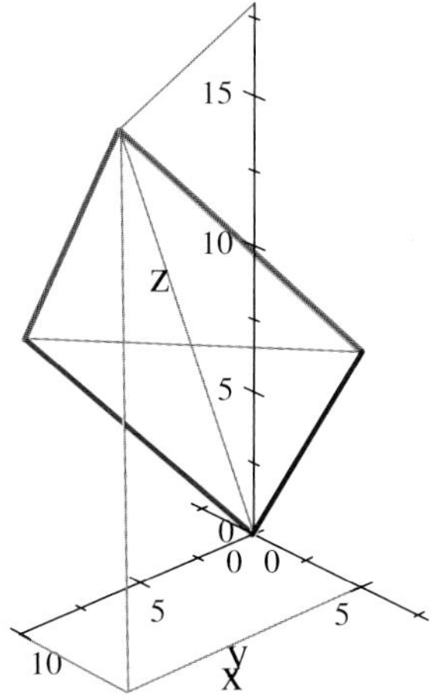

Figure 3.13 ...crossed

Problems

1. Figure 3.1 shows the addition of two displacement vectors. What are the directions (in degrees) of all the vectors in the figure?

2. Make a table that lists the x-component, the y-component, and the quadrant for the following vectors.

   a. $\vec{\mathbf{r}} = 4\,\hat{\mathbf{x}} + 7\,\hat{\mathbf{y}}$

   b. $\vec{\mathbf{r}} = 4\,\hat{\mathbf{x}} - 7\,\hat{\mathbf{y}}$

   c. $\vec{\mathbf{r}} = -4\,\hat{\mathbf{x}} + 7\,\hat{\mathbf{y}}$

   d. $\vec{\mathbf{r}} = -4\,\hat{\mathbf{x}} - 7\,\hat{\mathbf{y}}$

3. Calculate the magnitude and direction of the vectors in the previous problem. Make a sketch of each vector.

4. Determine the magnitude and direction of a generic vector $\vec{V}$ that has components $V_x = +5.9$ and $V_y = -8.4$.

5. Determine the magnitude and direction of a generic vector $\vec{V}$ that has components $V_x = -5.9$ and $V_y = +8.4$.

6. Show that the magnitude and direction of the position vector in Figure 3.3 are given by $r = \sqrt{x^2 + y^2}$ and $\theta = \tan^{-1}\dfrac{y}{x}$.

7. Vector $\vec{A}$ is 6.6 units long and points along the negative x-axis. Vector $\vec{B}$ is 8.5 units long and points at 45° to the positive x-axis.

   a. What are the x and y components of each vector?

   b. Determine the sum of the two vectors (magnitude and direction).

8. You go for a walk through the woods, walking 1.50 km due east, 0.75 km due south, and 2.2 km in a direction 35° north of west. Then you sprain your ankle and call me for help. What directions should I give the rescue helicopter?

9. A car is driven 215 km west and then 108 km southwest. What is the car's displacement (magnitude and direction)? Make a plot.

10. An airplane flies in a pattern described by three vectors $\vec{A}$, $\vec{B}$, and $\vec{C}$. The values for these vectors are $\vec{A} = 250$ km due north, $\vec{B} = 370$ km due east, and $\vec{C} = 170$ km in a direction that is 25° south of west.

    a. Make a graph of the plane's trip. Use the positive x-axis to represent "east".

    b. Calculate the components of the plane's final displacement.

    c. Using the component method, calculate the magnitude and direction of the plane's final displacement.

11. You walk 5 m due south, then make a right turn and walk 12 m due west. What is the magnitude and direction of your final position? Make a plot of your trip.

12. A hiker travels $4.47\,\mathrm{km}$ in a direction 63.4° south of east, then travels $2.24\,\mathrm{km}$ in a direction 63.4° north of east, and finally $2.24\,\mathrm{km}$ in a direction 26.6° north of east.

    a. Use the Component Method to find her final position.

    b. Use the Graphing Method to check your result.

13. A biker travels $7.0\,\mathrm{km}$ in a direction 66° north of west, then travels $4.0\,\mathrm{km}$ due east, and finally $5.0\,\mathrm{km}$ in a direction 26° south of west.

    a. Use the *SNB* method to find his final position.

    b. Use the Graphing Method to check your result.

14. Look at the procedure for plotting the components of a 3-d vector. Explain all those zeros.

15. Reproduce Figure 3.12.

16. Make a plot of the vector $-\overrightarrow{\boldsymbol{B}}$ and its components, where $\overrightarrow{\boldsymbol{B}} = \{5 \quad 3\}$. Measure the direction of $-\overrightarrow{\boldsymbol{B}}$, and compare your result to the expected value using Eqs. (3.4).

17. Show by direct calculation that the following identities are correct:

    a. $\widehat{x} \cdot \widehat{y} = 0$

    b. $\widehat{y} \cdot \widehat{z} = 0$

    c. $\widehat{x} \cdot \widehat{x} = 1$

18. Show by direct calculation that the following identities are correct:

    a. $\widehat{x} \times \widehat{y} = \widehat{z}$

    b. $\widehat{y} \times \widehat{z} = \widehat{x}$

    c. $\widehat{x} \times \widehat{z} = -\widehat{y}$

19. What are the coordinates of the four vertices of the parallelogram in Figure 3.13?

20. The force on an electric charge q moving with velocity $\overrightarrow{\boldsymbol{v}}$ through an electric field $\overrightarrow{\boldsymbol{E}}$ and a magnetic field $\overrightarrow{\boldsymbol{B}}$ is $\overrightarrow{\boldsymbol{F}} = q\left(\overrightarrow{\boldsymbol{E}} + \overrightarrow{\boldsymbol{v}} \times \overrightarrow{\boldsymbol{B}}\right)$. Find the force for the following situations.

    a. $\overrightarrow{\boldsymbol{E}} = E\,\widehat{x}$, $\overrightarrow{\boldsymbol{v}} = v\,\widehat{x}$, and $\overrightarrow{\boldsymbol{B}} = B\,\widehat{z}$

    b. $\overrightarrow{\boldsymbol{E}} = E\,\widehat{x}$, $\overrightarrow{\boldsymbol{v}} = v\,\widehat{x}$, and $\overrightarrow{\boldsymbol{B}} = B\,\widehat{y}$

    c. $\overrightarrow{\boldsymbol{E}} = E\,\widehat{x}$, $\overrightarrow{\boldsymbol{v}} = v\,\widehat{y}$, and $\overrightarrow{\boldsymbol{B}} = B\,\widehat{y}$

21. The angular momentum of a moving object is $\overrightarrow{\boldsymbol{L}} = m\,\overrightarrow{\boldsymbol{r}} \times \overrightarrow{\boldsymbol{v}}$, where $\overrightarrow{\boldsymbol{r}}$ is the position and $\overrightarrow{\boldsymbol{v}}$ is the velocity of the object. Find a general expression for the components of the angular momentum of an object whose motion is confined to the x-y plane.

4 Projectile Motion

The 1-dimensional kinematics of Chapter 2 allow us to describe horizontal or vertical motion in a straight line. Vectors allow us to analyze 2-dimensional objects. Now we combine the two and use vectors to describe 2-dimensional motion. A 2-d kinematics problem is really two 1-d kinematics problems and we use vectors to break the analysis into components.

One example of 2-d motion is **projectile motion**, where an object is launched with an initial velocity and then is influenced only by gravity and air resistance. The object's initial coordinates are x_0 and y_0. Its initial velocity has magnitude v_0 and direction θ_0 above the horizontal.

The horizontal component of the initial velocity is $v_{ox} = v_0 \cos\theta_0$ and the vertical component is $v_{oy} = v_0 \sin\theta_0$. This unusual notation is due to a limitation of *SNB*: the engine can't handle a double subscript unless the indices are either both numbers or both letters. So we'll use v_{ox} instead of v_{0x} and v_{oy} instead of v_{0y}. Here is a summary of the notation for the velocity and position of our projectile.

Quantity	*x*-component	*y*-component	Magnitude	Direction
Initial Velocity	v_{ox}	v_{oy}	v_0	θ_0
Final Velocity	v_x	v_y	v	θ_v
Initial Position	x_0	y_0	r_0	—
Final Position	x	y	r	θ

Table 4.1

With the equations of 1-dimensional kinematics and vectors, we can analyze projectile motion with and without air resistance. We will calculate the horizontal position x and the height y as functions of time directly and separately, and calculate the trajectory, the mathematical description of the path of the projectile that gives the projectile's height as a function of horizontal position. We will also calculate how far the projectile travels horizontally before landing, how much time it spends in the air, and its maximum height.

No Air Resistance

In the absence of air resistance, the only influence on the projectile is gravity. Near the Earth's surface, the acceleration due to gravity is constant so we can use Eqs. (2.5) and (2.6) to describe the projectile's motion. Since we will describe both vertical and horizontal motions, we need to alter those equations to distinguish the two motions.

Doing Physics with Scientific Notebook: A Problem-solving Approach, First Edition. Joseph Gallant.

Here are the general horizontal equations of motion.

$$x = x_0 + v_{ox}t + \tfrac{1}{2}a_x\,t^2 \tag{4.1a}$$
$$v_x = v_{ox} + a_x\,t \tag{4.1b}$$
$$v_x^2 = v_{ox}^2 + 2a_x\,(x - x_0) \tag{4.1c}$$

The subscript "x" on the velocity and acceleration indicates the horizontal components of those vectors.

A similar alteration produces the vertical equations of motion.

$$y = y_0 + v_{oy}t + \tfrac{1}{2}a_y\,t^2 \tag{4.2a}$$
$$v_y = v_{oy} + a_y\,t \tag{4.2b}$$
$$v_y^2 = v_{oy}^2 + 2a_y\,(y - y_0) \tag{4.2c}$$

These six equations describe all 2-dimensional motion with constant acceleration.

Gravity always pulls objects down, never sideways, never up. The acceleration due to gravity is constant, with a downward vertical component only

$$a_x = 0 \tag{4.3a}$$
$$a_y = -g \tag{4.3b}$$

where $g = 9.806\,7\,\mathrm{m/\,s^2}$. This condition is valid for motion in a vacuum or at speeds small enough that the effects of air resistance are negligible. Since the acceleration is constant, we can apply Eqs. (4.1a) and (4.2a) to the position of a projectile.

$$x = x_0 + v_{ox}\,t \tag{4.4a}$$
$$y = y_0 + v_{oy}\,t - \tfrac{1}{2}\,g\,t^2 \tag{4.4b}$$

You can use these equations to calculate the projectile's position (its x and y coordinates) or its displacement ($\Delta x = x - x_0$ and $\Delta y = y - y_0$) as functions of time.

Example 4.1 *And the pitch...*
How much time does a 95 mile-per-hour fastball (thrown horizontally) take to get to home plate? How far does the ball drop? Ignore air resistance and spin effects.
Solution. The ball's horizontal speed is constant $v_{ox} = 95\,\mathrm{mi/\,h} = 42.469\,\mathrm{m/\,s}$ and its horizontal displacement is $\Delta x = 60.5\,\mathrm{ft} = 18.440\,\mathrm{m}$. Using Eq. (4.4a), the time it takes the ball to get to home plate is $t = \Delta x / v_{ox}$. Create and Evaluate the following expression for the time.

$$t = \frac{18.440\,\mathrm{m}}{42.469\dfrac{\mathrm{m}}{\mathrm{s}}} = 0.434\,20\,\mathrm{s}$$

The ball was thrown horizontally, so its initial vertical velocity is zero ($v_{oy} = 0$). Use Eq. (4.4b) to create and Evaluate an expression for the ball's change in height.

$$\Delta y = -\tfrac{1}{2}\,g\,t^2 = -\tfrac{1}{2}\left(9.8067\frac{\mathrm{m}}{\mathrm{s}^2}\right)(0.434\,20\,\mathrm{s})^2 = -0.924\,43\,\mathrm{m}$$

If the pitcher releases the ball from a $2\,\mathrm{m}$ height, the ball crosses the plate at a height of $1.075\,6\,\mathrm{m}$ and should be called a strike. The batter has less than $\frac{1}{2}$ second to watch the ball, decide what to do and then to do it, during which time the ball drops $92.4\,\mathrm{cm}$. ◄

When there is no air resistance, the horizontal component of the acceleration is zero and the horizontal component of the projectile's velocity is constant.

$$v_x = v_{ox} \tag{4.5a}$$

$$v_y = v_{oy} - g\,t \tag{4.5b}$$

$$v_y^2 = v_{oy}^2 - 2\,g\,(y - y_0) \tag{4.5c}$$

The vertical component of the velocity changes at a constant rate due to gravity.

Example 4.2 *Strike one!*
If the batter does not hit the ball in the previous example, how fast is it moving when it gets to home plate?
Solution. The horizontal component of the velocity is constant at $v_x = 42.469\,\mathrm{m/\,s}$. To calculate the vertical component of the velocity, use Eq. (4.5b) to create an expression for v_y and Evaluate it.

$$v_y = 0 - \left(9.806\,7\frac{\mathrm{m}}{\mathrm{s}^2}\right)(0.434\,20\,\mathrm{s}) = -4.258\,1\frac{\mathrm{m}}{\mathrm{s}}$$

The ball's speed is the magnitude of its velocity. Let's create and Evaluate an expression for the magnitude of the velocity.

$$v = \sqrt{v_x^2 + v_y^2} = \sqrt{\left(42.469\frac{\mathrm{m}}{\mathrm{s}}\right)^2 + \left(-4.258\,1\frac{\mathrm{m}}{\mathrm{s}}\right)^2} = 42.682\frac{\mathrm{m}}{\mathrm{s}}$$

With no air resistance, the ball is moving slightly *faster* than when it started. The horizontal component of the velocity remained constant but the magnitude of the vertical component increased. In real life, air resistance slows the ball by about 8 mph. [1] ◀

There is an important and interesting property of projectile motion with no air resistance. The horizontal and vertical motions are independent, so changes in one do not affect the other. If you drop one object and throw another horizontally from the same height, they will land at exactly the same time.

Let's use Eqs. (4.4) to explore this. Suppose two balls start from $x_0 = 1\,\mathrm{m}$, $y_0 = 4.9\,\mathrm{m}$. You drop one ball from rest and you throw the other horizontally with an initial velocity of $v_{ox} = 5\,\mathrm{m/\,s}$. Both balls take about 1 second to reach the ground.

Figure 4.1 shows the path of each ball and its position every $\frac{1}{5}$ s.

They start from the same vertical position (y_0) with the same initial vertical velocity ($v_{oy} = 0$) and experience the same vertical acceleration ($-\,g$), so the vertical motions of the two balls are the same.

They always have the same height and they land at the same time.

Their vertical motions are identical!

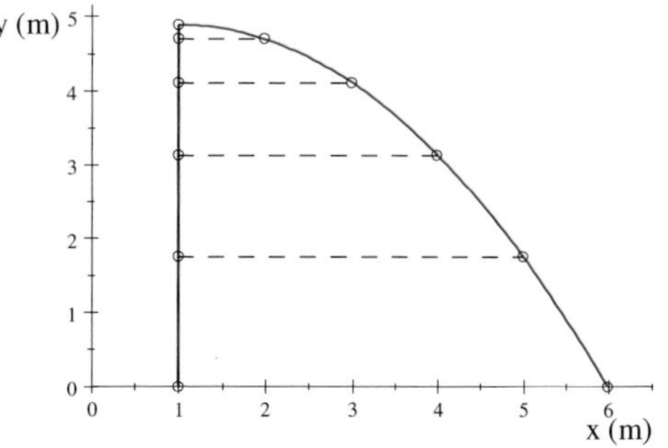

Figure 4.1 Mr. Galileo was right!

A fun and popular physics demonstration shows another property of projectile motion. In this demonstration, a projectile is launched at a high target that is released from rest at the same time the projectile is launched. This brings up an important question for the demonstrator: in what direction should the projectile be aimed initially to guarantee a mid-air collision?

The motions of both objects obey Eqs. (4.4), but with different initial conditions. The target falls from rest while the projectile has initial speed v_0 and a launch angle of θ_0. Suppose the projectile starts from position (x_0, y_0) and the target starts from rest at position (X_0, Y_0).

Here are the projectile's coordinates at a time t after launch.

$$\begin{aligned} x_{\mathrm{p}} &= x_0 + (v_0 \cos\theta_0)\, t \\ y_{\mathrm{p}} &= y_0 + (v_0 \sin\theta_0)\, t - \tfrac{1}{2} g t^2 \end{aligned}$$

Both coordinates depend on time.

Here are the target's coordinates at time t.

$$\begin{aligned} x_{\mathrm{t}} &= X_0 \\ y_{\mathrm{t}} &= Y_0 - \tfrac{1}{2} g t^2 \end{aligned}$$

The target's x-coordinate is constant because it falls straight down.

If the projectile and the falling target collide, then both their x and y coordinates must be the same at the time of impact. We can use that condition to find the direction of the initial velocity that produces a collision.

Create expressions setting $x_{\mathrm{p}} = x_{\mathrm{t}}$ and $y_{\mathrm{p}} = y_{\mathrm{t}}$, and place them in a 2-row, 1-column matrix. Leave the insertion point anywhere in the matrix and click the Solve Exact button. Enter $t, \sin\theta_0$ in the Variable(s) to Solve for window, then Factor the result for $\sin\theta_0$ in-place.

$$\begin{bmatrix} x_0 + (v_0 \cos\theta_0)\, t = X_0 \\ y_0 + (v_0 \sin\theta_0)\, t - \tfrac{1}{2} g t^2 = Y_0 - \tfrac{1}{2} g t^2 \end{bmatrix}, \text{ Solution is:}$$

$$\left[t = \frac{1}{v_0 \cos\theta_0} (X_0 - x_0), \sin\theta_0 = (\cos\theta_0) \frac{Y_0 - y_0}{X_0 - x_0} \right]$$

A little editing by-hand shows the direction of the initial velocity depends only on initial positions of the target and projectile.

$$\tan\theta_0 = \frac{Y_0 - y_0}{X_0 - x_0} \tag{4.6}$$

This is exactly the angle defined by a straight line connecting the two initial positions and the x-axis. The projectile should be aimed directly at the target!

The acceleration due to gravity is constant, so the target and the projectile experience the same acceleration. If that acceleration was zero, the dart would not fall, the projectile would move in a straight line (see Figure 4.2), and it would hit the target. Since the two experience the same acceleration, their paths deviate from that straight line in exactly the same way and they collide.

Example 4.3 *Le singe est sur la branche*
A sick monkey, hanging from a tree branch, is 8 m high and 10 m down range from a doctor with a dart gun. The gun has a muzzle speed of 12 m/ s and its end is 1 m off the ground. If the monkey lets go just when the gun fires, where should the doctor aim the medicinal darts? How much time does it take for the dart to hit the monkey? Where is the monkey when it is hit?
Solution. The good doctor should aim directly at the monkey. Use Evaluate Numerically and Eq. (4.6) to calculate the correct angle in degrees.

$$\theta_0 = \frac{180}{\pi} \arctan \frac{8-1}{10-0} = 34.992$$

Now use Evaluate Numerically to Substitute the appropriate values (using the **red** degree symbol for θ_0) into two expressions for the time.

$$t = \left[\frac{X_0 - x_0}{v_0 \cos\theta_0}\right]_{\{X_0=10\,\mathrm{m},x_0=0\,\mathrm{m},v_0=12\,\mathrm{m/\,s},\theta_0=34.992°\}} = 1.0172\,\mathrm{s}$$

$$t = \left[\frac{Y_0 - y_0}{v_0 \sin\theta_0}\right]_{\{Y_0=8\,\mathrm{m},y_0=1\,\mathrm{m},v_0=12\,\mathrm{m/\,s},\theta_0=34.992°\}} = 1.0172\,\mathrm{s}$$

They are the same! The medicinal dart hits the monkey 1.017 2 s after launch.

Create the following expression for the monkey's y-coordinate at the collision, and Substitute the appropriate values.

$$y_m = \left[Y_0 - \tfrac{1}{2}gt^2\right]_{\{Y_0=8\,\mathrm{m},g=9.8067\,\mathrm{m/\,s^2},t=1.0172\,\mathrm{s}\}} = 2.9265\,\mathrm{m}$$

The monkey falls straight down, so the dart impacts the monkey 10 m down range at a height of 2.926 5 m, just over 1 second after launch. Figure 4.2 shows the paths of the dart and monkey from launch to impact.

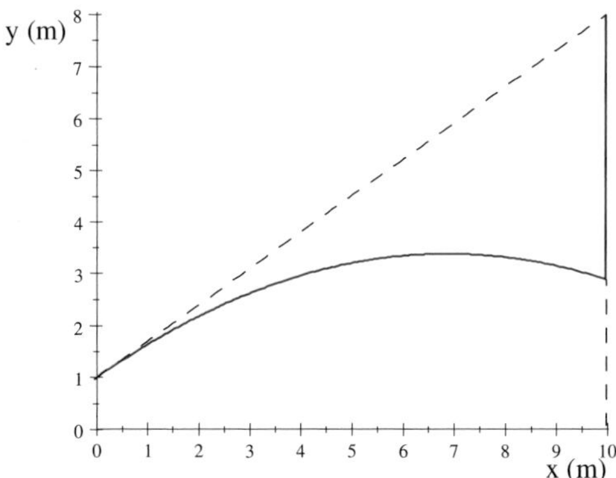

Figure 4.2 You always wanted a monkey

No monkeys were harmed during the writing of this book. ◀

Trajectory

It is often more useful to calculate the projectile's **trajectory**, its path through space giving its height y as a function of its horizontal position x. We need to eliminate the time t from Eqs. (4.4). First use Solve Exact to solve the x-equation for time.

$x = x_0 + v_{ox}t$, Solution is: $\frac{1}{v_{ox}}(x - x_0)$

Now Substitute (with Evaluate) the result for the time into the y-equation.

$$y = \left[y_0 + v_{oy}t - \tfrac{1}{2}\,g\,t^2\right]_{\left\{t=\frac{x-x_0}{v_{ox}}\right\}} = y_0 + \tfrac{1}{v_{ox}}v_{oy}(x - x_0) - \tfrac{1}{2}\tfrac{g}{v_{ox}^2}(x - x_0)^2$$

We can edit this output by-hand and put the trajectory in a more standard form.

$$y = y_0 + \frac{v_{oy}}{v_{ox}}(x - x_0) - \frac{g}{2v_{ox}^2}(x - x_0)^2$$

In terms of the magnitude and direction of the initial velocity, the trajectory is

$$y = y_0 + \tan\theta_0\,(x - x_0) - \frac{g}{2v_0^2\cos^2\theta_0}(x - x_0)^2 \tag{4.7}$$

This is the equation of a parabola.

The **range** (R) is the projectile's horizontal displacement from launch to impact, so its final horizontal position is $x = x_0 + R$. To find an expression for the range, write Eq. (4.7) in terms of R, and use Solve Exact to find R.

$y = y_0 + \frac{v_{oy}}{v_{ox}}R - \frac{g}{2v_{ox}^2}R^2$, Solution is: $\frac{1}{g}\left(v_{oy} + \sqrt{v_{oy}^2 + 2gy_0 - 2gy}\right)v_{ox}$

A little editing by-hand gives the range in terms of the components of the initial velocity.

$$R = \frac{v_{ox}\,v_{oy}}{g}\left(1 + \sqrt{1 - \frac{2g}{v_{oy}^2}(y - y_0)}\right)$$

We can replace those components with the initial velocity's magnitude and direction.

$$R = \frac{v_0^2}{2g}\left(1 + \sqrt{1 - \frac{2g}{v_0^2\sin^2\theta_0}(y - y_0)}\right)\sin 2\theta_0 \tag{4.8}$$

This expression gives the horizontal distance travelled by a projectile launched from height y_0 with initial velocity v_0 at angle θ_0 that lands at final height y. If the initial and final heights are equal then Eq. (4.8) reduces to

$$R = \frac{v_0^2}{g}\sin 2\theta_0 \tag{4.9}$$

and a launch angle of $\theta_0 = 45°$ produces the maximum range $R_{\text{max}} = v_0^2/g$. This is not the case when the initial and final heights are different (see Problem 18). When you attempt this derivation yourself, use Combine + Trig Functions to find the double-angle result $\sin\theta\cos\theta = \frac{1}{2}\sin 2\theta$.

There is an important question we need to answer: do these equations describe the motion of an actual projectile? The path of a batted baseball is a common and fun example of projectile motion that we can use as a test.

Example 4.4 *Going, going, gone!*
A routine major league fly ball typically leaves the bat with initial speed $v_0 = 44\,\mathrm{m/s}$ at an angle of $\theta_0 = 45°$, starting from an initial height of $y_0 = 1\,\mathrm{m}$ at $x_0 = 0\,\mathrm{m}$. Ignoring air resistance, plot the ball's trajectory and calculate its range.
Solution. Create and Evaluate Numerically an expression for the trajectory using Eq. (4.7) with the appropriate values without units so you can plot the expression.

$$y = 1 + x \tan 45° - \frac{9.8067}{2\,(44)^2 \cos^2 45°}\, x^2 = -5.0654 \times 10^{-3} x^2 + x + 1.0$$

To find the ball's range, Substitute the appropriate values into Eq. (4.8).

$$R = \left[\frac{v_0^2}{2g}\left(1 + \sqrt{1 - \frac{2g\,(y - y_0)}{v_0^2 \sin^2 \theta_0}}\right) \sin 2\theta_0\right]_{\left\{v_0 = 44\frac{\mathrm{m}}{\mathrm{s}},\, \theta_0 = 45°,\, y = 0,\, y_0 = 1\,\mathrm{m},\, g = 9.8067\frac{\mathrm{m}}{\mathrm{s}^2}\right\}}$$
$$= 198.41\,\mathrm{m}$$

Use the Plot 2D Rectangular button to plot the trajectory (Figure 4.3). Open the Plot Properties dialog and set the Interval for the variable x from 0 to 198.41 for Item 1.

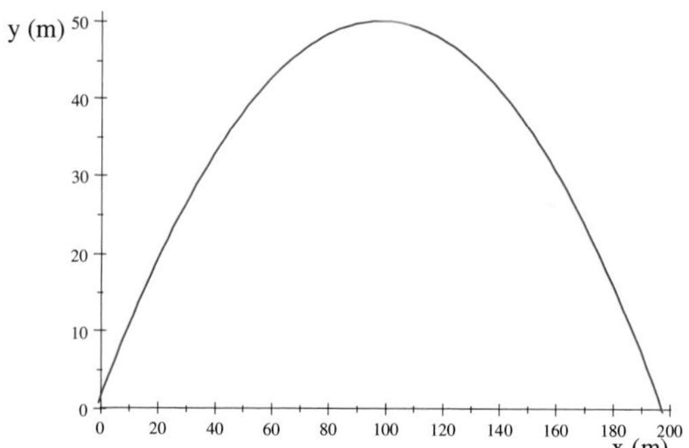

Figure 4.3 A wicked long fly ball

The trajectory of a baseball in the absence of air resistance is a parabola. You can also calculate the horizontal distance the ball traveled directly from the trajectory. Set the final height to zero ($y = 0$) and use Solve Exact to find x.

$0 = x - 5.0654 \times 10^{-3} x^2 + 1.0$, Solution is: $198.41, -0.99499$

Discard the negative solution, which tells you where the ball would land if it were projected back from its landing place with the same initial speed and angle. The ball travels $198.41\,\mathrm{m} = 650.95\,\mathrm{ft}$ and would be the longest home run in the history of baseball.

Ignoring air resistance in this example is not such a good idea. ◀

Time of Flight

The **time of flight** (T) is the amount of time it takes the projectile to arrive at its final position. To find an expression for the time of flight, write Eq. (4.4b) in terms of T and use Solve Exact to solve for T.

$y = y_0 + v_{oy}\,T - \frac{1}{2}\,g\,T^2$, Solution is: $\frac{1}{g}\left(v_{oy} + \sqrt{v_{oy}^2 + 2gy_0 - 2gy}\right)$

Now Expand the solution.

$$T = \frac{1}{g}\left(v_{oy} + \sqrt{2gy_0 - 2gy + v_{oy}^2}\right) = \frac{1}{g}v_{oy} + \frac{1}{g}\sqrt{v_{oy}^2 + 2gy_0 - 2gy}$$

In terms of the magnitude and direction of the initial velocity, the time of flight is

$$T = \frac{v_0}{g}\left(1 + \sqrt{1 - \frac{2g}{v_0^2\sin^2\theta_0}\,(y - y_0)}\right)\sin\theta_0 \tag{4.10}$$

This expression gives the time for the projectile launched from initial height y_0 with initial velocity v_0 at angle θ_0 to land at its final height y. When the final height equals the initial height, Eq. (4.10) reduces to

$$T = \frac{2v_0}{g}\sin\theta_0. \tag{4.11}$$

Aiming the projectile straight up ($\theta_0 = 90°$) produces the largest time of flight for a given initial speed.

Example 4.5 *Time of flight*
During our historic home run, how much time did the baseball stay in the air?
Solution. The ball's vertical displacement was not zero, so Substitute (with Evaluate Numerically) the numerical values into Eq. (4.10) to calculate the ball's time of flight.

$$T = \left[\frac{v_0}{g}\left(1 + \sqrt{1 - \frac{2g\,(y - y_0)}{v_0^2\sin^2\theta_0}}\right)\sin\theta_0\right]_{\{v_0=44\frac{\text{m}}{\text{s}},\theta_0=45°,y=0,y_0=1\,\text{m},g=9.8067\frac{\text{m}}{\text{s}^2}\}}$$
$$= 6.3772\,\text{s}$$

The ball hits the ground $6.3772\,\text{s}$ after it was batted. ◄

When the initial and final heights are equal, the trajectory is symmetric and there is another way to think about the time of flight. At the projectile's highest point, the vertical component of the velocity is zero ($v_y = 0$). In other words, when it stops going up, its going-up stops. Let's use Solve Exact to find how long that takes.

$0 = v_{oy} - g\,t_\uparrow$, Solution is: $\frac{1}{g}v_{oy}$

The projectile spends time $t_\uparrow = v_{oy}/g$ going up. Since the trajectory is symmetric, the total time of flight is twice this time $T = 2\,t_\uparrow$, reproducing Eq. (4.11).

Example 4.6 *Time of flight II*
How much time would the ball stay in the air if it hit a fence at a point 1 m above the ground? How far from home plate is the fence?
Solution. Since the initial and final heights are equal, you can use Eq. (4.11) and Evaluate Numerically to calculate the ball's time of flight.

$$T = 2 \times \frac{44\,\mathrm{m/\,s}}{9.806\,7\,\mathrm{m/\,s^2}} \sin 45° = 6.345\,2\,\mathrm{s}$$

Substitute (with Evaluate Numerically) this time into Eq. (4.4a) to find the distance to the fence.

$$x = [x_0 + v_0\,T\cos\theta_0]_{\{v_0=44\,\mathrm{m/\,s},\theta_0=45°,x_0=0\,\mathrm{m},T=6.3452\,\mathrm{s}\}} = 197.42\,\mathrm{m}$$

The fence is 0.99 m closer to home plate than the landing point of our home run. Since the initial and final heights are equal, you could also use Eq. (4.9).

$$R = \left[\frac{v_0^2}{g}\sin 2\theta_0\right]_{\{v_0=44\,\mathrm{m/\,s},\theta_0=45°,g=9.8067\,\mathrm{m/\,s^2}\}} = 197.42\,\mathrm{m}$$

The two answers are the same. ◀

Maximum Height

As we saw in the last section, the projectile reaches its **maximum height** (H) when the vertical component of the velocity is zero at time $t_\uparrow = v_{oy}/g$. The maximum height is the value of the y-coordinate at this time.

$$H = \left[y_0 + v_{oy}\,t_\uparrow - \tfrac{1}{2}\,g\,t_\uparrow^2\right]_{\{t_\uparrow=(v_0\sin\theta_0)/g,v_{oy}=v_0\sin\theta_0\}} = y_0 + \frac{1}{2g}v_0^2\sin^2\theta_0$$

A little editing by-hand gives the maximum height in terms of the initial velocity.

$$H = y_0 + \frac{v_0^2}{2g}\sin^2\theta_0 \tag{4.12}$$

Because the parabolic trajectory is symmetric, when the initial and final heights are equal the projectile reaches its maximum height at half the time of flight $t = T/2$ and half the range $x = R/2$. Aiming the projectile straight up ($\theta_0 = 90°$) produces the largest maximum height for a given initial speed.

Example 4.7 *Maximum height*
What was the maximum height attained by the ball during our historic home run?
Solution. Substitute (with Evaluate) the appropriate values into Eq. (4.12).

$$H = \left[y_0 + \frac{1}{2g}v_0^2\sin^2\theta_0\right]_{\{v_0=44\,\mathrm{m/\,s},\theta_0=45°,y_0=1\,\mathrm{m},g=9.8067\,\mathrm{m/\,s^2}\}} = 50.354\,\mathrm{m}$$

The ball reaches a maximum height of 50.354 m, doing so 3.172 6 s after launch. ◀

Example 4.8 *Dinger details*
When does the height of the home run equal $\frac{3}{4}$ of the maximum height? How high is the ball when its vertical velocity is $\frac{1}{2}$ the initial value? When does that occur?
Solution. Create and Evaluate this simple expression for $\frac{3}{4}$ of the maximum height.

$$y = \tfrac{3}{4}\,(50.354\,\mathrm{m}) = 37.766\,\mathrm{m}$$

Use Solve Exact and Eq. (4.4b) to find the time the ball reaches this height.

$$37.766\,\mathrm{m} = 1\,\mathrm{m} + \left(31.113\frac{\mathrm{m}}{\mathrm{s}}\right)t - \tfrac{1}{2}\left(9.8067\frac{\mathrm{m}}{\mathrm{s}^2}\right)t^2, \text{ Solution is: } 4.774\,9\,\mathrm{s}, 1.570\,3\,\mathrm{s}$$

Both solutions are physically meaningful. At $t = 1.570\,3\,\mathrm{s}$, use Evaluate Numerically to show the ball is going up so its vertical velocity is positive.

$$v_y = 44\frac{\mathrm{m}}{\mathrm{s}}\sin 45^\circ - \left(9.8067\frac{\mathrm{m}}{\mathrm{s}^2}\right)(1.570\,3\,\mathrm{s}) = 15.713\frac{\mathrm{m}}{\mathrm{s}}$$

At $t = 4.774\,9\,\mathrm{s}$, use Evaluate Numerically to show the ball is going down so its vertical velocity is negative.

$$v_y = 44\frac{\mathrm{m}}{\mathrm{s}}\sin 45^\circ - \left(9.8067\frac{\mathrm{m}}{\mathrm{s}^2}\right)(4.774\,9\,\mathrm{s}) = -15.713\frac{\mathrm{m}}{\mathrm{s}}$$

The parabolic trajectory is symmetric: at a given height the y-component of the velocity on the way up has the same magnitude but opposite sign as it does on the way down.

Use Evaluate Numerically to get the vertical component of the ball's initial velocity.

$$v_{oy} = 44\frac{\mathrm{m}}{\mathrm{s}}\sin 45^\circ = 31.113\frac{\mathrm{m}}{\mathrm{s}}$$

Use Solve Exact and Eq. (4.5c) to find the ball's height when $v_y = \frac{1}{2}v_{oy}$.

$$\left(\tfrac{1}{2}\left(31.113\frac{\mathrm{m}}{\mathrm{s}}\right)\right)^2 = \left(31.113\frac{\mathrm{m}}{\mathrm{s}}\right)^2 - 2\left(9.8067\frac{\mathrm{m}}{\mathrm{s}^2}\right)(y - 1\,\mathrm{m}), \text{ Solution is: } 38.016\,\mathrm{m}$$

When the ball is $38.016\,\mathrm{m}$ above the ground, the y-component of its velocity is half its initial value.

Use Solve Exact and Eq. (4.5b) to find the time when the ball's vertical velocity is half its initial value.

$$\tfrac{1}{2}\left(31.113\frac{\mathrm{m}}{\mathrm{s}}\right) = 31.113\frac{\mathrm{m}}{\mathrm{s}} - 9.8067\frac{\mathrm{m}}{\mathrm{s}^2}t, \text{ Solution is: } 1.586\,3\,\mathrm{s}$$

The y-component of the ball's velocity is half its initial value $1.586\,3\,\mathrm{s}$ after launch. ◄

When a baseball player hits a home run, the ball rarely lands at ground level. Usually it hits a seat, a fan, or a facade of some kind. The point of impact allows an accurate measurement of one point of the ball's trajectory, and we can use that information to estimate the ball's initial speed and how far it would have travelled unobstructed.

Example 4.9 *Ted Williams and his 502* ft *home run*

In 1946, Ted Williams hit the longest home run ever measured in Fenway Park. The ball hit a fan sitting 502 ft from home plate approximately 30 ft above ground level. Ignoring air resistance, estimate the ball's initial speed. How far would the ball have traveled if it hit the ground? Assume an initial height of 3 ft and a launch angle of 45° (which maximizes the range).

Solution. When the ball hit the fan, its vertical coordinate was $y = 30\,\text{ft}$ and its horizontal displacement was $\Delta x = 502\,\text{ft}$. In American units, the acceleration due to gravity is $g = 32.174\,\text{ft/s}^2$. Create an expression for the trajectory using Eq. (4.7) with the appropriate numerical values, and use Solve Exact to find the initial speed.

$$30 = 3 + 502\tan 45^\circ - \frac{32.174}{2v_0^2\cos^2 45^\circ}(502)^2, \text{ Solution is: } -130.65, 130.65$$

The ball leaves the bat with an initial speed of $v_0 = 130.65\,\text{ft/s}$. If the ball hits the ground, its vertical displacement would be $\Delta y = -3\,\text{ft}$. Create an expression for the range using Eq. (4.8) with the appropriate numerical values, and Evaluate it.

$$R = \frac{(130.65\,\text{ft/s})^2}{2\,(32.174\,\text{ft/s}^2)}\left(1 + \sqrt{1 - \frac{2\,(32.174\,\text{ft/s}^2)}{(130.65\,\text{ft/s})^2\sin^2 45^\circ}(-3\,\text{ft})}\right) = 533.52\,\text{ft}$$

The ball would hit the ground 533.52 ft from home plate. Ignoring air resistance again is not appropriate, as we see by converting the ball's initial speed into miles per hour.

$$130.65\,\text{ft/s} = v_0\,\text{mi/h}, \text{ Solution is: } 89.080$$

Major league home runs typically leave the bat with speeds around 110 mi/h. [1] ◄

Linear Air Resistance

If you've ridden an open roller coaster or watched a leaf spiral gently to the ground, you have experienced the effects of air resistance. These experiences tell us two properties of air resistance. The faster the object moves, the more air resistance it experiences. The less mass the object has, the more effect air resistance has on its motion.

We can model these effects with the following acceleration.

$$a_x = -\frac{b}{m}v_x \tag{4.13a}$$

$$a_y = -g - \frac{b}{m}v_y \tag{4.13b}$$

While this linear approach to air resistance is only an approximation, it produces analytical results that are consistent with the motion of real projectiles at low speeds.

This acceleration is not constant and depends on velocity, so we must use the calculus form of kinematics. As we did in the no-air-resistance case, we will alter the notation for velocity and acceleration to distinguish vertical and horizontal motion.

The definition for the components of the velocity follows from Eq. (2.7).

$$v_x = \frac{dx}{dt} \tag{4.14a}$$

$$v_y = \frac{dy}{dt} \tag{4.14b}$$

The definition for the components of the acceleration follows from Eqs. (2.8) and (2.9).

$$a_x = \frac{dv_x}{dt} = \frac{d^2x}{dt^2} \tag{4.15a}$$

$$a_y = \frac{dv_y}{dt} = \frac{d^2y}{dt^2} \tag{4.15b}$$

With these definitions and Eqs. (4.13), we can calculate the trajectory of a projectile under the influence of gravity and linear air resistance.

When we substitute these definitions into our acceleration modeling linear air resistance (with $k \equiv b/m$), we get the equations of motion for the our projectile.

$$\frac{d^2x}{dt^2} = -k\frac{dx}{dt} \tag{4.16a}$$

$$\frac{d^2y}{dt^2} = -g - k\frac{dy}{dt} \tag{4.16b}$$

These equations have exact solutions. To find the x-coordinate as a function of time, use Solve ODE Exact to solve Eq. (4.16a) and Expand the result in-place.

$$\begin{bmatrix} \frac{d^2x}{dt^2} = -k\frac{dx}{dt} \\ x(0) = x_0 \\ x'(0) = v_o \end{bmatrix}, \text{Exact solution is: } \left\{x_0 + \tfrac{1}{k}v_o - \tfrac{1}{k}v_o e^{-kt}\right\}$$

Notice I omitted the "x" from the subscript of the initial velocity, so *SNB* would not try to include it in the solution.

To find the y-coordinate as a function of time, repeat the process for Eq. (4.16b).

$$\begin{bmatrix} \frac{d^2y}{dt^2} = -g - k\frac{dy}{dt} \\ y(0) = y_0 \\ y'(0) = v_o \end{bmatrix}, \text{Exact solution is: } \left\{y_0 + \tfrac{g}{k^2} + \tfrac{1}{k}v_o - \tfrac{g}{k}t - \left(\tfrac{g}{k^2} + \tfrac{1}{k}v_o\right)e^{-kt}\right\}$$

We can edit these results by-hand, restore the double-subscripts on the initial velocity, and put them in a more traditional form.

$$x = x_0 + \frac{v_{ox}}{k}\left(1 - e^{-kt}\right) \tag{4.17a}$$

$$y = y_0 - \frac{g}{k}t + \left(\frac{v_{oy}}{k} + \frac{g}{k^2}\right)\left(1 - e^{-kt}\right) \tag{4.17b}$$

You can use these equations to calculate the projectile's position (its x and y coordinates) or its displacement ($\Delta x = x - x_0$ and $\Delta y = y - y_0$) as a function of time, including the effects of linear air resistance. You can also use $v_{ox} = v_0 \cos\theta_0$ and $v_{oy} = v_0 \sin\theta_0$ to express the projectile's position in terms of its initial speed and direction.

The value of the constant b depends on the object's size and shape and is determined experimentally. Table 4.2 lists some values for the mass and constant for several common projectiles moving through air.

Object	m (kg)	b (kg/s)	k (1/s)
Shot Put	7.26	0.475	0.0654
Skydiver	80.0	13.1	0.164
Baseball	0.142	0.0328	0.231
Tennis Ball	0.0567	0.0179	0.316
Ping Pong Ball	0.00249	0.00271	1.09

Table 4.2

The ratio $k = b/m$ indicates how much air resistance will affect the projectile's motion. The larger the ratio, the larger the effect air resistance has on the projectile's motion. The unit of k is "per second" $(1/\text{s})$ which is convenient since it's the same in both systems. The baseball value for b is based on experiments that suggest the resistive force on a baseball moving at speeds near $95\,\text{mi/h}$ is approximately equal to its weight. [1]

Let's revisit Example 4.1 and see how air resistance affects a baseball pitched at major league speeds.

Example 4.10 *Now the 0-1 pitch...*
Take air resistance into account and calculate how much time a fastball thrown horizontally with an initial velocity of $95\,\text{mi/h}$ takes to get to home plate. How far does the ball drop?
Solution. Let's use Evaluate Numerically to verify the ball's initial horizontal speed and horizontal displacement.

$$v_{ox} = 95\frac{\text{mi}}{\text{h}} = 42.469\frac{\text{m}}{\text{s}}$$
$$\Delta x = 60.5\,\text{ft} = 18.44\,\text{m}$$

Use these values, Eq. (4.17a), and Solve Exact to find the time of flight.

$$18.44\,\text{m} = 0\,\text{m} + \frac{42.469\,\text{m/s}}{0.231/\text{s}}\left(1 - e^{-\frac{0.231}{\text{s}}t}\right), \text{ Solution is: } 0.457\,55\,\text{s}$$

Since the ball was thrown horizontally, its initial vertical velocity is zero ($v_{oy} = 0$). Substitute the appropriate values into Eq. (4.17b) to find the ball's change in height.

$$\Delta y = \left[-\frac{9.8067\,\text{m/s}^2}{0.231/\text{s}}t + \frac{9.8067\,\text{m/s}^2}{(0.231/\text{s})^2}\left(1 - e^{-\frac{0.231}{\text{s}}t}\right)\right]_{t=0.45755\,\text{s}} = -0.991\,30\,\text{m}$$

Due to the effects of air resistance, the ball takes about 5.38% more time ($23.35\,\text{ms}$) to arrive at home plate, which gives the ball more time to fall an additional $6.687\,\text{cm}$. ◀

Now that we know the projectile's coordinates as functions of time, we can calculate its velocity by taking the derivative with respect to time. First we take the derivative of the x-component.

$$v_x = \frac{d}{dt}\left(x_0 + \frac{v_{ox}}{k}\left(1 - e^{-kt}\right)\right) = e^{-kt} v_{ox}$$

Then we do the same to the y-component.

$$v_y = \frac{d}{dt}\left(y_0 - \frac{g}{k}t + \left(\frac{v_{oy}}{k} + \frac{g}{k^2}\right)\left(1 - e^{-kt}\right)\right) = \frac{1}{k}\left(g e^{-kt} - g + k e^{-kt} v_{oy}\right)$$

Again we can edit these results by-hand to put them in a more traditional form.

$$v_x = v_{ox}\, e^{-kt} \tag{4.18a}$$

$$\begin{aligned} v_y &= \left(v_{oy} + \frac{g}{k}\right) e^{-kt} - \frac{g}{k} \\ &= v_{oy}\, e^{-kt} - \frac{g}{k}\left(1 - e^{-kt}\right) \end{aligned} \tag{4.18b}$$

You can see that including air resistance complicates our equations. The horizontal velocity is no longer constant and slows down exponentially. The vertical velocity tends toward a constant non-zero value over time.

Let's revisit Example 4.2 and see how air resistance affects the speed of a baseball pitched at major league speeds.

Example 4.11 *Strike two!*
If the batter does not hit the ball in the previous example, how fast is it moving when it gets to home plate?
Solution. Air resistance affects both components of the velocity. Use Eq. (4.18a) to create and Evaluate the following expression for the horizontal component of the velocity.

$$v_x = \left(42.469\frac{\text{m}}{\text{s}}\right) e^{-\left(\frac{0.231}{\text{s}}\right)(0.457\,55\,\text{s})} = 38.209\frac{\text{m}}{\text{s}}$$

Since $v_{oy} = 0$, the following expression gives the vertical component of the velocity.

$$v_y = -\left(\frac{(9.806\,7\,\text{m}/\,\text{s}^2)}{0.231/\,\text{s}}\right)\left(1 - e^{-\left(\frac{0.231}{\text{s}}\right)(0.457\,55\,\text{s})}\right) = -4.258\,1\frac{\text{m}}{\text{s}}$$

The ball's speed is the magnitude of its velocity vector, so create and Simplify the following expression.

$$v = \sqrt{v_x^2 + v_y^2} = \sqrt{\left(38.209\frac{\text{m}}{\text{s}}\right)^2 + \left(-4.258\,1\frac{\text{m}}{\text{s}}\right)^2} = 38.446\frac{\text{m}}{\text{s}}$$

Due to the effects of air resistance, the ball crosses the plate moving at $4.236\,\text{m}/\,\text{s}$ (about $9.47\,\text{mi}/\,\text{h}$) slower than when it was initially thrown. ◀

When there is no air resistance, a dropped object moves vertically in free-fall and its acceleration is $a = -\,g$. But if air resistance is considered, the acceleration is no longer constant. The acceleration defined in Eq. (4.13b) becomes

$$a = -\,g + k\,v \tag{4.19}$$

where v is the (positive) magnitude of the (vertical) velocity vector. As the object moves faster, its acceleration decreases. When the object reaches **terminal velocity** its acceleration is zero.

$$v_{\mathrm{T}} = \frac{mg}{b} = \frac{g}{k} \tag{4.20}$$

When the object's speed is v_{T}, its acceleration is zero and it continues to fall at that constant velocity.

Figure 4.4 shows the vertical speed as a function of time for an object dropped from rest ($v_{ox} = 0$) with and without air resistance.

With no air resistance, the speed

$$v = g\,t$$

increases linearly with time.

With air resistance, the falling speed

$$v = v_{\mathrm{T}}\left(1 - e^{-kt}\right)$$

approaches terminal velocity.

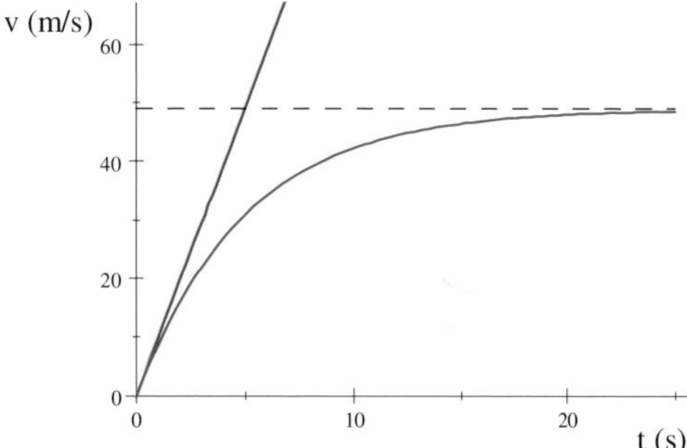

Figure 4.4 Terminal velocity

The plot uses $k = \frac{1}{5\,\mathrm{s}}$ (which is about the baseball value) so the terminal velocity is approximately $v_{\mathrm{T}} = 49\,\mathrm{m/\,s}$.

Trajectory

When we ignored air resistance, the calculation to find the trajectory was relatively easy and straightforward. Now the equations are more complicated, but the basic ideas are the same. Use Solve Exact and Eq. (4.17a) for the x-coordinate to find the time.

$$x = x_0 + \frac{v_{ox}}{k}\left(1 - e^{-kt}\right), \text{ Solution is: } -\frac{1}{k}\ln\frac{1}{v_{ox}}\left(kx_0 + v_{ox} - kx\right)$$

Now Substitute (with Evaluate) the time into Eq. (4.17b) for the y-coordinate, and then Simplify in-place the second term and Expand in-place the argument of the natural logarithm.

$$\begin{aligned} y &= \left[y_0 - \tfrac{g}{k}t + \left(\tfrac{v_{oy}}{k} + \tfrac{g}{k^2}\right)\left(1 - e^{-kt}\right)\right]_{\left\{t = -\frac{1}{k}\ln\frac{1}{v_{ox}}(kx_0 + v_{ox} - kx)\right\}} \\ &= y_0 - \left(\tfrac{1}{v_{ox}}\left(kx_0 + v_{ox} - kx\right) - 1\right)\left(\tfrac{1}{k}v_{oy} + \tfrac{g}{k^2}\right) + \tfrac{g}{k^2}\ln\tfrac{1}{v_{ox}}\left(kx_0 + v_{ox} - kx\right) \\ &= y_0 + \left(\tfrac{1}{v_{ox}}v_{oy} + \tfrac{g}{kv_{ox}}\right)(x - x_0) + \tfrac{g}{k^2}\ln\left(k\tfrac{x_0}{v_{ox}} - k\tfrac{x}{v_{ox}} + 1\right) \end{aligned}$$

We can edit this result by-hand and put the trajectory in a more standard form.

$$y = y_0 + \left(\frac{v_{oy}}{v_{ox}} + \frac{g}{kv_{ox}}\right)(x - x_0) + \frac{g}{k^2}\ln\left(1 - \frac{k(x - x_0)}{v_{ox}}\right)$$

Here is the trajectory in terms of the magnitude and direction of the initial velocity.

$$y = y_0 + \left(\tan\theta_0 + \frac{g}{kv_0\cos\theta_0}\right)(x - x_0) + \frac{g}{k^2}\ln\left(1 - \frac{k(x - x_0)}{v_0\cos\theta_0}\right) \tag{4.21}$$

This is *not* the equation of a parabola!

The expression analogous to Eq. (4.8) for the range of the projectile under the influence of linear air resistance cannot be expressed in terms of elementary functions. If you try to find the range with Solve Exact and Eq. (4.21), *SNB* cannot find an algebraic solution.

$$y = y_0 + \left(\tan\theta_0 + \frac{g}{kv_0\cos\theta_0}\right)R + \frac{g}{k^2}\ln\left(1 - \frac{kR}{v_0\cos\theta_0}\right), \text{ No solution found.}$$

Even so, we can find the range for particular problems using Eq. (4.21) and Solve Numeric.

Example 4.12 *A more realistic result*

For the same typical fly ball from Example 4.4, plot both trajectories (with and without air resistance) and calculate the baseball's range including the effects of air resistance.

Solution. To include the effects of air resistance on the ball's trajectory, use Eq. (4.21) with $k = 0.231/\,\mathrm{s}$. Create and Evaluate Numerically an expression for the trajectory with the appropriate values without units, since you're going to plot the expression.

$$y = 1 + \left(\tan 45^\circ + \frac{(9.8067)}{(0.231)(44)\cos 45^\circ}\right)x + \frac{(9.8067)}{(0.231)^2}\ln\left(1 - \frac{(0.231)\,x}{(44)\cos 45^\circ}\right)$$
$$= 2.3645x + 183.78\ln\left(1.0 - 7.4246\times 10^{-3}x\right) + 1.0$$

Calculate the horizontal distance traveled by the ball by setting the final height to zero ($y = 0$) and using Solve Numeric to find the range somewhere between zero and the no-air-resistance result.

$$\begin{bmatrix} 0 = 2.3645R + 183.78\ln\left(1.0 - 7.4246\times 10^{-3}R\right) + 1.0 \\ R \in \{0, 200\} \end{bmatrix},$$

Solution is: $\{[R = 95.463]\}$

Under these conditions, the ball travels $95.463\,\mathrm{m} \approx 313\,\mathrm{ft}$ and it would be a routine fly ball (and not a home run) in most major league parks.

The trajectory is quite different than the parabolic no-air-resistance trajectory. To add this expression to the plot of the no-air-resistance trajectory, select it with your mouse and drag it onto the plot. Open the Plot Properties dialog and set the Interval from 0 to 95.463 for Item 2.

When we include air resistance, the ball travels about half as far horizontally and 70% as high as the no-air-resistance case.

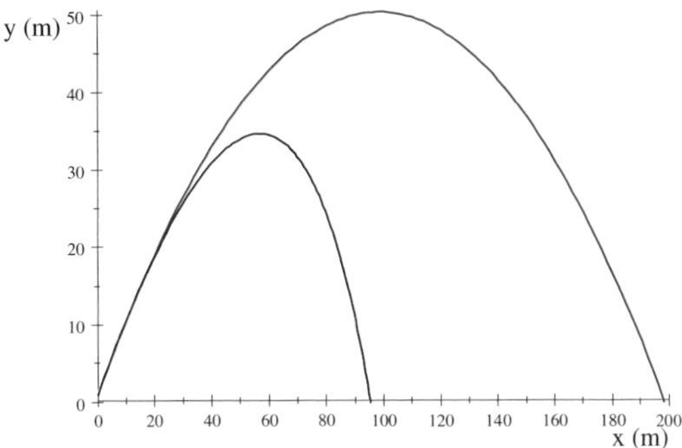

Figure 4.5 A tale of two trajectories

The trajectory is neither symmetric nor parabolic. Air resistance has a huge effect on the trajectory of a major league fly ball. ◀

In the following example, we visit Fenway Park, home of the 2004 and 2007 World Champion Boston Red Sox.

Example 4.13 *A little Wall ball...*

Would the ball hit in previous example be a home run if it was hit down the left field line in Fenway Park?

Solution. In Fenway Park, the left field Wall is 37 ft high and is officially listed as 310 ft from home plate. In **SI** units, the Wall is 11.3 m high and 94.5 m from home plate. Substitute (with Evaluate) that distance ($x = 94.5$) into the trajectory.

$$y = \left[2.364\,5x + 183.78 \ln\left(1.0 - 7.424\,6 \times 10^{-3} x\right) + 1.0\right]_{x=94.5} = 2.181\,1$$

When the ball gets to the Wall, it is 2.1811 m (about 7 ft) off the ground. Since the Wall is 11.3 m high, this ball is not a home run. It would either hit the Wall or be caught. ◀

Time of Flight and Range

The expression analogous to Eq. (4.10) for the time of flight for a projectile under the influence of linear air resistance cannot be expressed in terms of elementary functions. If you try to find the time with Solve Exact and Eq. (4.17b), *SNB* cannot find an algebraic solution.

The time of flight (T) and the range (R) are determined by Eqs. (4.17), which we can write in terms of R and T.

$$R = \frac{v_0 \cos\theta_0}{k}\left(1 - e^{-kT}\right) \tag{4.22a}$$

$$y = y_0 - \frac{g}{k}T + \left(\frac{v_0 \sin\theta_0}{k} + \frac{g}{k^2}\right)\left(1 - e^{-kT}\right) \tag{4.22b}$$

Use Eqs. (4.22) and Solve Numeric to find the range and time of flight simultaneously when you know the projectile's initial conditions and final height. If you know the projectile's final height, you can solve Eq. (4.22b) numerically for the time of flight.

Example 4.14 *More of the more realistic result*
For the same typical fly ball from Example 4.12, calculate the baseball's range and time of flight.
Solution. Create expressions using Eqs. (4.22) with the appropriate values (I used Evaluate Numerically in-place to reduce the expressions) and use Solve Numeric to find the range and time of flight between zero and the no-air-resistance results.

$$\begin{bmatrix} R = 134.69\left(1 - e^{-0.231\,T}\right) \\ 0 = 1 - 42.453T + (318.47)\left(1 - e^{-0.231\,T}\right) \\ R \in \{0, 198\} \\ T \in \{0, 6.37\} \end{bmatrix},$$

Solution is: $\{[T = 5.340\,6, R = 95.466]\}$

The ball lands $95.466\,\text{m}$ from home plate $5.340\,6\,\text{s}$ after it was hit. ◄

If you already know the range, then you can use Eq. (4.22a) and Solve Exact to find the time of flight.

$$R = \frac{v_0 \cos\theta_0}{k}\left(1 - e^{-kT}\right), \text{ Solution is: } -\frac{1}{k}\ln\frac{1}{v_0\cos\theta_0}\left(v_0\cos\theta_0 - Rk\right)$$

A little editing by-hand (and applying Expand in-place to the argument of the logarithm) gives us

$$T = -\frac{1}{k}\ln\left(1 - \frac{kR}{v_0\cos\theta_0}\right) \tag{4.23}$$

an expression for the time of flight for any horizontal displacement.

Example 4.15 *Time of Wall ball*
How much time did it take the ball in Example 4.13 to hit the Wall?
Solution. Since you know the distance to the Wall, you can use Evaluate Numerically and Eq. (4.23) to find the time.

$$T = -\frac{1}{0.231/\,\text{s}}\ln\left(1 - \frac{(0.231/\,\text{s})\,(94.5\,\text{m})}{(44\,\text{m}/\,\text{s})\cos 45°}\right) = 5.235\,5\,\text{s}$$

Let's compare that to the time it takes without air resistance.

$$T_{\text{noAR}} = \frac{R}{v_0\cos\theta_0} = \frac{94.5\,\text{m}}{(44\,\text{m}/\,\text{s})\cos 45°} = 3.037\,3\,\text{s}$$

With no air resistance, the ball hits the Wall just over $3\,\text{s}$ after it leaves the bat. Let's look at the percent deviation.

$$\text{pd} = 100\frac{5.2355\,\text{s} - 3.037\,3\,\text{s}}{3.037\,3\,\text{s}} = 72.373$$

That's an increase of over 72%! Air resistance has a huge effect on the time of flight of a major league fly ball. ◄

Maximum Height

As we saw in our analysis of no-air-resistance projectile motion, the projectile reaches its maximum height (H) when the vertical component of the velocity is zero. This is still the case, so we can use *SNB* to Solve Eq. (4.18b) with $v_y = 0$ for the time.

$$0 = \left(v_{oy} + \tfrac{g}{k}\right) e^{-kt} - \tfrac{g}{k}, \text{ Solution is: } -\frac{1}{k} \ln \frac{g}{g + kv_{oy}}$$

In terms of the magnitude and direction of the initial velocity, the amount of time the projectile spends going up is

$$t_\uparrow = \frac{1}{k} \ln \left(1 + \frac{kv_0 \sin\theta_0}{g}\right) \tag{4.24}$$

The trajectory is not symmetric, and the time going up does not generally equal the time coming down. We can Substitute (with Evaluate) this time into the y-coordinate (and Simplify in-place the second term) to get the maximum height.

$$\begin{aligned} H &= \left[y_0 - \tfrac{g}{k}t_\uparrow + \left(\tfrac{v_{oy}}{k} + \tfrac{g}{k^2}\right)\left(1 - e^{-kt_\uparrow}\right)\right]_{\{v_{oy}=v_0 \sin\theta_0, t_\uparrow = \frac{1}{k}\ln(1+(kv_0 \sin\theta_0)/g)\}} \\ &= y_0 + \tfrac{1}{k} v_0 \sin\theta_0 - \tfrac{g}{k^2} \ln\left(\tfrac{1}{g} k v_0 \sin\theta_0 + 1\right) \end{aligned}$$

A little editing gives the maximum height in terms of the initial velocity.

$$H = y_0 + \frac{v_0 \sin\theta_0}{k} - \frac{g}{k^2} \ln\left(1 + \frac{kv_0 \sin\theta_0}{g}\right) \tag{4.25}$$

A projectile shot straight up reaches the largest maximum height for a given initial speed.

Example 4.16 *A more realistic maximum height*
What is the maximum height attained by the baseball in Example 4.12? How much time elapsed before it reached its maximum height? How far down range is the ball?
Solution. To calculate the ball's maximum height, use Evaluate Numerically with the appropriate values for Eq. (4.25).

$$H = 1\,\text{m} + \frac{(44)\sin 45^\circ}{(0.231)}\,\text{m} - \frac{(9.8067)\ \text{m}}{(0.231)^2} \ln\left(1 + \frac{(0.231)(44)\sin 45^\circ}{(9.8067)}\right) = 34.648\,\text{m}$$

Calculate the elapsed time using Evaluate Numerically and Eq. (4.24).

$$t_\uparrow = \frac{1}{0.231/\,\text{s}} \ln\left(1 + \frac{(0.231)(44)\sin 45^\circ}{(9.8067)}\right) = 2.380\,0\,\text{s}$$

Use Eq. (4.17a) to create and Evaluate this expression for the ball's x-coordinate.

$$x = 0\,\text{m} + \frac{(44\,\text{m/s})\cos 45^\circ}{0.231/\,\text{s}}\left(1 - e^{-(0.231/\,\text{s})(2.38\,\text{s})}\right) = 56.962\,\text{m}$$

The ball reaches its maximum height $2.38\,\text{s}$ after it was hit, at a horizontal distance $59.962\,\text{m}$ from home plate. ◀

Example 4.17 *What goes up...*
How far from home plate is the baseball in Example 4.12 when its height is half the maximum? How much time elapsed before the ball reaches half its maximum height?
Solution. We need to solve Eqs. (4.17) simultaneously for the time and x-coordinate when the ball's height is $y = \frac{1}{2}H = 17.325\,\mathrm{m}$. Create expressions with the appropriate values, apply Evaluate Numerically in-place to reduce the expressions, and use Solve Numeric to find the horizontal distance and time.

$$\begin{bmatrix} x = 134.69\left(1 - e^{-0.231t}\right) \\ 17.325 = 1 - 42.453t + (318.47)\left(1 - e^{-0.231t}\right) \end{bmatrix},$$

Solution is: $\{[t = 0.627\,07, x = 18.163]\}$

When the ball is $18.163\,\mathrm{m}$ down range, $0.62710\,\mathrm{s}$ after it was hit, it is moving up at a height of $17.325\,\mathrm{m}$, half the maximum value.

To find the time and horizontal distance when the ball is coming down at that height, tell Solve Numeric to look for a solution between the "going up" and final results.

$$\begin{bmatrix} x = 134.69\left(1 - e^{-0.231t}\right) \\ 17.325 = 1 - 42.453t + (318.47)\left(1 - e^{-0.231t}\right) \\ x \in \{20, 100\} \\ t \in \{2.38, 5.3\} \end{bmatrix},$$

Solution is: $\{[x = 86.017, t = 4.406\,2]\}$

Notice how close the ball is to its maximum range when it is halfway down. The ball's descent is very steep, which we can see by using Evaluate Numerically to find the direction of the ball's motion.

$$\frac{v_y}{v_x} = \frac{\left(44\sin 45^\circ + \dfrac{9.8067}{0.231}\right)e^{-(0.231)(4.4061)} - \dfrac{9.8067}{0.231}}{44\left(\cos 45^\circ\right)e^{-(0.231)(4.4061)}} = -1.411\,2$$

$$\theta_v = \tan^{-1}\frac{v_y}{v_x} = \frac{180}{\pi}\tan^{-1}\left(-1.411\,2\right) = -54.678$$

The ball is moving at an angle of 54.678° below horizontal when it's halfway down. ◄

Turn Off the Air!

We started our discussion with an acceleration that models the effects of air resistance. That acceleration reduces to the no-air-resistance acceleration of Eqs. (4.3) when we "turn off" air resistance by setting $b = 0$ (or equivalently $k = 0$). If our results describing motion including the effects of air resistance are correct, they should also reduce to the appropriate no-air-resistance expressions.

In many of those results, the constant k appears in the denominator so we cannot directly Evaluate the expressions with $k = 0$. Instead, we can use *SNB* to take the limit as the constant k goes to zero.

The expression for the trajectory should reduce to Eq. (4.7).

$$y \to \lim_{k\to 0}\left(y_0 + \left(\tan\theta_0 + \frac{g}{kv_0\cos\theta_0}\right)x + \frac{g}{k^2}\ln\left(1 - \frac{kx}{v_0\cos\theta_0}\right)\right)$$
$$= y_0 + x\tan\theta_0 - \tfrac{1}{2}g\frac{x^2}{v_0^2\cos^2\theta_0}$$

Twice the amount of time it takes the projectile to reach its maximum height should reduce to the time of flight given by Eq. (4.11).

$$T \to 2\lim_{k\to 0}\left(\tfrac{1}{k}\ln\left(1 + \frac{kv_0\sin\theta_0}{g}\right)\right) = \frac{2}{g}v_0\sin\theta_0$$

The projectile's maximum height should reduce to Eq. (4.12).

$$H \to \lim_{k\to 0}\left(y_0 + \frac{v_{oy}}{k} - \frac{g}{k^2}\ln\left(1 + \frac{kv_{oy}}{g}\right)\right) = y_0 + \frac{1}{2g}v_{oy}^2$$

In all three cases, the results reduce to the appropriate no-air-resistance expression.

Turn Down the Air!

The equations for projectile motion with linear air resistance are complicated. It is often useful and instructive to expand complicated expressions in a power series. We can use a power series to approximate an expression as long as we can identify a small parameter. When the parameter is small enough, first-order corrections that are linear in that parameter often suffice.

Let's find the first-order corrections to no-air-resistance motion with the constant k as our parameter. These corrections allow us to estimate the effects of air resistance and are valid for low air resistance (small k), slow initial speeds or short distances.

Maximum Range

We don't have an analytical expression for the range to expand, so let's start with the trajectory of Eq. (4.21) for $y = y_0$ and $\Delta x = R$. Leave the insertion point anywhere in the expression and choose a Power Series with 2 terms, and Expand in Powers of k.

$$0 = \left(\tan\theta_0 + \tfrac{g}{kv_0\cos\theta_0}\right)R + \frac{g}{k^2}\ln\left(1 - \tfrac{kR}{v_0\cos\theta_0}\right)$$
$$= \left(R\tan\theta_0 - \tfrac{1}{2}R^2\frac{g}{v_0^2\cos^2\theta_0}\right) - \tfrac{1}{3}R^3g\frac{k}{v_0^3\cos^3\theta_0} + O\left(k^2\right)$$

Delete the $+O\left(k^2\right)$, factor out an R and divide by $\tan\theta_0$ by-hand, and use Rewrite + Sin and Cos in-place on the trigonometric functions. Then tell *SNB* to assume $(g > 0)$ and use Solve Exact on the resulting equation to find R.

The solution is not to first-order in k.

$$0 = 1 - \tfrac{1}{2}\frac{g}{v_0^2 \cos\theta_0 \sin\theta_0} R - \tfrac{1}{3} g \frac{k}{v_0^3 \cos^2\theta_0 \sin\theta_0} R^2,$$

Solution is: $\frac{1}{4\sqrt{gk}}\left(6v_0 (\cos\theta_0)\sqrt{\tfrac{1}{4}g + \tfrac{4}{3}kv_0 \sin\theta_0} - 3\sqrt{g}v_0 \cos\theta_0\right)$

We want a first-order correction, so let's lose the radical with a 3-term Power Series expansion in k of the solution. Delete the $+O\left(k^2\right)$, and Expand the rest in-place.

$$\begin{aligned} R &= \frac{1}{4\sqrt{gk}}\left(6v_0 (\cos\theta_0)\sqrt{\tfrac{1}{4}g + \tfrac{4}{3}kv_0 \sin\theta_0} - 3\sqrt{g}v_0 \cos\theta_0\right) \\ &= \frac{2}{g}v_0^2 \cos\theta_0 \sin\theta_0 - \frac{4}{3g^2}kv_0^3 \cos\theta_0 \sin^2\theta_0 \end{aligned}$$

A little editing by-hand gives us the projectile's range to first-order in the constant k.

$$R \approx \frac{v_0^2 \sin 2\theta_0}{g}\left(1 - \tfrac{4}{3}\frac{kv_0}{g}\sin\theta_0\right) \tag{4.26}$$

The lead term outside the parentheses is the no-air-resistance range. The term in the parentheses is less than one, so air resistance reduces the projectile's range.

When the effects of air resistance are small, the angle producing the maximum range should be slightly different than $45°$. But is it slightly larger or slightly smaller? We can find the maximizing angle by taking the derivative of Eq. (4.26) with respect to θ. Since we expect a small deviation from $45°$, let's Substitute $\theta = \frac{\pi}{4} + \varepsilon$ where ε is that small deviation. Then let's do a Power Series with 2 terms, and Expand in Powers of ε.

With *SNB*, we can do all that in one step.

$$\begin{aligned} 0 &= \left[\frac{d}{d\theta}\left(\frac{v_0^2 \sin 2\theta}{g}\left(1 - \tfrac{4}{3}\frac{kv_0}{g}\sin\theta\right)\right)\right]_{\theta=\frac{\pi}{4}+\varepsilon} \\ &= -\tfrac{2}{3}\frac{\sqrt{2}}{g^2}kv_0^3 - \frac{1}{3g^2}\varepsilon\left(12gv_0^2 - 10\sqrt{2}kv_0^3\right) + O\left(\varepsilon^2\right) \end{aligned}$$

Keep the terms that are linear in ε or k (the product εk is "second-order" and very small), and use Solve Exact to find ε.

$$0 = -\tfrac{2}{3}\frac{\sqrt{2}}{g^2}kv_0^3 - \frac{12}{3g^2}\varepsilon g v_0^2, \text{ Solution is: } -\tfrac{1}{6}\frac{\sqrt{2}}{g}kv_0$$

We now have the angle that maximizes the projectile's range to first-order in k.

$$\theta_{\max} \approx \frac{\pi}{4} - \frac{\sqrt{2}}{6}\frac{kv_0}{g} \tag{4.27}$$

Unlike the no-air-resistance case, the maximizing angle depends on the initial velocity and is less than (not equal to) $45°$. This compensates for the decreasing horizontal velocity, which does not happen in the absence of air resistance, by giving the projectile more horizontal velocity initially.

To calculate the maximum range to first-order in k, let's Substitute the value for $\theta_{\max}$ into Eq. (4.26).

$$R_{\max} = \left[\frac{v_0^2 \sin 2\theta_0}{g} \left(1 - \tfrac{4}{3} \frac{k v_0 \sin \theta_0}{g} \right) \right]_{\theta_0 = \frac{\pi}{4} - \frac{\sqrt{2}}{6} \frac{k v_0}{g}} = \frac{1}{g} v_0^2 - \tfrac{2}{3} \frac{\sqrt{2}}{g^2} k v_0^3 + O\left(k^2\right)$$

We now have the projectile's maximum range to first-order in the constant k.

$$R_{\max} \approx \frac{v_0^2}{g} \left(1 - \tfrac{2\sqrt{2}}{3} \frac{k v_0}{g} \right) \tag{4.28}$$

The lead term outside the parentheses is the maximum range with no air resistance. The term in the parentheses is less than one, so air resistance reduces the projectile's maximum range for a given initial speed.

Time of Flight

We have expressions for the time of flight and time it takes the projectile to reach its maximum height. We can expand these expressions to first order in the air resistance constant k. To find the first-order correction to the time of flight, leave the insertion point anywhere in Eq. (4.23), choose a Power Series with 2 terms, and Expand in Powers of k.

$$T = -\tfrac{1}{k} \ln \left(1 - \frac{kR}{v_0 \cos \theta_0} \right) = \frac{R}{v_0 \cos \theta_0} + \tfrac{1}{2} R^2 \frac{k}{v_0^2 \cos^2 \theta_0} + O\left(k^2\right)$$

The following expression gives the time of flight to first order in the constant k.

$$\begin{aligned} T &\approx \frac{R}{v_0 \cos \theta_0} \left(1 + \tfrac{1}{2} \frac{kR}{v_0 \cos \theta_0} \right) \\ &= T_{\text{noAR}} \left(1 + \tfrac{1}{2} k T_{\text{noAR}} \right) \end{aligned} \tag{4.29}$$

The subscript "noAR" refers to the corresponding result without air resistance. Air resistance increases the total time the projectile spends in the air for a given range, which is consistent with the result of Example 4.15.

To find the first-order correction to the time going up, leave the insertion point anywhere in Eq. (4.24), choose a Power Series with 2 terms, and Expand in Powers of k.

$$t_{\uparrow} = \tfrac{1}{k} \ln \left(1 + \frac{k v_0 \sin \theta_0}{g} \right) = \tfrac{1}{g} v_0 \sin \theta_0 - \tfrac{1}{2g^2} k v_0^2 \sin^2 \theta_0 + O\left(k^2\right)$$

The following expression gives the time the projectile spends going up, to first order in the constant k.

$$\begin{aligned} t_{\uparrow} &\approx \frac{v_0 \sin \theta_0}{g} \left(1 - \frac{k v_0 \sin \theta_0}{2g} \right) \\ &= t_{\text{noAR}} \left(1 - \tfrac{1}{2} k t_{\text{noAR}} \right) \end{aligned} \tag{4.30}$$

Air resistance reduces the time the projectile spends going up.

Maximum Height

We have an expression for the projectile's maximum height that we can expand to first order in the air resistance constant k. To find the first-order correction to the maximum height, leave the insertion point anywhere in Eq. (4.25), choose a Power Series with 2 terms, and Expand in Powers of k.

$$H = y_0 + \frac{v_0 \sin\theta_0}{k} - \frac{g}{k^2} \ln\left(1 + \frac{k v_0 \sin\theta_0}{g}\right)$$
$$= \left(y_0 + \frac{1}{2g} v_0^2 \sin^2\theta_0\right) - \frac{1}{3g^2} k v_0^3 \sin^3\theta_0 + O\left(k^2\right)$$

The maximum height of the projectile, to first order in the constant k, is less than the no-air-resistance result.

$$H \approx y_0 + \frac{v_0^2}{2g} \sin^2\theta_0 \left(1 - \tfrac{2}{3} \frac{k v_0}{g} \sin\theta_0\right) \tag{4.31}$$

The term in the parentheses is less than one, so air resistance reduces the projectile's maximum height. Notice all the fractional first-order corrections are proportional to kv_0/g. For these approximations to be valid, this dimensionless quantity must be small (meaning it's much less than one $\frac{kv_0}{g} \ll 1$).

Example 4.18 *Not so fast...*

Consider a baseball thrown at $12\,\mathrm{mi/h}$. Verify that the first-order approximation is valid, and then use it to calculate the angle that maximizes the ball's range. When thrown at that angle, how far does the ball travel and how long is it in the air?

Solution. At this speed, the dimensionless quantity kv_0/g is small enough for the first-order approximation to be valid.

$$\frac{kv_0}{g} = \frac{(0.231/\mathrm{s})\,(5.3645\,\mathrm{m/s})}{9.8067\,\mathrm{m/s^2}} = 0.12636$$

Use Eq. (4.27) and Evaluate Numerically to calculate the angle that maximizes the ball's range.

$$\theta_{\max} = \frac{180}{\pi}\left(\frac{\pi}{4} - \tfrac{\sqrt{2}}{6} \frac{(0.231/\mathrm{s})\,(5.3645\,\mathrm{m/s})}{9.8067\,\mathrm{m/s^2}}\right) = 43.294$$

To calculate the ball's range, use Eq. (4.28) and Evaluate Numerically.

$$R_{\max} = \frac{(5.3645\,\mathrm{m/s})^2}{9.8067\,\mathrm{m/s^2}}\left(1 - \tfrac{2\sqrt{2}}{3} \frac{(0.231/\mathrm{s})\,(5.3645\,\mathrm{m/s})}{9.8067\,\mathrm{m/s^2}}\right) = 2.5849\,\mathrm{m}$$

Use Eq. (4.27) and Evaluate Numerically to calculate the time the ball is in the air.

$$T = \frac{2.5849\,\mathrm{m}}{(5.3645\,\mathrm{m/s})\,(43.294°)}\left(1 + \tfrac{1}{2} \frac{(0.231/\mathrm{s})\,(2.5849\,\mathrm{m})}{(5.3645\,\mathrm{m/s})\,(43.294°)}\right) = 0.68466\,\mathrm{s}$$

The gently-tossed ball only travels about $8.5\,\mathrm{ft}$ and is in the air for less than $0.7\,\mathrm{s}$. ◄

Quadratic Air Resistance

The linear model of air resistance is valid for the motion of slow, smooth spheres. Because it allows exact solutions and gives reasonable results, it is a useful model. For most projectiles moving at reasonable speeds, a quadratic model where air resistance is proportional to v^2 is more appropriate.

The equations of motion for a projectile experiencing quadratic air resistance are

$$\frac{d^2x}{dt^2} = -\kappa v v_x \tag{4.32a}$$

$$\frac{d^2y}{dt^2} = -g - \kappa v v_y \tag{4.32b}$$

where $v = \sqrt{v_x^2 + v_y^2}$ is the magnitude of the velocity and the constant κ depends on the projectile's mass, size and aerodynamic properties.

These equations do not have an exact solution, so we must solve them numerically with Solve ODE + Numeric. The range, time of flight, maximum height, and trajectory cannot be calculated analytically. Notice there is no more independence of the vertical and horizontal motions, since each coordinate depends on both components of the velocity.

Example 4.19 *A better more realistic result*
Calculate and plot the trajectory of the typical fly ball in Example 4.12 using quadratic air resistance, and include the trajectory from that example in your plot. Estimate the range and time of flight of the ball.
Solution. For a baseball, the average value for the air-resistance constant is $\kappa = 0.00541/\,\mathrm{m}$. Create a 1-column, 6-row matrix for the two equations of motion and the four initial conditions. Leave the insertion point anywhere in the matrix and choose Solve ODE + Numeric.

$$\begin{bmatrix} x'' = -0.00541x'\sqrt{(x')^2 + (y')^2} \\ y'' = -9.8067 - 0.00541y'\sqrt{(x')^2 + (y')^2} \\ x(0) = 0 \\ y(0) = 1 \\ x'(0) = 44\cos 45° \\ y'(0) = 44\sin 45° \end{bmatrix}, \text{ Functions defined: } x, y$$

To estimate the ball's time of flight, Evaluate the defined function y for various times until the ball's height is near $y = 0$.

$$y\,(5.4026) = 2.3576 \times 10^{-4}$$

At $t = 5.4026\,\mathrm{s}$, the ball is less than $\frac{1}{4}$ of a millimeter off the ground. To find the ball's range, Evaluate the defined function x at that time.

$$x\,(5.4026) = 114.01$$

The ball lands $114.01\,\mathrm{m}$ (about $374\,\mathrm{ft}$) away from home plate $5.4026\,\mathrm{s}$ after it was hit. To plot the ball's trajectory, create the parametric expression $(x\,(t)\,, y\,(t))$.

Leave the insertion point anywhere in it and click the Plot 2D Rectangular button. Open the Plot Properties dialog and set the Interval for t from 0 to 5.4026 for Item 1.

To include the linear air-resistance trajectory, create the following parametric expression and add it to your plot.

Linear AR: $\left(134.70 - 134.70\, e^{-0.231\, t}, 319.50 - 318.50\, e^{-0.231\, t} - 42.456\, t\right)$

Set the Interval for the variable t from 0 to 5.3406 (see Example 4.14) for Item 2.

Figure 4.6 shows the trajectory for the linear and quadratic air resistance. For the speeds of major league fly balls, the quadratic model of air resistance produces results that are more consistent with actual trajectories. The ball goes further and the descent is not as steep as in the linear case.

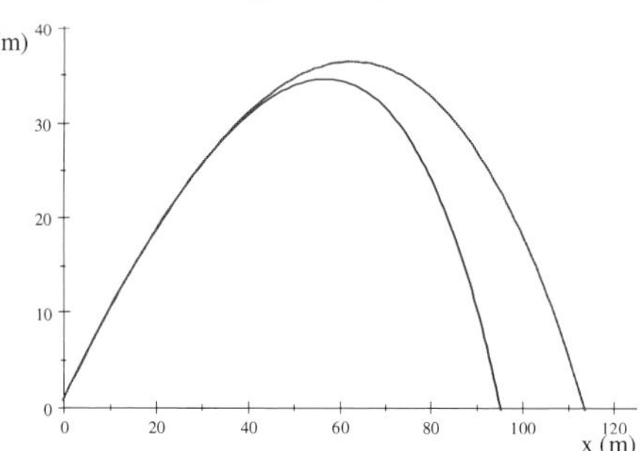

Figure 4.6 More Trajectories

◀

Height-Dependent Air Resistance

In real life, the air resistance encountered by a projectile decreases with altitude because of the reduction in air density. We can include this effect by letting the resistance parameter be a decreasing function of height. One model used in ballistics is $k(y) = k_0 e^{-y/h}$ where $h \ln 2$ is the height where the air resistance equals one-half its surface value. A typical value in ballistics is $h = 9650\,\mathrm{m}$. [15]

The equations of motion for a projectile experiencing a linear, height-dependent air resistance are

$$\frac{d^2x}{dt^2} = -k\frac{dx}{dt}e^{-y/h} \tag{4.33a}$$

$$\frac{d^2y}{dt^2} = -g - k\frac{dy}{dt}e^{-y/h} \tag{4.33b}$$

where k is the same constant listed in Table 4.2. Notice there is no independence of the vertical and horizontal motions, since the horizontal coordinate depends on the vertical coordinate.

As in the case of quadratic air resistance, these equations do not have an exact solution, and we must solve them numerically. We can use the numerical solutions to estimate the range, time of flight, maximum height, and trajectory.

Example 4.20 *Don't sit in the upper deck*
Calculate and plot the trajectory of the typical fly ball in Example 4.12 using a linear, height-dependent air resistance. Use the unrealistic value $h = 25\,\mathrm{m}$ to exaggerate the effects for the purpose of comparison to our baseball trajectories. Include the corresponding trajectories for the linear and no air resistance examples in your plot. Estimate the range and time of flight of the ball.

Solution. Use $k = 0.231/\,\mathrm{s}$ for the value of the air-resistance constant. Create a 1-column, 6-row matrix for the two equations of motion and the four initial conditions. Leave the insertion point anywhere in the matrix and choose Solve ODE + Numeric.

$$\begin{bmatrix} x'' = -0.231x'e^{-y/25} \\ y'' = -9.8067 - 0.231y'e^{-y/25} \\ x(0) = 0 \\ x'(0) = 44\cos 45^\circ \\ y(0) = 1 \\ y'(0) = 44\sin 45^\circ \end{bmatrix}, \text{Functions defined: } x, y$$

To estimate the ball's time of flight, Evaluate the defined function y for various times until the ball's height is near $y = 0$.

$$y\,(5.7032) = -2.7867 \times 10^{-4}$$

At $t = 5.7032\,\mathrm{s}$, the ball's height is less than $\frac{1}{3}$ of a millimeter below ground level. To find the ball's range, Evaluate the defined function x at that time.

$$x\,(5.7032) = 139.71$$

The ball lands $139.71\,\mathrm{m}$ (about $458\,\mathrm{ft}$) away from home plate $5.7032\,\mathrm{s}$ after it was hit.

We'll have to plot this trajectory parametrically, so create the expression $(x\,(t)\,, y\,(t))$ and click the Plot 2D Rectangular button. Add the following parametric expressions to your plot to include the linear and no air resistance trajectories.

Linear AR: $\left(134.70 - 134.70\,e^{-0.231\,t}, 319.50 - 318.50\,e^{-0.231\,t} - 42.456\,t\right)$

No AR: $\left(31.113\,t, 1 + 31.113\,t - \frac{1}{2}9.8067t^2\right)$

Adjust the three time Intervals accordingly (see Examples 4.5 and 4.14).

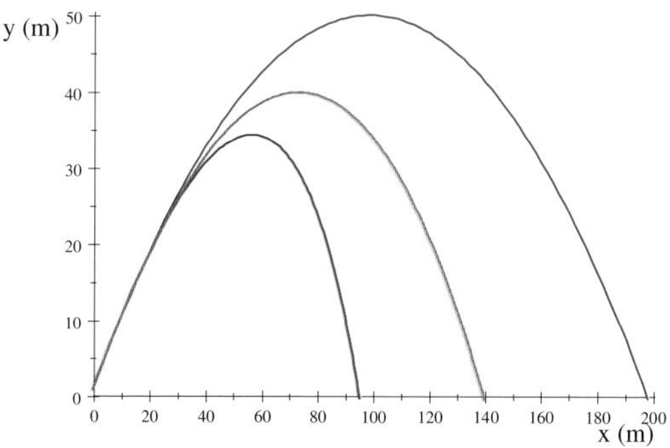

Figure 4.7 A tale of three trajectories

The reduced air resistance at higher altitude places the ball's range, time of flight, and maximum height between the linear and no air resistance results. ◀

Problems

1. Show that the acceleration due to gravity is $g = 32.174\,\mathrm{ft/\,s^2}$.

2. Playing second base for the Red Sox, you pick up a ground ball and throw it to first. You throw the ball horizontally with a speed of $65\,\mathrm{mi/\,h}$ and the ball is $6\,\mathrm{ft}$ above the ground when you release it. The first baseman catches the ball $0.57\,\mathrm{s}$ after you throw it.

   a. How far are you from the first baseman?

   b. How high is the ball when caught?

3. A bullet is shot from a rifle aimed directly at the center of a target. The bullet leaves the rifle horizontally with a speed of $650\,\mathrm{m/\,s}$ and it strikes the target $0.025\,\mathrm{m}$ below the center. What is the horizontal distance between the end of the rifle and the target?

4. A bullet is fired from a rifle that is held $1.6\,\mathrm{m}$ above the ground in a horizontal position. The initial speed of the bullet is $800\,\mathrm{m/\,s}$.

   a. Calculate the time it takes for the bullet to strike the ground.

   b. Find the horizontal distance travelled by the bullet.

5. In a friendly game of catch, the pitcher throws the ball with an initial speed of $25\,\mathrm{m/\,s}$ at an angle of $35°$ above the horizontal. The catcher catches the ball at the same height from which it was thrown. How long was the ball in the air? How far did the ball travel horizontally?

6. In a livelier game of catch, the pitcher releases the ball from an initial height of $2\,\mathrm{m}$ at $35\,\mathrm{m/\,s}$ at an angle of $35°$ above the horizontal. The catcher misses the ball and it hits the ground. What was the ball's impact velocity?

7. A projectile is launched with an initial speed of $55\,\mathrm{m/\,s}$ at an angle of $34°$ above the horizontal on a long flat firing range. Ignore air resistance and determine

   a. the maximum height achieved by the projectile,

   b. the total time it is in the air,

   c. the total horizontal distance covered, and

   d. the velocity of the projectile $1.5\,\mathrm{s}$ after firing.

8. Show that the collision time in Example 4.3 also equals $(Y_0 - y_0)\,/\,(v_0 \sin\theta_0)$.

9. Calculate and plot the slope of the line between the monkey and the dart in Example 4.3 for the entire time before the collision.

10. Golf balls have a mass of 45.9 grams and a terminal velocity of $74\,\mathrm{mi/\,h}$. Calculate the air resistance parameters b and k for a golf ball.

11. Repeat Problem 7 including the effects of linear air resistance with $k = \frac{1}{5\,\mathrm{s}}$.

12. Prove the following expression for a projectile moving without air resistance and interpret it.

$$1 - \frac{2g}{v_0^2 \sin^2 \theta_0} \Delta y = \frac{H - y}{H - y_0}$$

13. Show that Eq. (4.17b) can also be written as

$$y = y_0 + \frac{1}{k}\left(v_{oy} - gt + \frac{g}{k} - \left(\frac{g}{k} + v_{oy}\right) e^{-kt}\right)$$

14. Verify that the time of flight for a projectile experiencing linear air resistance

$$T = -\frac{1}{k} \ln\left(1 - \frac{kx}{v_{ox}}\right)$$

reduces to the appropriate result in the absence of air resistance.

15. Show that Eqs. (4.17) reduce to the appropriate results in the absence of air resistance.

16. Find the first-order correction to the horizontal and vertical position of a projectile due to the effects of air resistance.

17. Would the ball hit in Example 4.12 be a home run if it was hit down the right field line toward the Pesky Pole in Fenway Park? The wall there is $3\,\mathrm{ft}$ high and only $302\,\mathrm{ft}$ from home plate.

18. When a projectile not experiencing air resistance has a non-zero vertical displacement ($\Delta y \neq 0$), the launch angle (in degrees) that maximizes its range is

$$\theta_{\mathrm{max}} = \frac{180}{\pi} \tan^{-1} \frac{v_0}{\sqrt{v_0^2 - 2g\Delta y}}$$

where $\Delta y = y - y_0$ is the projectile's vertical displacement.

a. Show that this produces the correct result when the projectile's initial and final heights are equal.

b. Make a plot of θ_{max} versus final height for $v_0 = 25$, $g = 9.81$, and $y_0 = 0$. Let the final height vary from -70 to $+30$.

c. Calculate the maximizing angles for three different final heights.

d. Verify by direct calculation that these angles produce the maximum range.

19. How fast would you have to launch the ball in the more realistic Example 4.12 so it has the same range as the ball in Example 4.4? Try launch angles of $35°$ and $45°$. Give your answers in $\mathrm{mi/\,h}$.

20. How much time does it take for a baseball, dropped from rest and experiencing linear air resistance, to be moving half as fast as it would be if it were falling without air resistance?

21. A ping pong ball is launched straight up with a speed of $45\,\mathrm{mi/\,h}$. Find its maximum height and time of flight without and with linear air resistance.

22. Repeat the previous problem for a ball with $k = \frac{1}{5}\,\mathrm{s}^{-1}$.

23. A ping pong ball is launched at $37°$ with a speed of $45\,\mathrm{mi/\,h}$. Find its maximum height, time of flight, and range without and with linear air resistance.

24. Derive this relationship between the velocity and height of an object thrown straight up (you may find Eq. (2.16) useful).

$$y = y_0 - \frac{1}{k}(v - v_0) + \frac{g}{k^2}\ln\left(\frac{v + \frac{g}{k}}{v_0 + \frac{g}{k}}\right)$$

Show it reproduces Eq. (4.25) under the right circumstances.

25. The following table lists the impact coordinates for several of the most prodigious home runs in major league history.

Batter	x (ft)	y (ft)	**Location (Date)**
Ted Williams	502	30	Fenway Park (1946)
Mickey Mantle	374	118	Yankee Stadium (1963)
Reggie Jackson	415	132	Tiger Stadium (1971)
Mark McGwire	436	70	Cleveland (1997)

These numbers are of course only ball-park figures.

a. Ignoring air resistance, calculate the initial speed and the hit-the-ground distance for each home run. Assume $y_0 = 3\,\mathrm{ft}$ and $\theta_0 = 45°$.

b. Using linear air resistance, calculate the initial speed and the hit-the-ground distance for each home run. Assume $y_0 = 3\,\mathrm{ft}$ and $\theta_0 = 37°$.

26. An alternate approach to projectile motion with linear air resistance is to integrate the equations

$$\begin{aligned}\frac{dv_x}{dt} &= -k\,v_x \\ \frac{dv_y}{dt} &= -g - k\,v_y\end{aligned}$$

for the velocity and then integrate again to get the position.

a. Tell *SNB* the variables are positive and Evaluate the integral $\int_{v_o}^{v_x}\frac{dv}{v}$.

b. Use the result of part a. to find the horizontal velocity as a function of time.

c. Repeat part a. for the integral $\int_{v_o}^{v_y}\frac{dv}{g+k\,v}$.

d. Use the result of part c. to find the vertical velocity as a function of time.

27. Plot the first-order trajectory

$$y = y_0 + x\tan\theta_0 - \frac{1}{2}\frac{g}{v_0^2\cos^2\theta_0}x^2 - \frac{1}{3}\frac{kg}{v_0^3\cos^3\theta_0}x^3$$

for the major league fly ball discussed in the text, and compare it to the two trajectories in Figure 4.5. Comment on whether the baseball's kv_0/g is a "small" value.

28. Use the quadratic model of air resistance to estimate the hit-the-ground distance and initial velocity of the $502\,\mathrm{ft}$ home run hit by Ted Williams.

5 Newton's Laws of Motion

In the previous chapters on kinematics, we described motion in a quantitative sense. Once we know an object's acceleration, we can describe its motion in terms of position, velocity, and time. Knowing how its motion changes lets us calculate how far, how fast, and how long the object moves.

Now we move to a branch of physics known as dynamics. Dynamics is the study of the effects of forces on an object's motion. Dynamics explains changes in motion by relating the cause of the changes (forces) to the effect (acceleration). Newton's Second Law provides the rule relating the acceleration to the net force.

There are three ingredients to Newton's Second Law: the object's acceleration, **mass**, and the net force acting on it. As we discussed in Chapter 2, the object's acceleration is the rate of change in its velocity. Mass is a property of the object that determines how much change the net force produces. An object's mass tells you how difficult it is to change its velocity and how much matter it has.

A **force** is a push or a pull that can cause changes in motion. Forces are vectors, so they have magnitude and direction. This is consistent with your experience. When you exert a force on something, two things matter: how hard you push or pull and which way. Often objects have more than one force acting on them. The **net force** acting on an object is the vector sum of all the forces acting on it.

Newton's First Law

Newton's First Law tells us what happens when there is no net force acting on an object.

> **Newton's 1st Law**: An object will remain in a state of rest or continue in motion at a constant velocity unless compelled to change by a non-zero net force.

When there is no net force acting on the object, there is no change in the object's velocity. If it is at rest, it remains at rest. If it is moving, it keeps moving at constant velocity. Since velocity is a vector, constant velocity means no change in both speed and direction. Motion at constant velocity is motion in a straight line at a constant speed.

The First Law is actually a definition of inertia, the tendency of an object to maintain its current state of motion. The object's motion remains unchanged unless a non-zero net force acts on it. The First Law also explains the reason for wearing a seat belt. A seat belt attaches you to the car's brakes. The brakes stop the car, but anything not attached to the car keeps going with its initial speed and direction until something stops it.

Doing Physics with Scientific Notebook: A Problem-solving Approach, First Edition. Joseph Gallant.

Newton's Second Law for Constant Forces

The First Law tells us what happens when there is no net force. The Second Law tells us what happens when there is a net force.

> **Newton's 2nd Law**: The acceleration produced by a net force acting on an object is proportional to the net force, inversely proportional to the mass of the object, and is in the direction of the net force.

The traditional way to write the Second Law is that the net force equals mass times acceleration.

$$\vec{F} = m\,\vec{a} \tag{5.1}$$

A non-zero net force causes certain changes in the object's motion, and the Newton's Second Law allows us to calculate the effects of those changes. It is a recipe for calculating the effect - an acceleration - from the causes of the change in motion - a non-zero net force. This is a vector equation; the Second Law relates each component of the net force to the corresponding component of the acceleration.

$$F_x = m\,a_x \tag{5.2a}$$

$$F_y = m\,a_y \tag{5.2b}$$

To understand the cause-and-effect relationship between the net force and the acceleration, it is helpful to rearrange Newton's Second Law.

$$\vec{a} = \frac{\vec{F}}{m} \Longleftrightarrow \begin{cases} \text{For a given Mass} \\ \text{Larger net Force} \Longrightarrow \text{larger resulting Acceleration} \\ \\ \text{For a given net Force} \\ \text{Larger Mass} \Longrightarrow \text{smaller resulting Acceleration} \end{cases}$$

The metric unit of force is the **newton** (N) and the American unit of force is the **pound** (which *SNB* denotes as lbf).

$$1\,\mathrm{N} = 0.224\,81\,\mathrm{lbf}$$

$$1\,\mathrm{lbf} = 4.448\,2\,\mathrm{N}$$

The newton is a derived unit where $1\,\mathrm{N} = 1\,\mathrm{kg\,m/s^2}$. One newton is approximately one-quarter of a pound and is about the weight of a small apple.

The following example reminds you that the kinematic results from the previous chapters are still useful.

Example 5.1 *When push comes to throw*
How much force must a pitcher exert to accelerate a baseball horizontally from rest to $100\,\mathrm{mi/h}$ over a distance of $2\,\mathrm{m}$? Ignore air resistance.
Solution. The only horizontal force on the ball is exerted by the pitcher. Use Eq. (4.1c) to create and Evaluate an expression for the ball's horizontal acceleration.

$$a_x = \frac{\left(44.704\frac{\mathrm{m}}{\mathrm{s}}\right)^2}{2\,(2\,\mathrm{m})} = 499.61\frac{\mathrm{m}}{\mathrm{s}^2}$$

This acceleration is huge (about $50.9g$).

The horizontal force on the ball is the horizontal component of the net force.

$$ma_x = (0.142\,\mathrm{kg})\left(499.61\frac{\mathrm{m}}{\mathrm{s}^2}\right) = 70.945\frac{\mathrm{m}}{\mathrm{s}^2}\,\mathrm{kg}$$

The pitcher exerts a force of $70.945\,\mathrm{N}$ on the baseball. This is not a very big force, but it is applied to an object with a small mass, so the resulting acceleration is huge. ◄

When more than one force acts on an object, the net force is the vector sum of the individual forces. The horizontal component of the net force (ma_x) equals the vector sum of all the forces acting horizontally. The vertical component of the net force (ma_y) equals the vector sum of all the forces acting vertically.

We can write the 2nd Law in a more useful form that takes into account its vector nature.

$$ma_x = \text{Right} - \text{Left} \tag{5.3a}$$

$$ma_y = \text{Up} - \text{Down} \tag{5.3b}$$

This is the answer to *every* 2nd Law problem! You're very welcome.

So what does it mean? As with most equations, it has a simple interpretation.

The net force equals the net of the forces!

The "Right" term is the sum of all forces acting to the right (positive-horizontal) and "Left" is the sum of all forces acting to the left (negative-horizontal). The right-hand side of Eq. (5.3a) is the net of the horizontal forces. Similarly, "Up" is the sum of all forces acting up (positive-vertical) and "Down" is the sum of all forces acting down (negative-vertical). The right-hand side of Eq. (5.3b) is the net of the vertical forces. Each force you include in Eqs. (5.3) is a positive number. The minus signs in the equations account for the difference in direction.

Once you know the forces acting on the object, you can use Eqs. (5.3) to calculate its acceleration. For constant forces, this acceleration will be constant, and you can use it in Eqs. (2.5) to calculate the object's position or velocity. If the competing forces in a particular direction balance, then that component of the net force is zero. In that direction, there is no acceleration and the object moves at constant velocity.

Most problems with constant acceleration involve four forces: Weight, the Normal force, Tension, and Friction. Weight is the force due to gravity. It is a downward force with magnitude mg. The Normal force is a contact force that is always perpendicular to and points away from the surface. Its magnitude depends on the situation. The Normal force is sometimes called the "apparent weight" because its magnitude equals the reading on a scale. Tension always points along the rope and is the same throughout the rope. Tension always pulls, never pushes, and its magnitude depends on the situation. Let's ignore Friction for now.

To use the 2nd Law correctly, the trick is to identify **all** the forces acting on the object. The best way to do that uses a special diagram called the Free Body Diagram (FBD). An FBD contains the object and a labeled arrow for each force acting on the object. Each arrow corresponds to a term in Eqs. (5.3).

Suppose you pull a sled with a rope across an ice surface slippery enough that we can ignore friction. There are three forces acting on the sled: your pull (P), the sled's weight (mg), and the normal force (N). The pull force makes an angle θ to the horizontal. The FBD of the sled looks like this:

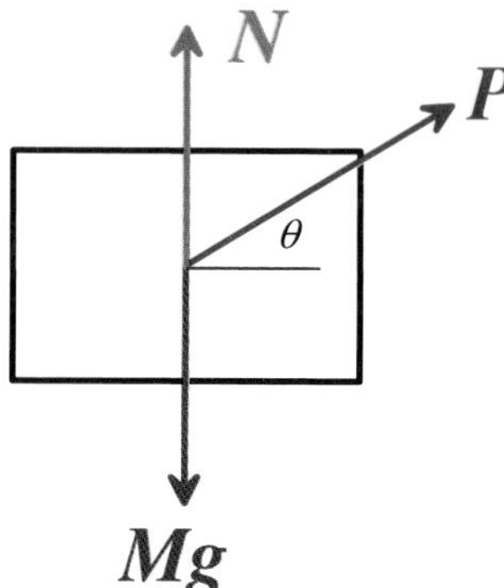

Since the pull force P does not point along either axis, we must resolve it into its x- and y-components. The FBD now looks like this:

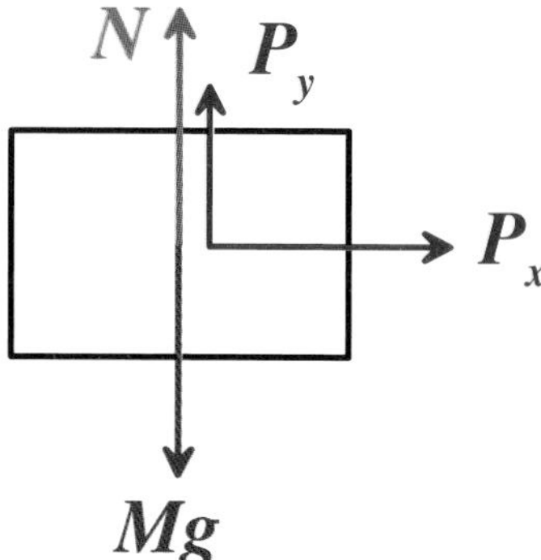

With the help of the FBD and Eqs. (5.3), the 2nd Law force equations are

$$\begin{aligned} ma_x &= P_x \\ 0 &= N + P_y - Mg \end{aligned}$$

where $P_x = P\cos\theta$ and $P_y = P\sin\theta$. There are 4 arrows in the FBD and there are 4 terms to the right of the equal signs. We set $a_y = 0$ because the sled slides along the ice surface so its vertical velocity is constant (zero).

Here is a summary of the problem-solving strategy using Newton's 2nd Law.

- Isolate the object the forces are acting on.
- Use a Free-Body Diagram to sketch the forces!
- Choose a convenient coordinate system.
  For example, you might pick the x-direction along the object's motion.
- Resolve the forces into components along the x-axis and y-axis.
- Apply Newton's 2nd Law using Eqs. (5.3).

Example 5.2 *Going up!*

A $65\,\mathrm{kg}$ physics student gets on an elevator with a built-in floor scale. As the elevator accelerates upward, the tension in the supporting cable is $9500\,\mathrm{N}$. As the elevator slows to a stop on the top floor, the tension in the supporting cable is $7000\,\mathrm{N}$. The combined mass of the student and elevator is $850\,\mathrm{kg}$. What does the scale read while the elevator is speeding up, slowing down, and moving at constant speed? What is the tension in the supporting cable while the elevator is moving at constant speed?

Solution. Construct two Free-Body Diagrams, one for the student and one for the elevator. The forces in this problem all act vertically. The Tension (T) in the cable supports the elevator, which has weight Mg. The Normal force (N) supports the student, who has weight mg.

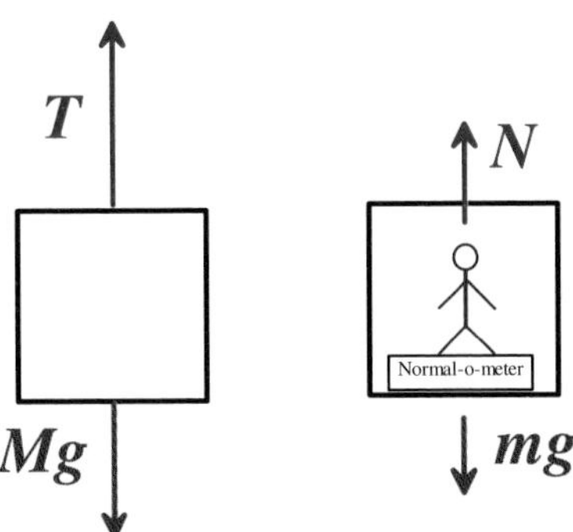

Use the diagrams and Eq. (5.3b) to write the Newton's 2nd Law force equations. The student and the elevator have the same acceleration.

$$Ma_y = T - Mg \quad \text{(Elevator)}$$

$$ma_y = N - mg \quad \text{(Student)}$$

Create the following expression using the elevator's force equation and use Solve Exact to find the elevator's acceleration while it is speeding up.

$$(850\,\mathrm{kg})\,a_y = 9500\,\mathrm{N} - (850\,\mathrm{kg})\left(9.8067\frac{\mathrm{m}}{\mathrm{s}^2}\right),\ \text{Solution is: } 1.369\,8\frac{\mathrm{m}}{\mathrm{s}^2}$$

The elevator's upward acceleration is $a_y = 1.369\,8\,\mathrm{m/\,s^2}$. The student's force equation tells you the Normal force on the student is $N = m\,(a_y + g)$. To calculate the Normal force on the student, Evaluate the following expression.

$$N = (65\,\mathrm{kg})\left(1.369\,8\frac{\mathrm{m}}{\mathrm{s}^2} + 9.8067\frac{\mathrm{m}}{\mathrm{s}^2}\right) = 726.47\frac{\mathrm{m}}{\mathrm{s}^2}\,\mathrm{kg}$$

While the elevator is speeding up, the Normal force on the student is larger than the student's weight. Create the following expression and use Solve Exact to find the elevator's acceleration while it is slowing down.

$$(850\,\mathrm{kg})\,a_y = 7000\,\mathrm{N} - (850\,\mathrm{kg})\left(9.8067\frac{\mathrm{m}}{\mathrm{s}^2}\right),\ \text{Solution is: } -1.571\,4\frac{\mathrm{m}}{\mathrm{s}^2}$$

The elevator's acceleration while slowing down is $a_y = -1.571\,4\,\mathrm{m/\,s^2}$.

To calculate the Normal force on the student, Evaluate the following expression.

$$N = (65\,\text{kg})\left(-1.571\,4\frac{\text{m}}{\text{s}^2} + 9.8067\frac{\text{m}}{\text{s}^2}\right) = 535.29\frac{\text{m}}{\text{s}^2}\,\text{kg}$$

While the elevator is slowing down, the Normal force on the student is less than the student's weight. The elevator's acceleration while moving at constant speed is zero. Create the following expression and use Solve Exact to find the Tension in the cable.

$$0 = T - (850\,\text{kg})\left(9.8067\frac{\text{m}}{\text{s}^2}\right), \text{ Solution is: } 8335.7\frac{\text{m}}{\text{s}^2}\,\text{kg}$$

When the acceleration is zero, the tension in the cable is $T = 8335.7\,\text{N}$ and equals the elevator's weight, and the Normal force acting on the student equals the student's weight, just like when the elevator is at rest. ◀

Most real-life problems involve friction. Friction is a contact force that is always parallel to the surface and opposes the object's motion. Its magnitude is proportional to the Normal force. There are two kinds of friction, static friction and kinetic friction.

Static friction prevents an object at rest from moving. The magnitude of static friction depends on the applied force. The minimum force required to move the static object is $f_s = \mu_s N$. An applied force that is less than f_s will not produce motion, so the acceleration is zero and the force of static friction equals the applied force.

If the applied force is larger than f_s, the object accelerates and feels kinetic friction, which acts to slow a moving object. The magnitude of kinetic friction is $f_k = \mu_k N$. The quantities μ_s and μ_k are the coefficients of static and kinetic friction, respectively. They are dimensionless numbers, determined experimentally, that depend on the surfaces involved.

Example 5.3 *When push comes to push*
A $25\,\text{kg}$ crate is pushed across the floor with a horizontal force of $150\,\text{N}$, and it slides with an acceleration of $1.5\,\text{m/s}^2$. Find the coefficient of kinetic friction between the crate and the floor, the normal force, and the frictional force acting on the crate,
Solution. Construct the crate's Free-Body Diagram, showing the four forces acting on it: the Push, Friction, the Normal Force and Weight. Use the diagram to write the Newton's 2nd Law equations for the crate.

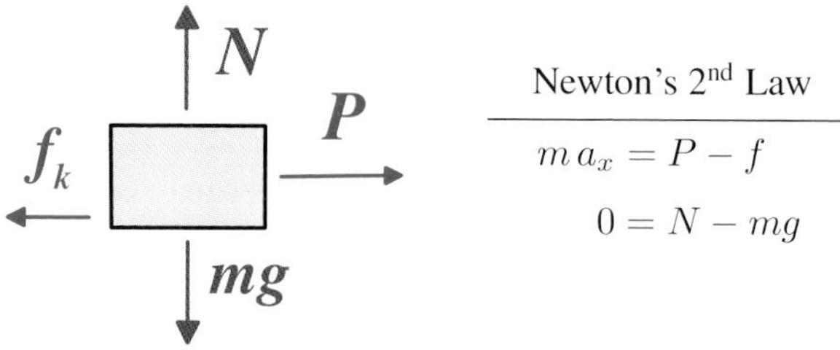

Since the crate slides along the floor, the vertical component of the acceleration is zero.

Create a 1-column, 2-row matrix and place one equation with the appropriate numerical values in each cell. Write the frictional force in terms of the normal force.

Leave the insertion point anywhere in the matrix and click the Solve Exact button.

$$\begin{bmatrix} (25\,\mathrm{kg})\left(1.5\dfrac{\mathrm{m}}{\mathrm{s}^2}\right) = 150\,\mathrm{N} - \mu_k N \\ 0 = N - (25\,\mathrm{kg})\left(9.8067\dfrac{\mathrm{m}}{\mathrm{s}^2}\right) \end{bmatrix},$$

$$\text{Solution is: } \left[N = 245.17\frac{\mathrm{m}}{\mathrm{s}^{2.0}}\,\mathrm{kg}, \mu_k = 0.458\,87\right]$$

The coefficient of kinetic friction is $\mu_k = 0.458\,87$ and the normal force on the crate is $N = 245.17\,\mathrm{N}$. To calculate the frictional force acting on the crate, create and Evaluate the following expression for $f_k = \mu_k N$.

$$f_k = (0.458\,87)\,(245.17\,\mathrm{N}) = 112.5\,\mathrm{N}$$

The force of kinetic friction acting on the crate is $112.5\,\mathrm{N}$. ◀

Example 5.4 *Up the incline*
A $7\,\mathrm{kg}$ block sits on a ramp connected to a $6\,\mathrm{kg}$ block hanging over the edge. The coefficient of kinetic friction between the block and the ramp is $\mu_k = 0.3$ and the incline makes a $\theta = 20°$ angle to the horizontal.

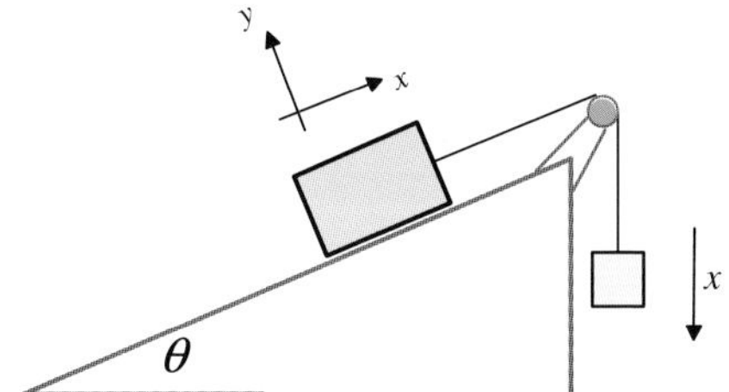

Find the acceleration of the two masses and the tension in the string.

Solution. In this problem, we choose the $+x$ direction as up the ramp (so $+x$ is down for the hanging block) and the $+y$ direction as perpendicular to the ramp. Construct an FBD for each mass, and use them to write the 2nd Law equations for each block.

Newton's 2nd Law

N, *mg* sin*θ*, *T*, f_k, *mg* cos*θ*

The block on the ramp

$$\begin{cases} m\,a_x = T - f_k - mg\sin\theta \\ 0 = N - mg\cos\theta \end{cases}$$

$$\left. M\,a_x = Mg - T \right\}$$

T, *Mg*

The hanging block

The block slides smoothly along the ramp so the vertical component of the acceleration is zero ($a_y = 0$) and both blocks have the same acceleration in the x-direction.

Create a 1-column, 3-row matrix and place one equation with the appropriate numerical values in each cell. Write the frictional force in terms of the normal force.

Leave the insertion point anywhere in the matrix and click the Solve Exact button.

$$\begin{bmatrix} (7\,\mathrm{kg})\,a_x = T - 0.3N - (7\,\mathrm{kg})\left(9.8067\frac{\mathrm{m}}{\mathrm{s}^2}\right)\sin 20^\circ \\ 0 = N - (7\,\mathrm{kg})\left(9.8067\frac{\mathrm{m}}{\mathrm{s}^2}\right)\cos 20^\circ \\ (6\,\mathrm{kg})\,a_x = (6\,\mathrm{kg})\left(9.8067\frac{\mathrm{m}}{\mathrm{s}^2}\right) - T \end{bmatrix},$$

Solution is: $\left[N = 64.507\frac{\mathrm{m}}{\mathrm{s}^{2.0}}\,\mathrm{kg}, T = 51.451\frac{\mathrm{m}}{\mathrm{s}^{2.0}}\,\mathrm{kg}, a_x = 1.231\,5\frac{\mathrm{m}}{\mathrm{s}^{2.0}}\right]$

The acceleration is $a_x = 1.231\,5\,\mathrm{m/\,s}^2$ and the tension is $T = 51.451\,\mathrm{N}$. You can also use Solve Exact on these equations (and Factor the solutions in-place) for the general case without numerical values. Enter N, T, a_x in the Solution Variable(s) window.

$$\begin{bmatrix} ma_x = T - \mu_k N - mg\sin\theta \\ 0 = N - mg\cos\theta \\ Ma_x = Mg - T \end{bmatrix}, \text{ Solution is:}$$

$$\left[N = gm\cos\theta, T = Mgm\frac{\sin\theta + \mu_k\cos\theta + 1}{M+m}, a_x = g\frac{M - m\sin\theta - m\mu_k\cos\theta}{M+m}\right]$$

Notice that friction reduces the acceleration and increases the tension. ◀

Example 5.5 *I'm not inclined to accelerate...*
How much mass should you place on the ramp so that the blocks do not move? The coefficient of static friction between the block and the ramp is $\mu_s = 0.4$.
Solution. Static friction will hold the blocks in place. Use the general result for the acceleration to create an expression using the appropriate numerical values. Set the acceleration equal to zero and replace kinetic friction with static. Leave the insertion point anywhere in the expression and click the Solve Exact button.

$$0 = \frac{6\,\mathrm{kg} - m\sin 20^\circ - m\,(0.4)\cos 20^\circ}{6\,\mathrm{kg} + m}, \text{ Solution is: } 8.357\,7\,\mathrm{kg}$$

If you put an $m = 8.357\,7\,\mathrm{kg}$ block on the ramp, static friction will hold the two blocks at rest. ◀

Example 5.6 *...but I am inclined to move*
With the $7\,\mathrm{kg}$ block back on the ramp, how much mass should you hang so that the $7\,\mathrm{kg}$ block moves up the ramp at constant velocity?
Solution. Since the motion is at constant velocity, the acceleration is zero. The blocks are moving, so use kinetic friction. Place the three force equations in a 1×3 matrix with the appropriate numerical values. Leave the insertion point anywhere in the matrix and click the Solve Exact button.

$$\begin{bmatrix} 0 = T - 0.3N - (7\,\mathrm{kg})\left(9.8067\frac{\mathrm{m}}{\mathrm{s}^2}\right)\sin 20^\circ \\ 0 = N - (7\,\mathrm{kg})\left(9.8067\frac{\mathrm{m}}{\mathrm{s}^2}\right)\cos 20^\circ \\ 0 = M\left(9.8067\frac{\mathrm{m}}{\mathrm{s}^2}\right) - T \end{bmatrix},$$

Solution is: $\left[M = 4.367\,5\,\mathrm{kg}, N = 64.507\frac{\mathrm{m}}{\mathrm{s}^{2.0}}\,\mathrm{kg}, T = 42.831\frac{\mathrm{m}}{\mathrm{s}^{2.0}}\,\mathrm{kg}\right]$

A hanging mass of $M = 4.367\,5\,\mathrm{kg}$ will produce the desired motion. ◀

Newton's Second Law for Varying Forces

In many physics problems, the net force is not constant but depends on time, velocity, or position. The approach of Eqs. (5.3) is still useful, but solving the equations produced by Newton's Second Law requires calculus. Mathematically, this section is very similar to Chapter 2, but the physics is different. In Chapter 2, we solved equations based on the definitions of kinematic quantities to describe motion. Here we solve one-dimensional dynamic equations based on Newton's Second Law to explain changes in motion.

Time-Dependent Forces

If the net force is a function of time, then we write the Second Law as

$$m\frac{dv}{dt} = F(t) \tag{5.4}$$

where $F(t)$ is the time-dependent net force. This is an equation we can solve directly by integrating to give the velocity as a function of time.

$$v(t) = v_0 + \frac{1}{m}\int_0^t F(t)\,dt \tag{5.5}$$

To calculate the position as a function of time, we integrate again.

$$x(t) = x_0 + \int_0^t v(t)\,dt \tag{5.6}$$

As long as the function $F(t)$ is integrable, we can determine the velocity and position. For the special case of a constant force $F(t) = F_0$, this becomes

$$v(t) = v_0 + \frac{F_0}{m}\,t \tag{5.7a}$$

$$x(t) = x_0 + v_0\,t + \tfrac{1}{2}\frac{F_0}{m}\,t^2 \tag{5.7b}$$

We recover the results of Eqs. (2.5) with the constant acceleration $a = F_0/m$.

Example 5.7 *An easy TDF example*
Find the position as a function of time when $F(t) = \alpha t^3$.
Solution. Use Eq. (5.5) to create an expression for the velocity as a function of time and Evaluate it.

$$v(t) = v_0 + \frac{1}{m}\int_0^t \alpha t^3\,dt = v_0 + \frac{1}{4m}t^4\alpha$$

Then use Eq. (5.6) to create an expression for the position as a function of time, Evaluate it, and Expand the last term in-place.

$$x(t) = x_0 + \int_0^t \left(v_0 + \frac{\alpha}{4m}t^4\right)dt = x_0 + tv_0 + \frac{1}{20m}t^5\alpha$$

This gives us the position as a function of time.

$$x(t) = x_0 + v_0t + \tfrac{1}{20}\frac{\alpha}{m}\,t^5$$

The details of this motion depend on the sign of α (see Problem 16). ◀

The following example would be much more difficult without *SNB*.

Example 5.8 *A not-so-easy TDF example*
Find the position as a function of time for an object experiencing this net force.

$$F(t) = F_0 \sin^4 \frac{\pi}{2} t$$

Solution. Use Eq. (5.5) to create an expression for the velocity as a function of time and Evaluate it.

$$v(t) = v_0 + \frac{F_0}{m} \int_0^t \sin^4 \frac{\pi}{2} t \, dt = v_0 + \frac{1}{16\pi m} F_0 \left(\sin 2\pi t - 8 \sin \pi t + 6\pi t\right)$$

Tell *SNB* the variables are positive.

$$\text{assume}\,(\text{positive}) = (0, \infty)$$

Then use Eq. (5.6) to create and Evaluate an expression for the position as a function of time.

$$x(t) = x_0 + \int_0^t \left(v_0 + \frac{F_0}{\pi m} \left(\tfrac{3}{8}\pi t - \tfrac{1}{2} \sin \pi t + \tfrac{1}{16} \sin 2\pi t\right)\right) dt$$
$$= x_0 + \frac{1}{32\pi^2 m} \left(16 F_0 \cos \pi t - 15 F_0 - F_0 \cos 2\pi t + 6\pi^2 t^2 F_0 + 32\pi^2 m t v_0\right)$$

Now Expand the result, then use Factor and Simplify in-place on the terms with F_0.

$$x(t) = x_0 + t v_0 + \frac{1}{32\pi^2 m} F_0 \left(6\pi^2 t^2 - 2\cos^2 \pi t + 16 \cos \pi t - 14\right)$$

Finally, edit the result by-hand to get the position as a function of time.

$$x(t) = x_0 + v_0\, t + \frac{F_0}{32\pi^2 m} \left(6\pi^2 t^2 - 2\cos^2 \pi t + 16 \cos \pi t - 14\right)$$

This time-dependent force models a short-duration impulse. Let's take a look at the force, position, and velocity for $F_0 = 1\,\text{N}$, $m = 1\,\text{kg}$, $x_0 = 0.5\,\text{m}$, and $v_0 = 0\,\text{m/s}$.

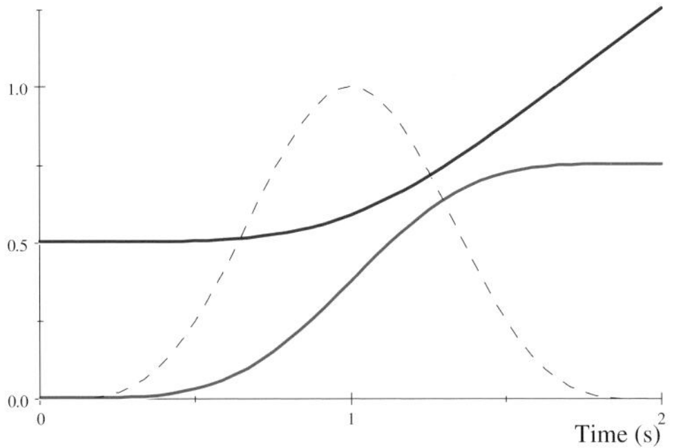

Figure 5.1 A short-duration impulse
The velocity (lower solid curve) changes only when the force (dashed line) is not zero and remains constant when the force is zero. ◄

Velocity-Dependent Forces

If the net force is a function of the velocity, then we write the Second Law as

$$m\frac{dv}{dt} = F(v) \tag{5.8}$$

where $F(v)$ is the velocity-dependent net force. We can integrate this equation as well.

$$\int_{v_0}^{v} \frac{dv}{F(v)} = \frac{1}{m}t \tag{5.9}$$

This gives us time as a function of velocity, an expression we can invert (solve for $v(t)$). Once you have the velocity as a function of time, use Eq. (5.6) to calculate the position as a function of time.

Example 5.9 *An easy VDF example*
Find the position as a function of time when $F(v) = -bv^2$. Such a force is often used to model the effects of air resistance.

Solution. First, tell *SNB* to assume the variables are positive so it can ignore the potential discontinuity at $v = 0$. Then Evaluate the integral on the left-hand side of Eq. (5.9) and Expand the result.

$$-\frac{1}{b}\int_{v_0}^{v} \frac{dv}{v^2} = -\frac{1}{bvv_0}(v - v_0) = \frac{1}{bv} - \frac{1}{bv_0}$$

Next, set the result equal to t/m, use Solve Exact to solve for v, and Expand the denominator in-place.

$$\frac{1}{bv} - \frac{1}{bv_0} = \frac{t}{m}, \text{ Solution is: } \frac{1}{\frac{1}{v_0} + \frac{b}{m}t}$$

Multiply both numerator and denominator by v_0. Then Simplify the numerator and Expand the denominator, both in-place.

$$v = \frac{1 \times v_0}{\left(\frac{1}{v_0} + \frac{b}{m}t\right) \times v_0} = \frac{v_0}{\frac{b}{m}tv_0 + 1}$$

This gives us the velocity as a function of time.

$$v(t) = \frac{v_0}{1 + \frac{bv_0}{m}t}$$

The resistive force slows the object. To calculate the position as a function of time, use Eq. (5.6) to create and Evaluate the following expression.

$$x(t) = x_0 + \int_0^t \frac{v_0}{1 + \frac{bv_0}{m}t}dt = x_0 + \frac{1}{b}(m\ln(m + btv_0) - m\ln m)$$

We can improve the appearance of the term in parentheses. Let's Factor it and then apply Combine + Logs and Expand in-place to the argument of the logarithm.

$$m\ln(m+btv_0)-m\ln m=m(-\ln m+\ln(m+btv_0))=m\ln\left(\frac{b}{m}tv_0+1\right)$$

In a more standard form, the position as a function of time is

$$x(t)=x_0+\frac{m}{b}\ln\left(1+\frac{bv_0}{m}t\right)$$

The object's displacement Δx is inversely proportional to the constant b, so the stronger the resistive force, the shorter the distance the object travels. ◀

The following two examples involve a more difficult problem that includes gravity.

Example 5.10 *It takes this much time...*

Find a general expression for the time needed for an object thrown straight up to reach its maximum height when it experiences the resistive force $-bv^2$.

Solution. On the way up, both gravity and air resistance exert a downward force on the object, which experiences a net downward force $F(v)=-mg-bv^2$. First, tell *SNB* the variables are positive.

$$\text{assume}(\text{positive})=(0,\infty)$$

Now Evaluate the integral on the left-hand side of Eq. (5.9).

$$-\int_{v_0}^{v}\frac{dv}{mg+bv^2}=\frac{1}{\sqrt{b}\sqrt{g}\sqrt{m}}\left(\arctan\frac{\sqrt{b}}{\sqrt{g}\sqrt{m}}v_0-\arctan\frac{\sqrt{b}}{\sqrt{g}\sqrt{m}}v\right)$$

Set the result equal to t/m, Solve for v, and apply Combine + Powers in-place to each group of constants.

$$\frac{1}{\sqrt{b}\sqrt{g}\sqrt{m}}\left(\arctan\frac{\sqrt{b}}{\sqrt{g}\sqrt{m}}v_0-\arctan\frac{\sqrt{b}}{\sqrt{g}\sqrt{m}}v\right)=\frac{t}{m},$$

$$\text{Solution is: }\sqrt{\frac{1}{b}gm}\tan\left(\arctan\frac{1}{\sqrt{\frac{1}{b}gm}}v_0-\sqrt{b\frac{g}{m}}t\right)$$

This give us the object's velocity as a function of time.

$$v(t)=\sqrt{\frac{mg}{b}}\tan\left(\arctan\sqrt{\frac{bv_0^2}{mg}}-\sqrt{\frac{bg}{m}}t\right)$$

At the maximum height the object's velocity is zero, and the argument of the tangent must be zero.

$$0=T\sqrt{\frac{bg}{m}}-\arctan\sqrt{\frac{bv_0^2}{mg}},\ \text{Solution is: }\frac{1}{\sqrt{b}\sqrt{g}}\sqrt{m}\arctan\frac{\sqrt{b}}{\sqrt{g}\sqrt{m}}v_0$$

The time needed to reach the highest point is

$$T = \sqrt{\frac{m}{bg}} \arctan \sqrt{\frac{bv_0^2}{mg}}$$

As the value b increases, the resistive force increases and reduces the time needed for the object to reach the highest point. ◀

Now that we know the time it takes the object to reach its highest point, we can calculate that maximum height.

Example 5.11 ... *to get this far off the ground*
Find a general expression for the maximum height attained by the object in the previous example.
Solution. The maximum height of the object is

$$H = y_0 + \int_0^T v(t)\,dt$$

where T is the time to the maximum height from the previous example. Let's help *SNB* by writing T and $v(t)$ in terms of simpler constants $A \equiv \sqrt{\frac{bv_0^2}{mg}}$ and $B \equiv \sqrt{\frac{bg}{m}}$.

$$\begin{aligned} v(t) &= \frac{v_0}{A} \tan(\arctan A - Bt) \\ T &= \frac{1}{B} \arctan A \end{aligned}$$

Tell *SNB* to assume the variables are positive. Then use the definite integral fudge and Expand the following integral, then Factor the result and use Combine + Logs on the numerator, both in-place.

$$\left[\int \frac{v_0}{A} \tan(\arctan A - Bt)\,dt\right]_{t=0}^{t=\frac{\arctan A}{B}} = \tfrac{1}{2} v_0 \frac{\ln(A^2+1)}{AB}$$

Now just Substitute (with Evaluate) the expressions for A and B.

$$\left[\frac{1}{2} v_0 \frac{\ln(A^2+1)}{AB}\right]_{\left\{A=\sqrt{bv_0^2/(mg)},B=\sqrt{bg/m}\right\}} = \frac{1}{2b} m \ln\left(\frac{b}{gm} v_0^2 + 1\right)$$

This gives us the object's maximum height for a given initial speed.

$$H = y_0 + \frac{m}{2b} \ln\left(1 + \frac{bv_0^2}{mg}\right)$$

As the value b increases, the resistive force increases and reduces the maximum height for a given mass and initial speed. A 2-term Power Series expansion in b shows this is the case for small values.

$$H = y_0 + \frac{m}{2b} \ln\left(1 + \frac{bv_0^2}{mg}\right) = \left(y_0 + \frac{1}{2g} v_0^2\right) - \tfrac{1}{4} \frac{b}{g^2 m} v_0^4 + O\left(b^2\right)$$

You should pick some reasonable values for the mass, initial speed and height, and confirm it for all b by graphing the maximum height as a function of b. ◀

Position-Dependent Forces

If the net force is a function of position only, then we write the Second Law as

$$m\frac{d^2x}{dt^2} = F(x) \tag{5.10}$$

where $F(x)$ is the position-dependent net force. For certain functions $F(x)$, you can use *SNB*'s Solve ODE + Exact command, with the appropriate initial conditions, to solve Eq. (5.10) for the position as a function of time.

This equation can be integrated to produce an alternative to the Second Law in terms of the velocity.

$$\frac{dx}{dt} = \pm\sqrt{\frac{2}{m}\int F(x)\,dx} \tag{5.11}$$

The choice of sign depends on whether the object moves to the right ("+") or left ("−").

For some functions $F(x)$, you can solve Eq. (5.11) by direct integration. To solve Eq. (5.11) directly, gather all the x-dependent terms on the same side.

$$\int_{x_0}^{x}\frac{dx}{\sqrt{\int_{x_0}^{x}F(x)\,dx}} = \pm\sqrt{\frac{2}{m}}\,t \tag{5.12}$$

If *SNB* can do the two integrals, then Eq. (5.12) will produce an equation you can Solve Exactly for the position as a function of time.

You have three ways to solve Newton's Second Law for position-dependent forces.

1. For certain functions $F(x)$, you can use Solve ODE + Exact to solve Eq. (5.10) to get an exact analytical solution for $x(t)$.

2. If *SNB* can do the two integrals, you can solve Eq. (5.12) by direct integration.

3. When no exact solution exists, use Solve ODE + Numeric to solve Eq. (5.10) to get a numerical approximation for $x(t)$.

One position-dependent force you can solve with Solve ODE + Exact is the linear force $F(x) \sim -x$, which applies to the motion of an object attached to a spring.

Example 5.12 *Spring forward*
Find the position and velocity as a function of time when $F(x) = -m\omega^2 x$ for an object with arbitrary initial position and velocity.
Solution. Place the equation in a 3×1 matrix with the appropriate force equation and initial conditions, leave the insertion point anywhere in the matrix and select Solve ODE + Exact.

$$\begin{bmatrix} m\dfrac{d^2x}{dt^2} = -m\omega^2 x \\ x(0) = x_0 \\ x'(0) = v_0 \end{bmatrix}, \text{ Exact solution is: } \left\{x_0\cos t\omega + \frac{1}{\omega}v_0\sin t\omega\right\}$$

The position as a function of time is $x(t) = x_0\cos\omega t + \dfrac{v_0}{\omega}\sin\omega t$.

The velocity is the time-derivative of the position. Create and Evaluate the following expression for the velocity as a function of time.

$$v(t) = \frac{d}{dt}\left(x_0 \cos \omega t + \frac{v_0}{\omega} \sin \omega t\right) = v_0 \cos t\omega - \omega x_0 \sin t\omega$$

The velocity as a function of time is $v(t) = v_0 \cos \omega t - \omega x_0 \sin \omega t$. ◀

The next example looks at the 1-dimensional motion for an attractive force that is proportional to the inverse-cube of the position.

Example 5.13 *An inverse-cubic force. Right!*
Find the position as a function of time when $F(x) = -k/x^3$ for an object released from rest from a positive initial position. How long does it take the object to reach the origin?

Solution. *SNB*'s Solve ODE + Exact command cannot solve Eq. (5.10) when you specify the initial conditions, so let's try Eq. (5.11) instead. First, useassume (positive) to tell *SNB* the variables are positive. Then create an expression for the integral inside the radical on the right-hand-side of Eq. (5.11) and Expand it.

$$\int F(x)\, dx = -k\int_{x_0}^{x} \frac{1}{x^3}\, dx = \frac{1}{2}\frac{k}{x^2} - \frac{1}{2}\frac{k}{x_0^2}$$

To use Eq. (5.12), gather all the x-dependent terms on the same side.

$$\frac{dx}{\sqrt{\frac{1}{x^2} - \frac{1}{x_0^2}}} = -\sqrt{\frac{k}{m}}dt$$

Notice I chose the negative root for the velocity, since the object will move to the left. Next, create an expression for the integral on the left-hand side and Evaluate it.

$$\int_{x_0}^{x} \frac{xdx}{\sqrt{1 - \frac{x^2}{x_0^2}}} = -x_0\sqrt{x_0^2 - x^2}$$

Create an expression equating this to the result of the trivial time integration, and use Solve Exact to find the position x.

$$x_0\sqrt{x_0^2 - x^2} = \sqrt{\frac{k}{m}}t, \text{ Solution is: } \frac{1}{\sqrt{m}x_0}\sqrt{-kt^2 + mx_0^4}$$

Edit the solution by-hand, and then Expand the expression inside the radical in-place.

$$x = \frac{1}{\sqrt{m}x_0}\sqrt{-kt^2 + mx_0^4} = x_0\sqrt{\frac{-kt^2 + mx_0^4}{\left(\sqrt{m}x_0^2\right)^2}} = x_0\sqrt{1 - \frac{k}{m}\frac{t^2}{x_0^4}}$$

This gives us the object's position as a function of time.

$$x(t) = x_0\sqrt{1 - \frac{k}{mx_0^4}t^2}$$

When the object reaches the origin its position is $x = 0$, which occurs when the term inside the radical is zero.

$$0 = 1 - \frac{k}{mx_0^4}t^2, \text{ Solution is: } \frac{1}{\sqrt{k}}\sqrt{mx_0^2}$$

The object reaches the origin at time $t = x_0^2\sqrt{m/k}$ after it is released from rest. For an inverse-cubic force, this time is proportional to the initial position squared. ◄

When you and *SNB* can't find an exact solution, try Solve ODE + Numeric to find a numerical solution to Eq. (5.10).

Example 5.14 *Approximately approximating the approximation*

Calculate and plot the position as a function of time when $F(x) = e^{\sin x}\cos x$ for an object with unit mass starting from rest at $x_0 = \pi$. Use the numerical solution to plot the velocity as a function of time.

Solution. You are given a position-dependent force for which there is no exact solution to either Eq. (5.10) or Eq. (5.11). You need to solve Eq. (5.10) numerically. Place the equation and initial conditions, with the appropriate numerical values, in the rows of a 3-row, 1-column matrix. Leave the insertion point anywhere in the matrix and choose Solve ODE + Numeric.

$$\begin{bmatrix} x'' = e^{\sin x}\cos x \\ x(0) = \pi \\ x'(0) = 0 \end{bmatrix}, \text{ Functions defined: } x$$

Use the Plot 2D Rectangular button and plot the solution $x(t)$, letting the time run from 0 to 13.20. Figure 5.2a shows the numerical solution $x(t)$ as the solid line and the function $\pi\cos^2\frac{t}{2.1}$ as the dashed line.

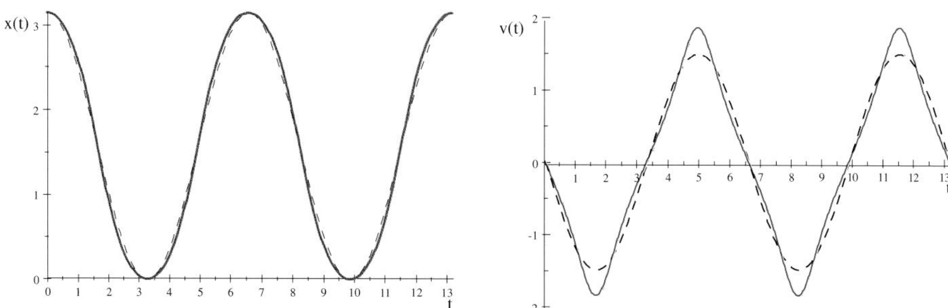

Figure 5.2a Approximating... **Figure 5.2b** ...the approximation

The function $\pi\cos^2\frac{t}{2.1}$ is a very good approximation to the numerical solution for the position.

There are two ways to calculate the velocity as a function of time. You can use the numerical solution $x(t)$ to calculate a numerical velocity $v(t)$ accurately with

$$v(t) \approx \frac{x(t+\varepsilon/2) - x(t-\varepsilon/2)}{\varepsilon}$$

as long as ε is a small number. Figure 5.2b uses $\varepsilon = 10^{-6}$. The solid line is the numerical velocity $v(t)$ and the dashed line is an approximation to the velocity.

$$v(t) \approx \frac{d}{dt}\left(\pi \cos^2 \frac{t}{2.1}\right) = -\frac{\pi}{2.1} \sin \frac{2t}{2.1}$$

As you can see, the approximation for the velocity is not nearly as good as the one for the position.

You can also use the definition of velocity from Eq. (2.7) and solve the differential equations for the position and velocity simultaneously.

$$\begin{bmatrix} v' = e^{\sin x} \cos x \\ v = x' \\ x(0) = \pi \\ v(0) = 0 \end{bmatrix}, \text{ Functions defined: } v, x$$

Figure 5.2b also include this velocity. The two numerical solutions for the velocity are indistinguishable. ◄

Newton's Third Law

The exertion of a force is an interaction between two objects. Newton's Third Law shows that the interaction affects both objects because it requires a force acting on both.

> **Newton's 3rd Law**: When one object exerts a force on another, the second object simultaneously exerts a force on the first object that is equal and opposite to the force it felt.

The "equal and opposite" refers to the equal magnitude and opposite direction of the force exerted by the second object. That force, also called the "reaction force", has equal magnitude and opposite direction of the original force. You may have heard the 3rd Law stated as "for every action there is an equal and opposite reaction". The "action" is the original force exerted **by** the first object, and the "reaction" is the force exerted **on** the first object by the second.

There are two important points here.

1. Forces come in pairs that have equal magnitudes and opposite directions.
2. The two forces act on different objects.

There are no isolated forces. Forces come in pairs which together comprise the interaction between two objects. In every interaction, forces act on each interacting object. For every force exerted by an object there is a force of equal magnitude but opposite direction which acts back on that object. An object can experience a non-zero net force and acceleration only if it interacts with some other object.

Example 5.15 *Cosmic litter bug*
While working in outer space, a 75 kg astronaut discards a 1 kg Illudium Q-36 explosive space modulator by literally throwing it away. During the throw, the astronaut exerts a constant force of 15 N for 0.5 s on the modulator. What happens to the astronaut? How far apart are the astronaut and modulator 5 minutes after the throw? What planet is the astronaut from?
Solution. By Newton's 3rd Law, the net force on the astronaut is also 15 N but in the direction opposite the throw. The astronaut pushed the modulator, but the modulator pushed back and in frictionless space the astronaut moves too. The astronaut and modulator feel the same amount of force, but their accelerations have different magnitudes and opposite directions.

Let's use Newton's 2nd Law and Evaluate Numerically to find the acceleration of the astronaut and modulator.

$$a_a = \frac{+15\,\mathrm{N}}{75\,\mathrm{kg}} = 0.2\frac{\mathrm{m}}{\mathrm{s}^2}$$

$$a_m = \frac{-15\,\mathrm{N}}{1\,\mathrm{kg}} = -15.0\frac{\mathrm{m}}{\mathrm{s}^2}$$

At the end of the half-second constant acceleration, they are both moving.

$$v_a = a_a\, t_{\text{throw}} = \left(+0.2\frac{\mathrm{m}}{\mathrm{s}^2}\right)(0.5\,\mathrm{s}) = 0.1\frac{\mathrm{m}}{\mathrm{s}}$$

$$v_m = a_m\, t_{\text{throw}} = \left(-15.0\frac{\mathrm{m}}{\mathrm{s}^2}\right)(0.5\,\mathrm{s}) = -7.5\frac{\mathrm{m}}{\mathrm{s}}$$

After the throw, there are no forces acting on either the astronaut or the modulator, so they move at a constant speed. Five minutes later, the astronaut's displacement is only 30 m.

$$\Delta x_a = v_a\, t = \left(+0.1\frac{\mathrm{m}}{\mathrm{s}}\right)(300\,\mathrm{s}) = 30.0\,\mathrm{m}$$

In that same time, the modulator travels much farther in the opposite direction.

$$\Delta x_m = v_m\, t = \left(-7.5\frac{\mathrm{m}}{\mathrm{s}}\right)(300\,\mathrm{s}) = -2250.0\,\mathrm{m}$$

After 5 minutes the astronaut and modulator are $30\,\mathrm{m} - (-2250\,\mathrm{m}) = 2280\,\mathrm{m}$ apart and still moving. The astronaut is a martian. ◀

Newton's 3rd Law is the science behind the motion of rockets. A rocket pushes exhaust gas out its back end, and by Newton's 3rd Law the gas pushes back. That reaction force is called thrust and it accelerates the rocket. Rocket science isn't rocket science. In the next chapter, we'll explore the motion of rockets in more detail.

Problems

1. How much time does it take the pitcher in Example 5.1 to accelerate the ball?

2. A $25\,\text{kg}$ crate is pushed across the floor with a horizontal force of $150\,\text{N}$. Ignoring friction, find the normal force acting on the crate and the acceleration of the crate. How do your answers compare with the results of Example 5.3?

3. Repeat Example 5.3, replacing the horizontal push with a pull of $150\,\text{N}$ at an angle of $25°$ to the horizontal. How do your answers compare with the results of the example?

4. A $7\,\text{kg}$ block sits on an inclined ramp connected to a $6\,\text{kg}$ block hanging over the edge. The ramp makes a $20°$ angle to the horizontal. Ignoring friction, find the acceleration of the two masses and the tension in the string. How do your answers compare with the results of Example 5.4?

5. To move a large crate across a rough floor, you push on it with a horizontal force. The mass of the crate is $27\,\text{kg}$ and the coefficient of static friction between the crate and the floor is 0.55. The coefficient of kinetic friction is 0.42.

   a. Calculate the normal force acting on the crate.

   b. Find the force necessary to start the crate moving.

   c. Find the force necessary to keep it moving at a constant speed.

   d. Find the acceleration if the applied force is $190\,\text{N}$.

6. A $m_1 = 4\,\text{kg}$ block is pulled along a frictionless table by the weight of a hanging mass ($m_2 = 2.2\,\text{kg}$) as shown in the figure.

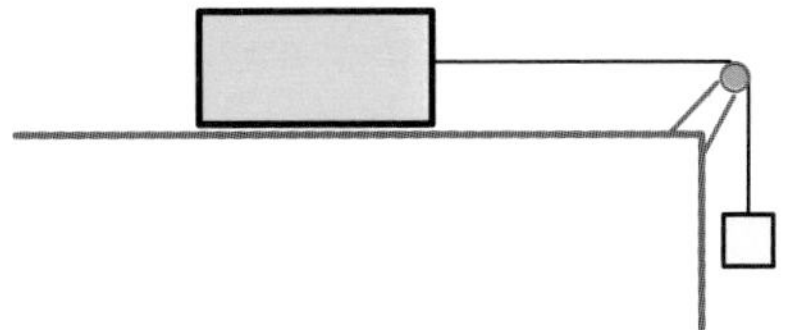

   a. Draw the free-body diagram for **each** block.

   b. Write the equations for Newton's 2nd law for each block.

   c. Calculate the acceleration of the blocks.

   d. Calculate the tension in the string connecting the two blocks.

7. A water skier whose mass is $89\,\text{kg}$ is being pulled with an acceleration of $\frac{1}{5}g$. The pulling force has a magnitude of $300\,\text{N}$ and points $10°$ above the horizontal. Find the magnitudes of the resistive and upward forces exerted on the skier.

8. A water skier whose mass is $89\,\text{kg}$ is being pulled at a constant velocity. The pulling force has a magnitude of $350\,\text{N}$ and points $15°$ above the horizontal. Find the magnitudes of the resistive and upward forces exerted on the skier.

9. A $75\,\mathrm{kg}$ block is sliding down a $25°$ ramp as shown in the figure. The coefficient of kinetic friction between block and the ramp is $\mu_k = 0.35$.

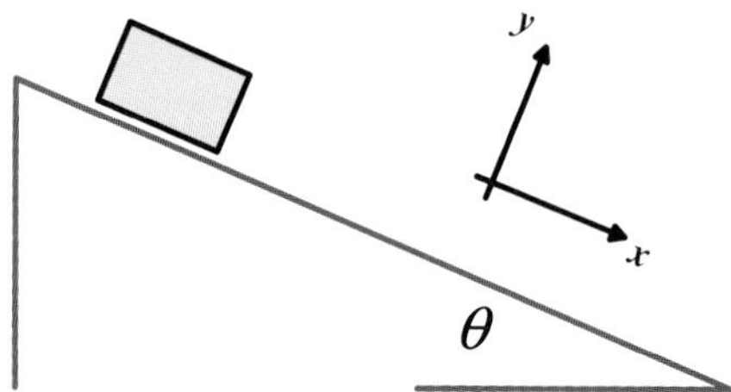

a. Draw the free-body diagram for the block.

b. Using Newton's 2nd law, write the equations for all the forces acting on the block.

c. Calculate the force of friction acting on the block.

d. Calculate the acceleration of the block.

10. A shopper pushes a $7.5\,\mathrm{kg}$ shopping cart up a $15°$ incline with a force that is parallel to the inclined surface. The cart experiences an acceleration of $1.41\,\mathrm{m/s^2}$. Ignore friction in this problem.

a. Draw the free-body diagram for the cart.

b. Using Newton's 2nd law, write the equations for all the forces acting on the cart.

c. Determine the force exerted by the shopper.

d. Calculate the normal force acting on the cart.

11. A small child pulls a $7.2\,\mathrm{kg}$ wagon up a $12°$ incline with a $35\,\mathrm{N}$ force that is parallel to the inclined surface. Ignore friction in this problem.

a. Draw the free-body diagram for the wagon.

b. Using Newton's 2nd law, write the equation for the horizontal forces acting on the wagon.

c. Determine the acceleration of the wagon as it is pulled by the child.

d. How much force is required to pull the wagon up the incline at a constant speed?

12. A four-kilogram box is released from rest at the top of a ramp that is $7\,\mathrm{m}$ long and inclined at an angle of $32°$ to the horizontal. The coefficient of kinetic friction between the box and the ramp is $\mu_k = \frac{1}{4}$.

a. What is the frictional force on the box as it slides down the ramp?

b. Calculate the acceleration of the box.

c. How much time does it take the box to reach the bottom of the ramp?

d. What is the speed of the box as it reaches the bottom of the ramp?

13. A five-kilogram box is given an initial speed of $10\,\mathrm{m/s}$ up a ramp that is $8\,\mathrm{m}$ long and inclined at an angle of $25°$ to the horizontal. The coefficient of kinetic friction between the box and the ramp is $\mu_k = \frac{1}{4}$.

    a. What is the frictional force on the box as it slides up the ramp?

    b. Calculate the acceleration of the box.

    c. How far up the ramp does the box go?

    d. How much time does it take the box to reach the highest point up the ramp?

14. Show that the coefficient of kinetic friction that produces motion at constant velocity on a ramp is

$$\mu_k = \frac{M - m\sin\theta}{m\cos\theta}$$

where M is the hanging mass and m is the mass on the ramp. What is the maximum angle for such motion when $m = 7\,\mathrm{kg}$ and $M = 6\,\mathrm{kg}$?

15. Repeat Example 5.6 for the $7\,\mathrm{kg}$ mass moving *down* the ramp at constant velocity.

16. Plot the results from Example 5.7 for the position and velocity with $x_0 = 1$, $v_0 = 5$, and $\alpha = \pm\frac{1}{5}$ for 5 seconds. Discuss the two motions, including if and when they return to the initial position and with what speed.

17. Repeat Example 5.8 for the TDF $F(t) = 4t\left(1 - \frac{1}{2}t\right)$.

18. Repeat Example 5.8 for the TDF $F(t) = 2\sin(\pi t/2)$.

19. Repeat Example 5.8 for the TDF $F(t) = 2\sin^2(\pi t/2)$.

20. Consider the force $F(t) = F_0\left(1 + \frac{t}{T}\right)^n$ where F_0 and T are constants.

    Find the velocity and position as functions of time for $n = -1$, -2, and $-1/2$.

21. Choose some physically reasonable numbers for the results of Example 5.9 and plot both $x(t)$ and $v(t)$ as a function of time.

22. Find $v(t)$ for $F(v) = -bv^n$ where $n \geq 0$. For $n < 0$, the object would experience an increasing resistive force as it slows down and an infinite force when it stopped.

23. Find $x(t)$ for $F(v) = -bv^n$. For what values of n is this expression invalid?

24. Choose some physically reasonable numbers for the previous problem and plot $x - x_0$ as a function of n for large values of t.

25. Show that the result for the time it takes the object in Example 5.10 to reach its maximum height reduces to the expected result in the absence of the resistive force.

26. Show that the result for the maximum height achieved by the object in Example 5.11 reduces to the expected result in the absence of the resistive force.

27. How fast is the object in Example 5.13 moving when it reaches the origin?

28. Use Expand and Factor to show that the result of Example 5.10 is equivalent to

$$v(t) = v_0 \frac{1 - \sqrt{\frac{mg}{bv_0^2}} \tan \sqrt{\frac{bg}{m}} t}{1 + \sqrt{\frac{bv_0^2}{mg}} \tan \sqrt{\frac{bg}{m}} t}$$

29. The object we launched in Examples 5.10 and 5.11 will come down. Show that its velocity and height as functions of time are as follows.

$$\begin{aligned} v(t) &= -\sqrt{\frac{mg}{b}} \tanh \sqrt{\frac{bg}{m}} t \\ y(t) &= H - \frac{m}{b} \ln \left(\cosh \sqrt{\frac{bg}{m}} t \right) \end{aligned}$$

Use these results to show the falling time and landing speed reduce to the expected results if you turn off air resistance.

30. Find the position as a function of time for the constant force $F(x) = -k$ for an object released from rest from an arbitrary initial position. How long does it take the object to reach the origin? What does the constant k mean?

31. For an object with initial position x_0 and initial velocity v_0, the position as a function of time for the position-dependent force $F(x) = -\kappa/x^3$ is

$$x(t) = x_0 \sqrt{1 + \frac{2v_0}{x_0} t + \left(\frac{v_0^2}{x_0^2} - \frac{\kappa}{m x_0^4} \right) t^2}$$

where κ is a positive constant. Use the numerical values $x_0 = 1$, $v_0 = 2$, and $\kappa = m = 1$. Make a plot of $x(t)$ from $t = 0$ to $t = 1.5486$. Describe the motion. When does the object reach the origin?

32. Show that the velocity as a function of time for the object in the previous problem is

$$v(t) = \frac{v_0 + \left(\frac{v_0^2}{x_0} - \frac{\kappa}{m x_0^3} \right) t}{\sqrt{1 + \frac{2v_0}{x_0} t + \left(\frac{v_0^2}{x_0^2} - \frac{\kappa}{m x_0^4} \right) t^2}}$$

Use the same numerical values and plot $v(t)$. When does the object stop? Show that both $x(t)$ and $v(t)$ reduce to the correct expressions when the force is very small.

33. An object dropped into a tunnel connecting the north and south poles falls with acceleration $a = -\frac{g}{R} y$ where R is the Earth's radius. How long does it take the object to reach the other pole? How fast is it going at the Earth's center?

34. Consider a tunnel connecting any two points on the Earth's surface. Choose your coordinate system so the y-axis bisects, and x-axis is parallel to the tunnel. An object dropped into such a tunnel falls at $a = -\frac{g}{R} x$ where R is the Earth's radius. How long does it take the object to traverse this tunnel? Does this time depend on its initial position? What does that mean? How fast is it going at the midpoint?

6 Conservation Laws

Conservation Laws are fundamentally important in physics. We say a quantity is "conserved" when its total amount never changes. Its initial value equals its final value equals its value at all times in between. In this chapter you'll meet two of the most important conserved quantities, Energy and Momentum.

Conservation laws are also powerful problem-solving tools. Even though they are consequences of Newton's Laws, the approach to using them is different than the cause-and-effect methods of the last chapter. When we used Newton's 2nd Law, we applied it to each object separately. When we use a conservation law, we'll look at the system as a whole, not at the individual parts, and keep track of the total amount of the conserved quantity. Still, our goal is the same: to understand and explain changes in motion.

Definitions

To understand and explain changes in motion, we need to consider both velocity and mass. Stopping a train moving at $5\,\mathrm{mi/h}$ is not the same as stopping an empty shopping cart moving at $5\,\mathrm{mi/h}$. They have the same speed, but not the same mass. Catching a baseball moving at $5\,\mathrm{mi/h}$ is not the same as catching one moving at $100\,\mathrm{mi/h}$. They have the same mass, but not the same speed.

We can define a new vector called **momentum** as the product of mass times velocity. This is such a useful and important concept that when Newton used the word "motion" in the *Principia*, he often meant what we today call momentum.

$$\overrightarrow{p} = m\overrightarrow{v} \tag{6.1}$$

The magnitude of the momentum is $p = mv$ and its direction is the same as the velocity vector. The metric unit of momentum is the $\mathrm{kg\,m/s}$.

Example 6.1 *Some typical values*
Calculate the momentum of a $1500\,\mathrm{kg}$ car moving at a speed of $15\,\mathrm{mi/h}$. Do the same for a baseball moving at a speed of $80\,\mathrm{mi/h}$. Which has more momentum?
Solution. Use Eq. (6.1) to create an expression for the car's momentum using the given values. Select the car's velocity and apply Evaluate Numerically in-place to convert it to meters per second, then Evaluate the expression.

$$p_{\text{car}} = (1500\,\mathrm{kg})\,(15\,\mathrm{mi/h}) = (1500\,\mathrm{kg})\left(6.705\,6\frac{\mathrm{m}}{\mathrm{s}}\right) = 10058.\frac{\mathrm{m}}{\mathrm{s}}\,\mathrm{kg}$$

Doing Physics with Scientific Notebook: A Problem-solving Approach, First Edition. Joseph Gallant.

Now let's repeat the process for the ball's momentum.

$$p_{\text{ball}} = (0.142\,\text{kg})\,(80\,\text{mi/h}) = (0.142\,\text{kg})\left(35.763\frac{\text{m}}{\text{s}}\right) = 5.0783\frac{\text{m}}{\text{s}}\,\text{kg}$$

One way to compare these momenta is to Evaluate their ratio.

$$\frac{p_{\text{car}}}{p_{\text{ball}}} = \frac{10058.\,\text{kg m/s}}{5.0783\,\text{kg m/s}} = 1980.6$$

The car has 1980.6 times as much momentum as the ball. ◀

In Chapter 5, I told you how Newton's 2nd Law related the net force acting on an object to the object's mass and resulting acceleration.

$$\vec{F} = m\vec{a}$$

There is another way to write the 2nd Law, in terms of momentum: the net force equals the rate of change of the momentum.

$$\vec{F} = \frac{\Delta\vec{P}}{\Delta t} \tag{6.2}$$

The change in momentum is $\Delta\vec{P} = \vec{P}_f - \vec{P}_i$ and Δt is the time interval in which the change occurred. This applies to individual objects and to systems of objects.

Example 6.2 *Stop!*
Which requires more force: stopping the car in the previous example in 3 seconds, or stopping the baseball in 1.6 milliseconds?
Solution. Use Eq. (6.2) and the result from the previous example to create and Evaluate an expression for the force on the car.

$$F_{\text{car}} = \frac{10058.\,\text{kg m/s}}{3\,\text{s}} = 3352.7\frac{\text{m}}{\text{s}^2}\,\text{kg}$$

It takes a force of 3352.7 N to stop the car. Repeat the process for the ball.

$$F_{\text{ball}} = \frac{5.0783\,\text{kg m/s}}{0.0016\,\text{s}} = 3173.9\frac{\text{m}}{\text{s}^2}\,\text{kg}$$

The two forces are approximately the same. ◀

Kinetic energy is the energy of motion. Objects in motion have **kinetic energy** (KE) and objects at rest have no kinetic energy. The kinetic energy of a moving object is

$$K = \tfrac{1}{2}mv^2 \tag{6.3}$$

where m is its mass and v is its speed. As the object moves faster, its kinetic energy increases dramatically. Doubling the object's velocity gives it four times as much kinetic energy. Kinetic energy is always positive, and like all forms of energy, it's a scalar. The metric unit of energy is the **joule** (J).

$$1\,\text{J} = 1\,\text{kg}\frac{\text{m}^2}{\text{s}^2} = 1\,\text{N m}$$

One joule is about the energy needed to lift an apple from your pocket to your mouth.

Example 6.3 *Some more typical values*
Calculate the kinetic energy of a 1500 kg car moving at a speed of 15 mi/h. Do the same for a baseball moving at a speed of 80 mi/h. Which has more kinetic energy?
Solution. Use Eq. (6.3) to create an expression for the car's kinetic energy using the given values. Select the car's velocity and apply Evaluate Numerically in-place to convert it to meters per second, then Evaluate the expression.

$$K_{\text{car}} = \tfrac{1}{2}\,(1500\,\text{kg})\left(6.705\,6\frac{\text{m}}{\text{s}}\right)^2 = 33724.\frac{\text{m}^2}{\text{s}^2}\,\text{kg}$$

The car has 33724 J of kinetic energy. Repeat the process for the ball.

$$K_{\text{ball}} = \tfrac{1}{2}\,(0.142\,\text{kg})\left(35.763\frac{\text{m}}{\text{s}}\right)^2 = 90.808\frac{\text{m}^2}{\text{s}^2}\,\text{kg}$$

The ball has 90.808 J of kinetic energy. Let's Evaluate this expression for the ratio of the car's kinetic energy to the ball's kinetic energy.

$$\frac{K_{\text{car}}}{K_{\text{ball}}} = \frac{33724\,\text{J}}{90.808\,\text{J}} = 371.38$$

The car has 371.38 times as much kinetic energy as the ball. ◀

Conservation of Energy

Energy is a conserved quantity, so the total amount of energy in the universe is constant. Another way to say this is "energy can neither be created nor destroyed" but energy can change from one form to another. Every form of energy is a scalar, so we specify energy with just a number and a unit.

Work

Suppose you throw a ball straight up in the air. When the ball leaves your hand, it has a large initial velocity, so it has a large initial kinetic energy. At the top of its trajectory the ball stops, so it has no kinetic energy. Where did the kinetic energy come from? Where did it go?

The answer to the first question is you. When you threw the ball, you exerted a force on it and you gave it kinetic energy. You did work on the ball, and energy was transferred from you to the ball. The work done equals the amount of energy transferred by the force. Work is the connection between Newton's forces and energy.

Constant Forces

The **work** done on an object by a constant force is

$$W = \vec{F}\cdot\Delta\vec{x} \tag{6.4a}$$

$$= F\Delta x\cos\theta \tag{6.4b}$$

where F and Δx are the magnitudes of the force and displacement vectors, and θ is the angle between those vectors. The metric unit of work is the Joule.

As the following example shows, the results from the previous chapters on kinematics and dynamics are still useful.

Example 6.4 *Pitching isn't that much Work*

How much work does a pitcher do to accelerate a baseball from rest to $100\,\mathrm{mi/h}$ over a distance of $2\,\mathrm{m}$? Assume the pitcher exerts a constant horizontal force on the ball.

Solution. As we saw in the previous chapter, the pitcher must exert a net force of $70.945\,\mathrm{N}$ on the baseball.

$$F = (0.142\,\mathrm{kg})\frac{\left(44.704\frac{\mathrm{m}}{\mathrm{s}}\right)^2}{2\,(2\,\mathrm{m})} = 70.945\frac{\mathrm{m}}{\mathrm{s}^2}\,\mathrm{kg}$$

The pitcher pushes the ball in the same direction as it moves, so the angle between the force and displacement vectors is $0°$. Use Eq. (6.4b) to create and Evaluate an expression for the work.

$$W = (70.945\,\mathrm{N})\,(2\,\mathrm{m})\cos 0° = 141.89\,\mathrm{N\,m}$$

The pitcher does $141.89\,\mathrm{J}$ of work to accelerate the baseball. ◀

Work has some interesting properties.

1. No Force $\Longrightarrow$ no Work

   If you didn't push on the object, then you did no work, no matter how far the object moves.

2. No Displacement $\Longrightarrow$ no Work

   If the object you pushed doesn't end up somewhere else, then you did no work no matter how hard you pushed.

3. Force perpendicular to Displacement $\Longrightarrow$ no Work

   If you didn't push along the object's motion, then you did no work no matter how hard you pushed.

You could push on a wall as hard as you can for as long as you can, but unless the wall moves you did no work. You could hold up a 200 pound weight while sliding on ice, but you did no work because the upward force you exert is perpendicular to your horizontal displacement. To do work, the force must have a component along the displacement, and the displacement must not be zero.

Work can be positive, negative, or zero. If the angle between the force and displacement vectors is between $\pm 90°$, then the work is positive and energy was added to the object. If the angle is between $+90°$ and $+270°$, then the work is negative and energy was removed. If the angle between the force and displacement vectors equals $\pm 90°$, then no work was done and no energy was transferred.

Let's see how work can remove energy from an object.

Example 6.5 *Catching is a lot of Work*

How much work does a catcher do to stop a baseball moving at $100\,\mathrm{mi/\,h}$ over a distance of $2.5\,\mathrm{cm}$? Assume the catcher exerts a constant horizontal force.

Solution. Create and Evaluate the following expression for the magnitude of the force exerted by the catcher.

$$F = (0.142\,\mathrm{kg})\,\frac{\left(44.704\dfrac{\mathrm{m}}{\mathrm{s}}\right)^2}{2\,(0.025\,\mathrm{m})} = 5675.6\frac{\mathrm{m}}{\mathrm{s}^2}\,\mathrm{kg}$$

To catch the ball, the catcher must push in the direction opposite the ball's velocity, so the angle between the force and displacement vectors is $180°$. Use Eq. (6.4b) to create and Evaluate an expression for the work done by the catcher.

$$W = \left(5675.6\frac{\mathrm{m}}{\mathrm{s}^2}\,\mathrm{kg}\right)(0.025\,\mathrm{m})\cos 180° = -141.89\frac{\mathrm{m}^2}{\mathrm{s}^2}\,\mathrm{kg}$$

To stop the ball, the catcher has to remove the same amount of energy ($141.89\,\mathrm{J}$) the pitcher gave it. ◀

When the force in Eqs. (6.4) is the net force, W is the total work done by all the forces. If more than one force acts on an object, the total work is the sum of the work done by each force.

$$W = W_1 + W_2 + W_3 + \cdots \tag{6.5}$$

Since work is a scalar, this is a simple algebraic sum.

For forces in more than 1-dimension, it is convenient to use the dot product to calculate the work. The work done is the dot product of the force and displacement vectors.

$$W = \vec{F}\cdot\Delta\vec{r} \tag{6.6a}$$

$$= F_x\Delta x + F_y\Delta y + F_z\Delta z \tag{6.6b}$$

You can also use the dot product to calculate the work in 1-dimension (where $F_y = F_z = 0$ and $F_x = F\cos\theta$).

Example 6.6 *What's our constant vector, Victor?*

A constant force $\vec{F} = 6\,\mathrm{N}\widehat{x} + 8\,\mathrm{N}\widehat{y} - 3\,\mathrm{N}\widehat{z}$ pushes an object from the origin to a final position of $\vec{r}_f = 3\,\mathrm{m}\,\widehat{x} + 4\,\mathrm{m}\,\widehat{y} - 2\,\mathrm{m}\,\widehat{z}$. How much work does the force do?

Solution. The displacement vector is $\Delta\vec{r} = \vec{r}_f - \vec{r}_i = \vec{r}_f$ since the object starts at the origin. Use Eq. (6.6a) to create and Evaluate the following expression for the work.

$$W = \begin{Bmatrix} 6\,\mathrm{N} \\ 8\,\mathrm{N} \\ -3\,\mathrm{N} \end{Bmatrix} \cdot \begin{Bmatrix} 3\,\mathrm{m} \\ 4\,\mathrm{m} \\ -2\,\mathrm{m} \end{Bmatrix} = 56\,\mathrm{N\,m}$$

The force does $56\,\mathrm{J}$ of work. ◀

Variable Forces

The work done on an object by a variable force is

$$W = \int \vec{F} \cdot d\vec{r} \tag{6.7a}$$

$$= \int_{x_0}^{x_f} F_x\, dx + \int_{y_0}^{y_f} F_y\, dy + \int_{z_0}^{z_f} F_z\, dz \tag{6.7b}$$

where each integral is over the appropriate component of the displacement. The dot product ensures that only the components of the force along a given direction contribute to the work. You can also calculate the work done by a varying force with

$$W = \vec{F}_{\text{ave}} \cdot \Delta\vec{r} \tag{6.8}$$

where $\vec{F}_{\text{ave}}$ is the average value of the force during the displacement. In 1-dimension, Eqs. (6.7) reduce to

$$W = \int_{x_0}^{x_f} F_x\, dx \tag{6.9a}$$

$$= \int_{x_0}^{x_f} F \cos\theta\, dx \tag{6.9b}$$

where θ is the angle between the force and the x-axis.

Example 6.7 *What's our varying vector, Victor?*

A variable force $\vec{F} = 2x^2\,\mathrm{N}\,\hat{x} + 6\sqrt{y}\,\mathrm{N}\,\hat{y} + 3z\,\mathrm{N}\,\hat{z}$ pushes an object from the origin to a final position of $\vec{r}_f = 3\,\mathrm{m}\,\hat{x} + 4\,\mathrm{m}\,\hat{y} - 2\,\mathrm{m}\,\hat{z}$. How much work does the force do?

Solution. Use Eq. (6.7b) to create the following expression for the work without units, and then apply Evaluate to it.

$$W = 2\,\mathrm{N\,m}\int_0^3 x^2 dx + 6\,\mathrm{N\,m}\int_0^4 \sqrt{y}\,dy + 3\,\mathrm{N\,m}\int_0^{-2} z dz = 56\,\mathrm{N\,m}$$

You can also use Eq. (6.7a) and Evaluate to calculate the work with vectors.

$$W = \begin{Bmatrix} \int_0^3 (2x^2\,\mathrm{N})\, dx \\ \int_0^4 (6\sqrt{y}\,\mathrm{N})\, dy \\ \int_0^{-2} (3z\,\mathrm{N})\, dz \end{Bmatrix} \cdot \begin{Bmatrix} 1\,\mathrm{m} \\ 1\,\mathrm{m} \\ 1\,\mathrm{m} \end{Bmatrix} \mathrm{m} = 56\,\mathrm{N\,m}$$

The force does 56 J of work, which is the same as in the previous example. ◄

Example 6.8 *The spring has sprung*

A slinky with a spring constant of $6\,\mathrm{N/\,m}$ hangs so that its unstretched length of $0.4\,\mathrm{m}$ runs along the y-axis. One end is attached to the wall and a mass that is free to slide along the x-axis is attached to the other end. How much work must you do to push the mass from $x = 1\,\mathrm{m}$ to $x = 5\,\mathrm{m}$?

Solution. The horizontal component of the force exerted on the mass by the spring is

$$F_x = -kx\left(1 - \left(1 + \frac{x^2}{L^2}\right)^{-1/2}\right)$$

where k is the spring constant and L is the unstretched length. To push the mass away from the origin along the x-axis, you must oppose this force. Using Eq. (6.9a), create and Evaluate the following expression for the work you do on the mass.

$$W = \int_1^5 6x\left(1 - \left(1 + \tfrac{x^2}{0.4^2}\right)^{-1/2}\right) dx = 62.547$$

It takes 62.547 J of work to push the mass from $x = 1\,\mathrm{m}$ to $x = 5\,\mathrm{m}$. ◂

The Work-Energy Theorem

The Work-Energy Theorem relates the total work done by the net force acting on an object to the resulting change in the object's kinetic energy. The total work done on the object equals the change in the object's kinetic energy.

$$W = \tfrac{1}{2}mv_f^2 - \tfrac{1}{2}mv_0^2 \tag{6.10}$$

The total work is the work done by the net force. How you calculate the left-hand side of Eq. (6.10) depends on whether or not the force is constant, but the right-hand side is always the change in kinetic energy. Keep in mind that while individual applied forces can do work, the Work-Energy Theorem relates the change in kinetic energy to the ***total*** work done by the ***net*** force.

Example 6.9 *Pitching is this much Work*
Verify that the work done by the pitcher in Example 6.4 equals the baseball's change in kinetic energy.
Solution. The ball starts from rest, so its initial kinetic energy is zero. Create and Evaluate the following expression for the change in the baseball's kinetic energy.

$$\Delta K = \tfrac{1}{2}\,(0.142\,\mathrm{kg})\left(44.704\tfrac{\mathrm{m}}{\mathrm{s}}\right)^2 = 141.89\tfrac{\mathrm{m}^2}{\mathrm{s}^2}\,\mathrm{kg}$$

The Work-Energy Theorem works! ◂

Example 6.10 *The spring springs back*
After you push the object in Example 6.8 to $x = 5\,\mathrm{m}$, you release it from rest and the slinky pulls it back toward the origin. If the object has a mass of $0.75\,\mathrm{kg}$, and you ignore friction, how fast is it going when it returns to $x = 1\,\mathrm{m}$?
Solution. To pull the object from $x = 5\,\mathrm{m}$ to $x = 1\,\mathrm{m}$, the slinky does the same amount of work you did. In the absence of friction, the force exerted by the slinky is the net force. Use the Work-Energy Theorem to create the following equation, and use Solve Exact to find the velocity.

$$141.89\,\mathrm{J} = \tfrac{1}{2}\,(0.75\,\mathrm{kg})\,v^2, \text{ Solution is: } -19.452\frac{\mathrm{m}}{\mathrm{s}}, 19.452\frac{\mathrm{m}}{\mathrm{s}}$$

The object is moving to the left at a speed of $19.452\,\mathrm{m/s}$. ◂

Potential Energy

We started our discussion of energy by explaining how the ball you threw acquired its initial kinetic energy. You did positive work on the ball and increased its kinetic energy. The total work you did on the ball equaled the change in the ball's kinetic energy.

Now let's answer the second question. When the ball stops at the top of its trajectory, where did all its kinetic energy go? While the ball was going up, gravity did negative work on the ball and decreased the ball's kinetic energy until it was zero. To understand what happened to the kinetic energy, we need to think about forces.

There are two kinds of forces, conservative forces and non-conservative forces. Both kinds of forces can do work, and the total work is the sum of the work done by both.

$$W = W_{\mathrm{c}} + W_{\mathrm{nc}} \tag{6.11}$$

Both kinds of forces can change an object's kinetic energy, but there is a significant difference. When a conservative force acts, the work done is stored as **potential energy** (PE). When a non-conservative force acts, the work done is not stored but converted into other forms of energy.

Potential energy is the energy of position or location. Potential energy comes from a conservative force doing work on an object so that the energy is stored and available to be converted into kinetic energy. The "potential" in potential energy is the potential to acquire kinetic energy.

The work done by a conservative force is recoverable, so you can get back any work you put in. The work over a closed path is zero for conservative forces. Two common examples of conservative forces are gravity and spring forces.

Example 6.11 *You call this a work out?*
During a vigorous exercise session, you repeatedly lift and lower a $30\,\mathrm{kg}$ mass through a vertical distance of $1.5\,\mathrm{m}$. How much work does gravity do when you lift the mass? Lower it? How much work does gravity do during one complete lifting and lowering?
Solution. When you lift the mass, gravity exerts a constant downward force mg over a distance y in the opposite direction of the weight's displacement. Use Eq. (6.4a) to create and Evaluate an expression for the work done by gravity.

$$W_{\mathrm{up}} = (30\,\mathrm{kg})\left(9.8067\frac{\mathrm{m}}{\mathrm{s}^2}\right)(1.5\,\mathrm{m})\cos 180° = -441.3\frac{\mathrm{m}^2}{\mathrm{s}^2}\,\mathrm{kg}$$

When you lower the weight, gravity exerts a downward force mg over the same distance but in the same direction of the weight's displacement.

$$W_{\mathrm{down}} = (30\,\mathrm{kg})\left(9.8067\frac{\mathrm{m}}{\mathrm{s}^2}\right)(1.5\,\mathrm{m})\cos 0° = 441.3\frac{\mathrm{m}^2}{\mathrm{s}^2}\,\mathrm{kg}$$

The total work done by gravity during one complete lifting and lowering is the sum.

$$W = W_{\mathrm{up}} + W_{\mathrm{down}} = -441.3\,\mathrm{J} + 441.3\,\mathrm{J} = 0$$

Gravity did no net work over the weight's round trip. Gravity is a conservative force. ◄

The work done by a conservative force is related to the change in potential energy.

$$W_c = -\Delta U \tag{6.12}$$

When a conservative force does positive work, the object's potential energy decreases. When a conservative force does negative work, the object's potential energy increases.

We can now say what happened to the kinetic energy you gave the ball. On the way up, gravity did negative work on the ball and converted its kinetic energy into potential energy. At its highest point, the ball's kinetic energy had been completely converted into potential. On the way down, gravity did positive work on the ball and converted its potential energy into kinetic energy.

Every conservative force has a potential energy associated with it. For a constant conservative force, you can use Eq. (6.4a) to calculate the change in potential energy.

$$\Delta U = -\overrightarrow{F} \cdot \Delta \overrightarrow{x} \tag{6.13}$$

For a varying conservative force, use Eq. (6.7a) to calculate the change in PE.

$$\Delta U = -\int \overrightarrow{F} \cdot d\overrightarrow{r} \tag{6.14}$$

In both cases, the change in potential energy is the negative of the work done by the conservative force.

In 1-dimension, if you know the force and want the potential energy as a function of position, integrate the force along the displacement.

$$U(x) - U(x_0) = -\int_{x_0}^{x} F(x)\, dx \tag{6.15}$$

Only changes in potential energy are physically meaningful, so the reference point x_0 is usually chosen so that $U(x_0) = 0$. If you know the potential energy and you want the force as a function of position, take the derivative of the potential energy with respect to the position.

$$F(x) = -\frac{dU}{dx} \tag{6.16}$$

For vertical problems, replace the "x" with "y" and for radial problems, replace the "x" with "r".

The work done by a non-conservative force cannot be expressed as potential energy and it is not recoverable. The work done by a non-conservative force is usually dissipated as heat energy. Two common examples of non-conservative forces are friction and air resistance.

Example 6.12 *Now this a work out!*
During a vigorous exercise session, you push a 30 kg sled a horizontal distance of 4 m to the right, and then you push it back to its original position. The floor is level and the coefficient of kinetic friction is 0.375. How much work does friction do when you push the sled to the right? To the left? How much total work did friction do?
Solution. As you push the sled to the right, friction exerts a constant force $f_k = \mu_k mg$ in the opposite direction of the sled's displacement. Use Eq. (6.4a) to create and Evaluate an expression for the work you did pushing the sled to the right.

$$W_{\text{right}} = (0.375)\,(30\,\text{kg})\left(9.8067\frac{\text{m}}{\text{s}^2}\right)(4\,\text{m})\cos 180° = -441.3\frac{\text{m}^2}{\text{s}^2}\,\text{kg}$$

As you push the sled to the left, friction exerts a force $f_k = \mu_k mg$ which is still in the opposite direction of the sled's displacement.

$$W_{\text{left}} = (0.375)\,(30\,\text{kg})\left(9.8067\frac{\text{m}}{\text{s}^2}\right)(4\,\text{m})\cos 180^\circ = -441.3\frac{\text{m}^2}{\text{s}^2}\,\text{kg}$$

The total work done by friction is the sum of the two.

$$W = W_{\text{right}} + W_{\text{left}} = -441.3\,\text{J} + (-441.3\,\text{J}) = -882.6\,\text{J}$$

Friction did $-882.6\,\text{J}$ of work against you while you pushed the sled back and forth along a closed path. Friction is a non-conservative force. ◀

Here is a simple way to remember the difference between the two kinds of forces:

Conservative forces store energy.
Non-conservative forces dissipate energy.

As you will soon see, there is a reason the two types of forces are given these names.

Mechanical Energy is Conserved

The **mechanical energy** is defined as the sum of the kinetic energy and the potential energy.

$$E = \tfrac{1}{2}mv^2 + U \tag{6.17}$$

The kinetic energy is always $\frac{1}{2}mv^2$, and the potential energy depends on the situation.

We can combine three previous results, namely $W = W_c + W_{nc}$, $W = \Delta K$, and $W_c = -\Delta U$ into one equation that defines the conditions for mechanical energy conservation.

$$W_{nc} = \Delta E \tag{6.18a}$$

$$= \Delta KE + \Delta PE \tag{6.18b}$$

When no non-conservative forces act, $W_{nc} = 0$ and the change in mechanical energy is zero. Mechanical energy is conserved when only conservative forces act. That's why we call them "*conservative*" forces. For an object moving from point-1 to point-2,

$$\tfrac{1}{2}mv_1^2 + U_1 = \tfrac{1}{2}mv_2^2 + U_2 \tag{6.19}$$

the sum of the object's kinetic energy and potential energy at point 1 equals the sum at point 2.

Example 6.13 *What goes up, revisited*
You throw a baseball straight up in the air with an initial speed of $22\,\text{m/s}$ (about $49\,\text{mi/h}$) from a height of $2\,\text{m}$. How high does the ball go? Ignore air resistance.
Solution. When the you release the ball, it has both kinetic ($\frac{1}{2}mv_1^2$) and gravitational potential energy (mgy_1). Use Eq. (6.17) and Evaluate to calculate the ball's initial mechanical energy.

$$\tfrac{1}{2}\,(0.142\,\text{kg})\left(22\frac{\text{m}}{\text{s}}\right)^2 + (0.142\,\text{kg})\left(9.8067\frac{\text{m}}{\text{s}^2}\right)(2\,\text{m}) = 37.149\frac{\text{m}^2}{\text{s}^2}\,\text{kg}$$

At its maximum height, the ball stops moving so it has no kinetic energy. Its final mechanical energy is all gravitational potential energy. Use Eq. (6.19) to create the following expression equating the ball's initial and final mechanical energy, and use Solve Exact to find the ball's final height.

$$37.149\,\mathrm{J} = (0.142\,\mathrm{kg})\left(9.8067\frac{\mathrm{m}}{\mathrm{s}^2}\right)y,\text{ Solution is: }26.677\,\mathrm{m}$$

The balls reaches a maximum height of $26.677\,\mathrm{m}$ above the ground (which is the same $24.677\,\mathrm{m}$ above its release point we found in Example 2.6). ◀

Example 6.14 *When I get to the bottom I go back to the top of the slide...*

A $25\,\mathrm{kg}$ child runs onto a water slide with an initial speed of $5\,\mathrm{m/\,s}$. The top of the slide is $7\,\mathrm{m}$ above the bottom. If we ignore friction, how fast is the child moving at the bottom of the slide?

Solution. Because the child ran onto the slide, initially she has both kinetic and potential energy. Use Eq. (6.17) and the Evaluate command to calculate her initial mechanical energy.

$$\tfrac{1}{2}(25\,\mathrm{kg})\left(5\tfrac{\mathrm{m}}{\mathrm{s}}\right)^2 + (25\,\mathrm{kg})\left(9.8067\frac{\mathrm{m}}{\mathrm{s}^2}\right)(7\,\mathrm{m}) = 2028.7\frac{\mathrm{m}^2}{\mathrm{s}^2}\,\mathrm{kg}$$

At the bottom of the slide, the child's mechanical energy is all kinetic. Use Eq. (6.19) to create the following expression equating the child's initial and final mechanical energy, and use Solve Exact to find her final speed.

$$2028.7\,\mathrm{J} = \tfrac{1}{2}(25\,\mathrm{kg})\,v^2,\text{ Solution is: }-12.740\frac{\mathrm{m}}{\mathrm{s}},\,12.740\frac{\mathrm{m}}{\mathrm{s}}$$

If we ignore friction, the child is moving at a speed of $12.740\,\mathrm{m/\,s}$. ◀

For 1-dimensional motion, we can use Eq. (6.17) to find the position as a function of time. We can apply the 1-d definition of velocity from Chapter 2 to Eq. (6.17).

$$\int_{x_0}^{x}\frac{dx}{\sqrt{E-U(x)}} = \pm\sqrt{\frac{2}{m}}\,t \qquad (6.20)$$

If *SNB* can do the integral, then Eq. (6.20) will produce an equation you can Solve Exactly for the position as a function of time. Notice this is similar to Eq. (5.12), with $\int F(x)\,dx = -U(x) +$ constant. The constant is the mechanical energy.

Example 6.15 *A potentially periodic perturbation*

Find an object's position as a function of time for $U(x) = \frac{1}{2}kx^2 + \frac{1}{2}\beta/x^2$. Express your answer both in terms of the energy and the initial position.

Solution. *SNB* cannot do the integral from Eq. (6.20) for this potential energy without your help. Start by telling *SNB* to assume the variables are positive. Then create an expression for the integral, leave the insertion point anywhere in it, and choose Calculus + Change of Variable... from the Compute menu.

Enter $u = x^2$ in the dialog box and click OK. Then Evaluate the resulting integral.

$$\int \frac{dx}{\sqrt{E - \frac{1}{2}kx^2 - \frac{1}{2}\frac{\beta}{x^2}}} = \int \frac{1}{2}\frac{\sqrt{2}}{\sqrt{-ku^2 + 2Eu - \beta}}\,du = -\frac{1}{2}\frac{\sqrt{2}}{\sqrt{k}} \arcsin \frac{E - ku}{\sqrt{E^2 - k\beta}}$$

Now replace the u with x^2 by-hand, set the result equal to $\sqrt{\frac{2}{m}}t$ and use Solve Exact to find x.

$$\sqrt{\frac{2}{m}}t = -\frac{1}{\sqrt{2k}} \arcsin \frac{E - kx^2}{\sqrt{E^2 - k\beta}}, \text{ Solution is: } \frac{1}{\sqrt{k}}\sqrt{E + \left(\sin 2\frac{\sqrt{k}}{\sqrt{m}}t\right)\sqrt{-k\beta + E^2}}$$

To relate the energy to the initial position, Substitute $t = 0$ into this expression and use Solve Exact to solve for E.

$$x_0 = \left[\sqrt{\frac{1}{k}\left(E + \sqrt{E^2 - k\beta}\sin 2\sqrt{\frac{k}{m}}t\right)}\right]_{t=0}, \text{ Solution is: } kx_0^2$$

Now Substitute $E = kx_0^2$ into the expression for the position.

$$x = \left[\sqrt{\frac{1}{k}\left(E + \sqrt{E^2 - k\beta}\,\sin 2\sqrt{\frac{k}{m}}t\right)}\right]_{E=kx_0^2}$$

$$= \frac{1}{\sqrt{k}}\sqrt{\left(\sin 2\frac{\sqrt{k}}{\sqrt{m}}t\right)\sqrt{k^2x_0^4 - k\beta} + kx_0^2}$$

A little editing by-hand produces our position as a function of time.

$$x(t) = x_0\sqrt{1 + \sqrt{1 - \frac{\beta}{kx_0^4}}\left(\sin 2\sqrt{\frac{k}{m}}t\right)}$$

Figure 6.1 shows $x(t)$ from $t = 0$ to $t = 2\pi$ with $x_0 = 2$ and $k = m = \beta = 1$.

The object oscillates between $x_{\min} = 2.805\,9$ at $t = \frac{1}{4}\pi$ and $x_{\max} = 0.356\,39$ at $t = \frac{3}{4}\pi$. It completes a cycle every π seconds.

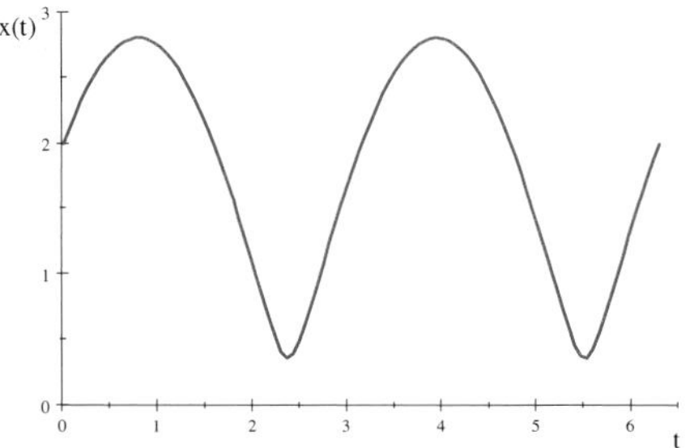

Figure 6.1 $x(t)$ for Example 6.15

◀

A Complete Bookkeeping

When non-conservative forces act on an object, its mechanical energy changes. The object's total energy is still conserved, but its mechanical energy is not. For an object moving from point-1 to point-2, this bookkeeping takes the form

$$\tfrac{1}{2}mv_1^2 + U_1 + W_{\text{nc}} = \tfrac{1}{2}mv_2^2 + U_2 \tag{6.21}$$

where W_{nc} is the energy lost to non-conservative forces along the way.

Example 6.16 *Not quite so high*
You throw yet another ball straight up in the air with an initial speed of $22\,\text{m/s}$ (about $49\,\text{mi/h}$) from a height of $2\,\text{m}$. On the way up, the ball experiences an average air resistance force of $0.27\,\text{N}$. How high does the ball go?
Solution. Air resistance opposes the ball's motion, so the angle between the force (down) and the ball's displacement (up) is $180°$. Let's use the average force and treat the air resistance as a constant. Use Eq. (6.4b) to create and Evaluate an expression for the work done by the non-conservative air resistance in terms in the unknown final height.

$$W_{\text{nc}} = F_{\text{ave}}\, y \cos\theta_{Fy} = (0.27\,\text{N})\, y \cos 180° = -0.27\,(\text{N})\, y$$

At its maximum height, the ball stops moving so its final mechanical energy is all potential energy. Use Eq. (6.21) to create the following expression, and use Solve Exact to find the ball's final height.

$$37.149\,\text{J} + (-0.27\,(\text{N})\, y) = (0.142\,\text{kg})\left(9.8067\frac{\text{m}}{\text{s}^2}\right) y, \text{ Solution is: } 22.345\,\text{m}$$

Air resistance reduced the ball's final height by over 19%. ◄

Even a really good slide has some friction.

Example 6.17 *...where I stop and I turn and I go for a ride*
When the child from Example 6.14 gets to the bottom of the slide, her actual speed there is $10\,\text{m/s}$. How much friction acted on the child?
Solution. According to Eq. (6.21), the energy conservation equation for the child is

$$\tfrac{1}{2}mv_1^2 + mgy_1 - f_k\, d = \tfrac{1}{2}mv_2^2$$

where $-f_k\, d$ is the work done by friction and d is the length of the slide. The child's initial mechanical energy is still 2028.7 J, so let's Evaluate her final kinetic energy.

$$\tfrac{1}{2}(25\,\text{kg})\left(10\frac{\text{m}}{\text{s}}\right)^2 = 1250\frac{\text{m}^2}{\text{s}^2}\,\text{kg}$$

Create the following expression, and use Solve Exact to find the force of friction.

$$2028.7\,\text{J} - f_k\,(25\,\text{m}) = 1250\,\text{J}, \text{ Solution is: } 31.148\frac{\text{m}}{\text{s}^2}\,\text{kg}$$

Friction exerts a force of $31.148\,\text{N}$ on the child. ◄

Example 6.18 *Slip sliding away*
A $0.17\,\mathrm{kg}$ hockey puck slides across a frictionless surface with a velocity of $+8\,\mathrm{m/s}$. At the origin, it encounters a ramp that exerts a position-dependent frictional force

$$f_k = \mu_0 mg\cos\theta\left(1 - e^{-x}\right)$$

where $\theta = 20^\circ$ is the ramp angle. How far along the ramp does the puck slide until it stops? Take $\mu_0 = 0.30$. (See Figure 6.5 in the Problems.)
Solution. The puck's initial energy is all kinetic, and its final energy is all gravitational potential. According to Eq. (6.21), the energy conservation equation for the puck is

$$\tfrac{1}{2}mv^2 + W_{\mathrm{nc}} = mgy$$

where W_{nc} is the work done by the varying frictional force.

Let's Evaluate the following expression for the work done by friction, which is not a constant force here.

$$W_{\mathrm{nc}} = \int \vec{f}_k \cdot d\vec{x} = -\mu_0 mg\cos\theta \int_0^x \left(1 - e^{-x}\right) dx = -gm\mu_0\left(\cos\theta\right)\left(x + e^{-x} - 1\right)$$

The exponential term suggests this problem will not have a simple algebraic solution. So let's solve it numerically, without units. Use Eq. (6.3) and Evaluate to calculate the initial kinetic energy.

$$\tfrac{1}{2}mv^2 = \tfrac{1}{2}\left(0.17\right)\left(8\right)^2 = 5.44$$

Let's apply Evaluate Numerically to the prefactor in the expression for W_{nc}.

$$\mu_0 mg\cos\theta = \left(0.30\right)\left(0.17\right)\left(9.8067\right)\cos 20.0^\circ = 0.469\,98$$

The final gravitational potential energy is $U_f = mgy = mgx\sin\theta$. Use Evaluate Numerically to calculate the potential energy as a function of the unknown distance.

$$U_f = \left(0.17\right)\left(9.8067\right)x\sin 20^\circ = 0.570\,20x$$

Use Eq. (6.21) and Solve Numeric to calculate the distance the puck slides.

$5.44 + \left(-0.469\,98\left(x + e^{-x} - 1\right)\right) = 0.570\,20x$, Solution is: $\{[x = 5.680\,1]\}$

The puck slides $5.680\,1\,\mathrm{m}$ along the ramp.

It's sometimes useful to look at the general equation, even if you can't solve it exactly.

$$\tfrac{1}{2}mv^2 - \mu_0 mg\cos\theta\left(x + e^{-x} - 1\right) = mgx\sin\theta$$

Notice the mass appears in every term. Let's use Solve Numeric on the energy conservation equation without the mass.

$\tfrac{1}{2}\left(8\right)^2 - \left(0.30\right)\left(9.8067\right)\left(\cos 20.0^\circ\right)\left(x + e^{-x} - 1\right) = \left(9.8067\right)\left(\sin 20.0^\circ\right)x$,
Solution is: $\{[x = 5.680\,2]\}$

The small difference is due to rounding errors. The distance the puck slides does not depend on its mass. ◄

Conservation of Momentum

There is an important and useful consequence to Eq. (6.2). When there is no net force, the change in momentum is zero and the total amount of momentum is constant. The total momentum is conserved. Since momentum is a vector, both its magnitude and direction are conserved.

Collisions in 1-Dimension

A collision is simply two objects crashing into one another. When they do, they exert a force on each other and each object experiences a change in its individual momentum and kinetic energy. Per Newton's 3rd Law, the forces have equal magnitude and opposite direction. So the vector sum of the forces is zero. If we consider the objects collectively as a system, the net force is zero and the total momentum is conserved. Every collision conserves the total momentum.

There are two kinds of collisions, elastic and inelastic. In both cases, the total momentum – the vector sum of the two momenta – stays constant.

Inelastic Collisions

Inelastic Collisions conserve only the total momentum. The total momentum before the collision equals the total momentum after the collision. Kinetic energy is not conserved and some is lost, converted into other forms of energy. The conservation equation for the momentum of two objects in a 1-dimensional inelastic collision is

$$m_1 v_1 + m_2 v_2 = m_1 v_1' + m_2 v_2' \tag{6.22}$$

where the "unprimed" variables indicate before the collision and "primed" variables indicate after the collision.

In a completely inelastic collision, the two objects stick together resulting in the maximum loss of total kinetic energy. In this case, the final mass is the sum of the individual masses.

$$m_1 v_1 + m_2 v_2 = (m_1 + m_2)\, v' \tag{6.23}$$

Even when the loss of kinetic energy is maximal, the final velocity of the objects is not necessarily zero.

Example 6.19 *A 1-dimensional inelastic collision*
A $1500\,\mathrm{kg}$ car moving to the right at $5.0\,\mathrm{m/s}$ crashes head-on into a $2500\,\mathrm{kg}$ van moving left at $1.8\,\mathrm{m/s}$. If the two vehicles stick together, what is their final speed?
Solution. Use Eq. (6.23) with $m_1 = 1500\,\mathrm{kg}$, $v_1 = 5.0\,\mathrm{m/s}$, $m_2 = 2500\,\mathrm{kg}$, and $v_2 = -1.8\,\mathrm{m/s}$ to create an equation conserving momentum for the collision. Place the insertion point anywhere in the equation and click the Solve Exact button.

$$(1500\,\mathrm{kg})\left(5.0\frac{\mathrm{m}}{\mathrm{s}}\right) + (2500\,\mathrm{kg})\left(-1.8\frac{\mathrm{m}}{\mathrm{s}}\right) = (1500\,\mathrm{kg} + 2500\,\mathrm{kg})\, v',$$

Solution is: $0.75\dfrac{\mathrm{m}}{\mathrm{s}}$

After the collision, the two vehicles move in the car's original (positive) direction with a speed of $v' = 0.75\,\mathrm{m/s}$. ◀

Elastic Collisions

Elastic Collisions conserve both the total momentum and the total kinetic energy. The total amount of each is unchanged. Here are the general conservation equations for the momentum and kinetic energy.

$$m_1 v_1 + m_2 v_2 = m_1 v_1' + m_2 v_2' \qquad (6.24a)$$

$$\tfrac{1}{2} m_1 v_1^2 + \tfrac{1}{2} m_2 v_2^2 = \tfrac{1}{2} m_1 \left(v_1'\right)^2 + \tfrac{1}{2} m_2 \left(v_2'\right)^2 \qquad (6.24b)$$

In a 1-d elastic collision, the total kinetic energy is conserved and the relative velocity of the two objects is reversed.

$$v_1 - v_2 = -\left(v_1' - v_2'\right) \qquad (6.25)$$

For example, if the two objects come together at a relative velocity of $20\,\mathrm{m/s}$ before the collision, they move apart at a relative velocity of $-20\,\mathrm{m/s}$ after the collision.

Example 6.20 *A 1-dimensional elastic collision*
On a frictionless surface, a puck with a mass of $2.0\,\mathrm{kg}$ and velocity of $22\,\mathrm{m/s}$ has an elastic collision with another puck at rest. After the collision, the first puck continues to move at one-quarter its original velocity. What is the mass and final velocity of the second puck?
Solution. Use the conservation equations for momentum and kinetic energy with $m_1 = 2.0\,\mathrm{kg}$, $v_1 = 22\,\mathrm{m/s}$, $v_1' = \frac{1}{4}(22\,\mathrm{m/s})$, and $v_2 = 0$. Place each equation in a row of a 2×1 matrix, leave the insertion point anywhere in the matrix, and click the Solve Exact button.

$$\begin{bmatrix} (2.0\,\mathrm{kg})(22\,\mathrm{m/s}) = (2.0\,\mathrm{kg})\frac{1}{4}(22\,\mathrm{m/s}) + m_2 v_2' \\ 22\,\mathrm{m/s} = -\left(\frac{1}{4}(22\,\mathrm{m/s}) - v_2'\right) \end{bmatrix},$$

Solution is: $\left[v_2' = 27.5\frac{\mathrm{m}}{\mathrm{s}^{1.0}}, m_2 = 1.2\,\mathrm{kg}\right]$

The second puck has a mass of $1.2\,\mathrm{kg}$ and moves off with a speed of $27.5\,\mathrm{m/s}$. ◀

Note The "prime" symbol is the same one we met in Chapter 1. It denotes a derivative in the context of differential equations but here it's just a label. You can create it with the APOSTROPHE key on your keyboard or with the Miscellaneous Symbols palette of the Symbols Panels.

You can use Solve Exact on Eqs. (6.24) to find general expressions for the two final velocities. Enter v_1', v_2' in the "Variable(s) to Solve for" box. Then Factor the first solution in-place.

$$\begin{bmatrix} m_1 v_1 + m_2 v_2 = m_1 v_1' + m_2 v_2' \\ \frac{1}{2} m_1 v_1^2 + \frac{1}{2} m_2 v_2^2 = \frac{1}{2} m_1 \left(v_1'\right)^2 + \frac{1}{2} m_2 \left(v_2'\right)^2 \end{bmatrix}, \text{ Solution is:}$$

$$\left[v_1' = \frac{m_1 v_1 - m_2 v_1 + 2 m_2 v_2}{m_1 + m_2}, v_2' = \frac{2 m_1 v_1 - m_1 v_2 + m_2 v_2}{m_1 + m_2}\right], \left[v_1' = v_1, v_2' = v_2\right]$$

The second solution is the no-collision, nothing-happens solution, which does indeed conserve momentum and kinetic energy but is not very interesting. The first solution gives the two final velocities in terms of the initial velocities and the masses.

The momentum conservation equations take on a much simpler form with the same solution if we use Eq. (6.25) instead of Eq. (6.24b).

$$\begin{bmatrix} m_1 v_1 + m_2 v_2 = m_1 v_1' + m_2 v_2' \\ v_1 - v_2 = -\left(v_1' - v_2'\right) \end{bmatrix}, \text{ Solution is: }$$

$$\left[v_1' = \frac{m_1 v_1 - m_2 v_1 + 2m_2 v_2}{m_1 + m_2}, v_2' = \frac{2m_1 v_1 - m_1 v_2 + m_2 v_2}{m_1 + m_2} \right]$$

Remember, Eq. (6.25) is valid for 1-dimensional, elastic collisions only.

We can write the general expressions for 1-d elastic collisions in a more standard form.

$$v_1' = \frac{m_1 - m_2}{m_1 + m_2} v_1 + \frac{2m_2}{m_1 + m_2} v_2 \tag{6.26a}$$

$$v_2' = \frac{2m_1}{m_1 + m_2} v_1 - \frac{m_1 - m_2}{m_1 + m_2} v_2 \tag{6.26b}$$

Usually, you won't need this general solution because you can use *SNB* to solve the conservation equations for your particular problem. But Eqs. (6.26) have some interesting special cases that are worth investigating.

Suppose the target is at rest ($v_2 = 0$) and the incident object is moving to the right (so v_1 is positive). Let's Substitute $v_2 = 0$ into Eqs. (6.26) to get the final velocity of each object.

$$v_1' = \left[\frac{m_1 - m_2}{m_1 + m_2} v_1 + \frac{2m_2}{m_1 + m_2} v_2 \right]_{v_2 = 0} = v_1 \frac{m_1 - m_2}{m_1 + m_2}$$

$$v_2' = \left[\frac{2m_1}{m_1 + m_2} v_1 - \frac{m_1 - m_2}{m_1 + m_2} v_2 \right]_{v_2 = 0} = 2m_1 \frac{v_1}{m_1 + m_2}$$

Since v_2' is a positive number times v_1, the target always moves off in the original direction of the incident object. The final direction of the incident object v_1' depends on which mass is larger. When the target mass is smaller than the incident mass ($m_2 < m_1$), v_1' is positive and the incident object keeps moving in its original direction. But if the target mass is larger than the incident mass ($m_2 > m_1$), then v_1' is negative and the incident object reverses direction.

When the masses are equal ($m_1 = m_2$) the two objects simply exchange velocities.

$$v_1' = \left[\frac{m_1 - m_2}{m_1 + m_2} v_1 + \frac{2m_2}{m_2 + m_1} v_2 \right]_{m_2 = m_1} = v_2$$

$$v_2' = \left[\frac{2m_1}{m_1 + m_2} v_1 - \frac{m_1 - m_2}{m_1 + m_2} v_2 \right]_{m_2 = m_1} = v_1$$

If the equal-mass target was stationary ($v_2 = 0$), then we're playing pool and the cue ball stops ($v_1' = 0$).

Consider a collision between two objects with very different masses. Let's set it up so mass–1 is much larger than mass–2 (so $m_1 \gg m_2$). When the smaller mass is at rest ($v_2 = 0$), the velocity of the massive object does not change but the small-mass object moves off in the same direction at twice that velocity!

$$v_1' \rightarrow \lim_{m_2 \rightarrow 0} \left(\frac{m_1 - m_2}{m_2 + m_1} v_1 \right) = v_1$$

$$v_2' \rightarrow \lim_{m_2 \rightarrow 0} \left(\frac{2m_1}{m_2 + m_1} v_1 \right) = 2v_1$$

When the small-mass hits a big stationary mass ($v_1 = 0$), the small-mass object bounces off with the exact opposite velocity while the massive object does not move.

$$v_1' \rightarrow \lim_{m_2 \rightarrow 0} \left(\frac{2m_2}{m_2 + m_1} v_2 \right) = 0$$

$$v_2' \rightarrow \lim_{m_2 \rightarrow 0} \left(\frac{m_2 - m_1}{m_2 + m_1} v_2 \right) = -v_2$$

Collisions in 2-Dimensions

Collisions that are not "head on" are 2-dimensional problems. In 2-d, we must conserve the total momentum by component. Each component is conserved as long as that component of the net force is zero. The initial momentum in the x-direction equals the final momentum in the x-direction, and the initial momentum in the y-direction equals the final momentum in the y-direction.

$$\begin{aligned} P_x &= P_x' \\ P_y &= P_y' \end{aligned}$$

The components of the vector $\overrightarrow{P}$ are given by

$$\begin{aligned} P_x &= P\cos\theta \\ P_y &= P\sin\theta \end{aligned}$$

where P is the vector's magnitude and θ is its direction.

All collisions conserve momentum, and the general momentum conservation equations for a 2-dimensional collision between two objects take the following form.

$$\begin{aligned} m_1 v_{1x} + m_2 v_{2x} &= m_1 v_{1x}' + m_2 v_{2x}' \\ m_1 v_{1y} + m_2 v_{2y} &= m_1 v_{1y}' + m_2 v_{2y}' \end{aligned} \tag{6.27}$$

If the two colliding masses stick together, then the collision is completely inelastic and the equations become

$$\begin{aligned} m_1 v_{1x} + m_2 v_{2x} &= (m_1 + m_2)\, v_x' \\ m_1 v_{1y} + m_2 v_{2y} &= (m_1 + m_2)\, v_y' \end{aligned} \tag{6.28}$$

where $m_1 + m_2$ is the total mass of the two objects. In all 2-d inelastic collisions, there are two conservation equations so you can only solve for two unknowns.

Example 6.21 *A 2-d inelastic collision*
A 2 kg stone with velocity $\overrightarrow{v}_1 = 11\frac{\text{m}}{\text{s}}\widehat{x}$ collides with a 10 kg blob of clay moving at $\overrightarrow{v}_2 = 3\frac{\text{m}}{\text{s}}\widehat{y}$. They collide at the origin and stick together. What is their final speed and direction?
Solution. The x-component of the initial momentum is carried by the stone.

$$P_x = (2\,\text{kg})\left(11\frac{\text{m}}{\text{s}}\right) = 22\frac{\text{m}}{\text{s}}\,\text{kg}$$

The y-component of the initial momentum is carried by the blob.

$$P_y = (10\,\text{kg})\left(3\frac{\text{m}}{\text{s}}\right) = 30\frac{\text{m}}{\text{s}}\,\text{kg}$$

The components of the final momentum are

$$\begin{aligned} P'_x &= (12\,\text{kg})\,v'\cos\theta \\ P'_y &= (12\,\text{kg})\,v'\sin\theta \end{aligned}$$

where v' is the speed and θ° is the direction of the stone-blob combination after the collision.

Put each of the two momentum conservation equations (without units) in a row of a 2×1 matrix, leave the insertion point anywhere in the matrix, and choose Solve Numeric. The θ° (with the **red** degree symbol) in the argument of the trigonometric functions ensures *SNB* returns the angle in degrees.

$$\begin{bmatrix} 22 = 12v'\cos\theta^\circ \\ 30 = 12v'\sin\theta^\circ \end{bmatrix}, \text{ Solution is: } \{[v' = 3.100\,2, \theta = 53.746]\}$$

You can also solve these equations directly by first dividing them,

$$\frac{30}{22} = \frac{12v'\sin\theta^\circ}{12v'\cos\theta^\circ} = \tan\theta^\circ$$

and then using Solve Numeric on the result.

$$\frac{30}{22} = \tan\theta^\circ, \text{ Solution is: } \{[\theta = 53.746]\}$$

This is the same as using Evaluate Numerically on the inverse-tangent function.

$$\theta = \frac{180}{\pi}\tan^{-1}\frac{30}{22} = 53.746$$

Once you know the angle, use either equation and Solve Exact to find the final velocity.

$$30 = 12v'\sin(53.746^\circ), \text{ Solution is: } 3.100\,2$$

$$22 = 12v'\cos(53.746^\circ), \text{ Solution is: } 3.100\,2$$

The stone-blob combination moves at speed $v' = 3.100\,2\,\text{m}/\text{s}$ at an angle of 53.746° relative to the positive x-axis. ◀

For a 2-d elastic collision, the total kinetic energy is also conserved so Eq. (6.24b) is a third conservation equation . Since there are three conservation equations, for 2-d elastic collisions you can solve for three unknowns.

Example 6.22 *A 2-d elastic collision*

Two rubber pucks slide along a frictionless surface and collide elastically. Puck 1 has a mass of $0.25\,\mathrm{kg}$ and an initial velocity $\overrightarrow{v}_1 = 1.45\frac{\mathrm{m}}{\mathrm{s}}\,\widehat{x}$. Puck 2 has a mass of $0.35\,\mathrm{kg}$ and an initial velocity $\overrightarrow{v}_2 = -1.35\frac{\mathrm{m}}{\mathrm{s}}\,\widehat{x}$. After the collision, Puck 2 moves in direction $\theta_2 = -137°$. Find both speeds and the direction of Puck 1 after the collision. Verify that the collision was elastic.

Solution. This is an elastic collision, so the total momentum and kinetic energy are both conserved. Let's Evaluate the x-component of the initial momentum.

$$P_x = (0.25\,\mathrm{kg})\left(1.45\frac{\mathrm{m}}{\mathrm{s}}\right) + (0.35\,\mathrm{kg})\left(-1.35\frac{\mathrm{m}}{\mathrm{s}}\right) = -0.11\frac{\mathrm{m}}{\mathrm{s}}\,\mathrm{kg}$$

Both pucks initially travel along the x-axis, so the y-component of the initial momentum is zero. Let's Evaluate the initial kinetic energy.

$$K = \tfrac{1}{2}(0.25\,\mathrm{kg})\left(1.45\frac{\mathrm{m}}{\mathrm{s}}\right)^2 + \tfrac{1}{2}(0.35\,\mathrm{kg})\left(-1.35\frac{\mathrm{m}}{\mathrm{s}}\right)^2 = 0.581\,75\frac{\mathrm{m}^2}{\mathrm{s}^2}\,\mathrm{kg}$$

The collision deflects Puck 2 in the $-\widehat{y}$ direction, so Puck 1 must deflect in the $+\widehat{y}$ direction to conserve momentum. Therefore its final direction is between $0°$ and $90°$.

Place the three conservation equations (without units) and a range for each unknown in a 6-row, 1-column matrix. Leave the insertion point anywhere in the matrix and choose Solve Numeric. The θ_1° (with the **red** degree symbol) in the argument of the trigonometric functions ensures *SNB* returns the angle in degrees.

$$\begin{bmatrix} -0.11 = (0.25)\,v_1'\cos(\theta_1^\circ) + (0.35)\,v_2'\cos(-137°) \\ 0 = (0.25)\,v_1'\sin(\theta_1^\circ) + (0.35)\,v_2'\sin(-137°) \\ 0.581\,75 = \tfrac{1}{2}(0.25)(v_1')^2 + \tfrac{1}{2}(0.35)(v_2')^2 \\ v_1' \in (0,5) \\ v_2' \in (0,5) \\ \theta_1 \in (0,90) \end{bmatrix},$$

Solution is: $\{[v_1' = 1.519\,8, v_2' = 1.294, \theta_1 = 54.388]\}$

The final speed of Puck 2 is $1.294\,\mathrm{m/s}$ and the final velocity of Puck 1 is $1.519\,8\,\mathrm{m/s}$ at $54.388°$. To verify that this collision was elastic, let's Evaluate expressions for the initial and final total kinetic energy.

$$K = \tfrac{1}{2}(0.25\,\mathrm{kg})\left(1.45\frac{\mathrm{m}}{\mathrm{s}}\right)^2 + \tfrac{1}{2}(0.35\,\mathrm{kg})\left(-1.35\frac{\mathrm{m}}{\mathrm{s}}\right)^2 = 0.581\,75\frac{\mathrm{m}^2}{\mathrm{s}^2}\,\mathrm{kg}$$

$$K' = \tfrac{1}{2}(0.25\,\mathrm{kg})\left(1.519\,8\frac{\mathrm{m}}{\mathrm{s}}\right)^2 + \tfrac{1}{2}(0.35\,\mathrm{kg})\left(1.294\frac{\mathrm{m}}{\mathrm{s}}\right)^2 = 0.581\,75\frac{\mathrm{m}^2}{\mathrm{s}^2}\,\mathrm{kg}$$

The total kinetic energy was conserved so the collision was elastic. ◀

Rockets

Until a successful warp drive is invented, rockets are the only means of propulsion that operate effectively in outer space. A rocket ejects exhausted fuel out its back end at high velocity. The rocket pushes the exhaust out and the exhaust pushes back on the rocket per Newton's 3rd Law. This reaction force is the thrust that pushes the rocket forward.

Newton's 2nd Law tells us the net force acting on the rocket is

$$m\overrightarrow{a} = \overrightarrow{T} + \overrightarrow{F}_{\text{app}} \tag{6.29}$$

where $\overrightarrow{T}$ is the thrust and $\overrightarrow{F}_{\text{app}}$ is any external applied force (such as gravity). This equation is valid at every moment in time, but it's not very useful to calculate the rocket's velocity or position. The rocket's mass changes as it ejects fuel and the thrust depends on the rate of that change. Instead, let's derive the rocket's equation of motion using momentum conservation.

When an external applied force $\overrightarrow{F}_{\text{app}}$ acts on the rocket, Newton's 2nd Law says

$$\frac{d\overrightarrow{P}}{dt} = \overrightarrow{F}_{\text{app}} \tag{6.30}$$

where $\overrightarrow{P}$ is the total momentum of the rocket/exhaust system. If the rocket ejects exhaust at a constant velocity $\overrightarrow{u}$ relative to the rocket, then this becomes

$$m\frac{d\overrightarrow{v}}{dt} = \frac{dm}{dt}\overrightarrow{u} + \overrightarrow{F}_{\text{app}} \tag{6.31}$$

where m is the rocket's decreasing mass and $\overrightarrow{v}$ is its velocity. The thrust exerted by the ejected mass on the rocket is $\frac{dm}{dt}\overrightarrow{u}$. The rocket's mass is decreasing, so the mass time-derivative is negative, but the exhaust velocity points backward so the thrust is forward.

Deep Space

A rocket in deep space is far away from any celestial objects so there are no external forces acting on it. In 1-dimension, the rocket's equation of motion is

$$m\frac{dv}{dt} = -u\frac{dm}{dt} \tag{6.32}$$

where u is the (positive) speed of the ejected mass relative to the rocket.

You can use *SNB* to solve this equation for the rocket's velocity. Cancel the explicit time dependence and arrange the equation as a differential equation for the velocity as a function of mass. Use Solve ODE + Exact with the appropriate initial condition in terms of the rocket's initial mass m_0. Then Factor and apply Combine + Logs in-place to the logarithm terms.

$$\begin{bmatrix} \frac{dv}{dm} = -\frac{u}{m} \\ v(m_0) = v_0 \end{bmatrix}, \text{ Exact solution is: } \left\{ v_0 + u\left(\ln\frac{1}{m}m_0\right) \right\}$$

Since the mass m is a function of time, this gives us an expression for the rocket's velocity as a function of time.

$$v = v_0 + u \ln \frac{m_0}{m} \tag{6.33}$$

This is valid for any physically reasonable time-dependent function for the mass.

To make the time-dependence explicit, we need to specify how the rocket's mass varies in time. Suppose the rocket burns fuel at a constant rate k (in $\mathrm{kg/\,s}$) so the thrust is $T = uk$. Then we can use *SNB* and Solve ODE + Exact to find the rocket's mass as a function of time.

$$\begin{bmatrix} \frac{dm}{dt} = -k \\ m(0) = m_0 \end{bmatrix}, \text{ Exact solution is: } \{m_0 - kt\}$$

This gives us an expression for the velocity of a rocket that ejects fuel at a constant rate and has no external forces acting on it.

$$v = v_0 - u \ln \left(1 - \frac{kt}{m_0}\right) \tag{6.34}$$

We can use Eqs. (2.10) and (6.34) to integrate the velocity and find the rocket's position as a function of time. Create and Expand the following expression for the position.

$$y = y_0 + v_0 t - u \left(\left[\int \ln \left(1 - \frac{kt}{m_0}\right) dt \right]_{t=0}^{t=t} \right)$$

$$= y_0 + tv_0 + tu - tu \ln \tfrac{1}{m_0}(m_0 - kt) + \frac{1}{k} u m_0 \ln \left(-\tfrac{1}{k}(m_0 - kt)\right) - \frac{1}{k} u m_0 \ln \left(-\tfrac{1}{k} m_0\right)$$

Using Expand in-place and a little algebra by-hand gives us an expression for the rocket's position as a function of time when no external forces act on it.

$$y = y_0 + (v_0 + u)\, t + \frac{m_0 u}{k} \left(1 - \frac{kt}{m_0}\right) \ln \left(1 - \frac{kt}{m_0}\right) \tag{6.35}$$

This expression is valid for a rocket that ejects propellant at a constant rate.

We can use Eqs. (2.8) and (6.34) to take the derivative of the velocity and find the rocket's acceleration as a function of time. Create and Evaluate the following expression for the acceleration.

$$a = \frac{d}{dt} \left(v_0 - u \ln \left(1 - \frac{kt}{m_0}\right) \right) = k \frac{u}{m_0 - kt}$$

This gives us an expression for the rocket's acceleration as a function of time when it is not experiencing any external forces.

$$a = \frac{uk}{m_0 - kt} \tag{6.36}$$

The rocket's acceleration is the thrust divided by its mass at time t. To specify the rocket's velocity, position, and acceleration, we need to know the initial values for the position and velocity, as well as its initial mass, exhaust speed, and mass rate of change.

Example 6.23 *Deep space mine*
The engine on a small rocket in deep space expels mass at a rate of $3\,\mathrm{kg/\,s}$ for 4 seconds. The exhaust exits the rocket at $70\,\mathrm{m/\,s}$ and the rocket starts at $20\,\mathrm{kg}$. Calculate the rocket's velocity, position, and acceleration 4 seconds after the engine starts.
Solution. Use Eq. (6.34) to create an expression with the appropriate numerical values for the rocket's velocity, and apply Evaluate Numerically to it.

$$v = -\left(70\frac{\mathrm{m}}{\mathrm{s}}\right)\ln\left(1 - \frac{(3\,\mathrm{kg/\,s})\,(4\,\mathrm{s})}{20\,\mathrm{kg}}\right) = 64.14\frac{\mathrm{m}}{\mathrm{s}}$$

At $t = 4\,\mathrm{s}$, the rocket has a velocity of $64.14\,\mathrm{m/\,s}$. Use Eq. (6.35) to create an expression with the appropriate numerical values for the rocket's position, and apply Evaluate Numerically to it.

$$y = \left(70\frac{\mathrm{m}}{\mathrm{s}}\right)(4\,\mathrm{s}) + \frac{(20\,\mathrm{kg})\,(70\,\mathrm{m/\,s})}{3\,\mathrm{kg/\,s}}\left(1 - \frac{(3\,\mathrm{kg/\,s})\,(4\,\mathrm{s})}{20\,\mathrm{kg}}\right)\ln\left(1 - \frac{(3\,\mathrm{kg/\,s})\,(4\,\mathrm{s})}{20\,\mathrm{kg}}\right)$$
$$= 108.96\,\mathrm{m}$$

At $t = 4\,\mathrm{s}$, the rocket is $108.96\,\mathrm{m}$ away from its starting point. Use Eq. (6.36) to create an expression with the appropriate numerical values for the rocket's acceleration, and apply Evaluate Numerically to it.

$$a = \frac{(70\,\mathrm{m/\,s})\,(3\,\mathrm{kg/\,s})}{20\,\mathrm{kg} - (3\,\mathrm{kg/\,s})\,(4\,\mathrm{s})} = 26.25\frac{\mathrm{m}}{\mathrm{s}^2}$$

At $t = 4\,\mathrm{s}$, the rocket's acceleration is $26.25\,\mathrm{m/\,s}^2$, which is about $2.7g$. ◀

A rocket's mass consists of fuel, payload, and the empty rocket. Once the fuel is exhausted only the payload and empty rocket mass remains. The mass ratio is the ratio of the final mass to the total initial mass.

$$\epsilon \equiv \frac{m_f}{m_0}$$

This limits the time the rocket engine can burn fuel. Let's use Solve Exact to find the burn time t_b and Simplify the result.

$$\epsilon m_0 = m_0 - kt_b, \text{ Solution is: } \frac{1}{k}(m_0 - \epsilon m_0) = -\frac{1}{k}m_0(\epsilon - 1)$$

The rocket can burn fuel for $t_b = \dfrac{m_0}{k}(1-\epsilon)$ seconds.

Let's calculate the rocket's speed at engine burnout. Use Evaluate to Substitute the burn time into Eq. (6.34).

$$v_f = \left[v_0 - u\ln\left(1 - \frac{kt}{m_0}\right)\right]_{t=m_0(1-\epsilon)/k} = v_0 - u\ln\epsilon$$

Decreasing the mass ratio increases the rocket's burn time and burnout speed.

To calculate the distance the rocket travelled while the engine ran, use Evaluate to Substitute the burn time into Eq. (6.35).

$$y_b = \left[(v_0 + u)\, t + u \frac{m_0}{k} \left(1 - \frac{kt}{m_0}\right) \ln \left(1 - \frac{kt}{m_0}\right) \right]_{t = \frac{m_0}{k}(1-\epsilon)}$$
$$= \frac{1}{k} u \epsilon m_0 \ln \epsilon - \frac{1}{k} m_0 (\epsilon - 1)(u + v_0)$$

Let's calculate the rocket's acceleration at engine burnout. Substitute (with Simplify) the burn time into Eq. (6.36).

$$a_b = \left[\frac{uk}{m_0 - kt} \right]_{t = m_0(1-\epsilon)/k} = k \frac{u}{\epsilon m_0}$$

Here is a summary of our results for a rocket in deep space.

$$t_b = \frac{m_0}{k}(1 - \epsilon) \tag{6.37a}$$

$$v_b = v_0 - u \ln \epsilon \tag{6.37b}$$

$$a_b = \frac{uk}{\epsilon m_0} \tag{6.37c}$$

$$y_b = \frac{m_0 u}{k}(1 - \epsilon + \epsilon \ln \epsilon) + \frac{m_0 v_0}{k}(1 - \epsilon) \tag{6.37d}$$

The mass ratio $\epsilon = m_f \,/\, m_0$ is always between zero and one.

Launch

To lift off a planet or moon, the rocket must overcome the force of gravity to lift its own weight. Near the surface, this force is constant and the applied force on the rocket is $F_{\text{app}} = -mg$. We can adjust the rocket's equation of motion to include this.

$$m \frac{dv}{dt} = -u \frac{dm}{dt} - mg \tag{6.38}$$

You can use *SNB* to solve this equation by direct calculation to find the rocket's velocity near the surface. Tell *SNB* to assume the variables in the problem are positive. Then use Eq. (6.38) to set up the appropriate integrals, Evaluate each one in-place, and use Combine + Logs in-place on the logarithm terms.

$$\int_{v_0}^{v} dv = u \left(-\int_{m_0}^{m} \frac{dm}{m} \right) - g \int_0^t dt \Longrightarrow v - v_0 = u \left(\ln \frac{1}{m} m_0 \right) - gt$$

Since the mass m is a function of time, this gives us an expression for the rocket's velocity as a function of time.

$$v = v_0 - gt + u \ln \frac{m_0}{m} \tag{6.39}$$

This is valid for any physically reasonable time-dependent function for the mass.

If the mass is ejected at a constant rate, then the velocity is

$$v = v_0 - gt - u\ln\left(1 - \frac{kt}{m_0}\right) \tag{6.40}$$

where k is the rate at which the exhaust mass is ejected (in $\mathrm{kg/\,s}$).

We can use this result to check the condition for a successful launch. Use Eq. (6.40) to create an expression for the velocity (with $v_0 = 0$). Leave the insertion point anywhere in it and choose Power Series. Then Expand in Powers of t, and pick 1 as your Number of Terms.

$$v = -gt - u\ln\left(1 - \frac{kt}{m_0}\right) = -t\left(g - k\frac{u}{m_0}\right) + O\left(t^2\right)$$

For a successful launch, the velocity must be positive for all times. Expand the coefficient (including the minus sign) of the time.

$$-\left(g - k\frac{u}{m_0}\right) = k\frac{u}{m_0} - g$$

This expression is positive as long as $uk > m_0 g$. In other words, the rocket's thrust must be greater than its initial weight!

Comparing Eqs. (6.34) and (6.40), we see the only difference is the "$-gt$" term. When we integrate to find the rocket's position, this will introduce an additional "$-\frac{1}{2}gt^2$" term. This produces the following expression for the rocket's height as a function of time when it is experiencing a constant gravitational force.

$$y = y_0 + (v_0 + u)\,t - \tfrac{1}{2}gt^2 + \frac{m_0 u}{k}\left(1 - \frac{kt}{m_0}\right)\ln\left(1 - \frac{kt}{m_0}\right) \tag{6.41}$$

This expression is valid for a rocket that ejects propellant at a constant rate.

We can use Eqs. (2.8) and (6.40) and take the derivative of the velocity to find the rocket's acceleration as a function of time. Create and Evaluate the following expression for the acceleration.

$$a = \frac{d}{dt}\left(v_0 - gt - u\ln\left(1 - \frac{kt}{m_0}\right)\right) = \frac{1}{m_0 - kt}\left(ku - gm_0 + gkt\right)$$

A little editing by-hand gives us the rocket's acceleration as a function of time when it is experiencing a constant gravitational force.

$$a = \frac{uk - (m_0 - kt)\,g}{m_0 - kt} = -g + \frac{uk}{m_0 - kt} \tag{6.42}$$

The rocket's acceleration is the thrust minus its time-dependent weight divided by its mass at time t, which also equals the deep-space acceleration minus the acceleration due to gravity. To specify the rocket's velocity, position, and acceleration, we need to know the initial values for the position and velocity, as well as its initial mass, exhaust speed, and mass rate of change.

Example 6.24 *Near Earth yours*

The same small rocket from the previous example is launched vertically from the Earth's surface. Its engine expels mass at a rate of $3\,\mathrm{kg/\,s}$ and runs for 4 seconds. The exhaust exits the rocket at a speed of $70\,\mathrm{m/\,s}$ and the rocket's initial mass is $20\,\mathrm{kg}$. Calculate the rocket's velocity, height, and acceleration at $t = 4$ seconds.

Solution. Use Eq. (6.40) to create an expression with the appropriate numerical values for the rocket's velocity, and apply Evaluate Numerically to it.

$$v = -\left(9.8067\frac{\mathrm{m}}{\mathrm{s}^2}\right)(4\,\mathrm{s}) - \left(70\frac{\mathrm{m}}{\mathrm{s}}\right)\ln\left(1 - \frac{(3\,\mathrm{kg/\,s})\,(4\,\mathrm{s})}{20\,\mathrm{kg}}\right) = 24.914\frac{\mathrm{m}}{\mathrm{s}}$$

At $t = 4\,\mathrm{s}$, the rocket has a velocity of $24.914\,\mathrm{m/\,s}$. Use Eq. (6.41) to create an expression with the appropriate numerical values for the rocket's height, and apply Evaluate Numerically to it.

$$\begin{aligned} y &= \left(70\frac{\mathrm{m}}{\mathrm{s}}\right)(4\,\mathrm{s}) - \tfrac{1}{2}\left(9.8067\frac{\mathrm{m}}{\mathrm{s}^2}\right)(4\,\mathrm{s})^2 \\ &\quad + \frac{(70\,\mathrm{m/\,s})\,(20\,\mathrm{kg})}{3\,\mathrm{kg/\,s}}\left(1 - \frac{(3\,\mathrm{kg/\,s})\,(4\,\mathrm{s})}{20\,\mathrm{kg}}\right)\ln\left(1 - \frac{(3\,\mathrm{kg/\,s})\,(4\,\mathrm{s})}{20\,\mathrm{kg}}\right) \\ &= 30.505\,\mathrm{m} \end{aligned}$$

At $t = 4\,\mathrm{s}$, the rocket is $30.505\,\mathrm{m}$ off the ground. Use Eq. (6.42) to create an expression with the appropriate numerical values for the rocket's acceleration, and apply Evaluate Numerically to it.

$$a = -9.8067\frac{\mathrm{m}}{\mathrm{s}^2} + \frac{(70\,\mathrm{m/\,s})\,(3\,\mathrm{kg/\,s})}{20\,\mathrm{kg} - (3\,\mathrm{kg/\,s})\,(4\,\mathrm{s})} = 16.443\frac{\mathrm{m}}{\mathrm{s}^2}$$

At $t = 4\,\mathrm{s}$, the rocket has an acceleration of $16.443\,\mathrm{m/\,s^2}$, which is $1.6767g$. Notice that all three answers were significantly reduced by the force of gravity. ◀

Once the rocket engine shuts down, the rocket becomes a projectile with gravity the only force acting on it. To calculate the elapsed time between launch and landing, we must divide the rocket's motion into two parts, engine on and engine off.

Example 6.25 *Heads up!*

What is the maximum height achieved by the rocket in the previous example? How much time elapsed before it hit the ground? Ignore air resistance.

Solution. During the 4 seconds that the rocket engine is running, the rocket's height as a function of time is given by Eq. (6.41).

$$\begin{aligned} y_{\mathrm{r}} &= \left[ut - \tfrac{1}{2}gt^2 + \frac{m_0 u}{k}\left(1 - \frac{kt}{m_0}\right)\ln\left(1 - \frac{kt}{m_0}\right)\right]_{g=9.8067,u=70,k=3,m_0=20} \\ &= 70.0t - 466.67\,(\ln(1.0 - 0.15t))\,(0.15t - 1.0) - 4.9034t^2 \end{aligned}$$

You can Substitute the appropriate values into this expression to verify the rocket's height when the engine stops.

$$\left[ut - \tfrac{1}{2}gt^2 + \frac{m_0 u}{k}\left(1 - \frac{kt}{m_0}\right)\ln\left(1 - \frac{kt}{m_0}\right)\right]_{g=9.8067,u=70,k=3,m_0=20,t=4} = 30.505$$

At $t = 4\,\mathrm{s}$, the engine shuts down and the rocket is $30.505\,\mathrm{m}$ off the ground.

Once the engine shuts down, the rocket becomes a simple projectile so we can use the results of Chapter 2. The "initial" values of this projectile motion are the rocket's velocity and height when the engine shuts down at time $t = t_b$ after launch. We can adjust the expressions so these "initial" values happen at $t = t_b$.

Let's use Eq. (2.6a) to create an expression for the rocket's height as a projectile with $v_0 = 24.914\,\mathrm{m/s}$ and $y_0 = 30.505\,\mathrm{m}$ from the previous example, adjusting the time to start at t_b. Then use Substitute to calculate the rocket's height as a function of time after the engine stops.

$$y_p = \left[y_0 + v_0(t - t_b) - \tfrac{1}{2}g(t - t_b)^2\right]_{y_0=30.505,v_0=24.914,g=9.8067,t_b=4}$$
$$= 24.914t - 4.9034(t-4)^2 - 69.151$$

At the rocket's maximum height, it stops going up and its velocity is zero. Use Eq. (2.6d) to create the following equation setting the rocket's velocity to zero, and use Solve Exact to find the maximum height.

$$0 = \left(24.914\frac{\mathrm{m}}{\mathrm{s}}\right)^2 - 2\left(9.8067\frac{\mathrm{m}}{\mathrm{s}^2}\right)(y - 30.505\,\mathrm{m}), \text{ Solution is: } 62.152\,\mathrm{m}$$

The rocket's maximum height is $62.152\,\mathrm{m}$. To calculate when the rocket hits the ground, set the projectile height y_p to zero and use Solve Exact to find the time.

$$0 = \left(24.914\frac{\mathrm{m}}{\mathrm{s}}\right)t - \left(4.9034\frac{\mathrm{m}}{\mathrm{s}^2}\right)(t - 4\,\mathrm{s})^2 - 69.151\,\mathrm{m}, \text{ Solution is: } 10.101\,\mathrm{s}$$

The rocket hits the ground $10.101\,\mathrm{s}$ after launch.

Figure 6.2 shows a graph of the rocket's height as a function of time for the entire flight.

The dashed lines indicate the rocket's height and the time when the engine shut down.

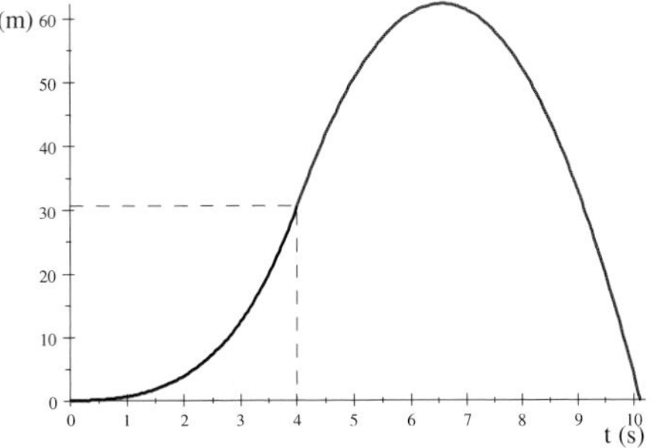

Figure 6.2 Near-Earth launch

◀

Including the effects of gravity does not change the properties of our rocket, so it still burns fuel for a time t_b. To calculate the rocket's speed at engine burnout, let's use Evaluate to Substitute the burn time into Eq. (6.40).

$$v_b = \left[v_0 - gt - u\ln\left(1 - \frac{kt}{m_0}\right)\right]_{t=m_0(1-\epsilon)/k} = v_0 - u\ln\epsilon + \frac{g}{k}m_0(\epsilon - 1)$$

To calculate the distance the rocket travelled while the engine ran, use Evaluate to Substitute the burn time into Eq. (6.41).

$$y_b = \left[y_0 + (v_0 + u)\,t - \tfrac{1}{2}gt^2 + \frac{m_0 u}{k}\left(1 - \frac{kt}{m_0}\right)\ln\left(1 - \frac{kt}{m_0}\right)\right]_{t=\frac{m_0}{k}(1-\epsilon)}$$
$$= y_0 - \frac{1}{k}m_0(\epsilon - 1)(u + v_0) - \tfrac{1}{2}\frac{g}{k^2}m_0^2(\epsilon - 1)^2 + \frac{1}{k}u\epsilon m_0\ln\epsilon$$

To calculate the rocket's acceleration at burnout, use Expand to Substitute the burn time into Eq. (6.42).

$$a_b = \left[\frac{uk}{m_0 - kt} - g\right]_{t=\frac{m_0}{k}(1-\epsilon)} = k\frac{u}{\epsilon m_0} - g$$

The burnout values for the velocity, height, and acceleration depend on the initial mass, the mass ejection rate, the mass ejection speed, the mass ratio (or equivalently the final mass), and gravity.

Here is a summary of our results for a rocket experiencing a constant gravitational force.

$$t_b = \frac{m_0}{k}(1 - \epsilon) \tag{6.43a}$$
$$v_b = v_0 - u\ln\epsilon - \frac{m_0 g}{k}(1 - \epsilon) \tag{6.43b}$$
$$a_b = -g + \frac{uk}{\epsilon m_0} \tag{6.43c}$$
$$y_b = y_0 + \tfrac{m_0 v_0}{k}(1 - \epsilon) + \tfrac{m_0 u}{k}(1 - \epsilon + \epsilon\ln\epsilon) - \tfrac{1}{2}\tfrac{m_0^2 g}{k^2}(1 - \epsilon)^2 \tag{6.43d}$$

As before, the mass ratio $\epsilon = m_f\,/\,m_0$ is always between zero and one.

Example 6.26 *Burning out his fuse up here alone*

The Redstone rocket that launched Alan Shepard into space in 1961 lifted $29700\,\mathrm{kg}$ off the Earth's surface. The engine burned for $144\,\mathrm{s}$ and when it stopped the rocket's velocity was $1970\,\mathrm{m/s}$ and its acceleration was $6.3g$. Calculate the rocket's thrust, mass, and height when the engine stopped.

Solution. Let's use Eqs. (6.43a-c) and Solve Numeric to calculate u, k, and ϵ. Put the equations, with the appropriate numerical values without units, in a 1-column, 3-row matrix.

Leave the insertion point anywhere in the matrix and choose Solve + Numeric.

$$\begin{bmatrix} 144 = \frac{29700}{k}(1-\epsilon) \\ 1970 = -u\ln\epsilon - \frac{(29700)(9.8067)}{k}(1-\epsilon) \\ 6.3(9.8067) = \frac{uk}{29700\epsilon} - 9.8067 \end{bmatrix},$$

Solution is: $\{[k = 176.26, u = 1754.0, \epsilon = 0.145\,41]\}$

Now we can Evaluate the rocket's mass when the engine stopped.

$$m_f = \epsilon m_0 = (0.145\,41)(29700\,\text{kg}) = 4318.7\,\text{kg}$$

Let's Evaluate the rocket's thrust.

$$uk = 1754.0\frac{\text{m}}{\text{s}} \times 176.26\frac{\text{kg}}{\text{s}} = 3.091\,6 \times 10^5 \frac{\text{m}}{\text{s}^2}\,\text{kg}$$

The thrust is $309.16\,\text{kN}$. This is less than the actual thrust of $347\,\text{kN}$, but Shepard's rocket had to overcome air resistance, which we ignored.

Let's use Evaluate and Eq. (6.43d) to calculate the rocket's height when the engine stopped.

$$\frac{(29700)(1754)}{176.26}(0.854\,59 + 0.145\,41\ln 0.145\,41) - \frac{1}{2}\frac{(29700)^2(9.8067)}{176.26^2}(0.854\,59)^2$$
$$= 68034.$$

The rocket was $68\,\text{km}$ above the Earth's surface when the rocket engine stopped. ◀

Air Resistance

Many planets and moons have an atmosphere, so our rocket must overcome both gravity and air resistance to fly successfully. We can include the effects of linear air resistance by changing the applied force on the rocket to $F_{\text{app}} = -mg - bv$. This will complicate our results, but don't worry. We'll let *SNB* do most of the work.

Here is the rocket's equation of motion for constant gravity and linear air resistance.

$$m\frac{dv}{dt} = -u\frac{dm}{dt} - mg - bv \tag{6.44}$$

We can't solve this for the velocity until we specify the time-dependent mass. Let's use our result when the mass is ejected at a constant rate.

$$(m_0 - kt)\frac{dv}{dt} = -uk - (m_0 - kt)g - bv$$

This is an equation *SNB* can solve. Place the equation and initial condition in a 2-row, 1-column matrix, leave the insertion point anywhere in the matrix.

Then choose Solve ODE + Exact.

$$\left[\begin{array}{c}(m_0 - kt)\dfrac{dv}{dt} = uk - (m_0 - kt)\,g - bv \\ v(0) = v_0\end{array}\right], \text{ Exact solution is:}$$

$$\left\{\frac{(k^2u + bgm_0 - bku - bgkt)}{bk - b^2} + \left(v_0 - \frac{(uk^2 - buk + bgm_0)}{bk - b^2}\right)\frac{(kt - m_0)^{\frac{b}{k}}}{(-m_0)^{\frac{b}{k}}}\right\}$$

A little algebra by-hand produces the rocket's velocity as a function of time.

$$\begin{aligned} v &= v_0\left(1 - \frac{kt}{m_0}\right)^{b/k} - \frac{gt}{1 - \frac{b}{k}} + \left(\tfrac{m_0 g}{k-b} + \tfrac{ku}{b}\right)\left(1 - \left(1 - \frac{kt}{m_0}\right)^{b/k}\right) \qquad (6.45) \\ &= C - \frac{gt}{1 - \frac{b}{k}} + (v_0 - C)\left(1 - \frac{kt}{m_0}\right)^{b/k} \end{aligned}$$

The constant $C \equiv \dfrac{m_0 g}{k - b} + \dfrac{uk}{b}$ has units of velocity.

We can use Eq. (6.45) to find the condition for a successful launch. Create an expression for the velocity, leave the insertion point in it and choose Power Series. Expand in Powers of t, and pick 2 as your Number of Terms. Then Simplify the coefficient of the time in-place.

$$\begin{aligned} v &= v_0\left(1 - \frac{kt}{m_0}\right)^{\frac{b}{k}} - \frac{gt}{1 - \frac{b}{k}} + \left(\frac{m_0 g}{k - b} + \frac{ku}{b}\right)\left(1 - \left(1 - \frac{kt}{m_0}\right)^{\frac{b}{k}}\right) \\ &= v_0 + t\left(-\frac{1}{m_0}(gm_0 + bv_0 - ku)\right) + O\left(t^2\right) \end{aligned}$$

A successful launch needs a positive velocity so the time coefficient must always be positive. This happens when $uk > m_0 g + bv_0$. In other words, the thrust must be greater than the initial weight and air-resistance combined! Unless the rocket is launched from a moving platform, the initial velocity is zero and the condition for a successful launch reduces to the thrust being larger than the initial weight.

There are two limits we need to check.

1. When we turn off the air resistance by taking the $b \to 0$ limit (with Evaluate), the rocket's velocity should reduce to the no-air-resistance result from the previous section.

$$\begin{aligned} v &\to \lim_{b\to 0}\left(v_0\left(1 - \frac{kt}{m_0}\right)^{b/k} - \frac{gt}{1 - \frac{b}{k}} + \left(\frac{m_0 g}{k - b} + \frac{ku}{b}\right)\left(1 - \left(1 - \frac{kt}{m_0}\right)^{b/k}\right)\right) \\ &= v_0 - u\ln\left(1 - k\frac{t}{m_0}\right) - gt \end{aligned}$$

The velocity does reduce to Eq. (6.40) which is the correct limit.

2. The velocity has terms that by themselves are infinite when $k = b$ so we should check that it is finite in the $b \to k$ limit. Use Evaluate to take the limit, then edit the result with Expand and Factor in-place.

$$v \to \lim_{b\to k}\left(v_0\left(1-\frac{kt}{m_0}\right)^{b/k} - \frac{gt}{1-\frac{b}{k}} + \left(\frac{m_0 g}{k-b}+\frac{ku}{b}\right)\left(1-\left(1-\frac{kt}{m_0}\right)^{b/k}\right)\right)$$

$$= v_0\left(1-\frac{kt}{m_0}\right) + \frac{uk}{m_0}t + \frac{m_0 g}{k}\left(1-\frac{kt}{m_0}\right)\ln\left(1-\frac{kt}{m_0}\right)$$

There are no inconvenient and unsightly infinities. As one last check, let's turn off the rocket engine by taking the $k \to 0$ limit.

$$v \to \lim_{k\to 0}\left(v_0\left(1-\frac{kt}{m_0}\right) + \frac{uk}{m_0}t + \frac{m_0 g}{k}\left(1-\frac{kt}{m_0}\right)\ln\left(1-\frac{kt}{m_0}\right)\right) = v_0 - gt$$

This result is an old friend from Chapter 2: the velocity of an object in free fall.

The calculation of the rocket's height as a function of time is straightforward but very cumbersome, especially if you try it all at once. We can use Eq. (2.10) to calculate the position from this time-dependent velocity. Here is the integral we need to calculate.

$$y - y_0 = \int_0^t \left(C - \frac{gt}{1-\frac{b}{k}} + (v_0 - C)\left(1-\frac{kt}{m_0}\right)^{b/k}\right) dt$$

Using the constant C makes the expression a little less complicated. Let's divide the integral into three terms. Create the following expressions and Evaluate the integrals in-place.

1. $\int_0^t C dt = Ct$

2. $-\frac{1}{1-\frac{b}{k}}\int_0^t g t dt = -\frac{1}{1-\frac{b}{k}}\left(\frac{1}{2}gt^2\right)$

3. $(v_0 - C)\left[\int\left(1-\frac{kt}{m_0}\right)^{b/k} dt\right]_{t=0}^{t=t} = (v_0 - C)\left(\frac{m_0}{b+k} - \frac{1}{m_0^{\frac{b}{k}}}\frac{(m_0 - kt)^{\frac{1}{k}(b+k)}}{b+k}\right)$

A little algebra by-hand produces the rocket's height as a function of time.

$$y = y_0 + C\,t - \frac{1}{2}\frac{gt^2}{1-\frac{b}{k}} + \frac{m_0(v_0 - C)}{k+b}\left(1-\left(1-\frac{kt}{m_0}\right)^{1+\frac{b}{k}}\right) \tag{6.46}$$

Remember, the constant C is the same one we defined in Eq. (6.45) for the velocity.

We can use Eqs. (2.8) and (6.45) and take the derivative of the velocity to find the rocket's acceleration as a function of time. Let's break it into three parts again. Create the following expression for the acceleration and Evaluate each derivative in-place.

$$a = \frac{dC}{dt} - \frac{1}{1-\frac{b}{k}}\frac{d}{dt}(gt) + (v_0 - C)\left(\frac{d}{dt}\left(1-\frac{kt}{m_0}\right)^{b/k}\right)$$

$$= 0 - \frac{1}{1-\frac{b}{k}}g + (v_0 - C)\left(-\frac{b}{m_0}\left(\frac{1}{m_0}(m_0 - kt)\right)^{\frac{1}{k}(b-k)}\right)$$

Using Expand in-place and a little algebra by-hand gives us an expression for the rocket's acceleration as a function of time when it is experiencing a constant gravitational force and a linear air resistance.

$$a = -\frac{g}{1-\frac{b}{k}} + \frac{b}{m_0}(C - v_0)\left(1-\frac{kt}{m_0}\right)^{(b-k)/k} \tag{6.47}$$

To turn off the air resistance, restore the constant C and take the $b \to 0$ limit. Create and Evaluate the following expression, and Expand the denominator in-place.

$$a \to \lim_{b\to 0}\left(-\frac{g}{1-\frac{b}{k}} + \frac{b}{m_0}\left(\frac{m_0 g}{k-b} + \frac{uk}{b} - v_0\right)\left(1-\frac{kt}{m_0}\right)^{(b-k)/k}\right) = -g - \frac{uk}{kt - m_0}$$

When we turn off the air resistance, this expression for the rocket's acceleration reduces to Eq. (6.42), the no-air-resistance result from the previous section.

Including the effects of gravity and air resistance does not change the properties of our rocket, so it still burns fuel for a time t_b. We can calculate the rocket's speed, height, and acceleration at engine burnout by substituting the burn time into Eqs. (6.45), (6.46), and (6.47), respectively. We've done similar calculations so there are no new *SNB* skills to learn here. Let me just quote you the results and encourage you to reproduce them.

$$t_b = \frac{m_0}{k}(1-\epsilon) \tag{6.48a}$$

$$v_b = v_0\epsilon^{b/k} + C\left(1-\epsilon^{b/k}\right) - \frac{m_0 g}{k-b}(1-\epsilon) \tag{6.48b}$$

$$a_b = -\frac{g}{1-\frac{b}{k}} + \frac{b}{m_0}(C-v_0)\,\epsilon^{(b-k)/k} \tag{6.48c}$$

$$y_b = y_0 + \frac{m_0 C}{k}(1-\epsilon) - m_0\frac{C-v_0}{k+b}\left(1-\epsilon^{(b+k)/k}\right) - \frac{1}{2}\frac{m_0^2 g}{k(k-b)}(1-\epsilon)^2 \tag{6.48d}$$

The constant is still $C \equiv \dfrac{m_0 g}{k-b} + \dfrac{uk}{b}$ and it still has units of velocity.

Example 6.27 *Near Earth yours II*

The same small rocket from Example 6.24 is launched vertically from the Earth's surface. Calculate the rocket's velocity, height, and acceleration when the engine shuts down. Include the effects of air resistance by letting $b = 2\,\mathrm{kg/\,s}$.

Solution. Let's use Evaluate Numerically to calculate the value of the constant C.

$$C = \frac{(20\,\mathrm{kg})\,(9.8067\,\mathrm{m/\,s^2})}{3\,\mathrm{kg/\,s} - 2\,\mathrm{kg/\,s}} + \frac{(70\,\mathrm{m/\,s})\,(3\,\mathrm{kg/\,s})}{2\,\mathrm{kg/\,s}} = 301.13\frac{\mathrm{m}}{\mathrm{s}}$$

Let's use Eq. (6.48a) and Solve Exact to find the mass ratio.

$$4\,\mathrm{s} = \frac{20\,\mathrm{kg}}{3\,\mathrm{kg/\,s}}\,(1-\epsilon),\ \text{Solution is: } \frac{2}{5}$$

Now we can use Eqs. (6.48) to create and apply Evaluate Numerically to expressions for the rocket's velocity, height, and acceleration when the engine stops.

$$v_b = 301.13\frac{\mathrm{m}}{\mathrm{s}}\left(1 - \tfrac{2}{5}^{2/3}\right) - \frac{(20\,\mathrm{kg})\,(9.8067\,\mathrm{m/\,s^2})}{3\,\mathrm{kg/\,s} - 2\,\mathrm{kg/\,s}}\,\left(1 - \tfrac{2}{5}\right) = 19.971\frac{\mathrm{m}}{\mathrm{s}}$$

$$y_b = \frac{20\,\mathrm{kg}\left(301.13\frac{\mathrm{m}}{\mathrm{s}}\right)}{3\frac{\mathrm{kg}}{\mathrm{s}}}\left(\tfrac{3}{5}\right) - \frac{20\,\mathrm{kg}\left(301.13\frac{\mathrm{m}}{\mathrm{s}}\right)}{3\frac{\mathrm{kg}}{\mathrm{s}}+2\frac{\mathrm{kg}}{\mathrm{s}}}\left(1 - \tfrac{2}{5}^{(2+3)/3}\right) - \frac{1}{2}\frac{(20\,\mathrm{kg})^2\left(9.8067\frac{\mathrm{m}}{\mathrm{s}^2}\right)}{3\frac{\mathrm{kg}}{\mathrm{s}}\left(3\frac{\mathrm{kg}}{\mathrm{s}}-2\frac{\mathrm{kg}}{\mathrm{s}}\right)}\left(\tfrac{3}{5}\right)^2$$
$$= 26.205\,\mathrm{m}$$

$$a_b = -\frac{9.8067\,\mathrm{m/\,s^2}}{1-\frac{2}{3}} + \frac{(2\,\mathrm{kg/\,s})}{20\,\mathrm{kg}}\left(301.13\frac{\mathrm{m}}{\mathrm{s}}\right)\tfrac{2}{5}^{(2-3)/3} = 11.450\frac{\mathrm{m}}{\mathrm{s}^2}$$

When the engine shuts down, the rocket is $26.205\,\mathrm{m}$ off the ground, moving up with a velocity of $19.971\,\mathrm{m/\,s}$, and has an acceleration of $11.45\,\mathrm{m/\,s^2}$ (about $1.17g$). All three results were significantly reduced by air resistance. ◀

Example 6.28 *Heads up II*

What is the maximum height achieved by the rocket in the previous example? How much time elapsed before it hit the ground?

Solution. After the engine shuts down, the rocket is a projectile. When we solved a similar problem in Example 6.25, we adjusted the expressions for projectile motion so the "initial" values happened at $t = t_b$. That wasn't necessary to solve the problem, so let's try a different approach here.

We'll need the rocket's final mass so let's Evaluate this expression.

$$m_f = \epsilon m_0 = \tfrac{2}{5}\,(20\,\mathrm{kg}) = 8\,\mathrm{kg}$$

After its engine stops, the rocket is an $8\,\mathrm{kg}$ projectile.

During the 4 seconds that the rocket engine is running, its height as a function of time is given by Eq. (6.46).

$$y_{\mathrm{r}} = \left[\begin{array}{c} \frac{m_0 g - uk + \frac{uk^2}{b}}{k-b} t - \frac{\frac{1}{2} g t^2}{1 - \frac{b}{k}} \\ - \frac{m_0 \left(m_0 g - uk + \frac{uk^2}{b} \right)}{(k^2 - b^2)} \left(1 - \left(1 - \frac{kt}{m_0} \right)^{1 + \frac{b}{k}} \right) \end{array} \right]_{g=9.8067, u=70, k=3, m_0=20, b=2}$$

$$= 301.13t + 1204.5 \left(1 - \tfrac{3}{20} t\right)^{\frac{5}{3}} - 14.71t^2 - 1204.5$$

After the engine shuts down, the rocket is a projectile experiencing the forces of gravity and linear air resistance. We can use Eqs. (4.17) and Eq. (4.18) unaltered as long as we recognize they start at $t = 4\,\mathrm{s}$. Use Eq. (4.17b) as is, with $v_0 = 19.971\,\mathrm{m/s}$ and $y_0 = 26.205\,\mathrm{m}$ from the previous example, and $k = b/m_f$, and Substitute to calculate the rocket's height as a function of time after the engine stops.

$$y_{\mathrm{p}} = \left[y_0 - \frac{g}{k} t + \left(\frac{v_0}{k} + \frac{g}{k^2} \right) \left(1 - e^{-kt} \right) \right]_{y_0=26.205, v_0=19.971, g=9.8067, k=2/8}$$

$$= 263.00 - 236.79 e^{-\frac{1}{4} t} - 39.227t$$

At the rocket's maximum height, its velocity is zero. Use Eq. (4.18b) to create the following equation setting the rocket's velocity to zero, and use Solve Numeric to find the time between the engine shutting down and the rocket stopping..

$$0 = \left[v_0 e^{-kt} - \frac{g}{k} \left(1 - e^{-kt} \right) \right]_{v_0=19.971, g=9.8067, k=2/8}, \text{ Solution is: } \{[t = 1.6461]\}$$

The rocket reaches its maximum height $1.6461\,\mathrm{s}$ after the engine stops and $4\,\mathrm{s} + 1.6461\,\mathrm{s} = 5.6461\,\mathrm{s}$ after launch.

Use Substitute to calculate the rocket's maximum height.

$$y_{\max} = \left[263.00 - 236.79 e^{-\frac{1}{4} t} - 39.227t \right]_{t=1.6461} = 41.522$$

The rocket's maximum height is $41.522\,\mathrm{m}$. To calculate how long after the engine stops the rocket hits the ground, set the projectile trajectory to zero and use Solve Numeric to find the time.

$$\left[\begin{array}{c} 0 = 263.00 - 236.79 e^{-\frac{1}{4} t} - 39.227t \\ t \in (4, 10) \end{array} \right], \text{ Solution is: } \{[t = 4.9560]\}$$

The rocket hits the ground $4.9560\,\mathrm{s}$ after the engine stops, and the total time of flight is $4\,\mathrm{s} + 4.9560\,\mathrm{s} = 8.956\,\mathrm{s}$.

Figure 6.3 shows a graph of the rocket's height as a function of time including the effects of air resistance and constant gravity.

The dashed lines indicate the rocket's height and time when the engine shut down, and the thin line is from Figure 6.2.

Gravity and air resistance both reduce the rocket's maximum height and time of flight.

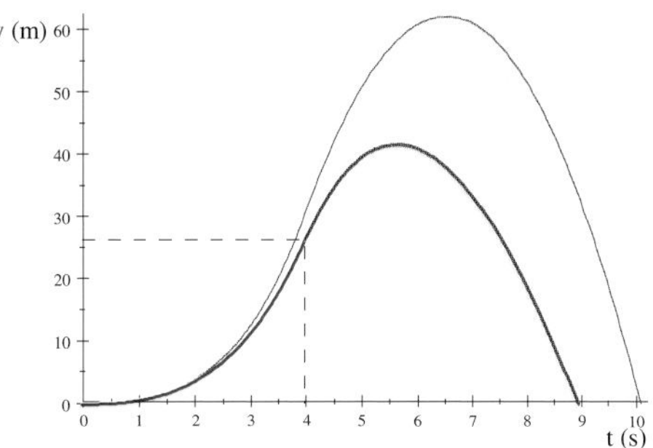

Figure 6.3 Launch with air resistance

◀

Varying Gravity and Air Resistance

As the rocket's height increases, the forces of gravity and air resistance acting on it both become weaker. Gravity weakens because the rocket is further away and the air resistance reduces due to the atmosphere's diminished density at altitude. We can incorporate these effects with the following adjustments.

$$g \rightarrow \frac{g}{\left(1+\dfrac{y}{R}\right)^2}$$

$$b \rightarrow b\,e^{-y/h}$$

The height-dependent air resistance is the same one we saw in Chapter 4. The height-dependent gravity is due to the inverse-square nature of the gravitational force. We'll see more about this in Chapter 9.

Adding these adjustments to Eq. (6.44), the equation of motion for the rocket's velocity becomes

$$m\frac{dv}{dt} = -u\frac{dm}{dt} - \frac{mg}{\left(1+\dfrac{y}{R}\right)^2} - bve^{-y/h} \tag{6.49}$$

where R is the radius of the celestial object and $h \ln 2$ is the height where the air resistance equals one-half its surface value. For Earth, these values are $R = 6378\,\mathrm{km}$ and $h = 9650\,\mathrm{m}$. The heights y and h are measured from the surface.

You can use Eq. (6.49) and the definition of velocity to calculate the rocket's height and velocity simultaneously.

$$v = \frac{dy}{dt}$$

You can also use this definition to write a single second-order differential equation for the rocket's height.

$$m\frac{d^2y}{dt^2} = -u\frac{dm}{dt} - \frac{mg}{\left(1+\dfrac{y}{R}\right)^2} - b\frac{dy}{dt}e^{-y/h} \tag{6.50}$$

Solve Eq. (6.50) if all you want is the rocket's height.

When the engine stops, the rocket becomes a projectile experiencing variable gravity and air resistance. Its equation of motion is then

$$m_f \frac{dv}{dt} = -\frac{m_f\, g}{\left(1+\frac{y}{R}\right)^2} - bve^{-y/h} \tag{6.51}$$

where m_f is the rocket's mass at engine cutoff. Neither Eq. (6.49) nor Eq. (6.51) has an exact solution, so we must solve them numerically with Solve ODE + Numeric.

We also need to add a third differential equation for the mass. These numerical calculations are not restricted to the constant mass-ejection rate, so we could experiment with varying rates that depend on time, height or velocity.

Example 6.29 *The higher, the weaker...*
Let's launch the same rocket from the previous examples from the surface of a small, dense moon. Suppose that moon has the same surface gravity as Earth, but a much smaller radius ($R = 500\,\mathrm{m}$) and an atmosphere that thins out at much lower altitudes ($h = 100\,\mathrm{m}$). (No such moon is known to exist, but these numbers produce a cool result.) Calculate and plot the rocket's height as a function of time. What is its time of flight? How fast is it going when it hits the ground?

Solution. When the rocket engine is on, you need to solve Eq. (6.49) numerically to find the rocket's height, velocity, and mass when engine stops. Create a 6-row, 1-column matrix containing 3 differential equations and the appropriate initial conditions without units, leave the insertion point anywhere in the matrix and choose Solve ODE + Numeric.

$$\begin{bmatrix} m\frac{dv}{dt} = 210 - \frac{9.8067m}{\left(1+\frac{y}{500}\right)^2} - 2ve^{-y/100} \\ v = \frac{dy}{dt} \\ \frac{dm}{dt} = -3 \\ v\left(0\right) = 0 \\ y\left(0\right) = 0 \\ m\left(0\right) = 20 \end{bmatrix}, \text{ Functions defined: } v, y, m$$

Use Evaluate or Evaluate Numerically to calculate the values for the height, velocity, and mass when the engine stops.

$$y_f = y\left(4\right) = 27.318$$

$$v_f = v\left(4\right) = 21.495$$

$$m_f = m\left(4\right) = 8.0$$

These values at engine shut-off are the initial values for the rocket's ensuing projectile motion.

When you use Solve ODE + Numeric to solve a differential equation, *SNB* assigns the most recent numerical solution to the variable. *SNB* treats uppercase and lowercase symbols as different variables, so you can avoid overwriting the previous solutions for y and v by using the uppercase letters Y and V for the projectile part of the solution.

Solve Eq. (6.51) numerically to find the time of flight and the rocket's impact velocity. Use $t = 4\,\mathrm{s}$ and the appropriate values for y_f, v_f, and m_f for the initial conditions.

$$\begin{bmatrix} 8\dfrac{dV}{dt} = -\dfrac{8\,(9.8067)}{\left(1+\dfrac{Y}{500}\right)^2} - 2Ve^{-Y/100} \\ V = \dfrac{dY}{dt} \\ V\,(4) = 21.495 \\ Y\,(4) = 27.318 \end{bmatrix}, \text{ Functions defined: } V, Y$$

Use trial and error (educated guesses really) to find the time of flight. Experiment with values for the time until you get the final height as close to zero as you can.

$$Y\,(9.9123) = -1.1849 \times 10^{-4}$$

$$V\,(9.9123) = -23.219$$

The total time of flight is about $9.9123\,\mathrm{s}$. The rocket has a downward speed of $23.219\,\mathrm{m/\,s}$ when it lands. This is less than the rocket's landing speed with no air resistance (about $35\,\mathrm{m/\,s}$) and slightly larger than the landing speed with linear air resistance ($22\,\mathrm{m/\,s}$).

Figure 6.4 uses these results to plot the rocket's height as a function of time including the effects of varying gravity and air resistance (the **green** curve). The dashed lines indicate the time and the rocket's height when the engine stopped, and the thin lines are the no-air-resistance (**blue**) and linear-air-resistance results (**red**) with constant gravity.

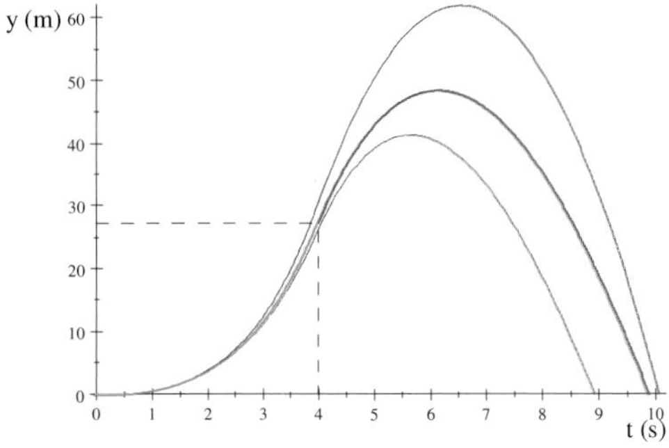

Figure 6.4 Varying variables

The reduced gravity and air resistance increased the rocket's maximum height and time to crash, as compared with the constant-gravity, linear-air-resistance result (**red**). ◀

Problems

1. Calculate the momentum of a baseball moving at $100\,\mathrm{mi/\,h}$. How much kinetic energy does it have? How quickly must it stop to impart a force of 1000 pounds?

2. Show that for an object with a constant mass, Eq. (6.2) reduces to the usual form of Newton's 2nd Law.

3. How much kinetic energy does a baseball have if it leaves the bat at a speed of $110\,\mathrm{mi/\,h}$?

4. The Impulse-Momentum Theorem states the change in momentum produce by a constant force is
$$\Delta\vec{P} = \vec{F}\,\Delta t$$
where Δt is the time interval over which it acts.

   a. Show that this is a direct result of the form of Newton's 2nd Law in Eq. (6.2).

   b. Calculate the change in momentum produced by a $1.333\,3\,\mathrm{N}$ force delivered over a $0.10\,\mathrm{s}$ interval.

5. The Impulse-Momentum Theorem states the change in momentum produce by a variable force is
$$\Delta\vec{P} = \int \vec{F}\,dt$$
where you integrate over the time interval.

   a. Show that this is a direct result of the form of Newton's 2nd Law in Eq. (6.30).

   b. Calculate the change in momentum produced by the force $F = 2\,\mathrm{N}\sin\dfrac{\pi t}{0.1\,\mathrm{s}}$ delivered over a $0.10\,\mathrm{s}$ interval.

6. Show that the magnitude of the force in Example 6.8 is
$$F = -kL\left(\left(1+\frac{x^2}{L^2}\right)^{1/2} - 1\right)$$
Use this result and Eq. (6.9b) to verify the result in the example.

7. Show that the work done by a varying force is
$$W = \vec{F}_{\text{ave}} \cdot \Delta\vec{r}$$
where $\vec{F}_{\text{ave}}$ is the average value of the force along the displacement.

8. Calculate the average value of the variable force $\vec{F} = 2x^2\,\mathrm{N}\widehat{x} + 6\sqrt{y}\,\mathrm{N}\widehat{y} + 3z\,\mathrm{N}\widehat{z}$ between the origin and the point $(3, 4, -2)$. Use the expression in the previous problem to verify the result of Example 6.7.

   Note: as you may recall, $f_{\text{ave}} = \dfrac{1}{x_2 - x_1}\displaystyle\int_{x_1}^{x_2} f(x)\,dx$.

9. Calculate the potential energy associated with the 1-dimensional force

$$F = -kx\left(1 - \left(1 + \frac{x^2}{L^2}\right)^{-1/2}\right)$$

from Example 6.8. Use the values for k and L from the example and plot the PE.

10. An object with unit mass, released from rest at $x_0 = \pi$, experiences the position-dependent force

$$F(x) = F_0\, e^{\sin x} \cos x$$

where $F_0 = 1\,\text{N}$.

a. Calculate and plot the potential energy as a function of position for the force.

b. Calculate the object's mechanical energy.

c. Add the mechanical energy to your plot.

11. Calculate the potential energy associated with the Yukawa force.

$$F\,(r) = -\frac{\kappa^2 e^{-mr}}{r^2}\,(1 + mr)$$

This interaction is used to describe the strong nuclear force.

12. The Lennard-Jones potential energy describes the interaction between two neutral molecules separated by a distance r.

$$U\,(r) = 4\epsilon\left(\frac{\rho^{12}}{r^{12}} - \frac{\rho^6}{r^6}\right)$$

a. Plot the potential energy as a function of separation between two molecules. Use $\epsilon = \rho = 1$ and let r range from 0.95 to 2.

b. Find the value of the separation for which the potential energy has a minimum.

c. Calculate the minimum value of the potential energy.

d. Calculate the force as a function of separation between two molecules.

13. What is the force acting on the object in Example 6.15?

14. Show that the velocity as a function of time for the object in Example 6.15 is

$$v\,(t) = v_0 \frac{\cos 2t\sqrt{\dfrac{k}{m}}}{\left(1 + \sqrt{1 - \dfrac{\beta}{kx_0^4}}\sin 2t\sqrt{\dfrac{k}{m}}\right)^{1/2}}$$

where v_0 is the initial velocity. Verify that the object moves back and forth between $x = 2.8059$ and $x = 0.35639$ by showing the velocity at those points is zero.

15. If the object in Example 6.15 was released from rest, show that its position as a function of time is

$$x\,(t) = x_0\left(\frac{1}{2}\left(1 + \frac{\beta}{kx_0^4} + \sqrt{1 - \frac{\beta}{kx_0^4}}\cos 2t\sqrt{\frac{k}{m}}\right)\right)^{1/2}$$

16. Show the result of the previous problem reduces to the correct result when $k = 0$.

17. Released from rest, a hockey puck slides down a frictionless ramp from an initial height of $2\,\mathrm{m}$. At the bottom of the ramp, the puck encounters a rough level surface. The coefficient of kinetic friction between the puck and the surface is $\mu_k = 0.35$.

    a. How fast is the puck moving at the bottom of the frictionless ramp?

    b. How far along the rough surface does the puck travel before stopping?

    c. Repeat the first two steps for an unspecified initial height h and coefficient μ_k. Verify your previous results.

18. A $0.163\,\mathrm{kg}$ hockey puck slides across a frictionless surface along the x-axis with a velocity of $+8\,\mathrm{m/\,s}$. At the origin, it encounters a 20° ramp that exerts a constant frictional force $f_k = \mu_k mg\cos\theta$, where $\mu_k = 0.30$.

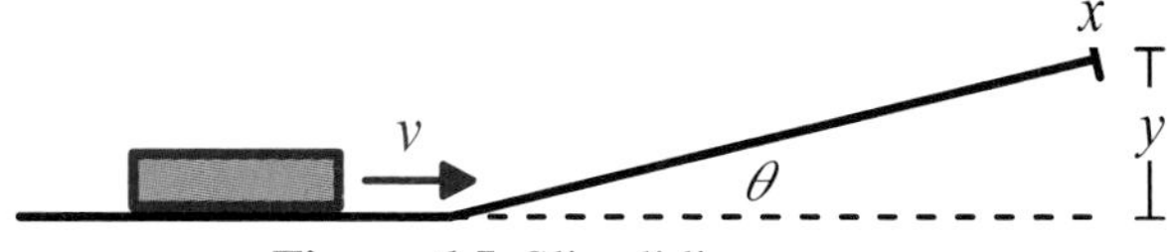

Figure 6.5 Slip sliding away

How far along the ramp does the puck slide until it stops? Compare your answer with the result of Example 6.18 and explain.

19. Starting from rest with an initial height of $0.8\,\mathrm{m}$, a $1.25\,\mathrm{kg}$ ball swings downward and strikes a $2.25\,\mathrm{kg}$ target ball that is at rest. The resulting collision is perfectly elastic.

    a. Calculate the speed of the $1.25\,\mathrm{kg}$ ball when it strikes the target ball.

    b. What is the velocity of each ball *immediately* after the collision?

    c. How high does the target ball swing after the collision?

20. Starting from rest with an initial height of $0.8\,\mathrm{m}$, a $1.25\,\mathrm{kg}$ stone swings downward and strikes a $2.25\,\mathrm{kg}$ target blob of clay that is at rest. The resulting collision is completely inelastic.

    a. Calculate the speed of the stone when it strikes the blob.

    b. What is the velocity of the stone-blob combo *immediately* after the collision?

    c. How high does the stone-blob combo swing after the collision?

21. A $^1/_2\,\mathrm{kg}$ ball, moving with a speed of $24\,\mathrm{m/\,s}$, has a perfectly elastic head-on collision with a $^3/_4\,\mathrm{kg}$ ball moving toward it at a speed of $20\,\mathrm{m/\,s}$. Calculate the final velocity of each ball.

22. A $^1/_2\,\mathrm{kg}$ ball, moving with a speed of $9\,\mathrm{m/\,s}$, has a head-on collision with a $^3/_4\,\mathrm{kg}$ ball moving toward it at a speed of $7\,\mathrm{m/\,s}$. Assuming the collision is perfectly elastic, calculate the final velocity of each ball.

23. A $^1/_4$ kg ball, moving with a speed of $22\,\mathrm{m/s}$ has a head-on collision with a $^1/_2$ kg ball moving toward it at a speed of $6\,\mathrm{m/s}$. The two balls collide elastically. Calculate the final velocity of each ball.

24. A $^1/_{10}$ kg ball, moving with a speed of $22\,\mathrm{m/s}$ has a collision with a $^1/_2$ kg ball moving in the same direction at a speed of $3\,\mathrm{m/s}$. The two balls collide elastically and in 1-dimension. Calculate the final velocity of each ball. Calculate the change of momentum for each ball, and verify that this collision conserves momentum.

25. A $50\,\mathrm{kg}$ skater is traveling due east at a speed of $4\,\mathrm{m/s}$. A $70\,\mathrm{kg}$ skater is moving due north at a speed of $6\,\mathrm{m/s}$. They collide and hold on to each other after the collision, managing to move off at some angle θ north of east with a speed of v_f. Assume that friction can be ignored. What is the final velocity (v_f) of the two skaters? Is this collision elastic? Use an energy calculation to support your answer.

26. A $0.50\,\mathrm{kg}$ ball is traveling due east at a speed of $8\,\mathrm{m/s}$. Another ball with a mass of $0.70\,\mathrm{kg}$ is moving due west at a speed of $6\,\mathrm{m/s}$. The two balls undergo an elastic head-on collision. Assume that friction can be ignored. Calculate the final velocity of each ball. Verify that the collision was elastic.

27. Prove that 1-d elastic collisions reverse the relative velocity between the two objects.

28. Explore a 1-dimensional elastic collision between two objects with much different masses, say $m_1 \gg m_2$, by letting the larger mass be very, very large.

29. A $2\,\mathrm{kg}$ mass moving at $4\,\mathrm{m/s}$ has an elastic collision with a second mass that is initially at rest. After the collision, the first mass moves off at a $90°$ angle at a speed of $2.31\,\mathrm{m/s}$. Calculate the mass and final velocity of the second mass.

30. A $^1/_4$ kg stone, moving with a speed of $22\,\mathrm{m/s}$ has a head-on collision with a $^1/_2$ kg blob of clay moving toward it at a speed of $6\,\mathrm{m/s}$. After the collision, the stone and clay blob merge into a single entity. Calculate the final velocity of the stone-blob combination. Calculate the total kinetic energy lost during this collision.

31. A $^1/_4$ kg stone, moving with a speed of $24\,\mathrm{m/s}$ to the right has a collision with a $^1/_2$ kg blob of clay moving to the left at a speed of $8\,\mathrm{m/s}$. The two objects collide head-on and stick together. Calculate the final velocity of the stone-clay combination. What is the ratio of the final kinetic energy to the initial kinetic energy of the system?

32. A $^1/_5$ kg stone, moving at $25\,\mathrm{m/s}$ has a collision with a $^1/_2$ kg lump of clay moving in the opposite direction at a speed of $9\,\mathrm{m/s}$. The two objects collide in 1-dimension and stick together.

    a. Calculate the change of momentum for each object. Is momentum conserved?

    b. Calculate the total change in the kinetic energy. Is this collision elastic?

33. Show the expression for the velocity of a rocket in deep space
$$v = v_0 - u \ln \left(1 - \frac{kt}{m_0}\right)$$
reduces to the expected result to first order in time.

34. How high would the rocket in Example 6.23 go if you launched it from the Moon's surface?

35. Use the burnout results of Eqs. (6.37) to verify the results of Example 6.23.

36. Suppose the rocket in Example 6.23 is modified so it can burn fuel for 5 seconds at the same rate. What is its mass ratio? Calculate its velocity, acceleration, and position when the engine shuts down.

37. How high would this modified rocket go if you launched it from the Earth's surface? Ignore air resistance.

38. Show the expression for the velocity of a rocket launched from the surface of a planet
$$v = v_0 - gt - u \ln \left(1 - \frac{kt}{m_0}\right)$$
reduces to the expected result to first order in time.

39. Use the variable mass $m = m_0 - kt$ and solve Eq. (6.38) exactly. Verify your answer agrees with Eq. (6.40).

40. Derive a general expression for the initial acceleration of a rocket launched from the Earth's surface. Calculate the rate of mass change for a rocket with an initial mass of 9×10^5 kg whose exhaust exits at $2000\,\mathrm{m/s}$ and has an initial acceleration of $\frac{1}{2}g$.

41. Derive a general expression for the acceleration as a function of time for a rocket launched near the Earth's surface, including the effects of air resistance when $b = k$.

42. Show that Eq. (6.46) for a rocket's height as a function of time including the effects of gravity and air resistance reduces to the expected result when you turn off the rocket engine and the effects of air resistance.

43. Prove that both terms in the constant $C \equiv \dfrac{m_0 g}{k - b} + \dfrac{uk}{b}$ have units of velocity.

44. Use Eqs. (6.45), (6.46), and (6.47) to verify the results of Example 6.27.

45. Verify that the burnout results of Eqs. (6.48) reduce to the correct expression when you turn off air resistance.

46. Use Examples 6.24 and 6.27 to reproduce the plots in Figures 6.2 and 6.3.

47. Verify the three landing speeds mentioned in Example 6.29.

7 Circular Motion

Many real-life motions include rotation. Merry-go-round riders move in circular paths, and the Earth rotates daily as it orbits the Sun in a nearly circular orbit. Tires and axles rotate as a car moves, and thrown and batted baseballs usually spin.

We could use an object's x and y coordinates to keep track of its position is it moves in a circle. It is easier to use the circle's radius (R) and the angle (θ) the object makes with the center of the circle and the x-axis (see Figure 7.1a). This coordinate system has the **radial** direction along the circle's radius, and the **tangential** direction is along the motion, perpendicular to the radius. The positive radial direction is "In" toward the center of the circle, and the negative radial direction is "Out" away from the center. The positive tangential direction is usually counterclockwise, and negative points clockwise.

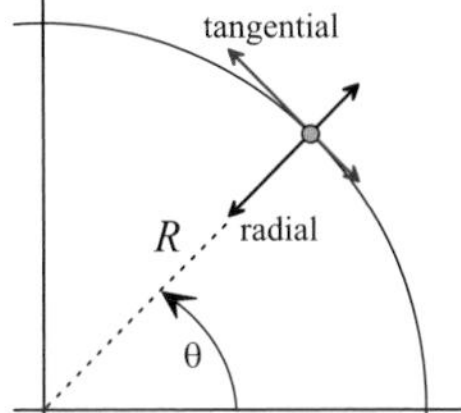

Figure 7.1a Angular Position

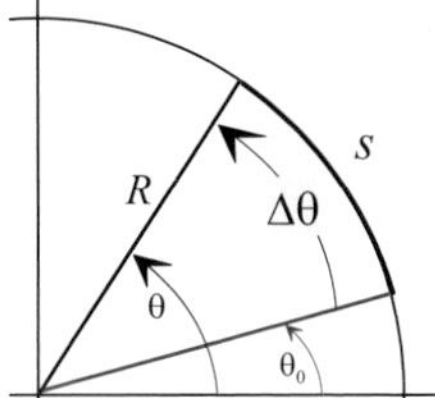

Figure 7.1b Angular Displacement

The **angular displacement** is the change in the object's angular position ($\Delta\theta$).

$$\Delta\theta = \theta - \theta_0 = \frac{s}{R} \tag{7.1}$$

The arc length s is the object's displacement as it moves along the circle from θ_0 to θ. When $\Delta\theta$ is positive, the net change in the object's angular position is counterclockwise (see Figure 7.1b) and it's negative when the net change in angular position is clockwise.

The ratio in Eq. (7.1) defines a unit called the **radian**, the standard unit for angles that *SNB* denotes rad and lists under Plane Angle in the Unit Name dialog box. The radian is a dimensionless unit, but you can think of it as a "meter-per-meter". It's a meter of displacement along the circular path per meter of radial distance from the circle's center.

The radian is one of three units for the angular displacement, along with degrees and revolutions. For one complete revolution, the arc length is the circle's circumference ($s = 2\pi R$) so Eq. (7.1) tells us there are $2\pi\,\mathrm{rad}$ in one revolution.

$$1\,\mathrm{rev} = 2\pi\,\mathrm{rad} = 360^\circ$$

SNB has two degree symbols but revolutions are not included in its built-in units. To create a user-defined unit, see Chapter 1 or *SNB*'s help.

Doing Physics with Scientific Notebook: A Problem-solving Approach, First Edition. Joseph Gallant.

Uniform Circular Motion

An object undergoing uniform circular motion (UCM) moves at a constant speed along a circular path. Circular motion is often due to the rotation of an object (such as a merry-go-round) and in UCM the object spins at a constant angular speed.

The **average angular speed** is the rate of change of angular displacement.

$$\omega = \frac{\Delta\theta}{\Delta t} \tag{7.2}$$

The metric unit for angular velocity is radians per second, although revolutions-per-second (rps) and revolutions per minute (rpm) are also common. When the problem asks "how many times...", you should give your answer for $\Delta\theta$ in revolutions or ω in revolutions per second or revolutions per minute.

The **period** (T) is the time to complete one revolution and is inversely related to the average angular velocity.

$$T = \frac{2\pi}{\omega} \tag{7.3}$$

The faster the object rotates, the less time it takes to complete one rotation. A careful look at the units will help you understand this. The period (in seconds per revolution) equals the number of radians per revolution divided by the number of radians per second.

One example of uniform circular motion is the rotation of a record. While the music is playing, a record rotates at a constant angular speed.

Example 7.1 *You say you want a revolution*

A local museum recently demonstrated an old phonograph by playing some vinyl LP records. What is the angular speed and period of the record's rotation? How many times did the record turn while each side played? How far does a point on the record's edge travel?

Solution. The old vinyl LPs were called "$33\frac{1}{3}$s" because their angular velocity was a constant $33\frac{1}{3}$ revolutions per minute while the music played. Let's use Evaluate Numerically to convert the angular velocity to radians per second.

$$\omega = 33\frac{1}{3}\frac{\text{rev}}{\text{min}} \times \frac{2\pi\,\text{rad}}{\text{rev}} \times \frac{1\,\text{min}}{60\,\text{s}} = 3.490\,7\frac{\text{rad}}{\text{s}}$$

Use Eq. (7.3) to create and Evaluate Numerically an expression for the period.

$$T = \frac{2\pi\,\text{rad}/\,\text{rev}}{3.490\,7\,\text{rad}/\,\text{s}} = 1.800\,0\frac{\text{s}}{\text{rev}}$$

The record completed one revolution every $1.8\,\text{s}$. Typically, each side of an LP played for about 22.5 minutes. Use Eq. (7.2) to create and Evaluate the following expression for the angular displacement.

$$\Delta\theta = \left(33\frac{1}{3}\frac{\text{rev}}{\text{min}}\right)(22.5\,\text{min}) = 750.0\,\text{rev}$$

The record turned 750 times per side. The vinyl LPs had a 12 inch diameter, so their radius was 6 in = $0.152\,4\,\mathrm{m}$. To calculate the distance travelled by a point on the record's edge, use Eq. (7.1) to create and Evaluate Numerically the following expression.

$$s = (0.152\,4\,\mathrm{m})\left(750.0\,\mathrm{rev} \times \frac{2\pi\,\mathrm{rad}}{\mathrm{rev}}\right) = 718.17\,\mathrm{m\,rad}$$

It's customary to drop the dimensionless "rad" for distance, so a point on the record's edge traveled $718.17\,\mathrm{m}$ while each side played. ◀

Since UCM is motion in a circle at constant speed, the speed equals the total distance traveled divided by the elapsed time. Along a circular path, the distance for one complete revolution is the circumference and the time is the period.

$$v = \frac{2\pi R}{T} \tag{7.4}$$

Again, a little unit analysis is helpful: the circumference is the number of meters per revolution and the period is the number of seconds per revolution.

The angular speed is related to the linear speed by the radius of the circular path.

$$v = R\omega \tag{7.5}$$

This relation quantifies what every merry-go-round rider knows: for a given angular speed, the further you are from the rotation center, the faster you go. In UCM, the speed (at a given radius), angular velocity, and period are all constant.

Example 7.2 *Take a sad song...*
The song "Hey Jude" lasts about 7 minutes and 8 seconds. How many times does the record turn while the song is playing? What is the linear speed of the record's edge?
Solution. Create the following expression, and use Evaluate and Evaluate Numerically to get the song's play time in minutes and then seconds.

$$t = \left(7.0 + \tfrac{8}{60}\right)\,\mathrm{min} = 7.133\,3\,\mathrm{min} = 428.00\,\mathrm{s}$$

Let's Evaluate this expression for the record's angular displacement.

$$\Delta\theta = \left(33\tfrac{1}{3}\frac{\mathrm{rev}}{\mathrm{min}}\right)(7.133\,3\,\mathrm{min}) = 237.78\,\mathrm{rev}$$

The record turned 237.78 times during the song. Use Eq. (7.5) to create and Evaluate an expression for the speed of a point on the record's edge.

$$v = (0.152\,4\,\mathrm{m})\left(3.490\,7\frac{\mathrm{rad}}{\mathrm{s}}\right) = 0.531\,98\frac{\mathrm{m}}{\mathrm{s}}\,\mathrm{rad}$$

Let's drop the "rad" so the speed is $v = 0.531\,98\,\mathrm{m/\,s}$. You could also use Eq. (7.4) to calculate the edge speed.

$$v = \frac{2\pi\,(0.152\,4\,\mathrm{m})}{1.8\,\mathrm{s}} = 0.531\,98\frac{\mathrm{m}}{\mathrm{s}}$$

◀

The Rotating Umbrella

Consider a wet umbrella held horizontally rotating at a constant angular velocity ω so that water drops leave all points along its edge and undergo projectile motion. [26] At what rotation angles do the launched drops achieve the maximum range and height?

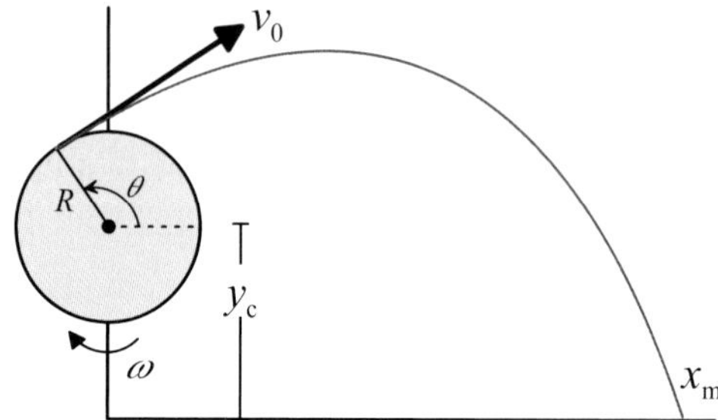

Figure 7.2 Raindrops keep falling on my head

This difficult problem combines uniform circular motion and projectile motion, and it is ideally suited for *SNB*.

Each drop is a projectile launched with initial speed $v_0 = R\omega$ from its initial position $x_0 = R\cos\theta$ and $y_0 = y_c + R\sin\theta$. The x- and y-coordinates of the drops along their trajectory are given by Eqs. (4.4)

$$x = R\cos\theta + R\omega t\sin\theta \tag{7.6a}$$

$$y = y_c + R\sin\theta - R\omega t\cos\theta - \tfrac{1}{2}gt^2 \tag{7.6b}$$

where θ is the rotation angle of the umbrella when the drop was launched. This angle is *not* the launch angle used in Chapter 4 (that's $\theta_0 = \theta - 90°$). We could use our previous results for projectile motion, but instead let's put *SNB*'s numerical capabilities to work.

Example 7.3 *Umbrella extrema*

A wet umbrella with a $1\,\mathrm{m}$ diameter, centered at $y_c = 1\,\mathrm{m}$, rotates with a constant angular speed of $10\,\mathrm{rad/\,s}$. What rotation angles produce trajectories with the maximum range and height?

Solution. We want the three rotation angles that maximize the range (left and right) and maximum height, so we need expressions for height and range as functions of angle. You can find the time of flight as a function of angle by setting the y-coordinate equal to zero and using Solve Exact. Enter t in the Variable(s) to Solve for box.

$0 = 1 + 0.5\sin\theta - 5t\cos\theta - \frac{1}{2}(9.8067)\,t^2$, Solution is:

$$0.101\,97\sqrt{12.5\cos 2\theta + 9.806\,7\sin\theta + 32.113} - 0.509\,86\cos\theta$$

Let's use this time, Eq. (7.6a), and New Definition to define a function for a drop's horizontal distance traveled as a function of angle.

$$x(\theta) = 0.5\cos\theta + 5\left(0.101\,97\sqrt{12.5\cos 2\theta + 9.806\,7\sin\theta + 32.113} - 0.509\,86\cos\theta\right)\sin\theta$$

Let's find the trajectory with the maximum range to the right, which has a rotation angle between $90°$ and $180°$ (see Figure 7.2).

Create an expression setting the derivative of the horizontal distance to zero and use Solve Numeric with this condition. The **red** degree symbol in the argument of $x(\theta°)$ ensures *SNB* returns the answer in degrees.

$$\begin{bmatrix} 0 = \dfrac{dx(\theta°)}{d\theta} \\ \theta \in (90, 180) \end{bmatrix}, \text{ Solution is: } \{[\theta = 119.82]\}$$

Because the drop's initial and final heights are not the same, this angle is ***not*** $135°$. Use Evaluate Numerically to find the maximum range to the right.

$$x_m = x(119.82°) = 3.4420$$

The drop with the maximum range to the right lands at $3.4420\,\mathrm{m}$. The trajectory with the maximum range to the left has a rotation angle between $180°$ and $270°$, so repeat the procedure with the new initial conditions.

$$\begin{bmatrix} 0 = \dfrac{dx(\theta°)}{d\theta} \\ \theta \in (180, 270) \end{bmatrix}, \text{ Solution is: } \{[\theta = 226.77]\}$$

This is not $225°$! Use Evaluate Numerically to find the maximum range to the left.

$$x_m = x(226.77°) = -3.4420$$

The drop with the maximum range to the left lands at $-3.4420\,\mathrm{m}$, so the maximum ranges are symmetric about the origin.

You can find the time it takes each drop to reach its maximum height (as a function of rotation angle) by setting the time-derivative of the y-coordinate equal to zero and using Solve Exact to find t.

$$0 = \frac{d}{dt}\left(1 + 0.5\sin\theta - 5t\cos\theta - \tfrac{1}{2}(9.8067)\,t^2\right), \text{ Solution is: } -0.50986\cos\theta$$

Now use this time, Eq. (7.6b), and New Definition to define the maximum height as a function of rotation angle.

$$y(\theta) = 1 + 0.5\sin\theta - 5(-0.50986\cos\theta)\cos\theta - \tfrac{1}{2}(9.8067)(-0.50986\cos\theta)^2$$

The maximizing trajectory will have a rotation angle between $90°$ and $180°$. Once again, let's use Solve Numeric.

$$\begin{bmatrix} 0 = \dfrac{dy(\theta°)}{d\theta} \\ \theta \in (90, 180) \end{bmatrix}, \text{ Solution is: } \{[\theta = 168.69]\}$$

Every trajectory has a maximum height, so this angle produces the maximum maximum-height. The answer is not $180°$!

The straight-up $\theta = 180°$ launch produces the maximum height only when $\sqrt{Rg} > v_0$ (see Problem 11). In our example here, $\sqrt{Rg} \approx 2.2\,\mathrm{m/s}$ is less than the initial launch speed $v_0 = (10/\mathrm{s})\left(\frac{1}{2}\,\mathrm{m}\right) = 5\,\mathrm{m/s}$. Use Evaluate Numerically to find the maximum maximum-height.

$$y_m = y\,(168.69°) = 2.3237$$

The highest projected water drop will be $2.3237\,\mathrm{m}$ off the ground. There is another solution between $0°$ and $90°$.

$$\begin{bmatrix} 0 = \dfrac{dy\,(\theta°)}{d\theta} \\ \theta \in (0, 90) \end{bmatrix}, \text{ Solution is: } \{[\theta = 11.311]\}$$

We'll see the significance of this second solution in the next example. ◀

Now that we know the maximum ranges and height and the launch angles that produced them, we can plot the trajectories.

Example 7.4 *Umbrella graphica*
Plot the trajectories calculated in the previous example.
Solution. We know the coordinates as functions of time, so let's use Plot 2D + Parametric to plot the maximal trajectories. We'll plot the general parametric expression

$$\left(R\cos\theta + v_0 t\sin\theta,\ y_c + R\sin\theta - v_0 t\cos\theta - \tfrac{1}{2}gt^2\right)$$

where $y_c = 1$, $R = 1$, $v_0 = 5$, $g = 9.8067$ and θ is one of the three maximizing angles we calculated in the previous example.

Let's start with the trajectory with the maximum range to the right, where $\theta = 119.82°$. Create and plot the following parametric expression with either Plot 2D + Rectangular or Plot 2D + Parametric.

$$\left(0.5\cos 119.82° + 5t\sin 119.82°, 1 + 0.5\sin 119.82° - 5t\cos 119.82° - \tfrac{1}{2}9.8067t^2\right)$$

To find the drop's time of flight, create an equation setting the y-coordinate equal to zero and apply Solve Exact to it.

$$0 = 1 + 0.5\sin 119.82° - 5t\cos 119.82° - \tfrac{1}{2}9.8067t^2, \text{ Solution is: } 0.85078$$

Set the Interval of the Variable t from 0 to 0.85078 for this Item. Add a thick black circle for the umbrella. Create the following parametric expression for a circle with radius $\frac{1}{2}$ centered at $(0, 1)$, add it to your plot, and for this Item let t run from 0 to 6.2832 (about 2π).

$$\left(0 + \tfrac{1}{2}\cos t, 1 + \tfrac{1}{2}\sin t\right)$$

Repeat the procedure with the other two maximizing angles ($\theta = 168.69°$ for the maximum height and $\theta = 226.77°$ for the maximum range to the left). Be sure to click the Equal Scaling Along Each Axis box on the Axes tab of the Plot Properties dialog, and include Labels for each axis.

Figure 7.3 shows the resulting plot with the three maximal trajectories and the umbrella.

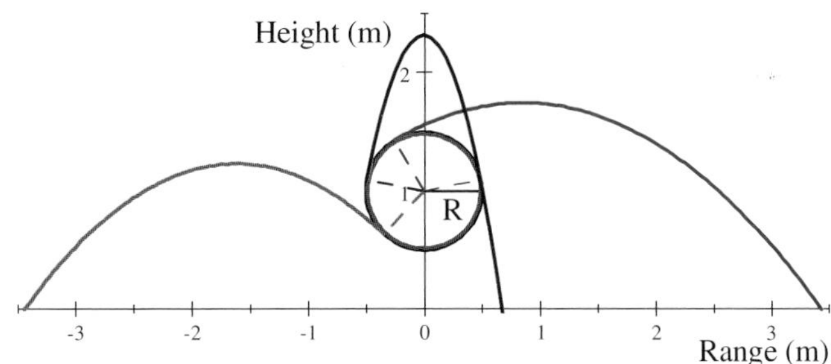

Figure 7.3 Call your shots

On its way down, the drop with the largest maximum height just skims the edge of the umbrella at $\theta = 11.311°$. If the umbrella was rotating in the opposite direction, then $11.311°$ would be the launch point for this trajectory. ◀

Rotational Kinematics

Not all rotational motion is uniform; in non-uniform circular motion, the angular speed changes. The **average angular acceleration** is the rate of change of the angular speed

$$\alpha = \frac{\Delta\omega}{\Delta t} \tag{7.7}$$

where $\Delta\omega = \omega - \omega_0$ is the change in angular speed. The usual unit for angular acceleration is the $\mathrm{rad/s^2}$, but we'll also use $\mathrm{rev / min^2}$.

From the definitions of angular displacement, speed, and acceleration, we can derive four basic equations which describe the angular position and speed of objects rotating with a constant acceleration.

$$\theta = \theta_0 + \omega_0 t + \tfrac{1}{2}\alpha t^2 \tag{7.8a}$$

$$\theta = \theta_0 + \omega_{ave}\, t \tag{7.8b}$$

$$\omega = \omega_0 + \alpha t \tag{7.8c}$$

$$\omega^2 = \omega_0^2 + 2\alpha\,(\theta - \theta_0) \tag{7.8d}$$

For constant angular acceleration, the average angular speed is $\omega_{ave} = \frac{1}{2}(\omega_0 + \omega)$.

Do these equations look familiar? These are the same as Eqs. (2.5) from Chapter 2. The connection between these angular quantities and their linear counterparts is the radius R of the circular path.

$$s = R\theta$$

$$v = R\omega$$

$$a = R\alpha$$

The radius is perpendicular to the displacement, velocity, and the component of the acceleration that changes the object's speed.

Example 7.5 *Turn that crap off*

After the LP record in our first two examples is finished, you turn it off and it slows to a stop after 10 seconds. Calculate the record's angular acceleration, assuming it to be constant. How many times does the record turn while it is slowing down?

Solution. Since we want $\Delta\theta$ in revolutions, let's use Eq. (7.7) to Evaluate the acceleration in revolutions and minutes.

$$\alpha = \frac{0 - 33\frac{1}{3}\dfrac{\text{rev}}{\text{min}}}{10\,\text{s} \times \dfrac{1\,\text{min}}{60\,\text{s}}} = -200\frac{\text{rev}}{\text{min}^2}$$

Now that we know the time and angular acceleration, we can use Eq. (7.8a) and Evaluate Numerically to find the number of times the record turns while it's slowing down.

$$\Delta\theta = \left(33\tfrac{1}{3}\frac{\text{rev}}{\text{min}}\right)\left(\tfrac{1}{6}\,\text{min}\right) + \tfrac{1}{2}\left(-200\frac{\text{rev}}{\text{min}^2}\right)\left(\tfrac{1}{6}\,\text{min}\right)^2 = 2.7778\,\text{rev}$$

The record turns about 2.8 times while it's slowing down. ◀

When the angular acceleration is not constant, you must use calculus to describe the object's rotational motion. The **angular velocity** is the rate of change of the angular displacement.

$$\omega = \frac{d\theta}{dt} \tag{7.9}$$

If you know the angular position as a function of time, you can calculate the angular speed.

The **angular acceleration** is the rate of change of angular speed.

$$\alpha = \frac{d\omega}{dt} \tag{7.10}$$

If you know the angular speed as a function of time, you can calculate the angular acceleration.

From these definitions we can derive four basic equations which describe the angular position and speed of objects rotating with a varying acceleration.

$$\theta(t) = \theta_0 + \int \omega(t)\,dt \tag{7.11a}$$

$$\theta(t) = \theta_0 + \omega_{ave}\,t \tag{7.11b}$$

$$\omega(t) = \omega_0 + \int \alpha(t)\,dt \tag{7.11c}$$

$$\omega^2(\theta) = \omega_0^2 + 2\int \alpha(\theta)\,d\theta \tag{7.11d}$$

For varying angular acceleration, the average angular velocity

$$\omega_{ave} = \frac{1}{T}\int_0^T \omega(t)\,dt$$

does not equal the arithmetic average of the initial and final angular velocities.

The Compact Disk

The rotational motion of a CD is an excellent example of nonuniform rotational motion with a time-dependent acceleration. To read the information on a record, a needle starts at the outer edge and follows a single spiral track while the disk rotates at a constant angular speed. To read the information on a CD, a laser starts at the inner radius and follows a single spiral track while the disk rotates at a decreasing angular speed. This ensures the data-read rate, which is proportional to the linear speed, is constant. [31]

The data on a typical CD is stored between $R_0 = 2.3\,\mathrm{cm}$ and $R_f = 5.8\,\mathrm{cm}$, and the spacing between tracks is $\Delta r = 1.6\,\mu\mathrm{m}/\,\mathrm{rev}$. We can use Evaluate Numerically to calculate the total number of times the disk rotates.

$$\Delta\theta = \frac{R_f - R_0}{\Delta r} = \frac{5.8\,\mathrm{cm} - 2.3\,\mathrm{cm}}{1.6\,\mu\mathrm{m}/\,\mathrm{rev}} = 21875.\,\mathrm{rev}$$

The arc length of the spiral track is approximately the number of rotations times the average circumference.

$$s \approx 2\pi R_{ave} \times \Delta\theta = 2\pi\,(4.05\,\mathrm{cm})\,(21875) = 5566.5\,\mathrm{m}$$

That's almost 3.5 miles! Since the linear speed is constant, it's just the length of the track divided by the total playing time of $T = 72\,\mathrm{min}$.

$$v = \frac{5566.5\,\mathrm{m}}{72\,\mathrm{min}} = 1.288\,5\frac{\mathrm{m}}{\mathrm{s}}$$

The laser moves along the track at a very pedestrian $v \approx 3\,\mathrm{mi}/\,\mathrm{h}$.

The radius of the spiral track as a function of angle is

$$r(\theta) = R_0 + \rho\theta \tag{7.12}$$

where $\rho = \Delta r/2\pi$ is the track spacing if the angle is measured in radians (the 2π is the revolutions-to-radians conversion factor). The arc length along the spiral track is

$$s = \int_{\theta_1}^{\theta_2} r(\theta)\,d\theta \tag{7.13}$$

where θ_1 and θ_2 are the rotation angles of the disk at the start and end of the track.

Since the track speed is constant, we can use $\omega = v/r(\theta)$ and Eq. (7.9) to create a differential equation, and then use Solve ODE + Exact to find the rotation angle as a function of time.

$$\begin{bmatrix} \dfrac{d\theta}{dt} = \dfrac{\omega_0 R_0}{R_0 + \rho\theta} \\ \theta(0) = 0 \end{bmatrix}, \text{ Exact solution is:}$$

$$\left\{ -\frac{1}{\rho}\left(R_0 + \sqrt{R_0^2 + 2t\rho\omega_0 R_0}\right), -\frac{1}{\rho}\left(R_0 - \sqrt{R_0^2 + 2t\rho\omega_0 R_0}\right) \right\}$$

Take the positive (second) solution and edit it by-hand to get an expression for the angle as a function of time (when $\theta_0 = 0$).

$$\theta(t) = \frac{R_0}{\rho}\left(\sqrt{1 + \frac{2\omega_0\rho}{R_0}t} - 1\right) \tag{7.14}$$

Now that we have the angle as a function of time, we can use Eqs. (7.12) and (7.14) to create an expression for the track radius as a function of time.

$$r(t) = R_0\sqrt{1 + \frac{2\omega_0\rho}{R_0}t} \tag{7.15}$$

Use Expand and Eq. (7.14) to calculate the angular speed by taking the derivative with respect to time.

$$\frac{d}{dt}\left(\frac{R_0}{\rho}\left(\sqrt{1 + \frac{2\omega_0\rho}{R_0}t} - 1\right)\right) = \frac{\omega_0}{\sqrt{2t\rho\frac{\omega_0}{R_0} + 1}}$$

The angular speed as a function of time is

$$\omega(t) = \frac{\omega_0}{\sqrt{1 + \frac{2\rho\omega_0}{R_0}t}} \tag{7.16}$$

where $\omega_0 = v/R_0$ is the initial angular speed.

Let's use Expand and Eq. (7.10) to calculate the CD's angular acceleration.

$$\frac{d}{dt}\left(\frac{\omega_0}{\sqrt{1 + \frac{2\rho\omega_0}{R_0}t}}\right) = -\rho\frac{\omega_0^2}{R_0\left(2t\rho\frac{\omega_0}{R_0} + 1\right)^{\frac{3}{2}}}$$

A little editing by-hand produces the CD's angular acceleration as a function of time.

$$\alpha(t) = -\frac{\frac{\rho\omega_0^2}{R_0}}{\left(1 + \frac{2\rho\omega_0}{R_0}t\right)^{3/2}}$$

The acceleration is negative because the CD is slowing down.

Example 7.6 *Average angular speed*
How much time does it take until the CD's actual angular speed equals the average? How many rotations?

Solution. We'll need the initial angular speed to calculate the average angular speed, so create and Evaluate the following expression. Let's work in revolutions and minutes.

$$\omega_0 = \frac{v}{R_0} = \frac{1.288\,5\frac{\text{m}}{\text{s}}}{0.023\,\text{m}} \times \frac{1\,\text{rev}}{2\pi} \times \frac{60\,\text{s}}{\text{min}} = 534.98\frac{\text{rev}}{\text{min}}$$

You'll also need to Evaluate the numerical prefactor multiplying the time.

$$\frac{2\rho\omega_0}{R_0} = \frac{2\left(1.6\times10^{-6}\dfrac{\mathrm{m}}{\mathrm{rev}}\right)\left(534.98\dfrac{\mathrm{rev}}{\mathrm{min}}\right)}{0.023\,\mathrm{m}} = \frac{7.443\,2\times10^{-2}}{\mathrm{min}}$$

Here's our expression for the angular speed as a function of time.

$$\omega\left(t\right) = \frac{535.59}{\sqrt{1+7.443\,2\times10^{-2}\,t}}$$

Use this expression and Evaluate Numerically to calculate the average angular speed.

$$\omega_{ave} = \frac{1}{T}\int_0^T \omega\left(t\right)dt = \frac{1}{72}\int_0^{72}\left(\frac{535.59}{\sqrt{1+7.443\,2\times10^{-2}\,t}}\right)dt = 304.16$$

The CD's average angular speed is $304.16\,\mathrm{rev}\,/\,\mathrm{min}$. Use Solve Exact to find the time.

$$304.16 = \frac{534.97}{\sqrt{7.443\,2\times10^{-2}t+1.0}},\ \text{Solution is: } 28.127$$

The actual angular speed equals the average value $28.127\,\mathrm{min}$ after the CD starts to play. With numerical values Substitute (with Evaluate).

$$\theta\left(t\right) = \left[\frac{R_0}{\rho}\left(\sqrt{1+\frac{2\omega_0\rho}{R_0}t}-1\right)\right]_{R_0=0.023,\rho=1.6\times10^{-6},\omega_0=534.98}$$
$$= 14375.\sqrt{7.443\,2\times10^{-2}t+1} - 14375.$$

Use this expression to Evaluate the number of revolutions.

$$\Delta\theta = \left[14375.\sqrt{7.443\,2\times10^{-2}t+1} - 14375.\right]_{t=28.127} = 10908.$$

By the time the actual and average angular speeds are equal, the CD has turned 10908 times. The following figures show the angular displacement (in revolutions) and speed (in rpm) for a CD's entire 72 min playing time.

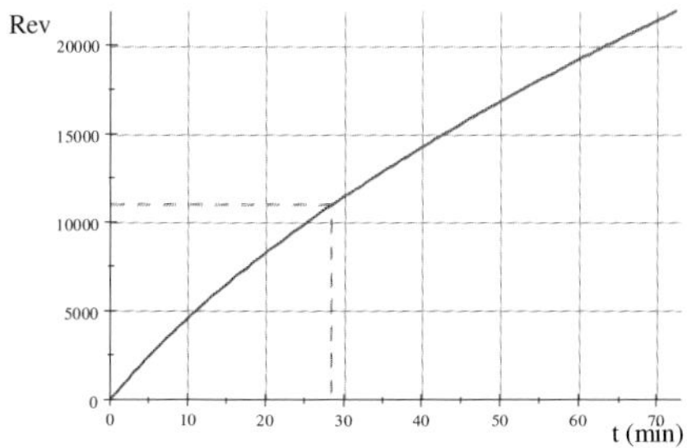

Figure 7.4a A CD's $\theta\left(t\right)$

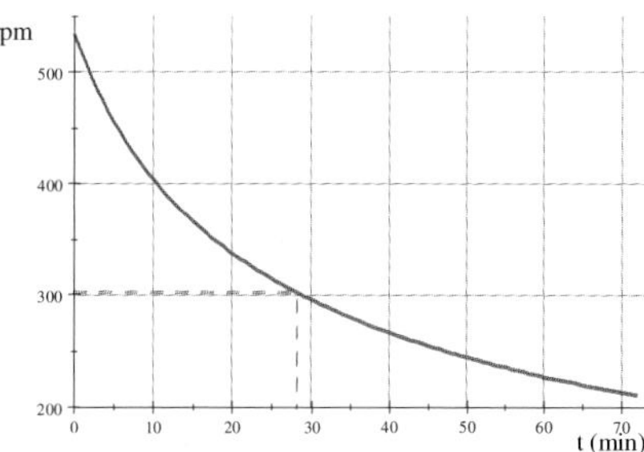

Figure 7.4b A CD's $\omega\left(t\right)$

◀

Example 7.7 *...and make it better*
How many times does a CD turn while the song "Hey Jude" is playing? What is the average angular speed of the disk during the song? How far does a point along the track travel?
Solution. Unlike with a record, all these answers depend on the song's location on the disk. We'll take the two limiting cases when the song is either first or last on the disk.

Let's start when the song is first. Use Eq. (7.11a) to create and Evaluate an expression for the number of rotations when the song is first on the disk.

$$\Delta\theta = \int_0^{7.133\,3} \left(\frac{535.59}{\sqrt{1 + 7.443\,2 \times 10^{-2}\,t}}\right) dt = 3415.3$$

The disk rotates 3415.3 times. Create and Evaluate the following expression for the average angular velocity during the disk's first $7.133\,3\,\mathrm{min}$ of playing time.

$$\omega_{ave} = \frac{1}{7.133\,3}\int_0^{7.133\,3} \left(\frac{535.59}{\sqrt{1 + 7.443\,2 \times 10^{-2}\,t}}\right) dt = 478.78$$

The disk's average angular velocity is $478.78\,\mathrm{rev}\,/\,\mathrm{min}$.

Eq. (7.13) gives the distance travelled by a point along the track (the arc length).

$$s = 2\pi \int_{\theta_1}^{\theta_2} (R_0 + \rho\theta)\, d\theta$$

Create this expression with all angles in revolutions and use Evaluate Numerically.

$$s = 2\pi \int_0^{3415.3} \left(0.023 + \left(1.6 \times 10^{-6}\right)\theta\right) d\theta = 552.19$$

The point along the spiral track travels $552.19\,\mathrm{m}$.

Now let's repeat our calculations when the song is last on the disk. The basic expressions are the same, but the limits of integration are different.

$$\Delta\theta = \int_{64.867}^{72} \left(\frac{535.59}{\sqrt{1 + 7.443\,2 \times 10^{-2}\,t}}\right) dt = 1548.0$$

$$\omega_{ave} = \frac{1}{7.133\,3}\int_{64.867}^{72} \left(\frac{535.59}{\sqrt{1 + 7.443\,2 \times 10^{-2}\,t}}\right) dt = 217.01$$

$$s = 2\pi \int_{20\,327}^{21\,875} \left(0.023 + \left(1.6 \times 10^{-6}\right)\theta\right) d\theta = 552.08$$

Since the disk is slowing down, the average angular speed and number of revolutions are smaller. But the linear speed is constant, so the distance along the spiral track is the same (within rounding error). ◀

Newton's Second Law and Circular Motion

A circular path is not a straight line. Even when the object's speed is constant, so its velocity vector has a constant magnitude, its direction changes along a circular path. Per Newton's 1st Law, there must be a non-zero net force acting on the object. To understand and explain circular motion, let's return to Newton's 2nd Law.

Uniform Circular Motion and the 2nd Law

The net force required to move an object in a circular path at constant speed points in the radial direction, toward the center of the circle. The resulting **centripetal** ("center seeking") **acceleration** also points "In" and has a magnitude

$$a_{\mathrm{c}} = \frac{v^2}{R} \tag{7.17}$$

that depends on the object's speed and the circle's radius.

The net force is still mass times acceleration, and in UCM it equals the net of the forces acting radially. Newton's 2nd Law takes the following form.

$$m\frac{v^2}{R} = \text{In} - \text{Out} \tag{7.18}$$

The strategy is the same as it was in Chapter 5. Use a free-body diagram to identify all the forces acting on the object, and then solve the 2nd Law.

Example 7.8 *Tote that barge, lift that bail, and swing it vertically!*
You want to use a $1.2\,\mathrm{m}$ rope to swing a $5\,\mathrm{kg}$ bucket of water in a vertical circle for a demonstration. What is the minimum angular speed you'll need? How much net force must you exert?
Solution. The minimum speed occurs when the tension in the string at the top of the circle is zero. Go slower and the object falls, go faster and the tension increases. At the top, both forces acting on the bucket (weight and tension) are "In" and there are no "Out" forces. Here are the bucket's FBD and 2nd Law equation, where F is the tension.

Newton's 2nd Law

$$m\frac{v^2}{R} = F + mg$$

FBD

F mg

Create an expression to Substitute $F = 0$ into Newton's 2nd Law and use Solve Exact to find v.

$$\left[m\frac{v^2}{R} = F + mg\right]_{F=0}, \text{ Solution is: } \sqrt{R}\sqrt{g}$$

The bucket's minimum speed is $v_{\min} = \sqrt{Rg}$.

Let's Evaluate the bucket's minimum speed $v_{\min}$ with the appropriate numerical values.

$$v_{\min} = \sqrt{Rg} = \sqrt{(1.2\,\mathrm{m})\left(9.8067\frac{\mathrm{m}}{\mathrm{s}^2}\right)} = 3.430\,5\frac{\mathrm{m}}{\mathrm{s}}$$

Now use Eq. (7.5) and Evaluate Numerically to calculate the minimum angular speed.

$$\omega_{\min} = \frac{v_{\min}}{R} = \frac{3.430\,5\frac{\mathrm{m}}{\mathrm{s}}}{1.2\,\mathrm{m}} \times \frac{1\,\mathrm{rev}}{2\pi} = 0.454\,98\frac{\mathrm{rev}}{\mathrm{s}}$$

It only takes about a half revolution-per-second. Let's Evaluate the minimum net force with the appropriate numerical values.

$$m\frac{v^2}{R} = (5\,\mathrm{kg})\frac{\left(3.430\,5\frac{\mathrm{m}}{\mathrm{s}}\right)^2}{1.2\,\mathrm{m}} = 49.035\frac{\mathrm{m}}{\mathrm{s}^2}\,\mathrm{kg}$$

The minimum angular speed does not depend on the bucket's mass, but the net force does. In fact, the minimum net force required to maintain the bucket in a vertical circle equals its weight. ◄

Example 7.9 *Tote that barge, lift that bail, and swing it horizontally!*
A tennis ball (mass $57\,\mathrm{g}$) is suspended by a string of length $L = 1\,\mathrm{m}$ and swung in a horizontal circle so that it makes an angle $\theta = 60\,°$ to the vertical (see Figure 7.5a).

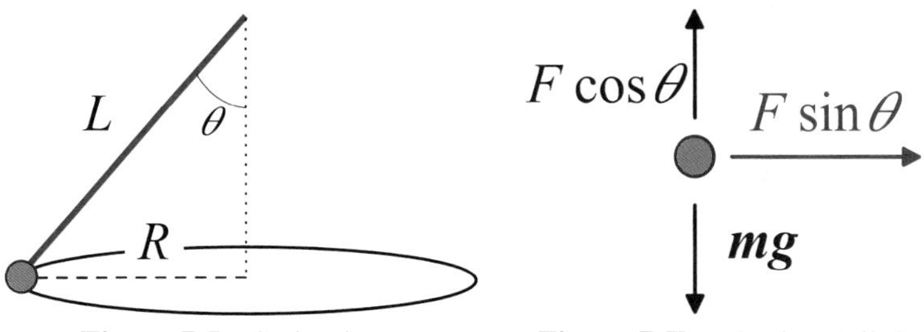

Figure 7.5a Swing it... **Figure 7.5b** ...horizontally!

What is the tension in the string? How much time does one orbit take?
Solution. Look at the FBD for the ball (Figure 7.5b). The vertical component of the tension balances the ball's weight and the horizontal component holds it in its circular path with speed $v = R\omega$ and radius $R = L\sin\theta$. We'll need both 2nd Law equations to find the tension and period.

$$mL\omega^2 = F \qquad \text{(Radial)}$$

$$0 = F\cos\theta - mg \qquad \text{(Vertical)}$$

We could use Solve Numeric and the numerical values, but there's something to learn from the general result. First, tell *SNB* the variables are positive.

$$\mathrm{assume}\,(\mathrm{positive}) = (0, \infty)$$

Then enter the two equations in a matrix and use Solve Exact to find F, ω.

$$\begin{bmatrix} mL\omega^2 = F \\ mg = F\cos\theta \end{bmatrix}, \text{ Solution is: } \left[F = g\frac{m}{\cos\theta}, \omega = \frac{1}{\sqrt{L}}\sqrt{g}\sqrt{\frac{1}{\cos\theta}} \right]$$

You can calculate the tension and angular speed using Substitute (with Simplify). A tennis ball has a 0.057 kg mass.

$$F = \left[\frac{mg}{\cos\theta}\right]_{m=0.057\,\mathrm{kg}, g=9.8067\,\mathrm{m/s^2}, \theta=60^\circ} = 1.1180\frac{\mathrm{m}}{\mathrm{s}^2}\,\mathrm{kg}$$

$$\omega = \left[\sqrt{\frac{g}{L\cos\theta}}\right]_{L=1\,\mathrm{m}, g=9.8067\,\mathrm{m/s^2}, \theta=60^\circ} = \frac{4.4287}{\mathrm{s}}$$

There is 1.11180 N of tension in the string and the ball has an angular speed of 4.4287 rad/ s. Use Eq. (7.3) and Evaluate Numerically to calculate the period.

$$T = \frac{2\pi\,\mathrm{rad/rev}}{4.4287\,\mathrm{rad/s}} = 1.4187\frac{\mathrm{s}}{\mathrm{rev}}$$

The ball completes one orbit every 1.4187 seconds.

Here's the promised something-to-learn. The angular speed has two interesting limits.

$$\omega = \begin{cases} \sqrt{\frac{g}{L}} & \text{if } \theta = 0^\circ \\ \infty & \text{if } \theta = 90^\circ \end{cases}$$

The $\theta = 0^\circ$ result is the angular frequency of a simple pendulum, and when $\theta = 90^\circ$ the angular speed is infinite. It's impossible to rotate an object in a horizontal circle without exerting some upward force to balance the object's weight. ◄

Non-Uniform Circular Motion and the 2nd Law

Now let's consider non-uniform circular motion (NCM) where the velocity's magnitude and direction both change. The net force has a radial and tangential component.

$$ma_{\mathrm{c}} = m\frac{v^2}{R} \tag{7.19a}$$

$$ma_{\mathrm{t}} = m\frac{dv}{dt} \tag{7.19b}$$

The radial component of the net force changes the object's direction, and the tangential component changes its speed.

The tangential component of the net force is the net of the forces acting tangentially. We'll call the forces that act in the direction of the motion "With" and those that oppose the motion "Against". Newton's 2nd Law takes the following form.

$$m\frac{dv}{dt} = \text{With} - \text{Against} \tag{7.20}$$

The strategy remains the same. Use a free-body diagram to identify all the forces acting on the object, and then solve the resulting 2nd Law equations.

Sliding on a Sphere

One common example of NCM is the motion of an object sliding on the surface of a sphere. An object is released (usually from rest) at the top of a hemisphere, and it slides along the surface. First we'll let our object slide on a smooth sphere with no friction. Then we'll explore a surprisingly more interesting problem and let the object slide on a rough surface where it encounters kinetic friction.

Sliding with no Friction

The only forces acting on the object are gravity and the normal force. The following figure shows the object's free-body diagram. Notice that positive is clockwise and $\theta = 0$ is the top of the sphere.

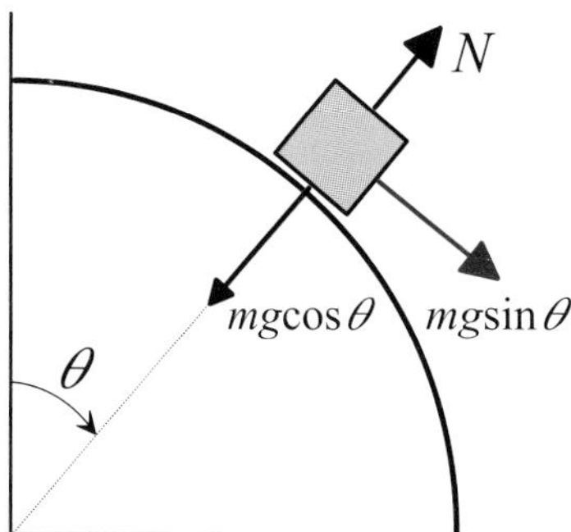

Figure 7.6 No-slip sliding away

Since there is no friction to slow the sliding object, it will speed up until it flies off the sphere's surface. We can calculate the object's fly-off angle (θ_f) and fly-off speed (v_f) with Newton's 2nd Law and energy conservation.

Here are the equations of motion per Newton's 2nd Law.

$$m\frac{v^2}{R} = mg\cos\theta - N \tag{7.21a}$$

$$m\frac{dv}{dt} = mg\sin\theta \tag{7.21b}$$

When the object leaves the surface, the normal force is zero. From the radial equation, we see this occurs when the object reaches its fly-off speed of $v_f = \sqrt{Rg\cos\theta_f}$. This is the maximum speed of the sliding object while it is on the sphere's surface.

We can use energy conservation to relate the object's speed and angle. For an object released from rest at the top of the sphere, the initial potential energy equals the final potential plus kinetic energy.

$$mgR = \tfrac{1}{2}mv^2 + mgR\cos\theta \tag{7.22}$$

To find the fly-off angle, Substitute the fly-off speed into this expression, use Solve Exact to find the fly-off angle, and apply Evaluate Numerically to the result.

$$\left[mgR = \tfrac{1}{2}mv^2 + mgR\cos\theta°\right]_{v=\sqrt{Rg\cos\theta°}}, \text{ Solution is: } -\frac{180}{\pi}\arccos\frac{2}{3} = -48.190$$

The block leaves the sphere's surface at $\theta_f = \pm 48.190°$.

This example shows you how to find the fly-off angle and speed simultaneously.

Example 7.10 *Don't fly-off the handle*
A block starts from rest at the top of a $1.5\,\mathrm{m}$-radius sphere. Where does it leave the sphere? How fast is it going when it does?
Solution. Set the normal force to zero, and apply Solve Numeric to Eqs. (7.21a) with $N = 0$ and (7.22) to calculate the block's fly-off angle and speed.

$$\begin{bmatrix} \dfrac{v_f^2}{1.5} = 9.8067\cos\theta_f^\circ \\ \frac{1}{2}v_f^2 + 1.5\,(9.8067)\cos\theta_f^\circ = 1.5\,(9.8067) \\ \theta_f \in (0, 90) \\ v_f \in (0, 10) \end{bmatrix},$$

Solution is: $\{[\theta_f = 48.190, v_f = 3.1316]\}$

The block leaves the sphere's surface at $\theta_f = 48.190^\circ$ moving at a speed of $v_f = 3.1316\,\mathrm{m/s}$. ◄

The following equations give the speed as a function of angle and the fly-off angle and speed for an object released from rest at the top of the sphere.

$$v = \sqrt{2Rg\,(1 - \cos\theta)} \tag{7.23a}$$

$$\theta_f = \cos^{-1}\left(\tfrac{2}{3}\right) \tag{7.23b}$$

$$v_f = \sqrt{\tfrac{2}{3}Rg} \tag{7.23c}$$

The fly-off angle does not depend on the object's mass, the sphere's radius, or the acceleration due to gravity. The fly-off speed depends on the sphere's radius and the acceleration due to gravity.

Example 7.11 *Do fly-off the handle*
A block starts from rest at the top of a $1.5\,\mathrm{m}$-radius sphere. Use Eqs. (7.23b) and (7.23c) to verify the results from the previous example for the block's fly-off angle and speed.
Solution. Create expressions for the two equations with the appropriate numerical values, and apply Evaluate Numerically to them.

$$\theta_f = \frac{180}{\pi}\cos^{-1}\left(\frac{2}{3}\right) = 48.190$$

$$v_f = \sqrt{\tfrac{2}{3}\,(1.5\,\mathrm{m})\left(9.8067\frac{\mathrm{m}}{\mathrm{s}^2}\right)} = 3.1316\frac{\mathrm{m}}{\mathrm{s}}$$

Our exact results agree with the numerical solutions. ◄

Releasing the object from rest at the top of the sphere produces one of many trajectories. The sliding object can also start from rest at a non-zero initial angle, or start from the top of the sphere with a non-zero initial speed. Energy conservation tells us

$$\tfrac{1}{2}mv_0^2 + mRg\cos\theta_0 = \tfrac{1}{2}mv^2 + mRg\cos\theta \tag{7.24}$$

where v_0 and θ_0 are the object's initial speed and angular position, respectively.

Eq. (7.24) tells us the speed as a function of angle and Newton's 2nd Law says the fly-off speed is still $v_f = \sqrt{Rg\cos\theta_f}$. Together they give us the fly-off angle and speed for an object that starts with speed v_0 at angle θ_0.

$$v = \sqrt{v_0^2 + 2Rg\left(\cos\theta_0 - \cos\theta\right)} \tag{7.25a}$$

$$\theta_f = \cos^{-1}\left(\tfrac{1}{3}\left(\frac{v_0^2}{Rg} + 2\cos\theta_0\right)\right) \tag{7.25b}$$

$$v_f = \sqrt{\tfrac{1}{3}\left(v_0^2 + 2Rg\cos\theta_0\right)} \tag{7.25c}$$

Example 7.12 *A little push*
A block starts from the top of a $1.5\,\mathrm{m}$-radius sphere with an initial speed of $2.5\,\mathrm{m/s}$. Where does it leave the sphere? How fast is it going when it does?
Solution. Let's apply Evaluate Numerically to Eqs. (7.25b) and (7.25c).

$$\theta_f = \frac{180}{\pi}\cos^{-1}\left(\frac{1}{3}\left(\frac{\left(2.5\frac{\mathrm{m}}{\mathrm{s}}\right)^2}{(1.5\,\mathrm{m})\left(9.8067\frac{\mathrm{m}}{\mathrm{s}^2}\right)} + 2\cos 0°\right)\right) = 36.07$$

$$v_f = \sqrt{\tfrac{1}{3}\left(\left(2.5\frac{\mathrm{m}}{\mathrm{s}}\right)^2 + 2\,(1.5\,\mathrm{m})\left(9.8067\frac{\mathrm{m}}{\mathrm{s}^2}\right)\cos 0°\right)} = 3.448\,2\frac{\mathrm{m}}{\mathrm{s}}$$

This block leaves the sphere's surface sooner and is moving faster than the one released from rest. ◀

Figure 7.7 uses Eq. (7.25a) to show the dimensionless speed $V \equiv v/\sqrt{Rg}$ as a function of angle for several trajectories. They all follow the same pattern: the object speeds up until it flies off the sphere's surface.

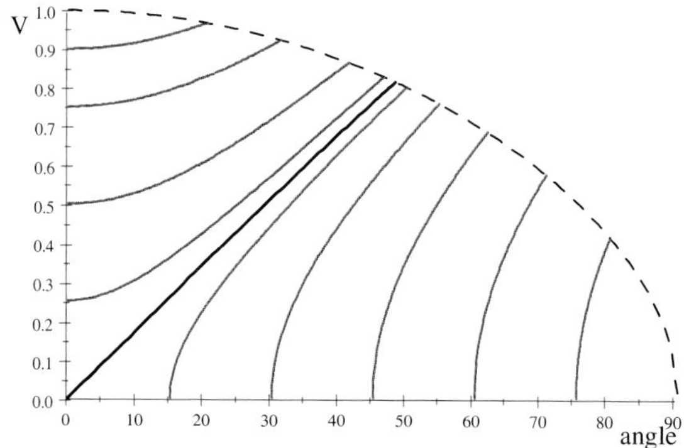

Figure 7.7 No friction
The thick curve starting from the origin describes the trajectory of an object released from rest at $\theta_0 = 0$. The curves above it describe non-zero initial speeds starting at $\theta_0 = 0$. Those below it describe trajectories released from rest at non-zero initial angles. The dashed line is the dimensionless fly-off speed $V_f = \sqrt{\cos\theta_f}$, an envelope that contains all trajectories on the sphere.

Speed and Position as Functions of Time

Now that we know the speed as a function of angle, we'd also like to know the speed and position as functions of time. We can use Eqs. (7.5) and (7.9) to recast the tangential equation of Newton's 2nd Law in terms of the angular position θ.

$$\frac{d^2\theta}{dt^2} = \frac{g}{R}\sin\theta \tag{7.26}$$

This equation has no exact solution, but we can use Solve ODE + Numeric to find approximate numerical solutions.

Example 7.13 *Infinity is a wicked long time*
Estimate how long it takes the block released from rest at the top of a sphere (radius $1.5\,\mathrm{m}$) to slide off the surface.
Solution. It takes an infinite amount of time for any object released from rest at the top of a sphere to slide off the surface. When $\theta_0 = 0$ and $v_0 = 0$, there is no net force acting on the object so it just sits there in unstable equilibrium.

Instead, let's start at a very small angle ($\theta_0 = 0.01°$) and apply Evaluate Numerically to Eq. (7.25a) to give our sliding block its initial speed.

$$v_0 = \sqrt{2\,(1.5)\,(9.8067)\,(1 - \cos 0.01°)} = 6.694\,0 \times 10^{-4}$$

Let's solve the equations for the angle and velocity as functions of time. Put Eq. (7.26) and Newton's 2nd Law tangential Eq. (7.21b) with the appropriate numerical values and initial conditions in a matrix. Leave the insertion point anywhere in the matrix and choose Solve ODE + Numeric.

$$\begin{bmatrix} \theta'' = \dfrac{9.8067}{1.5}\sin\theta \\ v' = 9.8067\sin\theta \\ v\,(0) = 6.694\,0 \times 10^{-4} \\ \theta\,(0) = \dfrac{0.01\pi}{180} \\ \theta'\,(0) = \dfrac{v\,(0)}{1.5} \end{bmatrix}, \text{ Functions defined: } \theta, v$$

Use Evaluate Numerically and a little trial-and-error to find the time it takes the block to reach $48.190°$.

$$\theta_f = \frac{180}{\pi}\theta\,(3.32245) = 48.19$$

Now that you have the time, use Evaluate Numerically to get the block's speed when it leaves the sphere's surface.

$$v_f = v\,(3.32245) = 3.131\,6$$

The block leaves the sphere's surface at $\theta_f = 48.19°$ about $t = 3.32245\,\mathrm{s}$ after it was released, moving at a speed $v_f = 3.131\,6\,\mathrm{m/\,s}$. This agrees with the results of our previous examples. ◀

Example 7.14 *Same little push*
A block starts from the top of a 1.5 m-radius sphere with an initial speed of 2.5 m/ s. How long does it take to leave the sphere? How fast is it going when it does?
Solution. Put Eqs. (7.26) and (7.21b) with the appropriate numerical values and initial conditions in a matrix, leave the insertion point anywhere in the matrix and choose Solve ODE + Numeric.

$$\begin{bmatrix} \theta'' = \dfrac{9.8067}{1.5}\sin\theta \\ v' = 9.8067\sin\theta \\ v(0) = 2.5 \\ \theta(0) = 0 \\ \theta'(0) = \dfrac{v(0)}{1.5} \end{bmatrix}, \text{ Functions defined: } \theta, v$$

We now have numerical approximations for $\theta(t)$ and $v(t)$, and we've already calculated the fly-off angle $\theta_f = 36.07°$. Use Evaluate Numerically on $\theta(t)$ and a little trial-and-error to find the time it takes the block to reach the fly-off angle.

$$\theta_f = \frac{180}{\pi}\theta(0.33572) = 36.070$$

Now that you have the time, use Evaluate Numerically on $v(t)$ to get the block's speed when it leaves the surface.

$$v_f = v(0.33572) = 3.448\,2$$

The block leaves the sphere 0.33572 s after it started, nearly 10 times sooner than the block released from rest. It does so at a speed of 3.448 2 m/ s, which agrees with our previous result. The following figure shows the block's speed as a function of time while it is on the sphere's surface.

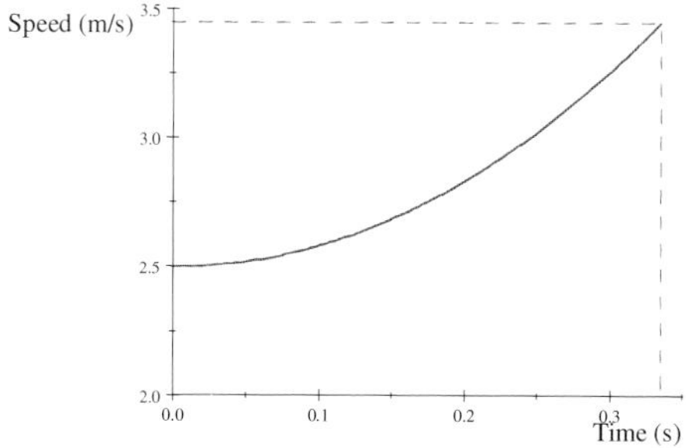

Figure 7.8 Weee!

In the absence of friction, the block's speed increases continually until it leaves the surface. You might find the following ratio is useful to compare this trajectory with those in Figure 7.7.

$$V_0 = \frac{v_0}{\sqrt{Rg}} = \frac{2.5\,\mathrm{m/\,s}}{\sqrt{(1.5\,\mathrm{m})(9.8067\,\mathrm{m/\,s^2})}} = 0.651\,83$$ ◀

Sliding with Friction

Let's complicate the issue slightly and include friction. [28] Gravity, the normal force, and now kinetic friction act on our sliding object. Figure 7.9 shows the object's free-body diagram. Positive is clockwise and $\theta = 0$ is the top of the sphere.

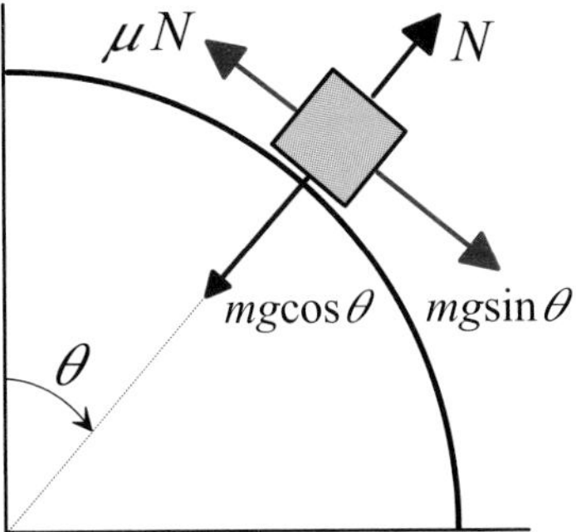

Figure 7.9 Not so much

The two 2nd Law equations are now

$$m\frac{v^2}{R} = mg\cos\theta - N \tag{7.27a}$$

$$m\frac{dv}{dt} = mg\sin\theta - \mu N \tag{7.27b}$$

where μ is the coefficient of kinetic friction. We can use the chain rule to change the tangential 2nd Law equation to a differential equation for the velocity as a function of angle.

$$\frac{dv}{dt} = \frac{d\theta}{dt}\frac{dv}{d\theta} = \omega\frac{dv}{d\theta} = \frac{v}{R}\frac{dv}{d\theta}$$

SNB prefers linear ODEs, so let's define a dimensionless variable $Z \equiv v^2/Rg$. After a little algebra, we can eliminate the Normal Force N

$$\frac{dZ}{d\theta} - 2\mu Z = 2(\sin\theta - \mu\cos\theta) \tag{7.28}$$

and produce a differential equation *SNB* can solve exactly. Place the equation and the appropriate initial condition in a matrix, and use Solve ODE + Exact.

$$\begin{bmatrix} \dfrac{dZ}{d\theta} - 2\mu Z = 2(\sin\theta - \mu\cos\theta) \\ Z(\theta_0) = Z_0 \end{bmatrix}$$, Exact solution is:

$$\left\{4\mu^2\frac{\cos\theta}{4\mu^2+1} - 6\mu\frac{\sin\theta}{4\mu^2+1} - 2\frac{\cos\theta}{4\mu^2+1} + Z_0\frac{e^{2(\theta\mu)}}{e^{2(\mu\theta_0)}} + 2\frac{e^{2(\theta\mu)}}{e^{2(\mu\theta_0)}}\frac{\cos\theta_0}{4\mu^2+1} - 4\mu^2\frac{e^{2(\theta\mu)}}{e^{2(\mu\theta_0)}}\frac{\cos\theta_0}{4\mu^2+1} + 6\mu\frac{e^{2(\theta\mu)}}{e^{2(\mu\theta_0)}}\frac{\sin\theta_0}{4\mu^2+1}\right\}$$

Using Simplify in-place and editing by-hand, we find the velocity as a function of angle

$$v(\theta) = \sqrt{v_0^2\, e^{2\mu\Delta\theta} + \frac{2Rg\left(\left(1-2\mu^2\right)\left(e^{2\mu\Delta\theta}\cos\theta_0 - \cos\theta\right) - 3\mu\left(\sin\theta - e^{2\mu\Delta\theta}\sin\theta_0\right)\right)}{1+4\mu^2}} \tag{7.29}$$

where $\Delta\theta = \theta - \theta_0$ is the sliding object's angular displacement.

The velocity is somewhat simpler when the initial angle is zero.

$$v(\theta) = \sqrt{v_0^2\, e^{2\mu\theta} + \frac{2Rg}{1+4\mu^2}\Big(\left(1-2\mu^2\right)\left(e^{2\mu\theta}-\cos\theta\right) - 3\mu\sin\theta\Big)} \tag{7.30}$$

The velocity as a function of angle depends on the sphere's radius, gravity, and the coefficient of kinetic friction.

You can use this and the 2nd Law radial equation to calculate the fly-off angle. The normal force is zero at the fly-off angle, so Eq. (7.27a) tells us $v_f^2 = Rg\cos\theta_f$. Eq. (7.30) tells us the same thing.

$$Rg\cos\theta_f = v_0^2\, e^{2\mu\theta_f} + \frac{2Rg}{1+4\mu^2}\Big(\left(1-2\mu^2\right)\left(e^{2\mu\theta_f}-\cos\theta_f\right) - 3\mu\sin\theta_f\Big) \tag{7.31}$$

This is an equation we'll have to solve numerically for θ_f.

There is a critical angle θ_c where both the velocity and its derivative are zero.

$$v(\theta_c) = 0 \tag{7.32a}$$

$$\left.\frac{dv}{d\theta}\right|_{\theta_c} = 0 \tag{7.32b}$$

There is a critical value for the coefficient of friction, and according to Eq. (7.28) that value is $\mu_c = \tan\theta_c$.

There are two possible outcomes for our sliding object. Either it stops on the surface or it flies off the sphere. This leads to four basic types of trajectories:

1. **Stop-and-Stay**

   The object comes to a stop at θ_c and stays there.

   Condition: $v(\theta_c) = 0$ and $\mu > \mu_c$

2. **Stop-and-Go**

   The object comes to a stop at θ_c then speeds up and flies off at θ_f.

   Condition: $v(\theta_c) = 0$ and $\mu = \mu_c$

3. **Speed-and-Go**

   The object starts from rest at θ_c, speeds up and flies off at θ_f.

   Condition: $v(\theta_c) = 0$ and $\mu < \mu_c$

4. **Slow-and-Go**

   The object slow downs to a minimum speed at θ_c but does not stop, and then speeds up and flies off at θ_f.

   Condition: $\left.\frac{dv}{d\theta}\right|_{\theta_c} = 0$ and $\mu > \mu_c$

The following four examples illustrate the different trajectories. Each uses $Rg = 1\,\mathrm{m}^2/\mathrm{s}^2$ which corresponds to a sphere with radius $R = 0.102\,\mathrm{m}$ on planet Earth.

Example 7.15 *Trajectory #1*: *Stop-and-Stay*!
Calculate the stopping angle when $v_0 = 0.687\,41\,\mathrm{m/s}$ and $\mu = 1.154\,7$. Plot the speed-versus-angle.
Solution. Substitute (with Evaluate) the appropriate values into Eq. (7.30).

$$v = \left[\sqrt{v_0^2 e^{2\mu\theta} + \frac{2}{1+4\mu^2}\left((1-2\mu^2)\left(e^{2\mu\theta} - \cos\theta\right) - 3\mu\sin\theta\right)}\right]_{\mu=1.154\,7, v_0=0.687\,41}$$
$$= \sqrt{0.526\,32\cos\theta - 1.093\,9\sin\theta - 5.378\,3\times 10^{-2}e^{2.\,3094\,\theta}}$$

Use Solve Numeric to find the stopping angle, and be sure to use the **red** degree symbol for θ so *SNB* returns the answer in degrees.

$$0 = \sqrt{0.526\,32\cos\theta^\circ - 1.093\,9\sin\theta^\circ - 5.378\,3\times 10^{-2}e^{2.\,3094\,\theta^\circ}},$$
Solution is: $\{[\theta = 20.0]\}$

The sliding objects stops at $\theta_c = 20.0°$, and stays there. The speed-versus-angle plot for this trajectory is labelled "1" on Figure 7.10. ◀

Example 7.16 *Trajectory #2*: *Stop-and-Go*!
Calculate and plot the speed-versus-angle for the Stop-and-Go trajectory when the stopping angle is $\theta_c = 45°$.
Solution. The Stop-and-Go trajectory occurs when $\mu = \tan\theta_c$, so for a stopping angle of $\theta_c = 45°$ the coefficient of friction is $\mu = \tan 45° = 1$. Substitute (with Evaluate) these vales into Eq. (7.30).

$$v = \left[v_0^2 e^{2\theta\mu} + \frac{2\left(1-2\mu^2\right)\left(e^{2\theta\mu} - \cos\theta\right) - 6\mu\sin\theta}{4\mu^2+1}\right]_{\theta=45°,\mu=1}$$
$$= e^{\frac{1}{2}\pi}v_0^2 - \tfrac{2}{5}e^{\frac{1}{2}\pi} - \tfrac{2}{5}\sqrt{2}$$

Use Solve Numeric to find the initial velocity.

$$\begin{bmatrix} 0 = e^{\frac{1}{2}\pi}v_0^2 - \frac{2}{5}e^{\frac{1}{2}\pi} - \frac{2}{5}\sqrt{2} \\ v_0 \in (0,1) \end{bmatrix}, \text{ Solution is: } \{[v_0 = -0.719\,44], [v_0 = 0.719\,44]\}$$

We need to know the fall-off angle for the plot, so let's find it with Eq. (7.31) and Solve Numeric.

$$\begin{bmatrix} \cos\theta^\circ = \left[v_0^2 e^{2\theta^\circ\mu} + \dfrac{2\left(1-2\mu^2\right)\left(e^{2\theta^\circ\mu} - \cos\theta^\circ\right) - 6\mu\sin\theta^\circ}{4\mu^2+1}\right]_{\mu=1, v_0=0.719\,44} \\ \theta \in (0,90) \end{bmatrix},$$
Solution is: $\{[\theta = 69.576]\}$

The sliding objects flies off the sphere's surface at $\theta_f = 69.576°$.

To create an expression for the object's speed, use Evaluate to Substitute the appropriate numerical values into Eq. (7.30).

$$v = \left[\sqrt{v_0^2 e^{2\theta\mu} + \frac{2\left(1-2\mu^2\right)\left(e^{2\theta\mu} - \cos\theta\right) - 6\mu\sin\theta}{4\mu^2+1}}\right]_{\mu=1, v_0=0.719\,44}$$

$$= \sqrt{0.117\,59 e^{2\theta} + \frac{2}{5}\cos\theta - \frac{6}{5}\sin\theta}$$

The complete expression for the speed-versus-angle is

$$v(\theta) = \sqrt{0.117\,59\, e^{2\theta} + \tfrac{2}{5}\cos\theta - \tfrac{6}{5}\sin\theta}$$

and the plot for this trajectory is labelled "2" on Figure 7.10. ◄

Example 7.17 *Trajectory #3*: *Speed-and-Go*!
Calculate and plot the speed-versus-angle for the Speed-and-Go trajectory of an object released from rest when $\theta_0 = 60°$. How fast is the object moving when it leaves the sphere?
Solution. The Speed-and-Go trajectory occurs when $\mu < \tan\theta_c$, so let's pick $\mu = \frac{1}{2}\tan 60°$. Use Simplify to Substitute these values into Eq. (7.30).

$$v = \left[\sqrt{\frac{2\left(\left(1-2\mu^2\right)\left(e^{2\mu(\theta-\theta_0)}\cos\theta_0 - \cos\theta\right) - 3\mu\left(\sin\theta - e^{2\mu(\theta-\theta_0)}\sin\theta_0\right)\right)}{1+4\mu^2}}\right]_{\mu=\frac{1}{2}\tan 60°, \theta_0=60°}$$

$$= \sqrt{\tfrac{1}{4}\cos\theta + e^{\sqrt{3}\theta - \frac{1}{3}\sqrt{3}\pi} - \tfrac{3}{4}\sqrt{3}\sin\theta}$$

Let's look between $60°$ and $90°$ to calculate the fly-off angle with Solve + Numeric.

$$\begin{bmatrix} \cos\theta° = \frac{1}{4}\cos\theta° + e^{\sqrt{3}\theta° - \pi/\sqrt{3}} - \frac{3\sqrt{3}}{4}\sin\theta° \\ \theta \in (60, 90) \end{bmatrix}, \text{ Solution is: } \{[\theta = 72.606]\}$$

The sliding objects flies off the sphere's surface at $\theta_f = 72.606°$.

To calculate the object's speed when it leaves the sphere, use Evaluate Numerically to Substitute the fly-off angle into the expression for the speed.

$$v_f = \left[\sqrt{\tfrac{1}{4}\cos\theta + e^{\sqrt{3}\theta - \pi/\sqrt{3}} - \tfrac{3\sqrt{3}}{4}\sin\theta}\right]_{\theta=72.606°} = 0.546\,78$$

The object's fly-off speed is $0.546\,78\,\mathrm{m/s}$.

The complete expression for the speed-versus-angle is

$$v(\theta) = \sqrt{\tfrac{1}{4}\cos\theta + e^{\sqrt{3}\theta - \pi/\sqrt{3}} - \tfrac{3\sqrt{3}}{4}\sin\theta}$$

and the plot for this trajectory is labelled "3" on Figure 7.10. ◄

Example 7.18 *Trajectory #4: Slow-and-Go!*
Calculate and plot the speed-versus-angle for the Slow-and-Go trajectory such that the object's minimum speed occurs at $\theta_c = 45°$. What is the value of the minimum speed?
Solution. The Slow-and-Go trajectory occurs when $\mu > \tan\theta_c$, so let's pick $\mu = 2\tan 45°$. Use Simplify to Substitute the values of μ and θ_c into the derivative of Eq. (7.30).

$$v' = \left[\frac{d}{d\theta}\left(v_0^2 e^{2\mu\theta} + \frac{2}{1+4\mu^2}\left(\left(1-2\mu^2\right)\left(e^{2\mu\theta} - \cos\theta\right) - 3\mu\sin\theta\right)\right)\right]_{\mu=2,\theta=45°}$$
$$= 4e^{\pi}v_0^2 - \frac{56}{17}e^{\pi} - \frac{13}{17}\sqrt{2}$$

Set the derivative equal to zero, and use Solve + Numeric to find the initial velocity.

$$0 = 4e^{\pi}v_0^2 - \tfrac{56}{17}e^{\pi} - \tfrac{13}{17}\sqrt{2}, \text{ Solution is: } \{[v_0 = -0.913\,90], [v_0 = 0.913\,90]\}$$

Look between $45°$ and $90°$ to calculate the fly-off angle in degrees with Solve + Numeric.

$$\left[\begin{array}{c} \cos\theta° = \left[v_0^2 e^{2\mu\theta°} + \dfrac{2\left(\left(1-2\mu^2\right)\left(e^{2\mu\theta°} - \cos\theta°\right) - 3\mu\sin\theta°\right)}{1+4\mu^2}\right]_{v_0=0.913\,90,\mu=2} \\ \theta \in (45, 90) \end{array}\right],$$
$$\text{Solution is: } \{[\theta = 58.505]\}$$

The sliding object flies off the sphere's surface at $\theta_f = 58.505°$. Use Substitute with Evaluate to create an expression for the object's speed.

$$v = \left[v_0^2 e^{2\mu\theta} + \tfrac{2}{1+4\mu^2}\left(\left(1-2\mu^2\right)\left(e^{2\mu\theta} - \cos\theta\right) - 3\mu\sin\theta\right)\right]_{v_0=0.91390,\mu=2}$$
$$= 1.168\,4\times10^{-2}e^{4\theta} + \tfrac{14}{17}\cos\theta - \tfrac{12}{17}\sin\theta$$

To calculate the object's minimum speed, use Evaluate Numerically to Substitute the value for θ_m into this expression for the object's speed.

$$v_{\min} = \left[\sqrt{1.168\,4\times10^{-2}e^{4\theta} + \tfrac{14}{17}\cos\theta - \tfrac{12}{17}\sin\theta}\right]_{\theta=45°} = 0.594\,61$$

The object's minimum speed is $0.594\,61\ \mathrm{m/s}$. The complete expression for the speed-versus-angle is

$$v(\theta) = \sqrt{1.168\,4\times10^{-2}e^{4\theta} + \tfrac{14}{17}\cos\theta - \tfrac{12}{17}\sin\theta}$$

and the speed-versus-angle plot for this trajectory is labelled "4" on Figure 7.10. The unlabeled curve is the frictionless trajectory of an object released from rest at the top of the sphere ($\theta_0 = 0$ and $v_0 = 0$). The dashed line is the dimensionless fly-off speed $V_f = \sqrt{\cos\theta}$. ◀

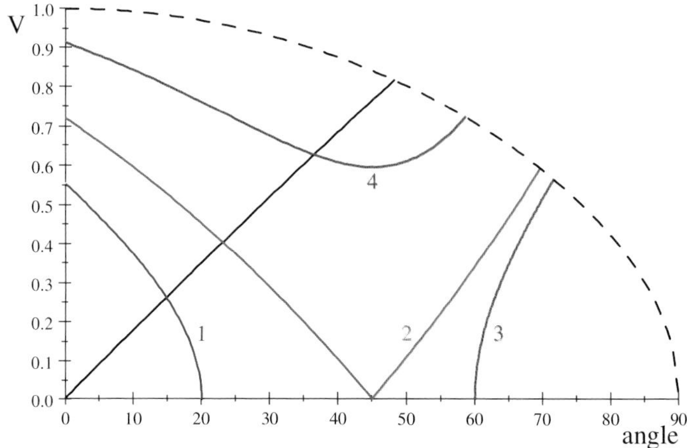

Figure 7.10 Four plots with friction

Speed and Position as Functions of Time

As in the no-friction case, we'd like to know the speed and position as functions of time. We can recast the tangential equation of Newton's 2nd Law in terms of the angular position θ and linear speed v.

$$\frac{d^2\theta}{dt^2} - \mu\left(\frac{d\theta}{dt}\right)^2 = \frac{g}{R}(\sin\theta - \mu\cos\theta) \tag{7.33a}$$

$$\frac{dv}{dt} - \frac{\mu}{R}v^2 = g(\sin\theta - \mu\cos\theta) \tag{7.33b}$$

Like the no-friction case, these equations have no exact solution, but we can use Solve ODE + Numeric to find approximate numerical solutions.

Example 7.19 *Same old same little push*

A block starts from the top of a $1.5\,\mathrm{m}$-radius sphere with an initial speed of $2.5\,\mathrm{m/s}$. The coefficient of kinetic friction between the block and sphere is 0.20. How long does it take to leave the sphere? How fast is it going when it does?

Solution. Use Evaluate to Substitute the numerical values for the coefficient of friction, initial speed, radius, and gravity into Eq. (7.30).

$$v = \left[\sqrt{v_0^2 e^{2\mu\theta} + \frac{2(1.5)(9.8067)}{1+4\mu^2}\left((1-2\mu^2)\left(e^{2\mu\theta}-\cos\theta\right)-3\mu\sin\theta\right)}\right]_{\mu=0.2,v_0=2.5}$$
$$= \sqrt{29.583e^{0.4\theta} - 15.217\sin\theta - 23.333\cos\theta}$$

Set the speed equal to the envelope, and use Solve Numeric to find the fly-off angle.

$$\left[\begin{array}{c}\sqrt{(9.8067)(1.5)\cos\theta^\circ} = \sqrt{29.583e^{0.4\theta^\circ} - 15.217\sin\theta^\circ - 23.333\cos\theta^\circ} \\ \theta\in(0,60)\end{array}\right],$$

Solution is: $\{[\theta = 39.617]\}$

Place Eqs. (7.33) in a matrix, and use Solve ODE + Numeric to calculate numerical approximations for the speed and angle as functions of time.

$$\begin{bmatrix} \theta'' - 0.20\,(\theta')^2 = \dfrac{9.8067}{1.5}\,(\sin\theta - 0.20\cos\theta) \\ v' - \dfrac{0.20}{1.5}v^2 = 9.8067\,(\sin\theta - 0.20\cos\theta) \\ v\,(0) = 2.5 \\ \theta\,(0) = 0 \\ \theta'\,(0) = \dfrac{v\,(0)}{1.5} \end{bmatrix}, \text{ Functions defined: } \theta, v$$

Use Evaluate Numerically and a little trial-and-error to find the time it takes the block to reach the fly-off angle 39.617°.

$$\theta_f = \frac{180}{\pi}\theta\,(0.38414) = 39.617$$

Now that you have the time, use Evaluate Numerically to get the block's speed when it leaves the surface.

$$v_f = v\,(0.38414) = 3.366\,2$$

The block leaves the sphere $0.38414\,\mathrm{s}$ after it was released, moving at $3.366\,2\,\mathrm{m/\,s}$.

Figure 7.11 shows the block's speed as a function of time while it is on the sphere's surface, along with the no-friction result for the same initial conditions.

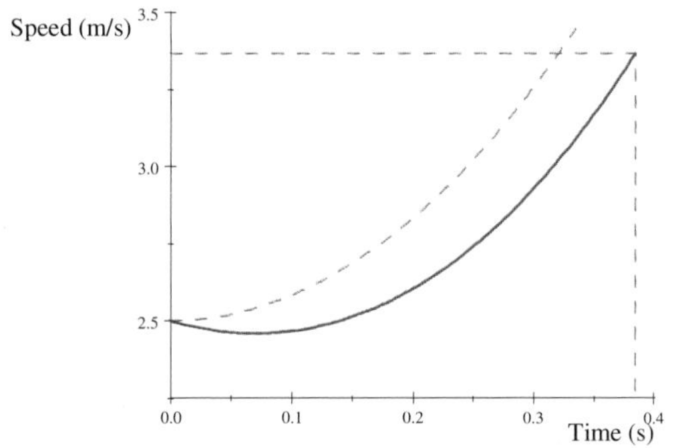

Figure 7.11 Slow-and-Go!

This is a Slow-and-Go trajectory. The block slows down initially, then speeds up and eventually flies off the sphere's surface. In Examples 7.12 and 7.14 we calculated the no-friction results for the same initial conditions.

$$\begin{aligned} t_f &= 0.33572\,\mathrm{s} \\ \theta_f &= 36.07\,^\circ \\ v_f &= 3.606\,5\frac{\mathrm{m}}{\mathrm{s}} \end{aligned}$$

With no friction to slow it down, the block just speeds up and flies off in less time, at a smaller angle, and moving faster. ◀

Problems

1. Repeat Example 7.1 for the song playing on a $45\,\mathrm{rev}\,/\,\mathrm{min}$ record with a 7 inch diameter.

2. Repeat Example 7.2 for the song playing on a $45\,\mathrm{rev}\,/\,\mathrm{min}$ record with a 7 inch diameter.

3. Repeat Example 7.1 for the song playing on a $78\,\mathrm{rev}\,/\,\mathrm{min}$ record with a 10 inch diameter.

4. Repeat Example 7.2 for the song playing on a $78\,\mathrm{rev}\,/\,\mathrm{min}$ record with a 10 inch diameter.

5. The album version of "In-A-Gadda-Da-Vida" lasts 17 minutes and 5 seconds. How many times does the record turn while the song is playing?

6. On side 2 of Abbey Road, there is a 14 second gap between "Her Majesty" and the previous song. How many times does the record turn during the gap?

7. The album version of "Thick as a Brick, Part 1" lasts 22 minutes and 40 seconds. How many times does the record turn while the song is playing? Life is a long song.

8. The landing x-coordinate of a drop from a rotating umbrella can be written in a remarkably simple form.
$$x^2 = R^2 - y_0^2 + \left(v_0^2 + gy_0\right)t^2 - \tfrac{1}{4}g^2t^4$$
(For a geometric interpretation of this result, see reference [26]).

   a. Show that the time that maximizes the x-coordinate is $t_m = \frac{1}{g}\sqrt{2\left(v_0^2 + gy_0\right)}$.

   b. Show that the maximum x-coordinate is $x_m = \sqrt{\frac{1}{g^2}\left(v_0^2 + gy_0\right)^2 + R^2 - y_0^2}$.

   c. Show that this is equivalent to $x_m = \frac{1}{g}\sqrt{v_0^4 + 2v_0^2gy_0 + R^2g^2}$.

9. Use the fact that the initial velocity v_0 is perpendicular to the umbrella's radius R and show that the difference between the umbrella's rotation angle and the drop's launch angle is $\theta - \theta_0 = 90°$.

10. Where does the drop with the maximum maximum-height in Figure 7.3 hit the ground? What is its maximum height?

11. Let's explore the claim that the umbrella's rotation angle for the maximum maximum-height trajectory depends on the ratio Rg/v_0^2.

    a. Show that a drop spends time $t_\uparrow = -\left(v_0 \cos\theta\right)/g$ going up.

    b. Use this result to create an expression for the maximum height as a function of launch angle.

    c. Show that the maximizing angle satisfies $0 = \left(Rg - v_0^2 \sin\theta_m\right)\cos\theta_m$.

    d. Use Solve Exact to find the two solutions.

12. Verify that the launch angle for a clockwise-rotating umbrella that produces the maximum maximum-height is

$$\theta_0 = \sin^{-1}\frac{Rg}{v_0^2} - \frac{\pi}{2}$$

when $v_0^2 \geq Rg$.

13. Prove that the linear speed of any point on a rotating CD is constant.

14. Use this expression for the arc length of a curve in polar coordinates

$$L = \int_{\theta_i}^{\theta_f} \sqrt{r^2 + \left(\frac{dr}{d\theta}\right)^2}\, d\theta$$

and Eq. (7.12) to verify the approximate result for the length of the spiral track on a CD.

15. Use Eq. (7.13) to verify my result for the arc length of the spiral track on a CD.

16. Show the average angular speed for rotational motion with a varying acceleration also equals

$$\omega_{ave} = \frac{\theta\,(T) - \theta_0}{T}$$

where T is the CD's total playing time.

17. Show that when the CD's average angular speed equals its actual angular speed, the CD has completed half the total number of rotations.

18. Show that the angular velocity of a CD as a function of rotation angle is

$$\omega\,(\theta) = \frac{\omega_0}{1 + \dfrac{2\pi\rho}{R_0}\theta}$$

where θ is given in revolutions.

19. Show that the angular acceleration of a CD as a function of rotation angle is

$$\alpha\,(\theta) = -\frac{\dfrac{2\pi\rho}{R_0}\omega_0^2}{\left(1 + \dfrac{2\pi\rho}{R_0}\theta\right)^3}$$

where θ is given in revolutions.

20. The CD version of "Thick as a Brick, Part 1" lasts 22 minutes and 40 seconds and it is the first song. How many times does the CD turn while the song is playing?

21. The CD version of "In-A-Gadda-Da-Vida" lasts 17 minutes and 5 seconds and it ends 36 minutes and 15 seconds after the first song starts. How many times does the CD turn while the song is playing?

22. On the Abbey Road CD, "Her Majesty" is the last song. The songs before it play for 46 minutes and 46 seconds, and there is a 14 second gap after them. How many times does the CD turn during the gap? Best album ever.

23. What is the CD's average angular speed during the $23\,\mathrm{s}$ "Her Majesty" is playing?

24. What is the tension in the string when the bucket of Example 7.8 is at the *top* of the circle if you swing it at twice the minimum angular speed?

25. What is the tension in the string when the bucket of Example 7.8 is at the *bottom* of the circle?

26. Repeat Example 7.8 for a bucket of mercury instead of water. Does your answer for the minimum speed surprise you?

27. A baseball is suspended by a string of length $L = 1.25\,\mathrm{m}$ and swung in a horizontal circle so that it makes an angle $\theta = 70°$ to the vertical. What is the tension in the string? How much time does one orbit take?

28. What angular speed is needed to swing the tennis ball of Example 7.9 in a horizontal circle that makes a $45°$ angle to the vertical? What is the tension in the string?

29. How do the results of Example 7.9 change if we replace the tennis ball with a baseball?

30. Find the speed as a function of angle using Solve ODE + Exact and Eq. (7.28) for an object sliding on a sphere *without* friction.

31. Derive the expressions for the fly-off speed and angle of an object sliding without friction on the surface of a sphere. Show they reduce to the expected result for an object released from rest at the top of the sphere.

32. Where does the speed of an object, released from rest very near the top of a frictionless sphere, equal half the fly-off speed?

33. Consider an object starting at the top of a frictionless sphere. What initial speed produces a fly-off angle of $45°$? What is the fly-off speed?

34. A block starts at the top of a sphere with a speed of $1\,\mathrm{m/s}$ and slides without friction. At what angle does the block fly off the sphere? How much time does it take?

35. A block starts at the top of a sphere with a speed of $1\,\mathrm{m/s}$. The coefficient of kinetic friction between the block and sphere is 0.20. At what angle does the block experience its minimum speed? Verify that this trajectory satisfies the Slow-and-Go conditions.

36. Calculate and plot the trajectory for a block sliding on a sphere starting from rest at $45°$. Use $\mu = 1$ for the kinetic friction coefficient. Compare your result with the appropriate curve on Figure 7.10.

8 Harmonic Motion

The study of Harmonic Motion, periodic motion that repeats itself at regular time intervals, is one of the most important and interesting in all of physics. Many physical systems experience such motion, and many cases are exactly solvable. Near a stable equilibrium point, the motion of any system can be approximated as harmonic motion.

There are three important properties of Harmonic Motion. The **period** (T) is the time needed to complete one oscillation. The SI unit is the second, which is the standard metric unit of time. The **frequency** (f) is the number of oscillations per unit time. The SI unit of frequency is the Hertz (Hz). An oscillating object moves back-and-forth about an equilibrium point. The **amplitude** (A) is the object's maximum displacement from equilibrium.

It is useful to think of the period as the number of seconds-per-oscillation and the frequency as the number of oscillations-per-second, so that $1\,\mathrm{Hz} = 1\,\mathrm{oscillation}/\mathrm{s}$. As their units suggest, the period and frequency have a reciprocal relationship.

$$T = \frac{1}{f} \tag{8.1}$$

When you consider oscillations, the period tells you how long, the frequency tells you how often, and the amplitude tells you how far.

Simple Harmonic Motion, Simply

Simple Harmonic Motion (SHM) is a special kind of periodic motion where the period and amplitude are constant and the period does not depend on the amplitude. SHM results when the net force acting on an object is proportional to, but in the opposite direction to the displacement from its equilibrium position. This position-dependent force, called **Hooke's Law**, always pushes the object toward equilibrium.

$$F(x) = -\,k\,x \tag{8.2}$$

A mass attached to a spring on a frictionless, horizontal surface provides such a linear restoring force. For this system, x is positive or negative when the spring is stretched or compressed. The positive constant k is the property of the spring (called the **spring constant**) that distinguishes shock absorbers (large k) from Slinkys (small k). The units of the spring constant are N/m.

Here is the equation of motion per Newton's 2nd Law for SHM in 1-dimension.

$$ma = -kx \tag{8.3}$$

Hooke and Newton in the same equation; neither would approve.

Doing Physics with Scientific Notebook: A Problem-solving Approach, First Edition. Joseph Gallant.

Like many humans, *SNB* prefers to solve equations with as few parameters as possible. Let's adopt a simpler equation of motion

$$a = -\omega_0^2\, x \tag{8.4}$$

where $\omega_0 \equiv \sqrt{k/m}$ is the **angular frequency** of the oscillations given in $\mathrm{rad/\,s}$.

The frequency of oscillations for a mass-spring system is

$$f = \frac{\omega_0}{2\pi} = \frac{1}{2\pi}\sqrt{\frac{k}{m}} \tag{8.5}$$

where $2\pi\,\mathrm{rad/oscillation}$ is the radians-oscillations conversion factor. Eq. (8.1) tells us the period of the motion.

$$T = \frac{2\pi}{\omega_0} = 2\pi\sqrt{\frac{m}{k}} \tag{8.6}$$

Example 8.1 *Spring forward...*
A $2\,\mathrm{kg}$ block is attached to a $47\,\mathrm{N/\,m}$ spring, arranged horizontally along a frictionless surface. The block is pulled $17\,\mathrm{cm}$ to the right and released from rest. How many oscillations does the block complete per second? How much time does it take for the block to return to equilibrium? What is the amplitude of these oscillations?
Solution. Use Eq. (8.5) to create and Evaluate Numerically an expression for the frequency.

$$f = \frac{1}{2\pi}\sqrt{\frac{47\,\mathrm{N/\,m}}{2\,\mathrm{kg}}} = \frac{0.771\,53}{\mathrm{s}}$$

The frequency is $0.77153\,\mathrm{Hz}$, so the block completes 0.77153 oscillations every second. Use Eq. (8.1) to create and Evaluate Numerically an expression for the period.

$$T = \frac{1}{0.77153\,\mathrm{Hz}} = 1.296\,1\,\mathrm{s}$$

The mass completes one oscillation every $1.296\,1\,\mathrm{s}$. The time it takes the mass to travel from the amplitude to equilibrium is one-quarter of the period.

$$t = \tfrac{1}{4} \times 1.296\,1\,\mathrm{s} = 0.324\,03\,\mathrm{s}$$

Because the block is released from rest, the amplitude is $A = 17\,\mathrm{cm}$. Neither the frequency nor the period depend on the amplitude. ◀

Frictionless surfaces are difficult for most labs to acquire. It's easier to suspend the spring vertically.

Example 8.2 *...Fall down*
The same $2\,\mathrm{kg}$ block is attached to the same $47\,\mathrm{N/\,m}$ spring, this time arranged vertically. The block is allowed to settle gently into equilibrium, then it's pulled down an additional $17\,\mathrm{cm}$ and released from rest. Where is the new equilibrium point? How many oscillations does the block complete per second? How much time does it take for the block to return to equilibrium? What is the amplitude of these oscillations?

Solution. The block's weight stretches the spring some distance x_{eq} until equilibrium is reached. At that point, the upward force exerted by the spring is balanced by the downward weight so $mg = kx_{eq}$. Create an expression for x_{eq}, and use Evaluate Numerically.

$$x_{eq} = \frac{(2\,\text{kg})\,(9.8067\,\text{m}/\,\text{s}^2)}{47\,\text{N}/\,\text{m}} = 0.417\,31\,\text{m}$$

The new equilibrium point is $0.417\,31\,\text{m}$ lower than it was before the block was attached. The answers for the frequency ($f = 0.77153\,\text{Hz}$), period ($T = 1.296\,1\,\text{s}$), time-to-equilibrium ($t = 0.324\,03\,\text{s}$), and amplitude ($A = 17\,\text{cm}$) are all the same as in Example 8.1. Hanging the spring vertically moves the equilibrium point, but the oscillations about that point remain the same. ◀

We can't solve Eq. (8.3) for the position as a function of time without calculus, and we can't use Eqs. (2.5) because the acceleration is not constant. But we can use the connection between SHM and Uniform Circular Motion. Consider an object moving in a horizontal circle of radius A at constant angular speed ω_0. As the object goes round and round, its shadow on each wall moves back-and-forth in SHM.

From this connection, we can derive the following equations that describe the position, speed, and acceleration as functions of time for objects released from rest into SHM.

$$x = A\cos\omega_0 t \tag{8.7a}$$

$$v = -\omega_0\,A\sin\omega_0 t \tag{8.7b}$$

$$a = -\omega_0^2\,A\cos\omega_0 t \tag{8.7c}$$

Notice that $a = -\omega_0^2\,x$, which agrees with Eq. (8.4). For an object released from rest, the amplitude equals the initial position.

Example 8.3 *Halfway isn't the half of it*
Once it's released, how long does it take the block in Example 8.1 to get halfway to equilibrium? How fast is it going when it gets there?
Solution. Tell *SNB* the angular frequency isn't zero.

$$\text{assume}\,(\omega_0 > 0) = (0, \infty)$$

Use Eq. (8.7a) and set the position equal to half the amplitude. Then use Solve Exact to calculate the time.

$$\tfrac{1}{2}A = A\cos\omega_0 t, \text{ Solution is: } \tfrac{1}{3}\frac{\pi}{\omega_0}$$

Use the definition of angular frequency to create an expression for the time, apply Evaluate Numerically in-place to the denominator to get the angular frequency, then use Evaluate Numerically.

$$t = \tfrac{1}{3}\frac{\pi}{\sqrt{\dfrac{47\,\text{N}/\,\text{m}}{2\,\text{kg}}}} = \tfrac{1}{3}\frac{\pi}{\dfrac{4.847\,7}{\text{s}}} = 0.216\,02\,\text{s}$$

Combine these results with Eq. (8.7b) and use Evaluate Numerically to calculate the block's velocity.

$$v = -\left(\frac{4.847\,7}{\mathrm{s}}\right)(0.17\,\mathrm{m})\sin\left(\frac{4.847\,7}{\mathrm{s}} \times 0.216\,02\,\mathrm{s}\right) = -0.713\,7\frac{\mathrm{m}}{\mathrm{s}}$$

The block is 0.085 m to the right of equilibrium and it's moving left at $0.713\,70\,\mathrm{m/s}$.

Let's look at the amplitude-to-midpoint time in terms of the period.

$$t = \tfrac{1}{3}\frac{\pi}{\omega_0} \times \frac{T}{2\pi/\omega_0} = \tfrac{1}{6}T$$

This is more than half the time it takes an object in SHM to go from amplitude to equilibrium $\left(\tfrac{1}{4}T\right)$. Even though the spring exerts a larger force when its stretched, the block starts from rest and spends more time outside the midpoint. ◀

Energy and SHM

The mechanical energy of an object in SHM is conserved, so the sum of its kinetic energy and potential energy is constant. As it moves, the oscillator's energy is transformed back and forth between the two. The oscillator's potential energy is

$$U = \tfrac{1}{2}k\,x^2 \tag{8.8}$$

where x is still its displacement.

At amplitude $(x = A)$ the oscillator stops and its energy is all potential. At equilibrium $(x = 0)$ the oscillator reaches its maximum speed and its energy is all kinetic. This gives us two expressions for the total energy.

$$E = \begin{cases} \tfrac{1}{2}k\,A^2 \\ \tfrac{1}{2}m\,v_{\text{max}}^2 \end{cases} \tag{8.9}$$

We can use these two expressions to relate the maximum speed to the maximum displacement – the amplitude.

$$v_{\text{max}} = \sqrt{\frac{k}{m}}\,A = \omega_0\,A \tag{8.10}$$

The further from equilibrium the oscillator starts, the faster it's going when it gets there.

Let's use energy conservation to equate the oscillator's energy at amplitude to its energy at some other position.

$$\tfrac{1}{2}mv^2 + \tfrac{1}{2}kx^2 = \tfrac{1}{2}kA^2 \tag{8.11}$$

This gives us the oscillator's velocity as a function of position.

$$v = \pm\sqrt{\frac{k}{m}\left(A^2 - x^2\right)} \tag{8.12a}$$

$$= \pm\,v_{\text{max}}\sqrt{1 - \frac{x^2}{A^2}} \tag{8.12b}$$

The positive and negative signs are for motion to the right and left, respectively.

Example 8.4 *Half-fast isn't either*
Where does the speed of a simple harmonic oscillator equal half its maximum value?
Solution. Use the definition of angular frequency and Eq. (8.12a) to create the following expression equating half the maximum speed to the speed at the unknown position x. Leave the insertion point anywhere in it, and click the Solve Exact button. Enter x in the Solution Variable(s) window and click OK.

$$\frac{1}{2}\sqrt{\frac{k}{m}}A = \sqrt{\frac{k}{m}\left(A^2 - x^2\right)}, \text{ Solution is: } -\tfrac{1}{2}\sqrt{3}A, \tfrac{1}{2}\sqrt{3}A$$

The half-maximum speed occurs at $x = \pm\frac{\sqrt{3}}{2}A \approx \pm 0.866\,03\,A$. The half-maximum speed does not occur at the midpoint between equilibrium and the amplitude. ◄

Not-Quite-as-Simple Harmonic Motion

Consider a mass attached to a horizontal spring on a frictionless surface. There are three ways to start the mass in SHM.

1. Move it away from equilibrium and release it from rest ($x_0 \neq 0$, $v_0 = 0$).
2. Leave it at equilibrium and give it a push ($x_0 = 0$, $v_0 \neq 0$).
3. Some push-and-release combination of the two ($x_0 \neq 0$, $v_0 \neq 0$).

The three have the same equation of motion but different initial conditions. Use Solve ODE + Exact to find the position as a function of time for general initial conditions.

$$\begin{bmatrix} \dfrac{d^2x}{dt^2} = -\omega_0^2\,x \\ x\,(0) = x_0 \\ x'\,(0) = v_0 \end{bmatrix}, \text{ Exact solution is: } \left\{ x_0 \cos t\omega_0 + \frac{1}{\omega_0} v_0 \sin t\omega_0 \right\}$$

This result gives the position of an object in SHM for any initial position and velocity. The amplitude of this motion

$$A = \sqrt{x_0^2 + \left(\frac{v_0}{\omega_0}\right)^2} \tag{8.13}$$

reduces to the initial position when the object is released from rest ($v_0 = 0$).

Let's use the definition of velocity to Evaluate the velocity as a function of time.

$$v = \frac{d}{dt}\left(x_0 \cos\omega_0 t + \frac{v_0}{\omega_0}\sin\omega_0 t\right) = v_0 \cos t\omega_0 - \omega_0 x_0 \sin t\omega_0$$

Let's use the definition of acceleration to Evaluate the acceleration as a function of time.

$$a = \frac{d}{dt}\left(v_0 \cos\omega_0 t - \omega_0 x_0 \sin\omega_0 t\right) = -\omega_0 v_0 \sin t\omega_0 - \omega_0^2 x_0 \cos t\omega_0$$

The following three equations describe the position, speed, and acceleration as functions of time of an object in SHM with arbitrary initial position and velocity.

$$x(t) = x_0 \cos \omega_0 t + \frac{v_0}{\omega_0} \sin \omega_0 t \quad (8.14a)$$

$$v(t) = v_0 \cos \omega_0 t - \omega_0 x_0 \sin \omega_0 t \quad (8.14b)$$

$$a(t) = -\omega_0 v_0 \sin \omega_0 t - \omega_0^2 x_0 \cos \omega_0 t \quad (8.14c)$$

The acceleration $a(t) = -\omega_0^2 x(t)$ agrees with Eq. (8.4) and all three equations reduce to their counterpart in Eqs. (8.7) for objects released from rest (so $v_0 = 0$ and $x_0 = A$).

Example 8.5 *Still the same*

The same $2\,\mathrm{kg}$ block is attached to the same $47\,\mathrm{N/m}$ spring, arranged horizontally along the same frictionless surface. The block is pulled $17\,\mathrm{cm}$ to the right and released such that it's moving $1.43\,\mathrm{m/s}$ to the right. How much time does it take for the block to return to equilibrium? What is the amplitude of these oscillations?

Solution. We've already calculated $\omega_0 = 4.8477\,\mathrm{rad/s}$. Use Eq. (8.14a) to create an expression setting the position equal to zero and use Solve Exact to find the time.

$$0 = (0.17\,\mathrm{m}) \cos\left(4.8477 \frac{\mathrm{rad}}{\mathrm{s}} t\right) + \frac{1.43\,\mathrm{m/s}}{4.8477/\mathrm{s}} \sin\left(4.8477 \frac{\mathrm{rad}}{\mathrm{s}} t\right),$$

Solution is: $0.54021\,\mathrm{s}$

Let's use Eq. (8.13) and Simplify to calculate the amplitude.

$$A = \left[\sqrt{x_0^2 + \left(\frac{v_0}{\omega_0}\right)^2}\right]_{x_0=0.17\,\mathrm{m}, v_0=1.43\,\mathrm{m/s}, \omega_0=4.8477/\mathrm{s}} = 0.34046\,\mathrm{m}$$

This amplitude is almost exactly twice that in Example 8.1.

Let's explore my claim that the period doesn't depend on the amplitude. Use Solve Exact to find the time it takes the block to go from this amplitude to equilibrium.

$$0 = (0.34046\,\mathrm{m}) \cos\left(4.8477 \tfrac{\mathrm{rad}}{\mathrm{s}} t\right) + \frac{0\,\mathrm{m/s}}{4.8477/\mathrm{s}} \sin\left(4.8477 \tfrac{\mathrm{rad}}{\mathrm{s}} t\right),$$

Solution is: $0.32403\,\mathrm{s}$

The period is quadruple the time it takes the block to travel from the amplitude to the equilibrium point.

$$T = 4 \times 0.32403\,\mathrm{s} = 1.2961\,\mathrm{s}$$

This is the same period we found in Example 8.1! The amplitude is larger and the block has further to travel, but the spring is stretched more and exerts more force at amplitude. The distance and speed are both proportionally bigger, so the time is the same. ◀

Energy and SHM, Again

Even though our oscillator now has some initial kinetic energy, it stops at amplitude so Eq. (8.9) and Eq. (8.10) are still valid. The amplitude is no longer the initial position, but is given by Eq. (8.13). We can use energy conservation to create a general expression equating the oscillator's initial energy to its energy at some other position.

$$\tfrac{1}{2}mv^2 + \tfrac{1}{2}kx^2 = \tfrac{1}{2}mv_0^2 + \tfrac{1}{2}kx_0^2 \tag{8.15}$$

This gives us the oscillator's velocity as a function of position.

$$v(x) = \pm\sqrt{v_0^2 + \frac{k}{m}\left(x_0^2 - x^2\right)} \tag{8.16}$$

When the initial speed is zero, the initial position is the amplitude and we recover Eq. (8.12a).

Let's verify that our previous results are valid for a non-zero initial velocity. If you Substitute with (Expand) the amplitude and angular frequency into Eq. (8.12a), the result is Eq. (8.16).

$$v = \left[\sqrt{\frac{k}{m}\left(A^2 - x^2\right)}\right]_{A=\sqrt{x_0^2+(v_0/\omega_0)^2},\,\omega_0=\sqrt{k/m}} = \sqrt{v_0^2 + \frac{k}{m}x_0^2 - \frac{k}{m}x^2}$$

For any initial conditions, you can use Eq. (8.13) to calculate the amplitude, then use Eqs. (8.10) and (8.12) for the speed.

In the previous section, *SNB* solved a differential equation for a general expression for the position of an object in SHM. Now let me show you how to use *SNB* to get the same result using basic calculus. Let's start with Eq. (8.12a) and the definition of velocity.

$$\frac{dx}{dt} = \pm\omega_0\sqrt{A^2 - x^2} \tag{8.17}$$

Tell *SNB* to assume $(A > 0)$ and recast this equation in terms of an integral *SNB* can Evaluate.

$$\omega_0 t = \int_{x_0}^{x} \frac{dx}{\sqrt{A^2 - x^2}} = \arcsin\frac{1}{A}x - \arcsin\frac{1}{A}x_0$$

We want to solve for the position, so let's take the *sine* of both sides and use Expand to eliminate the inverse trigonometric functions.

$$\sin\omega_0 t = \sin\left(\arcsin\frac{1}{A}x - \arcsin\frac{1}{A}x_0\right) = \frac{1}{A}x\sqrt{1 - \frac{1}{A^2}x_0^2} - \frac{1}{A}x_0\sqrt{1 - \frac{1}{A^2}x^2}$$

Now we can use Solve Exact to solve for x.

$$\sin\omega_0 t = \frac{1}{A}x\sqrt{1 - \frac{1}{A^2}x_0^2} - \frac{1}{A}x_0\sqrt{1 - \frac{1}{A^2}x^2},$$

Solution is: $(\sin t\omega_0)\sqrt{A^2 - x_0^2} + \frac{1}{2}\sqrt{2}x_0\sqrt{\cos 2t\omega_0 + 1}$

Create an expression for the position, replace the mess multiplying x_0 with $\cos\omega_0 t$. Tell *SNB* to assume$(v_0 > 0)$ and assume$(\omega_0 > 0)$. Then Substitute (with Simplify) the amplitude into your expression, and Expand the result in-place.

$$x = \left[(\sin\omega_0 t)\sqrt{A^2 - x_0^2} + x_0\cos\omega_0 t\right]_{A=\sqrt{x_0^2+(v_0/\omega_0)^2}} = x_0\cos t\omega_0 + \frac{1}{\omega_0} v_0 \sin t\omega_0$$

The answer is the same whether we use Newton's 2nd Law or energy conservation.

In SHM, both the kinetic and potential energies depend on time; only their sum is constant. It is sometimes useful to look at the time average of time-dependent quantities. Here is the average value of a time-dependent function over some time interval.

$$f_{\text{ave}} = \frac{1}{t_2 - t_1}\int_{t_1}^{t_2} f(t)\, dt \tag{8.18}$$

For SHM, the time interval is one period. The function's root-mean square (rms) value is a kind of average that prevents negative numbers from cancelling out.

$$f_{\text{rms}} = \sqrt{\frac{1}{t_2 - t_1}\int_{t_1}^{t_2} f^2(t)\, dt} \tag{8.19}$$

Root mean square quantities backwards named are. To find a function's rms value, first you square it, then average it and lastly take the square root of it.

The average kinetic energy is $K_{\text{ave}} = \frac{1}{2} m\, v_{\text{rms}}^2$ where v_{rms} is the rms velocity. Here's the expression for the average kinetic energy to Evaluate.

$$K_{\text{ave}} = \tfrac{1}{2} m \frac{\omega_0}{2\pi}\int_0^{2\pi/\omega_0} (v_0\cos\omega_0 t - \omega_0 x_0 \sin\omega_0 t)^2\, dt = \tfrac{1}{4} m\left(\omega_0^2 x_0^2 + v_0^2\right)$$

The average potential energy is $U_{\text{ave}} = \frac{1}{2} m\omega_0^2\, x_{\text{rms}}^2$ where $k = m\omega_0^2$ puts everything in terms of the angular frequency. Here's the expression for the average potential energy to Evaluate.

$$U_{\text{ave}} = \tfrac{1}{2} m\omega_0^2 \frac{\omega_0}{2\pi}\int_0^{2\pi/\omega_0} \left(x_0\cos\omega_0 t + \frac{v_0}{\omega_0}\sin\omega_0 t\right)^2 dt = \tfrac{1}{4} m\left(\omega_0^2 x_0^2 + v_0^2\right)$$

The average kinetic energy equals the average potential energy! We can verify that each average is half the total energy $E = \frac{1}{2} k A^2$ with Simplify.

$$\frac{K_{\text{ave}}}{E} = \frac{\frac{1}{4} m\left(\omega_0^2 x_0^2 + v_0^2\right)}{\frac{1}{2} m\omega_0^2\left(\sqrt{x_0^2 + (v_0/\omega_0)^2}\right)^2} = \frac{1}{2}$$

On average over a complete SHM cycle, half the energy is kinetic and half is potential. This is a special case of the Virial Theorem, which relates a system's average kinetic energy to the potential energy that binds it. SHM is the only case where the two averages are equal.

Damped Harmonic Motion

In real life, resistive forces are unavoidable so let's include one that is proportional to the velocity. There are two forces acting on the object and the resulting motion depends on their relative strength. Newton's 2nd Law says the equation of motion is

$$m\frac{d^2x}{dt^2} = -kx - b\frac{dx}{dt} \tag{8.20}$$

where the constant b is same one from Chapter 4 and k is the spring constant.

The general equation for the position is

$$\frac{d^2x}{dt^2} = -\omega_0^2\,x - 2\beta\frac{dx}{dt} \tag{8.21}$$

where $\omega_0 = \sqrt{k/m}$ is the natural angular frequency of the undamped oscillator and the **damping parameter** is $\beta \equiv b/(2m)$. The solution depends on the relative values of these two parameters (one for each force).

SNB can solve Eq. (8.21) exactly for arbitrary initial conditions (x_0 and v_0) but the solution is unwieldy and not very useful. Instead, let's try a trial solution (a fancy way of saying "inspired guess") of the form $x = Ae^{\kappa t}$. Create and Factor this form of Eq. (8.21).

$$0 = \frac{d^2\left(Ae^{\kappa t}\right)}{dt^2} + 2\beta\frac{d\left(Ae^{\kappa t}\right)}{dt} + \omega_0^2\left(Ae^{\kappa t}\right) = Ae^{t\kappa}\left(\kappa^2 + \omega_0^2 + 2\kappa\beta\right)$$

Use Solve Exact to find the values of κ that make our guess an actual solution.

$0 = \kappa^2 + \omega_0^2 + 2\kappa\beta$, Solution is: $\sqrt{\beta^2 - \omega_0^2} - \beta, -\beta - \sqrt{\beta^2 - \omega_0^2}$

The values we want are $\kappa = -\beta \pm \sqrt{\beta^2 - \omega_0^2}$. The resulting three types of motion depend on whether the term inside the square root is negative, zero, or positive.

Underdamped ($\beta^2 < \omega_0^2$)

Underdamped harmonic motion happens when the damping force is small. To avoid the unsightly general solution to Eq. (8.21), let's tell *SNB* the assumptions about the two force parameters. The two force parameters are positive.

$\text{assume}\,(\beta > 0) = (0, \infty)$

$\text{assume}\,(\omega_0 > 0) = (0, \infty)$

For underdamped motion, the damping parameter is smaller than the undamped frequency.

$\text{additionally}\,(\beta < \omega_0) = (0, \omega_0)$

Now we can solve Eq. (8.21).

Put the equation and the initial conditions in a matrix and use Solve ODE + Laplace.

$$\begin{bmatrix} \dfrac{d^2x}{dt^2} = -\omega_0^2\, x - 2\beta\dfrac{dx}{dt} \\ x\,(0) = x_0 \\ x'\,(0) = v_0 \end{bmatrix}, \text{ Laplace solution is:}$$

$$\left\{ x_0 e^{-t\beta}\left(\cos t\sqrt{\omega_0^2-\beta^2} - \left(\sin t\sqrt{\omega_0^2-\beta^2}\right)\frac{\beta - \frac{1}{x_0}\left(v_0 + 2\beta x_0\right)}{\sqrt{\omega_0^2-\beta^2}}\right)\right\}$$

A little editing by-hand and we have the position as a function of time.

$$x\,(t) = \left(x_0 \cos\omega t + \frac{\beta x_0 + v_0}{\omega}\sin\omega t\right) e^{-\beta t} \tag{8.22}$$

The underdamped oscillator does harmonic oscillations at an angular frequency

$$\omega \equiv \sqrt{\omega_0^2 - \beta^2}$$

that is smaller than the undamped oscillator, and its amplitude decreases exponentially.

Example 8.6 *Underdamped oscillators do*

A $1\,\mathrm{kg}$ block, attached to a $4\,\mathrm{N/\,m}$ spring, is subjected to a $-v$ resistive force. At the start of its motion, the block is $1\,\mathrm{m}$ and is moving at $2\,\mathrm{m/\,s}$ away from equilibrium. Show that the amplitude decreases exponentially.

Solution. We're given $\omega_0 = 2\,\mathrm{rad/\,s}$ and $\beta = \frac{1}{2}\,\mathrm{s}^{-1}$ so the block will undergo underdamped oscillations. Use Substitute to find the angular frequency of those oscillations.

$$\omega = \left[\sqrt{\omega_0^2 - \beta^2}\right]_{\omega_0=2,\beta=1/2} = \tfrac{1}{2}\sqrt{15}$$

The undamped frequency is $\omega = \sqrt{15}/2\,\mathrm{rad/\,s}$. Now use Substitute again to get your expression for the position as a function of time.

$$x = \left[\left(x_0 \cos\omega t + \frac{\beta x_0 + v_0}{\omega}\sin\omega t\right)e^{-\beta t}\right]_{x_0=1,v_0=2,\omega=\sqrt{15}/2,\beta=1/2}$$
$$= e^{-\frac{1}{2}t}\left(\cos\tfrac{1}{2}\sqrt{15}t + \tfrac{1}{3}\sqrt{15}\sin\tfrac{1}{2}\sqrt{15}t\right)$$

The solid line in Figure 8.1 is the position for this underdamped oscillator.

The period is constant but the amplitude decreases.

$$T = \frac{2\pi}{\omega} = 3.244\,6\,\mathrm{s}$$

Let's see if the decrease is an exponential one.

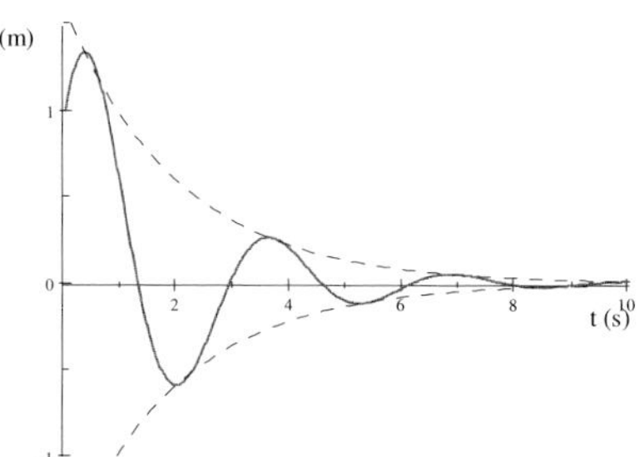

Figure 8.1 Underdamped

To find when the first maximum occurs, create an expression setting the time derivative of the position equal to zero, and use Solve Numeric to find the time.

$$0 = \frac{d}{dt}\left(e^{-t/2}\left(\cos\tfrac{\sqrt{15}}{2}t + \tfrac{\sqrt{15}}{3}\sin\tfrac{\sqrt{15}}{2}t\right)\right), \text{ Solution is: } \{[t = 0.340\,34]\}$$

To find where the first maximum occurs, Substitute (with Evaluate Numerically) that time into the position.

$$\left[e^{-t/2}\left(\cos\tfrac{\sqrt{15}}{2}t + \tfrac{\sqrt{15}}{3}\sin\tfrac{\sqrt{15}}{2}t\right)\right]_{t=0.340\,34} = 1.333\,7$$

Expressions of the form $\pm 1.333\,7\, e^{-(t-0.340\,34)/2}$ will pass through every extrema of an exponential decay. Add the exponentials to your plot and show them as dashed lines. As you can see, they pass through each maximum and minimum. ◄

Critically Damped ($\beta^2 = \omega_0^2$)

Critical damping occurs when the two force parameters are equal. Put Eq. (8.21) with $\beta^2 = \omega_0^2$ and the initial conditions in a matrix and use Solve ODE + Exact.

$$\begin{bmatrix} \dfrac{d^2x}{dt^2} = -\beta^2 x - 2\beta\dfrac{dx}{dt} \\ x(0) = x_0 \\ x'(0) = v_0 \end{bmatrix}, \text{ Exact solution is: } \left\{x_0 e^{-t\beta} + te^{-t\beta}(v_0 + \beta x_0)\right\}$$

Here is the position as a function of time for a critically-damped oscillator.

$$x(t) = \Big(x_0 + (\beta x_0 + v_0)\, t\Big)\, e^{-\beta t} \tag{8.23}$$

Notice there are no oscillations, only an exponential decay.

Example 8.7 *Critically-damped oscillators don't*

The block from the previous example is attached to the same spring, but now is subjected to a $-4v$ resistive force.

Solution. The frequency $\omega_0 = 2\,\text{rad/s}$ hasn't changed, but now $\beta = 2/\text{s}$ so this is a critically damped system. Use Substitute to get your expression for the position.

$$x = \left[(x_0 + (\beta x_0 + v_0)\, t)\, e^{-\beta t}\right]_{x_0=1, v_0=2, \beta=2} = e^{-2t}(4t+1)$$

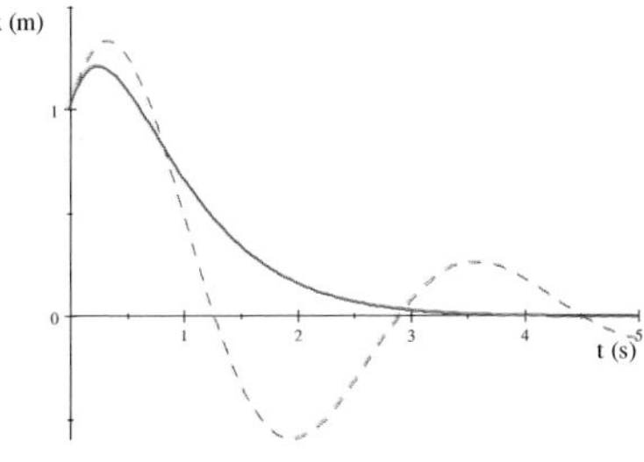

Figure 8.2 Critically damped

The solid line in Figure 8.2 is the position for this critically damped oscillator.

Any underdamped oscillator (the dashed line) gets to equilibrium sooner than the critically damped system. ◄

Overdamped ($\beta^2 > \omega_0^2$)

Overdamped motion happens when the damping force is large. Again, let's start by telling *SNB* the assumptions about the two force parameters.

$$\text{assume}\,(\beta > 0) = (0, \infty)$$
$$\text{assume}\,(\omega_0 > 0) = (0, \infty)$$
$$\text{additionally}\,(\beta > \omega_0) = (0, \beta)$$

Now solve Eq. (8.21) with Solve ODE + Laplace.

$$\begin{bmatrix} \dfrac{d^2x}{dt^2} = -\,\omega_0^2\, x - 2\beta\dfrac{dx}{dt} \\ x\,(0) = x_0 \\ x'\,(0) = v_0 \end{bmatrix}, \text{ Laplace solution is:}$$

$$\left\{ x_0 e^{-t\beta} \left(\cosh t\sqrt{\beta^2 - \omega_0^2} - \left(\sinh t\sqrt{\beta^2 - \omega_0^2} \right) \frac{\beta - \dfrac{1}{x_0}\left(v_0 + 2\beta x_0\right)}{\sqrt{\beta^2 - \omega_0^2}} \right) \right\}$$

The position as a function of time for an overdamped oscillator is

$$x\,(t) = \left(x_0 \cosh \omega t + \frac{\beta x_0 + v_0}{\omega} \sinh \omega t \right) e^{-\beta t} \tag{8.24}$$

where $\omega \equiv \sqrt{\beta^2 - \omega_0^2}$. The overdamped solution is similar in form to the underdamped, with the hyperbolic functions replacing their trigonometric counterparts. Like the critically damped system there are no oscillations here, and the overdamped oscillator is slower because of the larger resistive force.

Example 8.8 *Overdamped oscillators don't either*
The block from the previous example is still attached to the same spring, but now is subjected to a $-\,8v$ resistive force. Plot the position as a function of time, and compare it with the previous example.
Solution. The natural frequency $\omega_0 = 2\,\text{rad/s}$ remains the same, but now $\beta = 4\,\text{s}^{-1}$ so this is an overdamped system. Use Substitute (with Evaluate) to find ω.

$$\omega = \left[\sqrt{\beta^2 - \omega_0^2} \right]_{\omega_0 = 2, \beta = 4} = \sqrt{12}$$

Now use it again to get your expression for the position as a function of time.

$$\left[\left(x_0 \cosh \omega t + \frac{\beta x_0 + v_0}{\omega} \sinh \omega t \right) e^{-\beta t} \right]_{x_0 = 1, v_0 = 2, \omega = \sqrt{12}, \beta = 4}$$
$$= e^{-4t} \left(\cosh 2\sqrt{3}t + \sqrt{3} \sinh 2\sqrt{3}t \right)$$

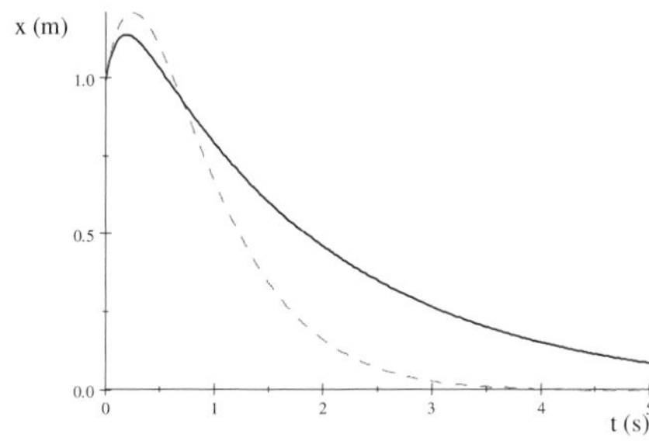

Figure 8.3 Overdamped

The solid line in Figure 8.3 is the position for this overdamped oscillator.
The dashed line is the critically damped result from the previous example.
The overdamped oscillator takes more time to approach equilibrium.
The critically damped system takes the shortest time to approach equilibrium without oscillation. ◀

Driven Harmonic Motion

Suppose we apply a time-dependent external force that pushes – drives – our damped oscillator. Newton's 2nd Law gives the equation of motion

$$m\frac{d^2x}{dt^2} = -kx - b\frac{dx}{dt} + F(t) \tag{8.25}$$

where $F(t)$ is the driving force. The general equation of motion is

$$\frac{d^2x}{dt^2} = -\omega_0^2 x - 2\beta\frac{dx}{dt} + f(t) \tag{8.26}$$

where $f(t) = F(t)/m$. Before we can solve this equation, we have to specify the driving force. We will consider constant and sinusoidal forces. Let's look at a few specific cases.

Constant Driving Force, no Damping

For a constant driving force with no damping, the equation of motion is

$$\frac{d^2x}{dt^2} = -\omega_0^2 x + f_0 \tag{8.27}$$

where f_0 is constant. *SNB* can solve this equation with Solve ODE + Exact.

$$\begin{bmatrix} \frac{d^2x}{dt^2} = -\omega_0^2 x + f_0 \\ x(0) = x_0 \\ x'(0) = v_0 \end{bmatrix}$$, Exact solution is:

$$\left\{ x_0 \cos t\omega_0 + \frac{1}{\omega_0^2} f_0 - \frac{1}{\omega_0^2} f_0 \cos t\omega_0 + \frac{1}{\omega_0} v_0 \sin t\omega_0 \right\}$$

A little editing by-hand and we have the position as a function of time.

$$x(t) = \frac{f_0}{\omega_0^2} + \left(x_0 - \frac{f_0}{\omega_0^2} \right) \cos \omega_0 t + \frac{v_0}{\omega_0} \sin \omega_0 t \tag{8.28}$$

As we saw in our vertical spring example, the constant force shifts the equilibrium point about which the object oscillates from the origin to f_0/ω_0^2. Notice that when $f_0 = \omega_0^2 x_0$ and $v_0 = 0$, the constant force exactly balances the force exerted by the spring and holds the object at that equilibrium point.

Example 8.9 *The boring oscillator*
A 1 kg object is connected to a 4 N/ m spring and placed at rest at the origin. The system has no damping and is driven by a constant 2 N force. What is the object's maximum distance from the origin? What is its maximum speed?

Solution. Let's start with the position. Substitute (with Evaluate) the appropriate numerical values into Eq. (8.28).

$$x = \left[\frac{f_0}{\omega_0^2} + \left(x_0 - \frac{f_0}{\omega_0^2}\right)\cos\omega_0 t + \frac{v_0}{\omega_0}\sin\omega_0 t\right]_{f_0=2,\omega_0=2,v_0=0,x_0=0} = \tfrac{1}{2} - \tfrac{1}{2}\cos 2t$$

Evaluate this expression to find the velocity.

$$v = \frac{d}{dt}\left(\tfrac{1}{2} - \tfrac{1}{2}\cos 2t\right) = \sin 2t$$

Since trigonometric functions max out at ± 1, the maximum distance from the origin is 1 m and the maximum speed is 1 m/ s. Here are the position and velocity graphs for this boring oscillator.

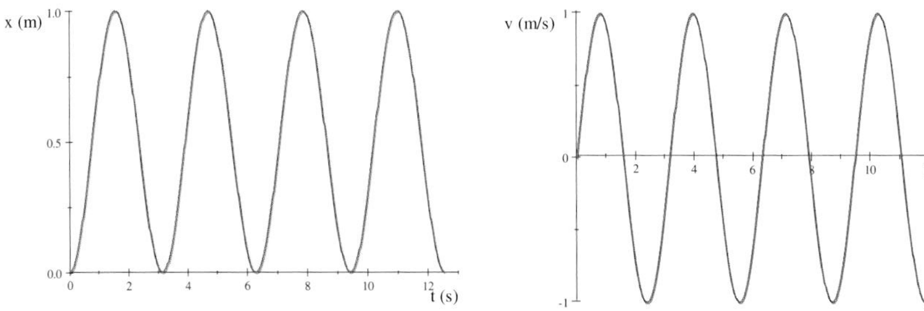

Figure 8.4a Position **Figure 8.4b** Velocity

The object does SHM about the equilibrium point $x = \frac{1}{2}$ with an amplitude of $\frac{1}{2}$ m at an angular frequency of 2 rad/ s. ◄

Sinusoidal Driving Force, no Damping

For a sinusoidal driving force with no damping, the equation of motion is

$$\frac{d^2x}{dt^2} = -\,\omega_0^2\,x + f_0\cos\omega_d t \tag{8.29}$$

where f_0 is constant and ω_d is the **driving frequency**. This is another equation *SNB* can solve exactly, giving a general solution that's too complicated to be useful. Instead, let's make some assumptions about the frequencies.

$$\text{assume}\,(\omega_d > 0) = (0,\infty)$$

$$\text{assume}\,(\omega_0 > 0) = (0,\infty)$$

The two frequencies are positive.

Now *SNB* can produce a useful solution with Solve ODE + Exact.

$$\begin{bmatrix} \dfrac{d^2x}{dt^2} = -\omega_0^2\, x + f_0 \cos \omega_d t \\ x(0) = x_0 \\ x'(0) = v_0 \end{bmatrix}, \text{ Exact solution is:}$$

$$\left\{ \frac{1}{\omega_0} v_0 \sin t\omega_0 - f_0 \frac{\cos t\omega_0}{\omega_0^2 - \omega_d^2} + f_0 \frac{\cos t\omega_d}{\omega_0^2 - \omega_d^2} + \omega_0^2 x_0 \frac{\cos t\omega_0}{\omega_0^2 - \omega_d^2} - \omega_d^2 x_0 \frac{\cos t\omega_0}{\omega_0^2 - \omega_d^2} \right\}$$

A little editing by-hand and we have the position as a function of time.

$$x(t) = x_0 \cos \omega_0 t + \frac{v_0}{\omega_0} \sin \omega_0 t + \frac{f_0}{\omega_0^2 - \omega_d^2} (\cos \omega_d t - \cos \omega_0 t) \tag{8.30}$$

There's a problem here as the next example shows.

Example 8.10 *The psycho oscillator*
A $1\,\mathrm{kg}$ object is connected to a $4\,\mathrm{N/\,m}$ spring and placed at rest at the origin. The system has no damping and is driven by a $2\,\mathrm{N}\cos 2t$ force. What happens?
Solution. We're given $\omega_0 = \omega_d = 2\,\mathrm{rad/\,s}$. There's no reason these two frequencies can't be equal, but we can't Substitute them into Eq. (8.30) because the denominator of the third term would be zero. Instead, let's use Solve ODE + Exact directly on Eq. (8.29) with the appropriate numerical values.

$$\begin{bmatrix} \dfrac{d^2x}{dt^2} = -4x + 2\cos 2t \\ x(0) = 0 \\ x'(0) = 0 \end{bmatrix}, \text{ Exact solution is: } \tfrac{1}{2} t \sin 2t$$

Here are the position and velocity graphs for this crazy oscillator.

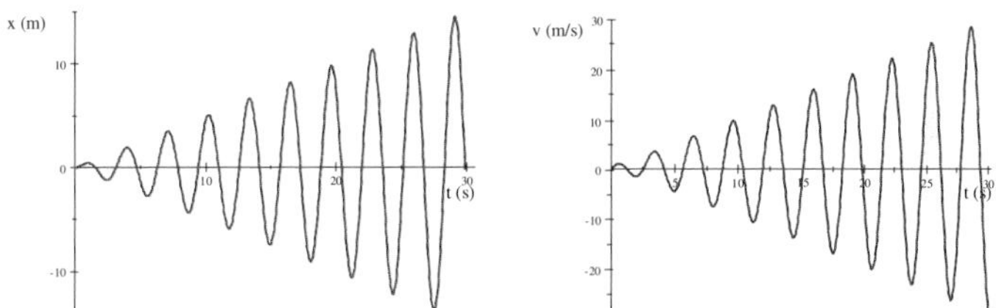

Figure 8.5a Position **Figure 8.5b** Velocity

When the driving and undamped frequencies are equal, the amplitude and maximum speed can become arbitrarily large. We avoid this unphysical situation in real life because there is always some damping force acting on the oscillator. ◀

Constant Driving Force with Damping

For a constant driving force with no damping, the equation of motion is

$$\frac{d^2x}{dt^2} = -\omega_0^2\, x - 2\beta \frac{dx}{dt} + f_0 \tag{8.31}$$

where f_0 is constant.

Even with simplifying assumptions, *SNB*'s solution to this equation is so ugly that I'm just going to tell you the answer. The position as a function of time is

$$x(t) = \frac{f_0}{\omega_0^2} + \left(x_0 - \frac{f_0}{\omega_0^2}\right) e^{-\beta t} \cos \omega t + \left(\frac{v_0}{\omega} + \frac{\beta}{\omega}\left(x_0 - \frac{f_0}{\omega_0^2}\right)\right) e^{-\beta t} \sin \omega t \quad (8.32)$$

where $\omega = \sqrt{\omega_0^2 - \beta^2}$. When $\beta^2 < \omega_0^2$, the object will oscillate and eventually reach static equilibrium at $x_f = f_0/\omega_0^2$, which is exactly the equilibrium point of the no-damping solution ($\beta = 0$).

Let's look at a numerical example.

Example 8.11 *The sane oscillator*

A $1\,\text{kg}$ object is connected to a $4\,\text{N/m}$ spring and placed at rest at the origin. The system has a damping parameter of $\frac{1}{5}\,\text{s}^{-1}$ and is driven by a constant $2\,\text{N}$ force. What is the object's maximum distance from the origin? What is its maximum speed?

Solution. It's easier to use Solve ODE + Exact and Eq. (8.31) directly to find the position as a function of time.

$$\begin{bmatrix} \dfrac{d^2x}{dt^2} = -4x - 2\left(\frac{1}{5}\right)\dfrac{dx}{dt} + 2 \\ x(0) = 0 \\ x'(0) = 0 \end{bmatrix}, \text{ Exact solution is:}$$

$$\left\{ -\frac{1}{2}\frac{\cos\frac{3}{5}\sqrt{11}t}{e^{\frac{1}{5}t}} - \frac{1}{66}\sqrt{11}\frac{\sin\frac{3}{5}\sqrt{11}t}{e^{\frac{1}{5}t}} + \frac{1}{2} \right\}$$

You can Evaluate the derivative of the position with respect to time to get the velocity as a function of time.

$$v = \frac{d}{dt}\left(\left(-\frac{1}{2}\cos\frac{3\sqrt{11}}{5}t - \frac{\sqrt{11}}{66}\sin\frac{3\sqrt{11}}{5}t\right)e^{-t/5} + \frac{1}{2}\right) = \frac{10}{33}\sqrt{11}\left(\sin\frac{3}{5}\sqrt{11}t\right)e^{-\frac{1}{5}t}$$

Let's look at plots of the position and velocity, so we can visualize our solutions.

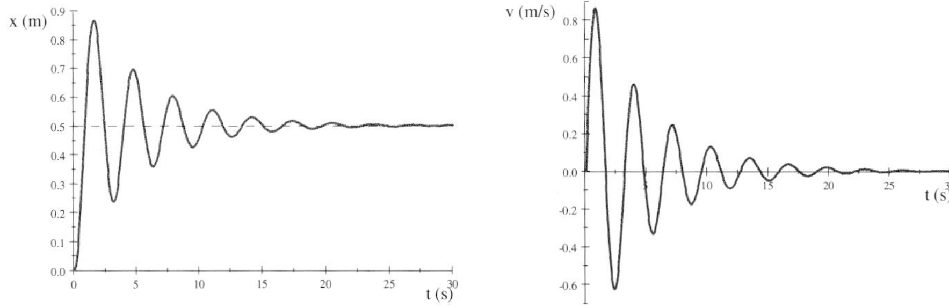

Figure 8.6a Position **Figure 8.6b** Velocity

The driving force adds energy into the system, but eventually the energy is lost as the damping force inhibits all motion and the system settles into static equilibrium.

Figure 8.6a tells you the maximum distance happens around $t = 2\,\mathrm{s}$, so use Solve + Numeric to look between $2\,\mathrm{s}$ and $4\,\mathrm{s}$ to find the time when the velocity is zero.

$$\left[\begin{array}{c}\frac{10\sqrt{11}}{33}\left(\sin\frac{3\sqrt{11}}{5}t\right)e^{-t/5}=0\\ t\in(1,3)\end{array}\right],\ \text{Solution is: }\{[t=1.578\,7]\}$$

To calculate that maximum distance, Substitute (with Evaluate Numerically) the value $t = 1.578\,7$ into the expression for the object's position as a function of time.

$$x_{\text{max}} = \left[\frac{1}{2}\left(1-\left(\cos\frac{3\sqrt{11}}{5}t+\frac{\sqrt{11}}{33}\sin\frac{3\sqrt{11}}{5}t\right)e^{-t/5}\right)\right]_{t=1.578\,7} = 0.864\,62$$

The object reaches its maximum distance from the origin of $x_{\text{max}} = 0.864\,62\,\mathrm{m}$ at time $t = 1.578\,7\,\mathrm{s}$.

Figure 8.6b tells you the maximum velocity happens around $t = 1\,\mathrm{s}$, so use Solve + Numeric to look between $\frac{1}{2}\,\mathrm{s}$ and $\frac{3}{2}\,\mathrm{s}$ to find the time when the acceleration is zero.

$$\left[\begin{array}{c}\frac{d}{dt}\left(\frac{10\sqrt{11}}{33}\left(\sin\frac{3\sqrt{11}}{5}t\right)e^{-t/5}\right)=0\\ t\in\left(\frac{1}{2},\frac{3}{2}\right)\end{array}\right],\ \text{Solution is: }\{[t=0.739\,02]\}$$

To calculate the maximum speed, Substitute (with Evaluate Numerically) this value for the time into the expression for the velocity as a function of time.

$$v_{\text{max}} = \left[\frac{10\sqrt{11}}{33}\left(\sin\frac{3\sqrt{11}}{5}t\right)e^{-t/5}\right]_{t=0.739\,02} = 0.862\,6$$

The object reaches its maximum speed of $v_{\text{max}} = 0.862\,6\,\mathrm{m/\,s}$ at time $t = 0.739\,02\,\mathrm{s}$. ◀

Sinusoidal Driving Force with Damping

For a sinusoidal driving force with damping, the equation of motion is

$$\frac{d^2x}{dt^2} = -\,\omega_0^2\,x - 2\beta\frac{dx}{dt} + f_0\cos\omega_d t \tag{8.33}$$

where f_0 is constant and ω_d is the driving frequency. Once again, *SNB*'s exact general solution is too messy to be useful so we'll solve this equation for a particular example.

Let's take the system in the previous example and let the driving force oscillate.

Example 8.12 *The steady-state oscillator*

A $1\,\mathrm{kg}$ object is connected to a $4\,\mathrm{N/\,m}$ spring and placed at rest at the origin. The system has a damping parameter of $\frac{1}{5}\,\mathrm{s}^{-1}$ and is driven by a $2\,\mathrm{N}\cos t$ force. What happens?

Solution. Let's use Solve ODE + Exact to solve Eq. (8.33) with the appropriate numerical values and initial conditions.

$$\begin{bmatrix} \dfrac{d^2x}{dt^2} = -4x - 2\left(\frac{1}{5}\right)\dfrac{dx}{dt} + 2\cos t \\ x(0) = 0 \\ x'(0) = 0 \end{bmatrix}, \text{ Exact solution is:}$$

$$\left\{ \tfrac{150}{229}\cos t + \tfrac{20}{229}\sin t - \tfrac{150}{229}\frac{\cos\frac{3}{5}\sqrt{11}t}{e^{\frac{1}{5}t}} - \tfrac{250}{7557}\sqrt{11}\frac{\sin\frac{3}{5}\sqrt{11}t}{e^{\frac{1}{5}t}} \right\}$$

We're given that the driving frequency is $\omega_d = 1\,\mathrm{rad/\,s}$ and the natural frequency is still $\omega_0 = 2\,\mathrm{rad/\,s}$, so let's use Evaluate Numerically to calculate the damped frequency.

$$\omega = \sqrt{\omega_0^2 - \beta^2} = \sqrt{2^2 - \tfrac{1}{5}^2} = \tfrac{3\sqrt{11}}{5} = 1.9900$$

Let's take a closer look at our solution (after a little editing by-hand).

$$x(t) = \tfrac{10}{229}(15\cos t + 2\sin t) - \tfrac{150}{229}\left(\cos\tfrac{3\sqrt{11}}{5}t + \tfrac{5\sqrt{11}}{99}\sin\tfrac{3\sqrt{11}}{5}t\right)e^{-t/5}$$

The first part is just SHM at the same frequency as the driving force. The second term looks like underdamped harmonic motion (SHM at the damped frequency ω multiplying the decaying exponential $e^{-\beta t}$). The system starts out in a combination of the two motions, but the second term quickly dies off.

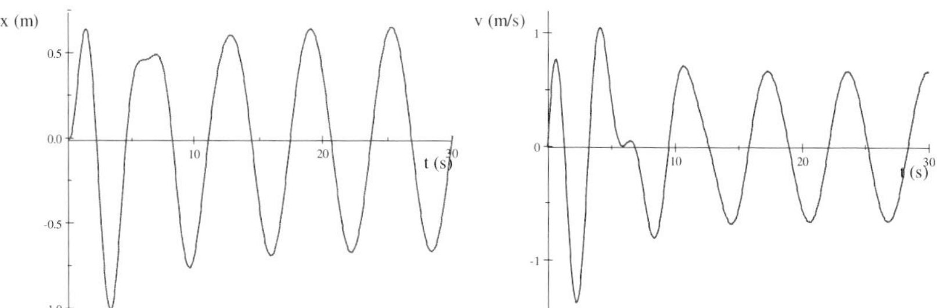

Figure 8.7a Position **Figure 8.7b** Velocity

The driving force adds energy to the system, but eventually the rate of energy lost to damping balances that gain and the system settles into steady SHM at the driving frequency. ◄

Resonance

Our direct just-solve-the-equation approach doesn't always work. The general solution to Eq. (8.33) is too messy for any useful analysis. We can work around that using a complex trial solution and a complex exponential rather than a cosine for the force term. Keep in mind that the real part of the solution has physical significance.

Let's try a complex trial solution of the form $x_c = A_c e^{i\omega t}$. Leave the insertion point anywhere in that expression and click New Definition. Then create and Factor the following form of Eq. (8.33).

$$0 = \frac{d^2x_c}{dt^2} + 2\beta\frac{dx_c}{dt} + \omega_0^2 x_c - f_0\, e^{i\omega t} = e^{it\omega}\left(-f_0 - \omega^2 A_c + \omega_0^2 A_c + 2i\beta\omega A_c\right)$$

We want to know how our solution depends on the driving frequency, so Substitute $\omega = \omega_d$ and use Solve Exact to find the complex amplitude A_c.

$$0 = \left[-f_0 - \omega^2 A_c + \omega_0^2 A_c + 2i\beta\omega A_c\right]_{\omega=\omega_d}, \text{ Solution is: } \frac{f_0}{\omega_0^2 - \omega_d^2 + 2i\beta\omega_d}$$

The physical amplitude A is the magnitude of this complex number. Before we calculate that, let's tell *SNB* the force parameters are real and let's make f_0 positive.

SNB uses absolute value notation for the magnitude of a complex number, so you can use the Brackets button on the Math Objects toolbar or CTRL + BACKSLASH (the key with the vertical bar). Create an expression for the amplitude and then use Rewrite + Rectangular, *SNB*'s command to put a complex number in standard form.

$$A = \left|\frac{f_0}{\omega_0^2 - \omega_d^2 + 2i\beta\omega_d}\right| = \frac{f_0}{\sqrt{\left(\omega_0^2 - \omega_d^2\right)^2 + 4\beta^2\omega_d^2}}$$

The amplitude depends on the driving frequency. Figure 8.8 shows a plot of the amplitude versus the driving frequency for $\omega_0 = 2\,\mathrm{rad/\,s}$, $\beta = \frac{1}{5}\,\mathrm{s}^{-1}$, and $f_0 = 2\,\mathrm{N/\,kg}$. The amplitude equals $f_0 \,/\, \omega_0^2$ when the driving frequency is zero, and the amplitude goes to zero as the driving frequency becomes large.

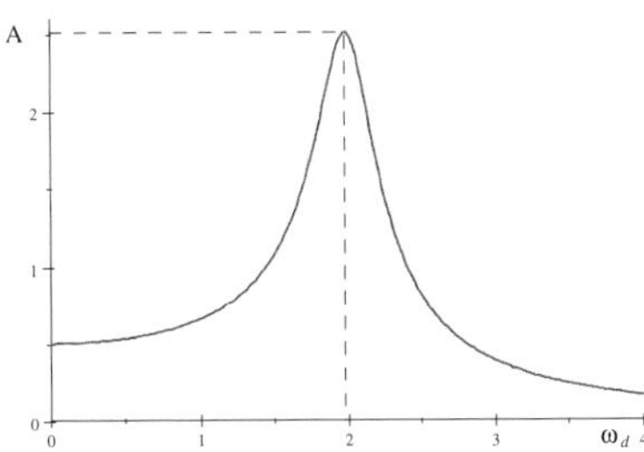

Figure 8.8 Resonance

To find the frequency that produces the maximum amplitude, Evaluate the derivative of the amplitude and then Factor the numerator of the result in-place.

$$\frac{d}{d\omega_d}\frac{f_0}{\sqrt{\left(\omega_0^2 - \omega_d^2\right)^2 + 4\beta^2\omega_d^2}} = -\frac{2\omega_d f_0\left(2\beta^2 - \omega_0^2 + \omega_d^2\right)}{\left(4\beta^2\omega_d^2 + \omega_0^4 - 2\omega_0^2\omega_d^2 + \omega_d^4\right)^{\frac{3}{2}}}$$

Set the numerator equal to zero and use Solve Exact to find ω_d.

$$0 = 2\beta^2 - \omega_0^2 + \omega_d^2, \text{ Solution is: } \sqrt{\omega_0^2 - 2\beta^2}$$

The maximum amplitude occurs when the driving frequency is $\omega_R = \sqrt{\omega_0^2 - 2\beta^2}$. This is called the **resonance frequency**. Think of the amplitude as the system's response to the driving force; resonance is the maximum response. We say a system driven at the resonance frequency is "in resonance" with the driving force.

This is exactly what happens when you ride a child's swing. To generate a significant response from the swing, you use your legs to apply a periodic driving force at just the right frequency. To swing higher, you don't move your legs faster, you push them harder. You increase F_0 but keep your driving frequency at ω_R.

Small Oscillations

An object bound in any potential well can oscillate, but the motion is not generally Simple Harmonic Motion. Stable equilibrium occurs at a local minimum in the potential energy. For small oscillations near the minimum, the motion is approximately SHM.

Here are the first three terms in the Taylor Series expansion of a generic potential energy $U(x)$ about the equilibrium point x_0.

$$U(x) = U(x_0) + \left.\frac{dU}{dx}\right|_{x=x_0} (x - x_0) + \tfrac{1}{2!} \left.\frac{d^2U}{dx^2}\right|_{x=x_0} (x - x_0)^2 + \cdots \tag{8.34}$$

The first term is constant and the second term is zero because the first derivative vanishes at equilibrium. We can approximate any potential energy as

$$U(x) \approx U(x_0) + \tfrac{1}{2}\, k_{\text{eff}}\, (x - x_0)^2 \tag{8.35}$$

where k_{eff} is the effective spring constant.

$$k_{\text{eff}} \equiv \left.\frac{d^2U}{dx^2}\right|_{x=x_0}$$

This is the potential energy of SHM! This SHM approximation is valid when k_{eff} is positive and $x - x_0$ is small. Physically, this says motion near a stable equilibrium point is SHM. Mathematically, it means you can always fit a parabola in a local extremum.

There are two ways we can calculate SHM approximations. The first uses *SNB*'s Power Series command, and the second is a more detailed hands-on procedure.

Example 8.13 *Quick and easy*

Find the SHM approximation for the potential energy

$$U(x) = -e^{-x^2/c^2}\left(U_0 - \tfrac{1}{2}bx^2\right)$$

where b, c, and U_0 are all positive constants.

Solution. Create an expression for the potential energy. Leave the insertion point anywhere in the expression and choose Power Series. In the Series Expansion of f(x) box, choose 3 as the Number of Terms and Expand in Powers of x.

$$U(x) = -e^{-x^2/c^2}\left(U_0 - \tfrac{1}{2}bx^2\right) = -U_0 + x^2\left(\tfrac{1}{2}b + \frac{1}{c^2}U_0\right) + O\left(x^3\right)$$

The effective spring constant is double the coefficient of x^2.

$$k_{\text{eff}} = 2\left(\tfrac{1}{2}b + \frac{1}{c^2}U_0\right) = b + \frac{2}{c^2}U_0$$

Here is our SHM approximation to this potential.

$$U(x) \approx -U_0 + \tfrac{1}{2}\left(b + \frac{2U_0}{c^2}\right)x^2$$

The effective spring constant is positive since all the constants are positive.

Figure 8.9 shows this potential energy (solid curve) and its SHM approximation (dashed) for $U_0 = \frac{1}{2}$, $b = 2$, and $c = 1$ (so $k_{\text{eff}} = 3$).

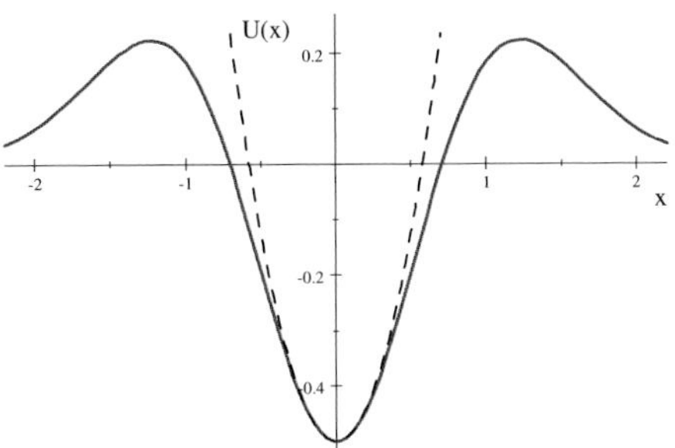

Figure 8.9 Small oscillations

The equilibrium point is $x_0 = 0$. This is not generally the case. ◄

Here is the step-by-step procedure for calculating the SHM approximation for any potential energy:

1. Find the equilibrium point x_0 as the solution to $\dfrac{dU}{dx} = 0$.

2. Calculate the effective spring constant $k_{\text{eff}} = \left.\dfrac{d^2U}{dx^2}\right|_{x=x_0}$.

3. Calculate any constant term in the potential energy $U_0 = U(x_0)$.

It is often useful to graph the potential energy and its SHM approximation.

Let's use this method and do another example.

Example 8.14 *A more hands-on approach*

Find the SHM approximation for the potential energy $U(x) = \dfrac{3}{x} + x^3$.

Solution. Use Solve Exact on the equilibrium condition to find the equilibrium point.

$$0 = \frac{d}{dx}\left(\frac{3}{x} + x^3\right), \text{ Solution is: } 1, i, -1, -i$$

A look at the plot (see Figure 8.10) tells us $x_0 = 1$ is the solution we want. Use this value and Substitute (with Evaluate) to calculate the constant term in the SHM approximation.

$$U_0 = \left[\frac{3}{x} + x^3\right]_{x=1} = 4$$

Use Substitute (with Evaluate) to calculate the effective spring constant.

$$k_{\text{eff}} = \left[\frac{d^2}{dx^2}\left(\frac{3}{x} + x^3\right)\right]_{x=1} = 12$$

Finally, use Expand and Eq. (8.35) to find the SHM approximation.

$$U(x) \approx 4 + \tfrac{1}{2}(12)(x-1)^2 = 6x^2 - 12x + 10$$

You can also use the Power Series method and Expand in Powers of $x - 1$ to three terms.

$$U = \frac{3}{x} + x^3 = 4 + 6(x-1)^2 + O\left((x-1)^3\right)$$

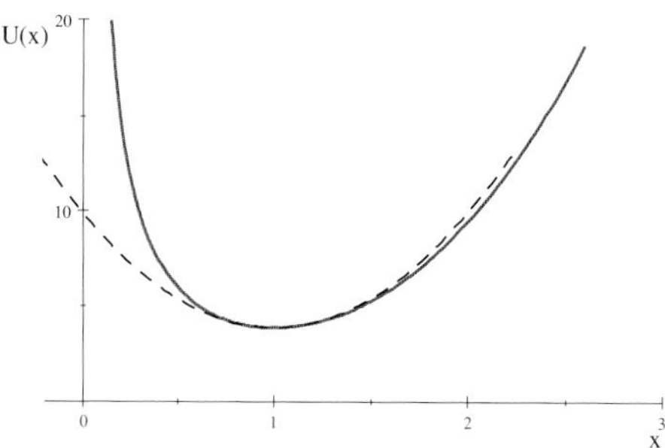

Figure 8.10 shows this potential energy (the solid curve) and its SHM approximation (dashed). ◀

Figure 8.10 On or off?

Not-so-Simple Harmonic Motion

The linear restoring force of Hooke's Law produces SHM with a period that is independent of the amplitude. Let's generalize this to the position-dependent force

$$F(x) = -kx^n \tag{8.36}$$

where the exponent n is unspecified ($n = 1$ for SHM). This suggests an interesting question: What is the period of the oscillatory motion for this force? [24]

SNB's exact ODE solver isn't much help here, so we can't use Newton's 2nd Law. Let's try energy conservation. The potential energy associated with this restoring force is

$$U(x) = \begin{cases} \dfrac{1}{1+n} k\, x^{1+n} & n \neq -1 \\ k \ln x & n = -1 \end{cases} \tag{8.37}$$

where k is a positive constant.

For an object released from rest, energy conservation tells us

$$\tfrac{1}{2}mv^2 + \frac{1}{1+n}kx^{1+n} = \frac{1}{1+n}kA^{1+n} \tag{8.38}$$

where A is the amplitude. This leads us to a general version of Eq. (8.17).

$$\frac{dx}{dt} = \pm A^{(1+n)/2}\sqrt{\frac{k}{m}}\sqrt{\frac{2}{1+n}}\sqrt{1 - \frac{x^{1+n}}{A^{1+n}}} \tag{8.39}$$

This equation is valid for every exponent except $n = -1$. We'll deal with that special case later.

An object released from rest at $x = +A$ moves left to equilibrium, arriving one-quarter of a period later. The equation we have to solve is

$$\int_A^0 \frac{dx}{\sqrt{1-\dfrac{x^{1+n}}{A^{1+n}}}} = -A^{(1+n)/2}\sqrt{\frac{k}{m}}\sqrt{\frac{2}{1+n}}\frac{T_n}{4} \tag{8.40}$$

where T_n is the period we seek.

Let's replace $\sqrt{k/m} = 2\pi/T_{\text{shm}}$ and let $u = x/A$ so $dx = Adu$.

$$\frac{T_n}{T_{\text{shm}}} = A^{(1-n)/2}\frac{1}{\pi}\sqrt{2\,(1+n)}\int_0^1 \frac{du}{\sqrt{1-u^{1+n}}} \tag{8.41}$$

Let's define a dimensionless period $\tau_n \equiv T_n\,/T_{\text{shm}}$ and write our expression in a simple form that isolates the amplitude dependence.

$$\tau_n = A^{(1-n)/2}\,I_n \tag{8.42}$$

We've solved the problem if we can evaluate the definite integral in I_n.

$$I_n \equiv \frac{1}{\pi}\sqrt{2\,(1+n)}\int_0^1 \frac{du}{\sqrt{1-u^{1+n}}} \tag{8.43}$$

SNB can do this integral as long as it knows the sign of $1+n$. Let's start with exponents that leave $1+n$ positive.

$$\text{assume}\,(n > -1) = (-1, \infty)$$

We can Evaluate the integral for exponents larger than -1.

$$I_n = \frac{1}{\pi}\sqrt{2\,(1+n)}\int_0^1 \frac{du}{\sqrt{1-u^{1+n}}} = \frac{\sqrt{2}}{\sqrt{\pi}\sqrt{n+1}}\frac{\Gamma\left(\dfrac{1}{n+1}\right)}{\Gamma\left(\dfrac{n+3}{2\,(n+1)}\right)}$$

Don't let the gamma functions freak you out. They're just a more general version of factorial that includes non-integers. You can see this with Rewrite + Gamma.

$$z! = \Gamma\,(z+1)$$

For the exponents less than -1, let $m = -\,(1+n)$ be a positive quantity.

$$\text{assume}\,(m > 0) = (0, \infty)$$

Rearrange the definite integral in I_n in terms of the positive m and Evaluate it.

$$I_n = \frac{1}{\pi}\sqrt{2m}\int_0^1 \frac{u^{m/2}\,du}{\sqrt{1-u^m}} = \frac{\sqrt{2}}{\sqrt{\pi}\sqrt{m}}\frac{\Gamma\left(\dfrac{m+2}{2m}\right)}{\Gamma\left(\dfrac{m+1}{m}\right)}$$

Now use Substitute (with Evaluate) to write I_n in terms of n.

$$\left[\frac{\sqrt{2}}{\sqrt{\pi}\sqrt{m}}\frac{\Gamma\left(\frac{m+2}{2m}\right)}{\Gamma\left(\frac{m+1}{m}\right)}\right]_{m=-(1+n)} = \frac{\sqrt{2}}{\sqrt{\pi}\sqrt{-n-1}}\frac{\Gamma\left(\frac{1}{2}\frac{n-1}{n+1}\right)}{\Gamma\left(\frac{n}{n+1}\right)}$$

What about $n = -1$? In that case the dimensionless period is proportional to the amplitude, and $I_{-1} = \sqrt{2/\pi}$. I've left the details to you in a problem (see Problem 34).

We have all the pieces to complete our general expression for the dimensionless period of this not-so-Simple Harmonic Motion (NSSHM).

$$\tau_n = A^{\frac{1-n}{2}} \begin{cases} \sqrt{\frac{2}{\pi}}\sqrt{\frac{1}{1+n}}\frac{\Gamma\left(\frac{1}{1+n}\right)}{\Gamma\left(\frac{1}{2}+\frac{1}{1+n}\right)} & \text{if} \quad n > -1 \\ \sqrt{\frac{2}{\pi}} & \text{if} \quad n = -1 \\ \sqrt{\frac{2}{\pi}}\sqrt{\frac{-1}{1+n}}\frac{\Gamma\left(\frac{1}{2}-\frac{1}{1+n}\right)}{\Gamma\left(1-\frac{1}{1+n}\right)} & \text{if} \quad n < -1 \end{cases} \tag{8.44}$$

For a given exponent, doubling the amplitude changes the period by a factor of $2^{(1-n)/2}$. Only in the case of SHM $(n = 1)$ is the period independent of the amplitude.

The following figure shows a plot of this expression for three different amplitudes. The **red** curve is $A = 2$, the **green** curve is $A = 1$, and the **blue** curve is $A = \frac{1}{2}$. All three reproduce the SHM result (dashed line) when $n = 1$.

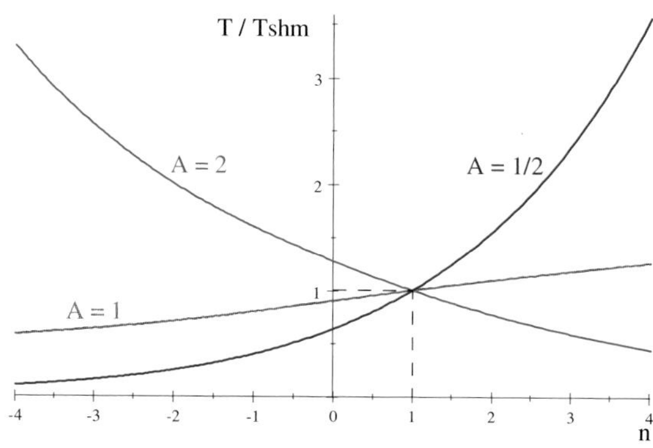

Figure 8.11 NSSHM

When $n > 1$, increasing the amplitude reduces the period, as if the spring's strength increased as it stretched. Similarly when $n < 1$, increasing the amplitude increases the period, as if the spring's strength decreased as it stretched. The $A = 1$ curve is just the quantity I_n from Eq. (8.43).

Problems

1. A 47 g block is attached to a 2.7 N/ m spring, arranged horizontally along a frictionless surface. The block is pulled 7 cm to the left and released from rest. How many oscillations does the block complete per second? How much time does it take for the block to return to equilibrium? What is the amplitude of these oscillations?

2. Prove an object in SHM spends twice as much time outside the midpoints $x = \pm\frac{1}{2}A$ as it does inside.

3. Show that when an oscillator released from rest is halfway to equilibrium, its speed is $v = \frac{\sqrt{3}}{2} v_{\max}$ where $v_{\max}$ is its maximum speed.

4. Show that the position, velocity, and acceleration for SHM starting from equilibrium with an arbitrary initial velocity are
$$\begin{aligned} x(t) &= \frac{v_0}{\omega_0} \sin \omega_0 t \\ v(t) &= v_0 \cos \omega_0 t \\ a(t) &= -\omega_0 v_0 \sin \omega_0 t \end{aligned}$$
where v_0 is the initial velocity.

5. When an object undergoes SHM, where does its kinetic energy equal its potential energy?

6. Let's use *SNB* to derive the expression for the amplitude of SHM with arbitrary initial position and speed.

   a. Show that the time to reach the amplitude is
   $$t = \frac{1}{\omega_0} \tan^{-1} \frac{v_0}{\omega_0 x_0}.$$

   b. Use Eq. (8.14a) to verify this expression for the amplitude.
   $$A^2 = \left(x_0 \cos \left(\tan^{-1} \frac{v_0}{\omega_0 x_0} \right) + \frac{v_0}{\omega_0} \sin \left(\tan^{-1} \frac{v_0}{\omega_0 x_0} \right) \right)^2$$

   c. Use Simplify and then Expand to finish the derivation.

7. Repeat Example 8.5 for a block initially moving at 1.43 m/ s to the left.

8. Show that Eq. (8.16) for the velocity as a function of position
$$v(x) = \pm\sqrt{v_0^2 + \frac{k}{m}\left(x_0^2 - x^2\right)}$$
is equivalent to $v = \pm\sqrt{\frac{k}{m}\left(A^2 - x^2\right)}$ when the amplitude is given by Eq. (8.13).

9. A 0.2 kg block is attached to a 4.7 N/ m spring, arranged horizontally along a frictionless surface. The block is pulled 4.7 cm to the left and released such that it's moving 0.85 m/ s to the right. How much time does it take for the block to return to equilibrium? What is the amplitude of these oscillations?

10. Use Eq. (8.17) to verify by direction calculation that the period of SHM is independent of the period.

11. Verify the "mess multiplying x_0" in the derivation following Eq. (8.17) does indeed equal $\cos \omega_0 t$.

12. Verify that for objects in SHM the sum of the average potential and kinetic energies equals the total energy.

13. Consider an object in SHM.

    a. Show that the object's average position is zero.

    b. Calculate the object's rms position.

    c. Use this result to verify the average potential energy found in the text.

14. Consider an object in SHM.

    a. Show that the object's average velocity is zero.

    b. Calculate the object's rms velocity.

    c. Use this result to verify the average kinetic energy found in the text.

15. Just for the fun of it, use *SNB* to find the exact solution to Eq. (8.21) for arbitrary initial conditions. Take a good look at the solution, and notice where the relative strength of the two forces plays a role.

16. Find the times when two consecutive minima occur for the oscillator in Example 8.6 and use them to verify the period of the motion.

17. How long does it take the block in Example 8.6 to first reach equilibrium?

18. Repeat Example 8.6 for a block that is initially

    a. at rest, and b. moving at $2\,\mathrm{m/\,s}$ toward equilibrium.

19. How long does it take the oscillator in Example 8.9 to complete one oscillation? How many oscillations does it complete in 5 seconds?

20. Show that the amplitude at resonance is $f_0/\left(2\beta\omega\right)$ where ω is the damped frequency.

21. Repeat Example 8.12 using the driving force $2\,\mathrm{N}\cos\omega_R t$ where ω_R is the resonance frequency. Plot the position as a function of time for the first 30 seconds. Compare your result to Figure 8.8 and confirm they are consistent. Verify that your amplitude agrees with the prediction of the previous problem.

22. Show that the amplitude of a damped oscillator with a sinusoidal driving force can be written as
$$A\left(\omega_d\right) = \frac{f_0}{\sqrt{\left(\omega_d^2 - \omega_R^2\right)^2 + 4\beta^2\omega^2}}$$
where ω_R is the resonance frequency and ω is the damped frequency.

23. Show that the position as a function of time for a damped oscillator driven by a constant force reduces to the expected result when the damping is turned off.

24. Show that the position as a function of time for an undamped oscillator driven by the force $F(t) = F_0 \sin \omega t$ is

$$x(t) = x_0 \cos \omega_0 t + \left(\frac{v_0}{\omega_0} - \frac{\omega f_0}{\omega_0 \left(\omega_0^2 - \omega^2\right)} \right) \sin \omega_0 t + \frac{f_0}{\omega_0^2 - \omega^2} \sin \omega t$$

where $f_0 = F_0/m$.

25. Use Solve ODE + Exact to show that for a constant driving force with damping, released from rest at the origin, the position as a function of time is

$$x(t) = \frac{f_0}{\omega_0^2} \left(1 - e^{-\beta t} \left(\cos \omega t + \frac{\beta}{\omega} \sin \omega t\right)\right)$$

26. Prove that resonance can only occur to a damped oscillator for a driving force proportional to $\cos \omega_d t$ if the system is underdamped.

27. Let's look at a system with no spring, that includes linear resistance and an oscillating driving force.

a. Verify that this system's equation of motion is $m\frac{d^2x}{dt^2} = -b\frac{dx}{dt} + F_0 \cos \omega t$.

b. Show that when $m = 1$, $b = 2$, $F_0 = 1$, and $\omega = 1$, the position as a function of time is

$$x(t) = x_0 + \tfrac{1}{2} v_0 \left(1 - e^{-2t}\right) + \tfrac{1}{5} \left(2 \sin t - \cos t + e^{-2t}\right)$$

where x_0 and v_0 are the initial position and velocity, respectively.

c. Plot the position for $x_0 = +1$ and $v_0 = -1$. Comment on the behavior of this system after $t = 3\,\mathrm{s}$.

28. Use the step-by-step procedure to find the SHM approximation to

$$U(x) = -e^{-x^2/c^2} \left(U_0 - \tfrac{1}{2} b x^2\right)$$

where U_0, b, and c are positive constants. Check your answer against Example 8.13.

29. Find the SHM approximation for the potential energy $U(x) = \dfrac{a}{x} + bx$. What is the frequency of the small oscillations about the equilibrium point?

30. Find the SHM approximation for the potential energy $U(x) = \dfrac{bx^2}{1 + \frac{x^2}{a^2}}$. What is the period of the small oscillations about the equilibrium point?

31. Find the SHM approximation for the potential energy $U(x) = -U_0\, e^{-x^2/a^2}$. What is the frequency of the small oscillations about the equilibrium point?

32. Find the SHM approximation for the potential energy $U(x) = -\dfrac{U_0}{1 + e^{x^2/a^2}}$. What is the period of the small oscillations about the equilibrium point?

33. Verify by direct calculation that the potential energy for the position-dependent force $F(x) = -kx^n$ is

$$U(x) = \begin{cases} \dfrac{1}{1+n} k\, x^{1+n} & n \neq -1 \\ k \ln x & n = -1 \end{cases}$$

34. Let's fill in the gaps in the derivation of the period of NSSHM for the case $n = -1$.

a. Start with energy conservation

$$\tfrac{1}{2}mv^2 + k\ln x = k\ln A$$

and show that $\tau_{-1} = A\dfrac{\sqrt{2}}{\pi}\displaystyle\int_0^1 \frac{du}{\sqrt{-\ln u}}$.

b. Use Calculus + Change Variable... and the substitution $u = e^{-w}$ to show that

$$\int_0^1 \frac{du}{\sqrt{-\ln u}} = \int_0^\infty \frac{e^{-w}}{\sqrt{w}}\,dw$$

c. Evaluate the new integral and show that $\tau_{-1} = A\sqrt{2/\pi}$.

35. The differential equation given by Newton's 2nd Law for the position-dependent force $F(x) = -kx^n$ is

$$\frac{d^2x}{dt^2} = -\frac{k}{m}x^n$$

where both the mass and force constants are positive. Set both constants equal to 1, use $n = 3$ for the exponent, and let the motion start from rest at $x_0 = 2$.

a. Use Solve ODE + Numeric to find a numerical solution for the position. Plot the solution from $t = 0$ to $t = 4$.

b. Use the graph and Evaluate Numerically to estimate the period of this motion.

c. Compare the period to that of the corresponding SHM. Is your result consistent with Figure 8.11?

36. Repeat the previous problem for $x_0 = \frac{1}{2}$. Compare the two part-b results. Is your result consistent with Eq. (8.43)?

37. Show that when $n > -1$, the expression for the period of not-so-simple harmonic motion has the correct limit as the exponent approaches -1 from above.

38. Show that when $n < -1$, the expression for the period of not-so-simple harmonic motion has the correct limit as the exponent approaches -1 from below.

9 Central Forces

Central forces play an important role in physics. Many interactions in nature are central forces. The orbits of planets, a child's balloon sticking to a wall, and the attraction between molecules are all explained with central forces.

A **central force** always points in the radial direction, either toward or away from the force center, with a magnitude that depends only on the radial distance.

$$\overrightarrow{F} \equiv -F(r)\,\widehat{r} \tag{9.1}$$

Central forces have no tangential $\left(\widehat{\phi}\right)$ or polar $\left(\widehat{\theta}\right)$ components.

Equations of Motion

In vector form, Newton's 2nd Law says

$$\frac{d^2\overrightarrow{r}}{dt^2} = \frac{\overrightarrow{F}}{m} \equiv -f(r)\,\widehat{r} \tag{9.2}$$

where m is the mass of the orbiting body and $f(r)$ is its acceleration. For bound orbits, $f(r)$ is positive so the force is always attractive.

Orbits tend to be round, and central-force orbits always lie in a plane, so it's easier to use 2-dimensional polar coordinates. There are two components of the acceleration.

$$a_r = \frac{d^2r}{dt^2} - r\left(\frac{d\phi}{dt}\right)^2 \tag{9.3a}$$

$$a_\phi = \frac{1}{r}\frac{d}{dt}\left(r^2\frac{d\phi}{dt}\right) \tag{9.3b}$$

For a central force, the radial acceleration is $a_r = -f(r)$ and the tangential acceleration is zero $(a_\phi = 0)$. We can use Eq. (9.3b) to define a constant ℓ called the angular momentum per unit mass.

$$\ell \equiv r^2\frac{d\phi}{dt} \tag{9.4}$$

When the position and velocity vectors are perpendicular $\ell = rv$.

We now have two coupled differential equations for the orbit.

$$\frac{d^2r}{dt^2} = -f(r) + \frac{\ell^2}{r^3} \tag{9.5a}$$

$$\frac{d\phi}{dt} = \frac{\ell}{r^2} \tag{9.5b}$$

This is Newton's 2nd Law for central forces.

Doing Physics with Scientific Notebook: A Problem-solving Approach, First Edition. Joseph Gallant.

The two solutions (the radial $r(t)$ and angular $\phi(t)$ positions as functions of time) require 4 initial conditions:

Radial Position	**Angular Momentum**	**Angular Position**	**Radial Speed**
$R_0 \equiv r(0)$	ℓ	$\phi_0 = \phi(0)$	$\left.\frac{dr}{dt}\right\|_{t=0}$

For bound orbits, we'll usually start our orbits on the x-axis ($\phi_0 = 0$) with zero radial speed. Every stable bound orbit has at least two such points, so this simplification is no limitation.

The two 2nd Law equations can be combined into a single equation for a central orbit as a function of angle.

$$\frac{d^2r}{d\phi^2} - \frac{2}{r}\left(\frac{dr}{d\phi}\right)^2 - r = -\frac{r^4}{\ell^2} f(r) \tag{9.6}$$

The solution $r(\phi)$ requires 3 initial conditions: radial position R_0, angular momentum ℓ, and radial derivative (usually zero).

It is sometimes more convenient to recast Eq. (9.6) in terms of the reciprocal variable $u \equiv 1/r$.

$$\frac{d^2u}{d\phi^2} + u = \frac{1}{u^2\ell^2} f(u) \tag{9.7}$$

This equation is simpler than Eq. (9.6) and it has more known exact solutions. If you know the force $f(r)$, you can solve Eq. (9.7) for the orbit $r(\phi)$. If you know the orbit $r(\phi)$, you can use Eq. (9.7) to find the force $f(r)$.

Example 9.1 *Cotes' spirals 1*

Analyze and plot the motion of an object subjected to the central force $f(r) = \beta/r^3$.

Solution. For an inverse-cubic force $f(u) = \beta u^3$ so Eq. (9.7) takes a simple form.

$$\frac{d^2u}{d\phi^2} + \left(1 - \frac{\beta}{\ell^2}\right) u = 0$$

There are three possible solutions, depending on the ratio β/ℓ^2. When $\beta < \ell^2$ the quantity $k^2 \equiv 1 - \dfrac{\beta}{\ell^2}$ is a positive number.

$$\text{assume}\,(k > 0) = (0, \infty)$$

Put Eq. (9.7) with the appropriate initial conditions in a matrix and choose Solve ODE + Exact.

$$\begin{bmatrix} \dfrac{d^2u}{d\phi^2} + k^2u = 0 \\ u(0) = \dfrac{1}{R_0} \\ u'(0) = 0 \end{bmatrix}, \text{ Exact solution is: } \left\{\frac{1}{R_0}\cos k\phi\right\}$$

When $\beta > \ell^2$ the quantity $k^2 \equiv \dfrac{\beta}{\ell^2} - 1$ is a positive number. Adjust Eq. (9.7) accordingly, and then use Solve ODE + Laplace (because the answer looks better).

$$\begin{bmatrix} \dfrac{d^2u}{d\phi^2} - k^2u = 0 \\ u(0) = \dfrac{1}{R_0} \\ u'(0) = 0 \end{bmatrix}, \text{ Laplace solution is: } \left\{ \frac{1}{R_0} \cosh k\phi \right\}$$

Now let's repeat the process for $\beta = \ell^2$.

$$\begin{bmatrix} \dfrac{d^2u}{d\phi^2} = 0 \\ u(0) = \dfrac{1}{R_0} \\ u'(0) = 0 \end{bmatrix}, \text{ Exact solution is: } \left\{ \frac{1}{R_0} \right\}$$

The resulting orbit is a circle. Here's a summary of these results, a collection of orbits known as Cotes' spirals.

$$r(\phi) = \begin{cases} \dfrac{R_0}{\cos k\phi} & k = \sqrt{1 - \dfrac{\beta}{\ell^2}} & (\beta < \ell^2) \\ R_0 & k = 1 & (\beta = \ell^2) \\ \dfrac{R_0}{\cosh k\phi} & k = \sqrt{\dfrac{\beta}{\ell^2} - 1} & (\beta > \ell^2) \end{cases}$$

Let's take a look at an example of each (using Plot 2D + Polar) with $R_0 = 1$ and $k = \frac{1}{\sqrt{10}}$ for the non-circular orbits.

Condition	Orbit
$\dfrac{\beta}{\ell^2} = \frac{9}{10}$	**Spiral out**
$\dfrac{\beta}{\ell^2} = 1$	**Circle**
$\dfrac{\beta}{\ell^2} = \frac{11}{10}$	**Spiral in**

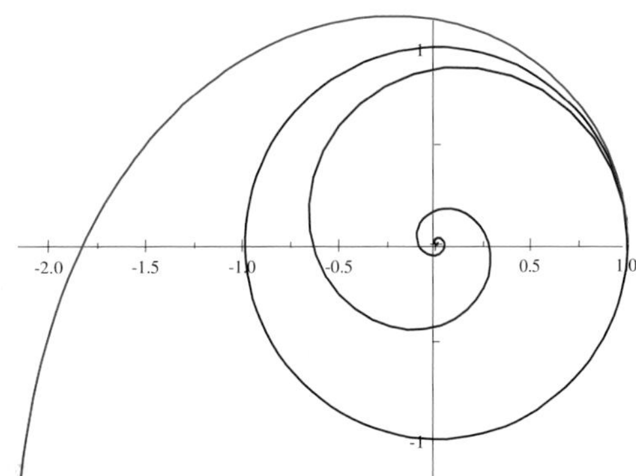

Figure 9.1 Cotes' spirals

If no force acts on the object, it moves off to infinity in a straight line at speed $v = \ell/R_0$. The attractive force pulls the object into a curved path. Depending on the strength of that pull, the object either spirals off to infinity, spirals into the force center, or is precariously balanced in an unstable circular orbit. ◄

Example 9.2 *Generalized cardioid*
Find the force law that produces the orbit $r(\phi) = A(1+\cos n\phi)^{-1/n}$ where $n \neq 0$.
Solution. Even though it's a small fib, let's tell *SNB* all the variables are positive.

$$\text{assume}(\text{positive}) = (0, \infty)$$

Create the following expression for the reciprocal variable u, leave the insertion point anywhere in it, and click the New Definition button.

$$u = \frac{1}{A}(1+\cos n\phi)^{1/n}$$

Use Eq. (9.7) to create and Simplify an expression for the force.

$$u^2\ell^2\left(\frac{d^2u}{d\phi^2}+u\right) = -\frac{1}{A^3}\frac{\ell^2}{(\cos n\phi+1)^{\frac{1}{n}(n-3)}}(n-2)$$

Now Substitute (with Simplify) $\cos n\phi = A^n r^{-n} - 1$ into the force.

$$f(r) = \left[-\frac{1}{A^3}\frac{\ell^2}{(\cos n\phi+1)^{\frac{1}{n}(n-3)}}(n-2)\right]_{\cos n\phi = A^n r^{-n}-1} = -\frac{1}{A^n}r^{n-3}\ell^2(n-2)$$

The desired force law requires $n < 2$ for bound orbits.

$$f(r) = \frac{(2-n)\,\ell^2}{A^n}r^{n-3}$$

Notice $n \neq 2$ since that makes the force zero. One special case of these orbits is the cardioid $r(\phi) = A(1+\cos\phi)$ where $n = -1$.

$$f(r) = \left[\frac{(2-n)\,\ell^2}{A^n}r^{n-3}\right]_{n=-1} = 3\frac{A}{r^4}\ell^2$$

This orbit passes through the origin where the force blows up. ◄

Besides defining angular momentum, Eq. (9.5b) has other uses. When you know the orbit in terms of time, you can use Eq. (9.5b) to calculate the rotation angle during some time interval.

$$\Delta\phi = \ell\int_{t_0}^{t_f}\frac{dt}{r^2(t)} \tag{9.8a}$$

When you know the orbit in terms of angle, you can use Eq. (9.5b) to calculate the elapsed time between two angular positions.

$$\Delta t = \frac{1}{\ell}\int_{\phi_0}^{\phi_f} r^2(\phi)\,d\phi \tag{9.8b}$$

In either case, you need to know the angular momentum.

Example 9.3 *Cotes' spirals 2*
How long does it take each orbit in the first example to complete half an orbit? Use $\beta = 4\pi^2$ (so the periods are in years, as you'll soon see).
Solution. We can use the given values of the ratio β/ℓ^2 and β to Evaluate (and then Evaluate Numerically) the angular momentum.

$$\textbf{Spiral out:} \quad \ell = \sqrt{\tfrac{10}{9} \times 4\pi^2} = \tfrac{1}{3}\sqrt{40}\pi = 6.623\,1$$
$$\textbf{Circle:} \quad \ell = \sqrt{1 \times 4\pi^2} = 2\pi = 6.283\,2$$
$$\textbf{Spiral in:} \quad \ell = \sqrt{\tfrac{10}{11} \times 4\pi^2} = \sqrt{\tfrac{40}{11}}\pi = 5.990\,8$$

For the Cotes orbits *SNB* can do the integrals of Eqs. (9.8) exactly (see Problems 6 and 7) but let's just use Evaluate Numerically to approximate the time for all three orbits.

$$\textbf{Spiral out:} \quad \Delta t = \frac{1}{\sqrt{\frac{40}{9}}\pi} \int_0^\pi \frac{d\phi}{\cos^2 \frac{\phi}{\sqrt{10}}} = 0.733\,02$$
$$\textbf{Circle:} \quad \Delta t = \frac{1}{2\pi} \int_0^\pi \frac{d\phi}{1^2} = 0.5$$
$$\textbf{Spiral in:} \quad \Delta t = \frac{1}{\sqrt{\frac{40}{11}}\pi} \int_0^\pi \frac{d\phi}{\cosh^2 \frac{\phi}{\sqrt{10}}} = 0.400\,56$$

The time for the Cotes semicircle is the same as the inverse-square semicircle, but that's just a coincidence because $R_0 = 1$ (see Problem 4). ◄

Eq. (9.5b) also lets you change variables between time and angle.

$$d\phi = \frac{\ell}{r^2}\, dt \tag{9.9}$$

This lets you change between radial position as a function of time or angle.

$$\frac{dr}{dt} = \frac{dr}{d\phi}\frac{d\phi}{dt} = \frac{\ell}{r^2}\frac{dr}{d\phi} \tag{9.10}$$

$\boldsymbol{r(\phi) \to r(t)}$: If you know $r(\phi)$ and you can express $\dfrac{dr}{d\phi}$ as a function of r – let's call it $D_\phi(r)$ – then you can calculate the radial position as a function of time.

$$\ell\, t = \int_{R_0}^{r} \frac{r^2}{D_\phi(r)}\, dr \tag{9.11}$$

After you do the integral, solve the resulting equation for r.

$\boldsymbol{r(t) \to r(\phi)}$: If you know $r(t)$ and you can express $\dfrac{dr}{dt}$ as a function of r – let's call it $D_t(r)$ – then you can calculate the radial position as a function of angle.

$$\frac{\phi}{\ell} = \int_{R_0}^{r} \frac{1}{r^2 D_t(r)}\, dr \tag{9.12}$$

After you do the integral, solve the resulting equation for r.

Example 9.4 *Cotes' spirals 3*
Find the radial position as a function of time for an inward Cotes' spiral. How long does the total inward spiral last?
Solution. From Example 9.1 we know the radial position as a function of angle.

$$r(\phi) = \frac{R_0}{\cosh k\phi}$$

Let's Evaluate the derivative with respect to angle.

$$\frac{d}{d\phi}\left(\frac{R_0}{\cosh k\phi}\right) = -k\frac{R_0}{\cosh^2 k\phi}\sinh k\phi$$

Use the identity $\cosh^2 x - \sinh^2 x = 1$ and Substitute (with Simplify) to express this derivative as a function of r, then edit the result by-hand.

$$\begin{aligned} D_\phi(r) &= -kR_0\left[\frac{\sinh k\phi}{\cosh^2 k\phi}\right]_{\sinh k\phi=\sqrt{\cosh^2 k\phi-1},\cosh k\phi=R_0/r} \\ &= -k\frac{r^2}{R_0}\sqrt{\frac{1}{r^2}\left(R_0^2-r^2\right)} \\ &= -\frac{k}{R_0}r\sqrt{R_0^2-r^2} \end{aligned}$$

SNB does the substitutions in the order you list them in the Subscript, so make sure you put the r-dependence last. Setup and Evaluate the integral in Eq. (9.11) using the definite integral fudge.

$$-\frac{R_0}{k}\left[\int\frac{r\,dr}{\sqrt{R_0^2-r^2}}\right]_{r=R_0}^{r=r} = \frac{1}{k}R_0\sqrt{R_0^2-r^2}$$

Create an expression for Eq. (9.11) and use Solve Exact to find r.

$$\ell t = \frac{1}{k}R_0\sqrt{R_0^2-r^2}, \text{ Solution is: } -\frac{1}{R_0}\sqrt{R_0^4-k^2t^2\ell^2}, \frac{1}{R_0}\sqrt{R_0^4-k^2t^2\ell^2}$$

Take the positive solution and do a little editing by-hand.

$$r(t) = R_0\sqrt{1-\frac{k^2\ell^2}{R_0^4}t^2}$$

The inward spiral ends when $r = 0$, so create an expression setting the stuff inside the radical equal to zero and use Solve Exact to find t.

$$0 = 1-\frac{k^2\ell^2}{R_0^4}t^2, \text{ Solution is: } -\frac{1}{k\ell}R_0^2, \frac{1}{k\ell}R_0^2$$

It takes an object in this orbit time $T_{\text{in}} = \frac{R_0^2}{k\ell}$ to spiral all the way to the origin. ◀

Newtonian Gravitation

Once Newton had his three Laws of Motion, he applied them to gravity and the orbital motion of the Moon around the Earth. He theorized the force of gravity holding the Moon in orbit, and then generalized his analysis to apply to any two objects. The result is called the Law of Universal Gravitation.

According to Newton, the magnitude of the gravitational force between two objects is

$$F = G\frac{m_1 m_2}{r^2} \tag{9.13}$$

where m_1 and m_2 are the masses of the two objects. The separation r is measured from the center of one object to the center of the other. Newton's gravity is an inverse-square force, so its magnitude depends on the reciprocal of the separation squared.

The constant G determines the strength of gravity and ensures the units work out.

$$G = 6.6726 \times 10^{-11}\frac{\mathrm{N\,m^2}}{\mathrm{kg^2}} \tag{9.14}$$

Gravity is a very weak force and it's always attractive, so the direction of gravity is towards the other object.

In our various equations of orbital motion, Newtonian gravity has a radial acceleration

$$f(r) = \frac{GM}{r^2} \tag{9.15}$$

where M is the mass of the central body around which the other object orbits.

One well-known orbit is the Earth's orbit around the Sun, so I'll occasionally use it for comparison. Just to remind you, the Earth orbits the Sun once per one year at an average distance of one astronomical unit.

$$a = 1\,\mathrm{AU} = 1.4960 \times 10^{11}\,\mathrm{m} \tag{9.16a}$$

$$T = 1\,\mathrm{y} = 3.1557 \times 10^{7}\,\mathrm{s} \tag{9.16b}$$

Here is the numerator in Eq. (9.15) for orbits around the Sun.

$$GM_S = 1.3271 \times 10^{20}\frac{\mathrm{m^3}}{\mathrm{s^2}} \tag{9.17a}$$

$$= 4\pi^2\frac{\mathrm{AU^3}}{\mathrm{y^2}} \tag{9.17b}$$

For planetary orbits, AUs and years are usually more convenient.

Newton's gravitational force explains weight. Since separation is measured center-to-center, when you're "on the surface" of a celestial body you're one radius away. Your weight on the surface is the gravitational force you feel at a distance of one radius.

$$mg = G\frac{mM}{R^2} \tag{9.18}$$

The acceleration due to gravity on the surface of celestial body X is

$$g_\mathrm{x} = \frac{GM_\mathrm{x}}{R_\mathrm{x}^2} \tag{9.19}$$

where M_x and R_x are the mass and radius of celestial body X. This is how the Apollo astronauts knew the Moon's surface gravity before they got there.

Example 9.5 *Gliese 581g*
A recently discovered (but unconfirmed) exoplanet called Gliese 581g may be able to sustain liquid water on its surface, and may be a candidate to harbor life. Preliminary data suggest it has 3.7 times the mass of the Earth and its radius is 1.65 times the Earth's. What is its surface gravity?
Solution. Look up the values for the Earth's radius and mass, and use Eq. (9.19) to create an expression for the surface gravity of Gliese 581g that you can Evaluate Numerically.

$$g_g = \frac{\left(6.6726 \times 10^{-11} \dfrac{\mathrm{N\,m^2}}{\mathrm{kg}^2}\right)(3.7 \times 5.9737 \times 10^{24}\,\mathrm{kg})}{(1.65 \times 6.3781 \times 10^6\,\mathrm{m})^2} = 13.316 \frac{\mathrm{m}}{\mathrm{s}^2}$$

Let's use Evaluate and convert that into units of Earth's g.

$$g_g = 13.316 \frac{\mathrm{m}}{\mathrm{s}^2} \times \frac{1g}{9.8067 \dfrac{\mathrm{m}}{\mathrm{s}^2}} = 1.3578\,g$$

The surface gravity on Gliese 581g is about 36% stronger than the Earth's. The "g" in the planet's name means it was the sixth planet (but seventh object including the star) discovered in the Gliese 581 system. ◄

Kepler's Laws

Kepler based his three laws of planetary motion on his detailed and painstaking analysis of Tycho Brahe's data on the motions of the planets. His results refuted 2000 years of wrong ideas about planetary motion.

1. The orbit of every planet is an ellipse with the Sun at one focus.
2. A line joining a planet and the Sun sweeps out equal areas during equal time intervals.
3. The square of the planet's orbital period is directly proportional to the cube of its average distance from the Sun.

Kepler's 2nd Law works for any central force, but the 1st and 3rd apply only to the inverse-square force.

Kepler's Laws are empirical. Even though they are based on observational data, they only describe the planetary orbits, they do not explain them. Newton used his three laws of motion and Universal Gravitation to explain Kepler's Laws theoretically. This agreement between theory and observation was convincing evidence that Newton was right.

We have an advantage over Kepler, because we know Newton's Laws and his Law of Universal Gravitation, and we have *SNB* to help us.

Kepler's 1st Law: The Ellipse Law

Let's use Eq. (9.7) with Newton's gravity $f(u) = GM\,u^2$ and *SNB*'s Solve ODE + Laplace to find the orbit.

$$\begin{bmatrix} \dfrac{d^2u}{d\phi^2} + u = \dfrac{GM}{\ell^2} \\ u(0) = \dfrac{1}{R_0} \\ u'(0) = 0 \end{bmatrix}, \text{ Laplace solution is:}$$

$$\left\{ G\frac{M}{\ell^2} + \frac{1}{\ell^2 R_0}(\cos\phi)\left(\ell^2 - GMR_0\right) \right\}$$

A little editing by-hand gives us the equation of an off-center ellipse.

$$r(\phi) = \frac{\dfrac{\ell^2}{GM}}{1 - \left(1 - \dfrac{\ell^2}{GMR_0}\right)\cos\phi} \tag{9.20}$$

The term multiplying the cosine is called the eccentricity (ε). We can write this orbit in the slightly simpler form

$$r(\phi) = \frac{R_0(1-\varepsilon)}{1 - \varepsilon\cos\phi} \tag{9.21}$$

where $\varepsilon \equiv 1 - \dfrac{\ell^2}{GMR_0}$. Circular orbits have zero eccentricity and free-fall zero-angular-momentum orbits have the maximum $\varepsilon = 1$.

For my choice of coordinate system, the maximum distance is the initial distance.

$$r_{\max} = \left[\frac{\dfrac{\ell^2}{GM}}{1 - \varepsilon\cos\phi} \right]_{\phi=0} = -\frac{1}{GM}\frac{\ell^2}{\varepsilon - 1}$$

The minimum distance occurs at $\phi = \pi$.

$$r_{\min} = \left[\frac{\dfrac{\ell^2}{GM}}{1 - \varepsilon\cos\phi} \right]_{\phi=\pi} = \frac{1}{GM}\frac{\ell^2}{\varepsilon + 1}$$

We can Simplify the average of these two extremes to get the average radial distance.

$$a = \frac{1}{2}\frac{\ell^2}{GM}\left(\frac{1}{1-\varepsilon} + \frac{1}{1+\varepsilon}\right) = \frac{1}{GM}\frac{\ell^2}{1-\varepsilon^2}$$

We can also write the eccentricity in terms of the average distance.

$$\varepsilon = \sqrt{1 - \frac{\ell^2}{GMa}} \tag{9.22}$$

Example 9.6 *OMG 3!*

The third planet in the Omagahd system starts at an initial distance of 3 AU and moves in an elliptical orbit around a 1-solar-mass star such that its maximum distance is three times the minimum. Find the eccentricity, angular momentum, and average distance of this orbit. Plot the orbit.

Solution. Let's Evaluate the ratio of the maximum distance to the minimum.

$$\frac{r_{\max}}{r_{\min}} = \frac{\left[\dfrac{\ell^2/(GM)}{1-\varepsilon\cos\phi}\right]_{\phi=0}}{\left[\dfrac{\ell^2/(GM)}{1-\varepsilon\cos\phi}\right]_{\phi=\pi}} = -\frac{1}{\varepsilon-1}(\varepsilon+1)$$

Set the ratio equal to 3 and use Solve Exact to find the eccentricity.

$$3 = \frac{1+\varepsilon}{1-\varepsilon}, \text{ Solution is: } \tfrac{1}{2}$$

Use the result for the minimum distance to create and Evaluate an expression for the angular momentum.

$$\ell = \sqrt{\left(4\pi^2\tfrac{\mathrm{AU}^3}{\mathrm{y}^2}\right)\left(1+\tfrac{1}{2}\right)1\,\mathrm{AU}} = \sqrt{6}\frac{\pi}{\mathrm{y}}\,\mathrm{AU}^2$$

Do the same for the average distance.

$$a = \frac{1}{4\pi^2\frac{\mathrm{AU}^3}{\mathrm{y}^2}}\,\frac{\left(\sqrt{6}\pi\frac{\mathrm{AU}^2}{\mathrm{y}}\right)^2}{1-\frac{1}{2}^2} = 2\,\mathrm{AU}$$

Now use Plot 2D + Polar to plot the orbit.

Elliptical Orbit

$$r(\phi) = \frac{\frac{3}{2}}{1-\frac{1}{2}\cos\phi}$$

$$r_{\max} = 3\,\mathrm{AU}$$
$$r_{\min} = 1\,\mathrm{AU}$$

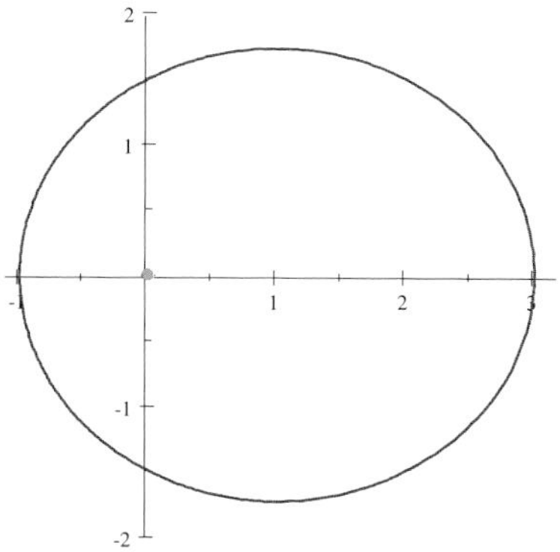

Figure 9.2 Kepler's 1st Law

The star (shown not to scale) is at the origin but it's not at the center of the orbit. The distance between the star's location at a focus and the orbit's center is $\varepsilon a = 1\,\mathrm{AU}$. ◀

Kepler's 2nd Law: The Area Law

Consider the orbital motion of an object under the influence of any central force.

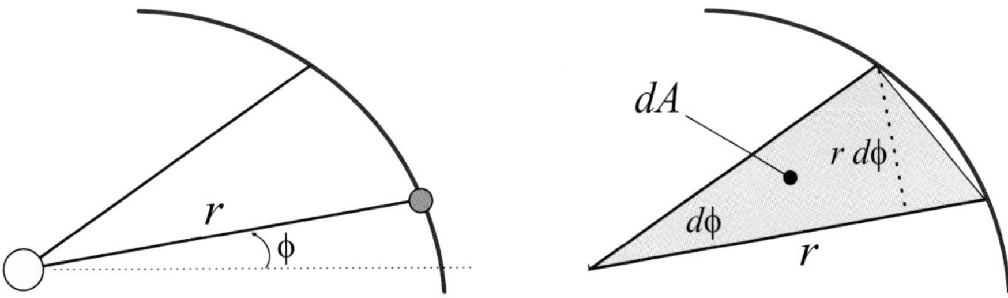

Figure 9.3 Orbital motion and geometry

As the object moves through a small angle $d\phi$, its radius sweeps out an area dA that is approximately triangular. A little geometry, some calculus algebrations, and Eq. (9.4) lead us directly to Kepler's result.

$$dA = \tfrac{1}{2}r^2\,d\phi \tag{9.23a}$$

$$= \frac{1}{2}\,r^2\frac{d\phi}{dt}\,dt \tag{9.23b}$$

$$= \tfrac{1}{2}\ell\,dt \tag{9.23c}$$

If we integrate over a finite time interval, we end up with the area of a "slice" of the orbit and Kepler's 2nd Law.

$$\Delta A = \tfrac{1}{2}\ell\,\Delta t \tag{9.24}$$

Since the angular momentum is constant, the radius vector sweeps equal areas in equal times. For periodic orbits, we can integrate over the entire orbit and relate the enclosed area to the orbital period.

$$A = \tfrac{1}{2}\ell T \tag{9.25}$$

The Area Law is a direct consequence of angular momentum conservation, so it's valid for any central force.

Example 9.7 *Time to get the ℓ out of here*

Verify Kepler's 2nd Law for our previous outward Cotes' spiral by direct calculation, and then compare your answers with Kepler's prediction.

Solution. Kepler's 2nd Law applies to "equal times" so we need to pick equal time intervals and calculate the corresponding angles. Let's calculate the area for the time intervals from $t_0 = 0$ to $t_1 = \frac{1}{3}$ and from $t_2 = \frac{2}{3}$ to $t_3 = 1$ so that each interval is $\Delta t = \frac{1}{3}$.

We'll need the position as a function of both time and angle for $R_0 = 1$, $\ell = \sqrt{40/9}$ and $\beta = 4\pi^2$.

$$r(\phi) = \frac{1}{\cos\frac{\phi}{\sqrt{10}}}$$

$$r(t) = \sqrt{1 + \tfrac{4}{9}\pi^2 t^2}$$

As before, we're working in units of AU and years.

Use Evaluate and Evaluate Numerically to calculate the exact and approximate radius and angle for t_1.

$$r_1 = \sqrt{1+\tfrac{4}{9}\pi^2\left(\tfrac{1}{3}\right)^2} = \sqrt{\tfrac{4}{81}\pi^2+1} = 1.2196$$

$$\phi_1 = \sqrt{10}\cos^{-1}\frac{1}{\sqrt{\frac{4}{81}\pi^2+1}} = 1.9273$$

Set up and Evaluate the first area integral.

$$\Delta A_1 = \tfrac{1}{2}\int_0^{\sqrt{10}\cos^{-1}\frac{1}{\sqrt{1+4\pi^2/81}}} \frac{d\phi}{\cos^2\frac{\phi}{\sqrt{10}}}$$

$$= \tfrac{1}{2}\sqrt{10}\left(\sin\left(2\arccos\frac{9}{\sqrt{4\pi^2+81}}\right)\right)\left(\tfrac{2}{81}\pi^2+\tfrac{1}{2}\right)$$

That's an ugly expression, but if you look closely you'll see the complicated middle-bit is just the sine of twice some number. Edit the expression by-hand, write $\sin 2x$ as $2\sin x\cos x$ and then Simplify.

$$\Delta A_1 = \tfrac{\sqrt{10}}{2}\left(2\sin\left(\cos^{-1}\tfrac{9}{\sqrt{4\pi^2+81}}\right)\cos\left(\cos^{-1}\tfrac{9}{\sqrt{4\pi^2+81}}\right)\right)\left(\tfrac{2\pi^2}{81}+\tfrac{1}{2}\right) = \tfrac{1}{9}\sqrt{10}\pi$$

Of course, it's easier to calculate the approximate area with Evaluate Numerically.

$$\Delta A_1 \approx \tfrac{1}{2}\int_0^{\sqrt{10}\cos^{-1}\frac{1}{\sqrt{1+4\pi^2/81}}} \frac{d\phi}{\cos^2\frac{\phi}{\sqrt{10}}} = 1.1038$$

Repeat the process for the other time interval.

$$\Delta A_3 = \tfrac{1}{2}\int_{\sqrt{10}\cos^{-1}\frac{1}{\sqrt{1+16\pi^2/81}}}^{\sqrt{10}\cos^{-1}\frac{1}{\sqrt{1+4\pi^2/9}}} \frac{d\phi}{\cos^2\frac{\phi}{\sqrt{10}}} = \tfrac{1}{9}\sqrt{10}\pi = 1.1038$$

Apply Evaluate and Evaluate Numerically to Eq. (9.24) and verify this is the area predicted by Kepler.

$$\Delta A = \tfrac{1}{2}\left(\sqrt{\tfrac{40}{9}}\pi\right)\left(\tfrac{1}{3}\right) = \tfrac{1}{9}\sqrt{10}\pi = 1.1038$$

Kepler's 2nd Law works for this decidedly non-elliptical orbit.

Here's a table of approximate numerical results and a plot of three equal-area orbital slices for our outward Cotes' spiral.

Time	Angle	Radius
0	0	1
$\frac{1}{3}$	1.927 3	1.219 6
$\frac{2}{3}$	3.001 9	1.717 4
1	3.558 6	2.320 9
(y)	(rad)	(AU)

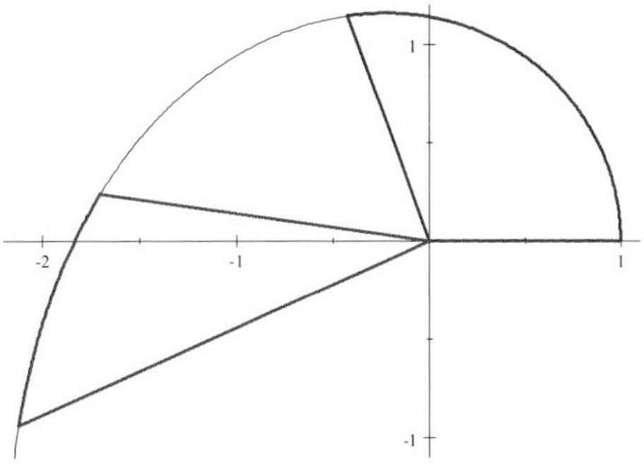

Figure 9.4 Kepler's 2nd Law

Each $\frac{1}{3}$-year slice of this orbit has an approximate area of $1.1038\,\mathrm{AU}^2$ in complete agreement with Kepler's 2nd Law. ◄

Kepler's 3rd Law: The Period Law

Since we know the orbit $r(\phi)$ produced by Newtonian Gravity, we can use Eq. (9.8a) to calculate the time for one complete orbit.

$$T = \frac{\left(a\left(1-\varepsilon^2\right)\right)^2}{\sqrt{aGM\left(1-\varepsilon^2\right)}} \int_0^{2\pi} \left(\frac{1}{1-\varepsilon\cos\phi}\right)^2 d\phi \tag{9.26}$$

I used the results from Kepler's 1st Law to write R_0 and ℓ in terms of a and ε. The current version of *SNB* can't do this integral analytically, but an older version can (with a little help from assume $(\varepsilon > 0)$ and additionally $(\varepsilon < 1)$).

$$\int_0^{2\pi} \frac{d\phi}{(1-\varepsilon\cos\phi)^2} = \frac{2\pi}{\left(1-\varepsilon^2\right)^{3/2}} \tag{9.27}$$

You can explore this later (see Problem 22). In the meantime you can use Evaluate Numerically to verify it for any particular value of the eccentricity.

$$\frac{\displaystyle\int_0^{2\pi} \frac{1}{\left[(1-\varepsilon\cos\phi)^2\right]_{\varepsilon=11/37}} d\phi}{\left[\dfrac{2\pi}{\left(1-\varepsilon^2\right)^{3/2}}\right]_{\varepsilon=11/37}} = 1.0$$

Let's square both sides of Eq. (9.26) and use Simplify.

$$T^2 = \left(\frac{\left(a\left(1-\varepsilon^2\right)\right)^2}{\sqrt{aGM\left(1-\varepsilon^2\right)}}\right)^2 \times \left(\frac{2\pi}{\left(1-\varepsilon^2\right)^{3/2}}\right)^2 = 4\frac{\pi^2}{GM}a^3$$

This is an extraordinary result! The period is related to the average size of the orbit in a very simple way.

Kepler's 3rd Law follows from Newton's 2nd Law and Universal Gravitation:

$$T^2 = \frac{4\pi^2}{GM} a^3 \tag{9.28}$$

Newton not only provided the proportionality constant, he also added a center-of-mass correction which depends on the orbiting mass.

$$T^2 = \frac{4\pi^2}{GM\left(1+\frac{m}{M}\right)} a^3 \tag{9.29}$$

Let's Evaluate the mass ratio for the Earth and Sun.

$$\frac{m_E}{M_S} = \frac{5.9742\times10^{24}}{1.9891\times10^{30}} = 3.0035\times10^{-6}$$

So for the Earth-Sun system, $GM_S = 4\pi^2\ \mathrm{AU}^3 / \mathrm{y}^2$ is accurate to one part in $332,940$.

Example 9.8 *Ruh-roh!*

If the Earth stopped its orbital motion, how long would it take to fall into the Sun?

Solution. We can approximate long-distance free fall motion as the inward trip along a very thin ellipse. The fall-in time is half an orbital period and the average distance is half the initial distance. Since we know the average distance $\left(a = \frac{1}{2}\ \mathrm{AU}\right)$ we can use Kepler's 3rd Law (and Evaluate Numerically) to calculate the time in years.

$$t_{\text{fall}} = \tfrac{1}{2}\times\tfrac{1}{2}^{3/2} = 0.17678$$

Let's use Evaluate and convert this into days.

$$t_{\text{fall}} = 0.17678\,\mathrm{y}\times\frac{365.24\,\mathrm{d}}{\mathrm{y}} = 64.567\,\mathrm{d}$$

According to Kepler, the Earth will crash into the Sun in just over 2 months. ◀

The Effective Potential

In 2-dimensional polar coordinates, the energy per unit mass of a moving object is

$$E = \tfrac{1}{2}\left(\frac{dr}{dt}\right)^2 + \tfrac{1}{2}r^2\left(\frac{d\phi}{dt}\right)^2 + V(r) \tag{9.30}$$

where $V(r)$ is the potential energy per unit mass associated with the central force.

$$V(r) \equiv -\frac{1}{m}\int \vec{F}\cdot d\vec{r} = \int f(r)\,dr \tag{9.31}$$

With the definition of angular momentum from Eq. (9.4), we can rewrite the energy of the orbiting object.

$$E = \tfrac{1}{2}\left(\frac{dr}{dt}\right)^2 + V_{\text{eff}}(r) \tag{9.32}$$

The units of E are usually J/kg or $\mathrm{AU}^2/\mathrm{y}^2$.

The **effective potential energy** contains all the radial-position dependence.

$$V_{\text{eff}}(r) \equiv V(r) + \tfrac{1}{2}\frac{\ell^2}{r^2} \tag{9.33}$$

The angular momentum term accounts for the effects of the angular motion on the radial motion.

The effective potential is a useful tool to analyze the radial part of orbital motion when you don't explicitly know the orbit. Circular orbits minimize the effective potential and are subject to two equivalent conditions.

$$\left.\frac{dV_{\text{eff}}}{dr}\right|_{r=r_{\text{cir}}} = 0 \quad \text{or} \quad f(r_{\text{cir}}) = \frac{\ell^2}{r_{\text{cir}}^3} \tag{9.34}$$

For non-circular orbits, the radial motion oscillates between two radial extremes which are solutions to $E = V_{\text{eff}}(r)$. Conversely, the energy equals the effective potential at those extremes.

$$E = V_{\text{eff}}(r_{\text{max}}) = V_{\text{eff}}(r_{\text{min}}) \tag{9.35}$$

The power law $f(r) = \beta r^n$ is an important family of central forces. For these orbits, we'll use the somewhat contrived $\beta = 4\pi^2\,\text{AU}^{1-n}/\text{y}^2$ so the units work out. Two special cases are Newton's inverse-square gravity $(n = -2)$ and the 3-d isotropic harmonic oscillator $(n = +1)$.

	Inverse-Square	3-d HO	Power Law
Force $f(r)$	$\dfrac{GM}{r^2}$	βr	βr^n
Potential $V(r)$	$-\dfrac{GM}{r}$	$\frac{1}{2}\beta r^2$	$\frac{1}{n+1}\beta r^{n+1}$

Here's a plot of the effective potential and energy for both special cases. Each has an initial position of $R_0 = 1\,\text{AU}$ and an angular momentum that's half the value for a circular orbit.

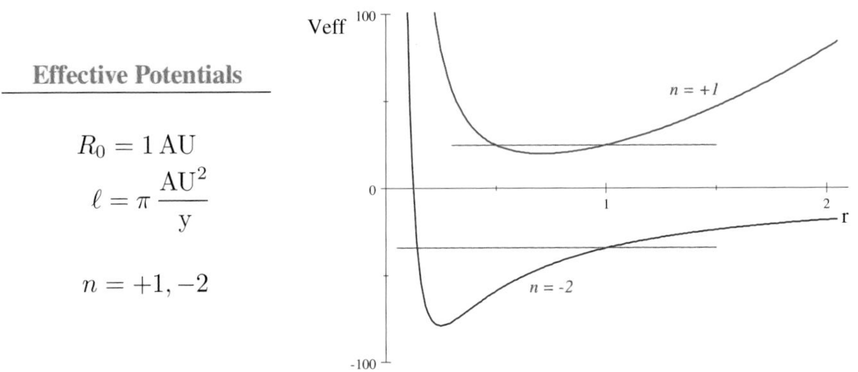

Figure 9.5 Effective potentials

For stable bound orbits, the effective potential must have a local minimum. The inverse-square curve is typical for forces with exponents $-3 < n < -1$ where bound orbits have negative energy. The harmonic oscillator curve is typical for exponents $n > -1$ with positive-energy bound orbits.

Example 9.9 *An affected effective potential*

Consider the orbit of an object under the effective potential

$$V(r) = 4\pi^2 \left(\frac{e^{\mu/r}}{r} + \frac{2}{\mu} \left(1 - e^{\mu/r} \right) \right)$$

where all quantities are in AU and years. The orbit starts at one radial extreme of 3 AU with angular momentum $\ell = 6.2794\,\mathrm{AU}^2\,/\,\mathrm{y}$. Use $\mu = 1\,\mathrm{AU}$.

Solution. Since we know one radial extreme, we can Substitute (with Evaluate Numerically) that value into Eq. (9.35) to calculate the energy.

$$E = \left[4\pi^2 \left(\frac{e^{1/r}}{r} + 2\left(1 - e^{1/r}\right) \right) + \frac{(6.2794)^2}{2r^2} \right]_{r=3} = -10.68$$

A plot of the effective potential and the energy (see Figure 9.6) shows that 3 AU is the radial maximum. Now that we know the energy, let's use Solve Numeric to find the orbit's radial minimum.

$$\left[\begin{array}{c} -10.68 = 4\pi^2 \left(\frac{e^{1/r}}{r} + 2\left(1 - e^{1/r}\right) \right) + \frac{(6.2794)^2}{2r^2} \\ r \in (1, 2) \end{array} \right], \text{ Solution is: } \{[r = 1.061\,3]\}$$

Let's use Solve Numeric and Eq. (9.34) to look between the two radial extremes for the circular orbit.

$$\left[\begin{array}{c} 0 = \frac{d}{dr} \left(4\pi^2 \left(\frac{e^{1/r}}{r} + 2\left(1 - e^{1/r}\right) \right) + \frac{(6.2794)^2}{2r^2} \right) \\ r \in (1.061\,3, 3) \end{array} \right], \text{ Solution is: } \{[r = 1.516\,5]\}$$

Here's a plot of this effective potential.

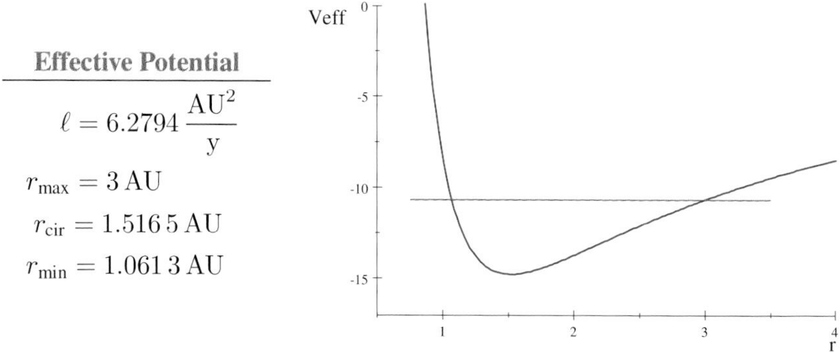

Figure 9.6 Effective potential

This effective potential mimics some relativistic effects for small values of μ. ◀

You can also use the effective potential to calculate the elapsed time or rotation angle between any two radial positions.

$$\Delta\phi = \ell\int_{r_1}^{r_2}\frac{dr}{r^2\sqrt{2\left(E-V_{\text{eff}}\left(r\right)\right)}} \tag{9.36a}$$

$$\Delta t = \int_{r_1}^{r_2}\frac{dr}{\sqrt{2\left(E-V_{\text{eff}}\left(r\right)\right)}} \tag{9.36b}$$

These expressions give you the same information as Eqs. (9.8) but you don't need to know the orbit.

Example 9.10 *Ruh-roh redux!*

If the Earth stopped in its orbit, how long would it *really* take to crash into the Sun?

Solution. When we used Kepler's 3rd Law, we calculated the time it takes the center of the Earth to fall from $r = 1\,\text{AU}$ to $r = 0$. But the crash occurs when the Earth's surface hits the Sun's surface, so the fall is from $1\,\text{AU} - R_E$ to R_S (where R_E and R_S are the Earth's and Sun's radii). Let's start by using Evaluate to convert the two radii into AU.

$$R_S = 6.955\times10^8\,\text{m}\times\frac{1\,\text{AU}}{1.4960\times10^{11}\,\text{m}} = 4.6491\times10^{-3}\,\text{AU}$$

$$R_E = 6.371\times10^6\,\text{m}\times\frac{1\,\text{AU}}{1.4960\times10^{11}\,\text{m}} = 4.2587\times10^{-5}\,\text{AU}$$

Free-fall is a zero-angular-momentum orbit so $V_{\text{eff}}\left(r\right) = -4\pi^2/r$. The Earth's initial energy is all potential so $E = -GM/R_0 = -4\pi^2$. Use Eq. (9.36b) and Evaluate Numerically to calculate the fall time in days.

$$\Delta t = 365.24\int_{0.0046491}^{1-0.000042587}\frac{dr}{\sqrt{2\left(-4\pi^2+\dfrac{4\pi^2}{r}\right)}} = 64.021$$

Not that it would matter much to our fate as a species, but the time to the imminent demise of our planet would be $0.546\,\text{d}$ shorter than Kepler's 3rd Law prediction. ◀

Once you know the maximum and minimum radial distances, you can use the effective potential to calculate two important orbital properties. The **apsidal angle** is defined as the angle between consecutive radial extrema.

$$\Phi_A = \ell\int_{r_{\text{min}}}^{r_{\text{max}}}\frac{dr}{r^2\sqrt{2\left(E-V_{\text{eff}}\left(r\right)\right)}} \tag{9.37a}$$

The angle between consecutive radial maxima is $2\Phi_A$. A periodic orbit with apsidal angle $\Phi_A = \frac{p}{q}\pi$ has q-lobes and repeats itself every p-orbits. Once you know the apsidal angle, you can calculate the **period**.

$$T = \frac{2\pi}{\Phi_A}\int_{r_{\text{min}}}^{r_{\text{max}}}\frac{dr}{\sqrt{2\left(E-V_{\text{eff}}\left(r\right)\right)}} \tag{9.37b}$$

The period is the time for one lobe times the number of lobes per orbit (q/p).

Two Special Forces

There are two important special cases of the power-law force: Newton's inverse-square force $(n = -2)$ and the 3-dimensional harmonic oscillator $(n = +1)$. Both allow closed, periodic elliptical orbits for any initial conditions. Even so, their apsidal angles, periods, and the orbits themselves are different.

The 3-d Harmonic Oscillator

The effective potential for the 3-d HO is

$$V_{\text{eff}}(r) = \tfrac{1}{2}\beta r^2 + \frac{\ell^2}{2r^2} \tag{9.38}$$

where β is a positive constant for bound orbits. Let's use the dummy variable $\rho = r^2$ and Solve Exact to find the radial extremes.

$$E = \tfrac{1}{2}\beta\rho + \frac{\ell^2}{2\rho}, \text{ Solution is: } \frac{1}{\beta}\left(E + \sqrt{E^2 - \beta\ell^2}\right), \frac{1}{\beta}\left(E - \sqrt{E^2 - \beta\ell^2}\right)$$

A little editing by-hand yields the following results (here the energy is always positive).

$$r_{\min}^2 = \frac{E}{\beta} - \sqrt{\frac{E^2 - \beta\ell^2}{\beta^2}} \tag{9.39a}$$

$$r_{\max}^2 = \frac{E}{\beta} + \sqrt{\frac{E^2 - \beta\ell^2}{\beta^2}} \tag{9.39b}$$

Apsidal Angle

Now that we know the maximum and minimum radial distances, let's tell *SNB* the variables are positive and calculate the apsidal angle. Even with the assumption, *SNB* can't do the integral in Eq. (9.37a) directly. Choose Calculus + Change Variables... and enter $\sigma = 1/r^2$ in the Substitution input box. Then Evaluate the resulting integral.

$$\ell \int \frac{dr}{r^2\sqrt{2\left(E - \frac{1}{2}\beta r^2 - \frac{1}{2}\frac{\ell^2}{r^2}\right)}} = \ell \int \left(-\frac{1}{2\sqrt{-\sigma^2\ell^2 + 2E\sigma - \beta}}\right) d\sigma$$

$$= \tfrac{1}{2}\arcsin\frac{E - \sigma\ell^2}{\sqrt{E^2 - \beta\ell^2}}$$

Use Evaluate at Endpoints ($r_{\min}$ and $r_{\max}$) to create and Simplify the expression.

$$\Phi_A = \left[\tfrac{1}{2}\arcsin\frac{E - \sigma\ell^2}{\sqrt{E^2 - \beta\ell^2}}\right]_{\sigma=1/r^2, r=\left(E/\beta - \sqrt{(E^2-\beta\ell^2)/\beta^2}\right)^{1/2}}^{\sigma=1/r^2, r=\left(E/\beta + \sqrt{(E^2-\beta\ell^2)/\beta^2}\right)^{1/2}} = \tfrac{1}{2}\pi$$

The $\Phi_A = 90^\circ$ angle between the minimum and maximum radial positions does not depend on the values of those positions.

Period

SNB also can't do the integral in Eq. (9.37b) directly. Choose Calculus + Change Variables... and enter $\rho = r^2$ in the Substitution input box. Then Evaluate the result.

$$\int \frac{rdr}{\sqrt{2r^2\left(E - \frac{1}{2}\beta r^2 - \frac{1}{2}\frac{\ell^2}{r^2}\right)}} = \int \frac{1}{2\sqrt{-\beta\rho^2 + 2E\rho - \ell^2}}\,d\rho$$

$$= -\frac{1}{2\sqrt{\beta}} \arcsin \frac{E - \beta\rho}{\sqrt{E^2 - \beta\ell^2}}$$

Use Evaluate at Endpoints ($r_{\min}$ and $r_{\max}$) to create and Simplify an expression for the period.

$$T = \frac{2\pi}{\frac{1}{2}\pi}\left[-\frac{1}{2\sqrt{\beta}} \arcsin \frac{E - \beta\rho}{\sqrt{E^2 - \beta\ell^2}}\right]_{\rho=r^2, r=\left(E/\beta - \sqrt{(E^2-\beta\ell^2)/\beta^2}\right)^{1/2}}^{\rho=r^2, r=\left(E/\beta + \sqrt{(E^2-\beta\ell^2)/\beta^2}\right)^{1/2}} = 2\frac{\pi}{\sqrt{\beta}}$$

The period is constant and independent of the initial conditions! As in the 1-d case, the period of 2-d harmonic-oscillator motion does not depend on the initial position. This isn't surprising since this 2-d motion is a combination of two 1-d harmonic oscillators with the same spring constant.

Orbit as a function of Angle

Exact solutions exist for the HO orbit, but *SNB* can't handle the differential equations so we'll use the effective potential instead. The procedure is simple – do the integral in Eq. (9.36a) from R_0 to r, and solve the resulting expression for the orbit $r(\phi)$.

We've already done the integral (in terms of $\sigma = 1/r^2$) for the apsidal angle calculation, so create an expression to Evaluate at Endpoints and use Solve Exact to find σ.

$$\phi = \left[\tfrac{1}{2} \arcsin \frac{E - \sigma\ell^2}{\sqrt{E^2 - \beta\ell^2}}\right]_{\sigma=\sigma_0}^{\sigma=\sigma} \text{, Solution is:}$$

$$\frac{1}{\ell^2}\left(E - \left(\sin\left(2\phi + \arcsin \frac{E - \sigma_0\ell^2}{\sqrt{E^2 - \beta\ell^2}}\right)\right)\sqrt{E^2 - \beta\ell^2}\right)$$

Edit this result by-hand, restoring the explicit r-dependence, to get an expression for the orbit as a function of angle.

$$r^2 = \frac{\frac{\ell^2}{E}}{1 - \sqrt{1 - \frac{\beta\ell^2}{E^2}}\,\sin\left(2\phi + \arcsin \frac{1 - \frac{\ell^2}{ER_0^2}}{\sqrt{1 - \frac{\beta\ell^2}{E^2}}}\right)}$$

This is a complicated expression, but we can simplify it with a good choice of coordinate system.

When the initial radial position is the maximum $(R_0 = r_{\max})$ we can Simplify the angle-dependence.

$$\left[\sin\left(2\phi + \arcsin\frac{1-\frac{\ell^2}{ER_0^2}}{\sqrt{1-\frac{\beta\ell^2}{E^2}}}\right)\right]_{R_0=\left(E/\beta+\sqrt{(E^2-\beta\ell^2)/\beta^2}\right)^{1/2}} = \cos 2\phi$$

A little editing by-hand yields the following result.

$$r^2(\phi) = \frac{\frac{\ell^2}{E}}{1-\sqrt{1-\frac{\beta\ell^2}{E^2}}\cos 2\phi} \tag{9.40}$$

This is the equation of an elliptical orbit whose center is at the origin.

Orbit as a function of Time

The procedure here is similar – do the integral in Eq. (9.36b) from R_0 to r, and solve the resulting expression for the orbit $r(t)$ – but the math can get a little ugly.

We've already done the integral (in terms of $\rho = r^2$) for the period calculation, so create an expression to Evaluate at Endpoints and use Solve Exact to find ρ.

$$t = \left[-\frac{1}{2\sqrt{\beta}}\arcsin\frac{E-\beta\rho}{\sqrt{E^2-\beta\ell^2}}\right]_{\rho=\rho_0}^{\rho=\rho}$$, Solution is:

$$\frac{1}{\beta}\left(E + \left(\sin 2\sqrt{\beta}\left(t - \frac{1}{2\sqrt{\beta}}\arcsin\frac{E-\beta\rho_0}{\sqrt{E^2-\beta\ell^2}}\right)\right)\sqrt{E^2-\beta\ell^2}\right)$$

Edit this result by-hand, restoring the explicit r-dependence, to get an expression for the orbit as a function of time.

$$r^2 = \frac{E}{\beta}\left(1 + \sqrt{1-\frac{\beta\ell^2}{E^2}}\sin\left(2\sqrt{\beta}t - \arcsin\frac{1-\frac{\beta R_0^2}{E}}{\sqrt{1-\frac{\beta\ell^2}{E^2}}}\right)\right)$$

When the initial radial position is the maximum $(R_0 = r_{\max})$ we can Simplify the angle-dependence in this expression.

$$\sin\left(2\sqrt{\beta}t - \arcsin\frac{1-\frac{\beta}{E}\left(\frac{E+\sqrt{E^2-\beta\ell^2}}{\beta}\right)}{\sqrt{1-\frac{\beta\ell^2}{E^2}}}\right) = \cos 2t\sqrt{\beta}$$

A little editing by-hand yields the following result.

$$r^2(t) = \frac{E}{\beta}\left(1 + \sqrt{1-\frac{\beta\ell^2}{E^2}}\cos 2\sqrt{\beta}t\right) \tag{9.41}$$

We now have exact expressions for the HO orbit as functions of angle and time.

The Inverse-Square Force

The effective potential for the inverse-square force is

$$V_{\text{eff}}(r) = -\frac{\beta}{r} + \frac{\ell^2}{2r^2} \tag{9.42}$$

where β is a positive constant for bound orbits. Let's use Solve Exact to find the radial extremes.

$$E = -\frac{\beta}{r} + \frac{\ell^2}{2r^2}, \text{ Solution is: } -\frac{1}{2E}\left(\beta + \sqrt{\beta^2 + 2E\ell^2}\right), -\frac{1}{2E}\left(\beta - \sqrt{\beta^2 + 2E\ell^2}\right)$$

A little editing by-hand yields the following results.

$$r_{\min} = -\frac{\beta}{2E}\left(1 - \sqrt{1 + \frac{2E\ell^2}{\beta^2}}\right) \tag{9.43a}$$

$$r_{\max} = -\frac{\beta}{2E}\left(1 + \sqrt{1 + \frac{2E\ell^2}{\beta^2}}\right) \tag{9.43b}$$

Remember here the energy is negative for bound orbits.

Apsidal Angle

Now that we know the maximum and minimum radial distances, let's tell *SNB* about the variables and calculate the apsidal angle.

$\text{assume}(\ell > 0) = (0, \infty)$ $\qquad$ $\text{assume}(\beta > 0) = (0, \infty)$

$\text{assume}(r > 0) = (0, \infty)$ $\qquad$ $\text{assume}(E < 0) = (-\infty, 0)$

With these assumptions *SNB* can do the integral in Eq. (9.37a) directly but the result is simpler if we change variables. Choose Calculus + Change Variables... and enter $u = 1/r$ in the Substitution input box. Then Evaluate the resulting integral.

$$\ell \int \frac{dr}{r^2\sqrt{2\left(E + \frac{\beta}{r} - \frac{\ell^2}{2r^2}\right)}} = \ell \int \left(-\frac{1}{\sqrt{-u^2\ell^2 + 2\beta u + 2E}}\right) du$$

$$= \arcsin \frac{\beta - u\ell^2}{\sqrt{\beta^2 + 2E\ell^2}}$$

Use Evaluate at Endpoints ($r_{\min}$ and $r_{\max}$) to create and Simplify the expression.

$$\Phi_A = \left[\arcsin \frac{\beta - u\ell^2}{\sqrt{\beta^2 + 2E\ell^2}}\right]_{u=1/r, r=-\beta/(2E)\left(1-\sqrt{1+2E\ell^2/\beta^2}\right)}^{u=1/r, r=-\beta/(2E)\left(1+\sqrt{1+2E\ell^2/\beta^2}\right)} = \pi$$

The angle between the minimum and maximum radial positions is $\Phi_A = 180°$ for all initial conditions.

Period

With the previous assumptions *SNB* can do the integral in Eq. (9.37b) directly.

$$\int \frac{rdr}{\sqrt{2r^2\left(E+\frac{\beta}{r}-\frac{\ell^2}{2r^2}\right)}} = \frac{1}{2E}\sqrt{2Er^2+2\beta r-\ell^2} - \frac{1}{4}\sqrt{2}\frac{\beta}{(-E)^{\frac{3}{2}}}\arcsin\frac{1}{2}\frac{2\beta+4rE}{\sqrt{\beta^2+2E\ell^2}}$$

The first term vanishes at both extrema so it doesn't contribute to the period.

$$\left[\frac{1}{2E}\sqrt{2Er^2+2\beta r-\ell^2}\right]_{r=-\frac{\beta}{2E}\left(1+\sqrt{1+\frac{2E\ell^2}{\beta^2}}\right)} = 0$$

$$\left[\frac{1}{2E}\sqrt{2Er^2+2\beta r-\ell^2}\right]_{r=-\frac{\beta}{2E}\left(1-\sqrt{1+\frac{2E\ell^2}{\beta^2}}\right)} = 0$$

Let's Simplify the second term, then Expand the result in-place.

$$T = \frac{2\pi}{\pi}\left[-\frac{\sqrt{2}}{4}\frac{\beta}{(-E)^{3/2}}\arcsin\frac{\beta+2rE}{\sqrt{\beta^2+2E\ell^2}}\right]_{r=-\frac{\beta}{2E}\left(1-\sqrt{1+\frac{2E\ell^2}{\beta^2}}\right)}^{r=-\frac{\beta}{2E}\left(1+\sqrt{1+\frac{2E\ell^2}{\beta^2}}\right)} = \frac{1}{2}\sqrt{2}\pi\frac{\beta}{(-E)^{\frac{3}{2}}}$$

The period depends on the initial conditions!

Let's use Solve Exact to relate the energy E to the average distance by averaging the minimum and maximum radial positions.

$$a = \frac{1}{2}\left(\frac{-\beta+\sqrt{\beta^2+2E\ell^2}}{2E}+\frac{-\beta-\sqrt{\beta^2+2E\ell^2}}{2E}\right), \text{ Solution is: } -\frac{1}{2a}\beta$$

The energy has a simple inverse-dependence on the average distance. Create and Simplify the following expression for the period of an inverse-square orbit.

$$T = \left[\frac{1}{2}\sqrt{2}\pi\frac{\beta}{(-E)^{\frac{3}{2}}}\right]_{E=-\beta/(2a)} = 2\pi\frac{a^{\frac{3}{2}}}{\sqrt{\beta}}$$

This is an old friend.

$$T^2 = \left[\left(\frac{1}{2}\sqrt{2}\pi\frac{\beta}{(-E)^{\frac{3}{2}}}\right)^2\right]_{E=-\beta/(2a)} = 4\pi^2\frac{a^3}{\beta}$$

It's Kepler's 3rd Law!

Orbit as a function of Angle

We've already seen a solution for the inverse-square orbit using Newton's 2nd Law, but let's explore it again with the effective potential. The procedure is simple: do the integral in Eq. (9.36a) from R_0 to r, and solve the resulting expression for the orbit $r(\phi)$.

We've already done the integral (in terms of $u = 1/r$) for the apsidal angle calculation, so create an expression to Evaluate at Endpoints and use Solve Exact to find r.

$$\phi = \left[\arcsin \frac{\beta - u\ell^2}{\sqrt{\beta^2 + 2E\ell^2}}\right]_{u=1/R_0}^{u=1/r}, \text{ Solution is: } \frac{\ell^2}{\beta - \left(\sin\left(\phi + \arcsin \frac{\beta - \frac{\ell^2}{R_0}}{\sqrt{\beta^2+2E\ell^2}}\right)\right)\sqrt{\beta^2+2E\ell^2}}$$

Edit this result by-hand to get an expression for the orbit as a function of angle.

$$r = \frac{\frac{\ell^2}{\beta}}{1 - \sqrt{1 + \frac{2E\ell^2}{\beta^2}} \sin\left(\phi + \arcsin \frac{1 - \frac{\ell^2}{\beta R_0}}{\sqrt{1 + \frac{2E\ell^2}{\beta^2}}}\right)}$$

When the initial radial position is the maximum $(R_0 = r_{\text{max}})$ we can Simplify the angle-dependence in this expression.

$$\left[\sin\left(\phi + \arcsin \frac{1 - \dfrac{\ell^2}{\beta R_0}}{\sqrt{1 + \dfrac{2E\ell^2}{\beta^2}}}\right)\right]_{R_0 = -\beta/(2E)\left(1+\sqrt{1+2E\ell^2/\beta^2}\right)} = \cos\phi$$

A little more editing by-hand yields the following result.

$$r(\phi) = \frac{\dfrac{\ell^2}{\beta}}{1 - \sqrt{1 + \dfrac{2E\ell^2}{\beta^2}} \cos\phi} \tag{9.44}$$

This is the same off-center ellipse from Eq. (9.20)!

Orbit as a function of Time

By now you see the method to my madness, but here we run into trouble. When we do the integral in Eq. (9.36b) for the inverse-square force, the resulting equation has no simple algebraic solution, even if we choose the simplest case where $R_0 = r_{\text{max}}$.

$$t = -\frac{\sqrt{2Er^2 + 2\beta r - \ell^2}}{2E} + \frac{1}{2\sqrt{2}}\beta(-E)^{-3/2}\left(\frac{\pi}{2} + \arcsin \frac{\beta + 2Er}{\sqrt{\beta^2 + 2E\ell^2}}\right) \tag{9.45}$$

There is no way to solve this equation for r, so inverse-square orbits have no closed-form solution $r(t)$.

Example 9.11 *A dual duel*

Calculate the orbit as a function of angle and time for inverse-square (IS) and harmonic oscillator (HO) elliptical orbits with the same maximum and minimum radial positions.

Solution. Let's start with the orbit from Example 9.6, where $r_{\max} = 3$ and $r_{\min} = 1$. Use Solve Exact on Eq. (9.35) for each radius to find the corresponding values for the energy and angular momentum of the harmonic oscillator orbit.

$$\begin{bmatrix} E = \frac{1}{2}\left(4\pi^2\right)3^2 + \frac{1}{2}\frac{\ell^2}{3^2} \\ E = \frac{1}{2}\left(4\pi^2\right)1^2 + \frac{1}{2}\frac{\ell^2}{1^2} \end{bmatrix}, \text{ Solution is: } \left[E = 20\pi^2, \ell = 6\pi\right]$$

Substitute (with Simplify) these values (along with $\beta = 4\pi^2$) into the expression for the HO orbit as a function of angle.

$$r_{\mathrm{HO}} = \left[\sqrt{\frac{\frac{\ell^2}{E}}{1 - \sqrt{1 - \frac{\beta\ell^2}{E^2}}\cos 2\phi}}\right]_{\beta=4\pi^2, E=20\pi^2, \ell=6\pi} = 3\sqrt{-\frac{1}{4\cos 2\phi - 5}}$$

Repeat the substitution for the HO orbit as a function of time.

$$r_{\mathrm{HO}} = \left[\sqrt{\frac{E}{\beta}\left(1 + \sqrt{1 - \frac{\beta\ell^2}{E^2}}\cos 2\sqrt{\beta}t\right)}\right]_{\beta=4\pi^2, E=20\pi^2, \ell=6\pi} = \sqrt{4\cos 4\pi t + 5}$$

We'll need the period of the HO orbit.

$$T = \left[\frac{2\pi}{\sqrt{\beta}}\right]_{\beta=4\pi^2} = 1$$

We already know the inverse-square orbit from Example 9.6.

$$r_{\mathrm{IS}} = \frac{\frac{3}{2}}{1 - \frac{1}{2}\cos\phi}$$

Since the inverse-square orbit has no closed-form solution $r(t)$ let's use Solve ODE + Numeric on Eq. (9.5a) $\left(\text{with } \ell = \sqrt{6}\pi\right)$ for a numerical approximation.

$$\begin{bmatrix} r'' = -\frac{4\pi^2}{r^2} + \frac{6\pi^2}{r^3} \\ r(0) = 3 \\ r'(0) = 0 \end{bmatrix}, \text{ Functions defined: } r$$

The period of the inverse-square orbit is $T = 2^{3/2} = 2.828\,4$ per Kepler's 3rd Law.

Figure 9.7 shows the two orbits and the two radial positions as functions of time.

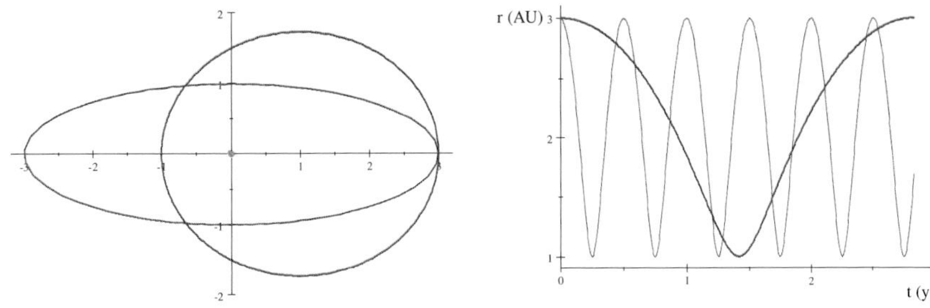

Figure 9.7a Angle **Figure 9.7b** Time

Both orbits are period-1 ellipses, but with different time and angle dependence. The HO orbit (with its smaller apsidal angle) cycles between the radial extrema much quicker than the inverse-square orbit.

Let's find the position of the inverse-square orbiter when the HO has completed one orbit. We'll need the energy of the IS orbiter, so Evaluate the following expression.

$$E = -\frac{4\pi^2}{1} + \frac{1}{2}\frac{6\pi^2}{1^2} = -\pi^2$$

Next Substitute the appropriate values into the right side of Eq. (9.45) and use Solve Numeric (with $t = 1$) to find the value of r.

$$1 = \left[-\frac{\sqrt{2Er^2+2\beta r-\ell^2}}{2E} + \frac{\beta(-E)^{-3/2}}{2\sqrt{2}} \left(\frac{\pi}{2} + \arcsin \frac{\beta+2Er}{\sqrt{\beta^2+2E\ell^2}} \right) \right]_{E=-\pi^2, \ell=\sqrt{6}\pi, \beta=4\pi^2},$$

Solution is: $\{[r = 1.843\,9]\}$

Take a look at Figure 9.7b and decide if you agree. ◄

Numerical Stuff

Many central force problems do not have an exact solution. Even the inverse-square force needs numerical approximations. We can use *SNB*'s numeric capabilities to analyze the orbits produced by these forces.

Example 9.12 *Periodicity*
Calculate the orbit for the force $f(r) = r^{-2.735}$ with angular momentum $\ell = 0.914\,06$. What are the maximum and minimum radial positions? Calculate the apsidal angle, average radial position, and period of this orbit.
Solution. This force allows no exact solution, so let's use Solve ODE + Numeric and Eq. (9.6) to calculate the orbit as a function of angle.

$$\begin{bmatrix} r'' - \frac{2}{r}(r')^2 - r = -\frac{r^4}{(0.914\,06)^2}\left(r^{-2.735}\right) \\ r(0) = 1 \\ r'(0) = 0 \end{bmatrix}, \text{ Functions defined: } r$$

Let's use Plot 2D + Polar to take a look at this unusual orbit.

It's a periodic orbit with one lobe that repeats every two revolutions.

Periodic orbits are possible for many force laws, but only under specific initial conditions.

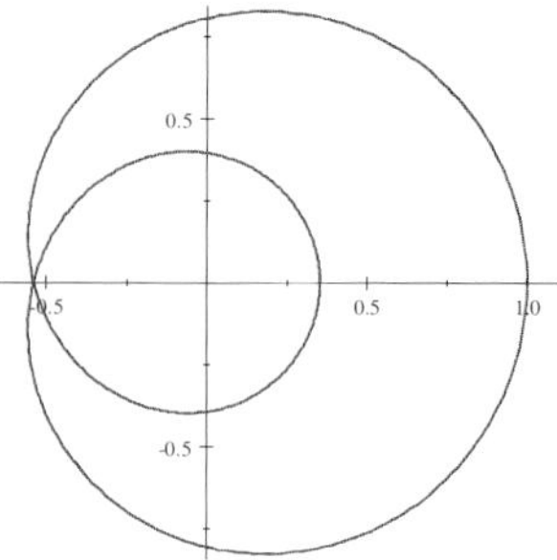

Figure 9.8 A $\Phi_A = 2\pi$ orbit

Let's use Evaluate Numerically to find the maximum and minimum radii.

$$r(2\pi) = 0.353\,63$$
$$r(4\pi) = 1.000\,0$$

Since we know the orbit, we can use Eq. (9.8b) and Evaluate Numerically to calculate the period.

$$T = \frac{1}{0.914\,06}\int_0^{4\pi} r^2(\phi)\,d\phi = 5.719\,1$$

Use Evaluate Numerically once more to get the average radial position.

$$a = \frac{1}{4\pi}\int_0^{4\pi} r(\phi)\,d\phi = 0.605\,13$$

To calculate the apsidal angle we need $r(t)$ so let's use Solve ODE + Numeric on Eq. (9.5a).

$$\begin{bmatrix} r'' = -r^{-2.735} + \dfrac{(0.914\,06)^2}{r^3} \\ r(0) = 1 \\ r'(0) = 0 \end{bmatrix}, \text{Functions defined: } r$$

Let's verify the two radial extrema with Evaluate Numerically and the period.

$$r(5.719\,1) = 1.000\,0$$
$$r\left(\tfrac{5.719\,1}{2}\right) = 0.353\,63$$

Use Eq. (9.8a) and Evaluate Numerically in-place to calculate the apsidal angle.

$$\Phi_A = \left(\frac{0.914\,06}{\pi}\int_0^{5.719\,1/2}\frac{dt}{r^2(t)}\right)\pi = 2.0\,\pi$$

The apsidal angle is $\Phi_A = 2\pi$ as expected. ◄

Problems

1. Show by direct calculation that $GM_S = 4\pi^2\,\mathrm{AU}^3\,/\,\mathrm{y}^2$.

2. Define the Moon Unit (MU) to be the average Earth-Moon distance (about $384400\,\mathrm{km}$) and the month (mon) to be the Moon's orbital period (about $27.322\,\mathrm{d}$). Show that $GM_E \approx 4\pi^2\,\mathrm{MU}^3\,/\,\mathrm{mon}^2$. Why isn't this as accurate as the Sun's value?

3. Show that for an object in a circular orbit, Eq. (9.5a) is consistent with the radial equation of uniform circular motion per Newton's 2nd Law.

4. For an object subjected to the central force $f(r) = \beta r^n$, show that the period of a circular orbit is proportional to $R^{(1-n)/2}$ where R is the circle's radius.

5. Let's explore the instability of the circular Cotes' orbit by giving it a small radial tweak. Set up the equation of motion with $\beta = \ell^2$ as in Example 9.1, but set $u'(0) = \delta$. Use Solve ODE + Exact and Plot 2D + Polar to plot the solution for $\delta = 0$, $+0.01$, and -0.01. Let each solution complete 5 orbits. What happens?

6. Consider an object moving along a Cotes' spiral. Show that the time it takes to spiral in from $\phi_0 = 0$ is
$$T_{\text{in}} = \frac{R_0^2}{k\ell}\tanh k\phi_f$$
where ϕ_f is the final angle. How long does it take to spiral all the way in?

7. Consider an object moving along a Cotes' spiral. Show that the time it takes to spiral out from $\phi_0 = 0$ is
$$T_{\text{out}} = \frac{R_0^2}{k\ell}\tan k\phi_f$$
where ϕ_f is the final angle. How long does it take to spiral all the way out?

8. Use the results of the two previous problems to verify the results of Example 9.3.

   The next 5 problems involve a circular orbit of radius a that contains but is not centered about the force center (the origin). [30]

9. Let's follow Example 9.2 and examine a circular orbit centered along the x-axis at $(x_c, 0)$.

   a. Show that $r(\phi) = x_c\cos\phi + \left(a^2 - x_c^2\left(1-\cos^2\phi\right)\right)^{1/2}$.

   b. Define the reciprocal variable $u = \dfrac{-x_c\cos\phi + \left(a^2 - x_c^2\left(1-\cos^2\phi\right)\right)^{1/2}}{a^2 - x_c^2}$.

   c. Simplify $u^2\ell^2\left(\dfrac{d^2u}{d\phi^2} + u\right)$ and notice the $\cos 2\phi$ terms.

   d. Show that $f(r) = \dfrac{8a^2\ell^2 r}{\left(r^2 + a^2 - x_c^2\right)^3}$.

   Note: you may find the identity $\cos 2\phi = 2\cos^2\phi - 1$ useful.

10. Let's examine a circular orbit centered somewhere within the orbit at (x_c, y_c). The expression for this orbit is

$$r(\phi) = x_c \cos\phi + y_c \sin\phi + \left(a^2 - x_c^2 - y_c^2 + (x_c \cos\phi + y_c \sin\phi)^2\right)^{1/2}$$

where a is the circle's radius. This problem requires more direct input from you.

a. Show that the reciprocal variable can be written as

$$u = \frac{-\Phi + \left(\rho^2 + \Phi^2\right)^{1/2}}{\rho^2}$$

where $\Phi \equiv x\cos\phi + y\sin\phi$ and $\rho \equiv \left(a^2 - x^2 - y^2\right)^{1/2}$.

b. Show that $\dfrac{du}{d\phi} = -\dfrac{2u^2}{1+u^2\rho^2}\dfrac{d\Phi}{d\phi}$.

c. Show that $\dfrac{d^2u}{d\phi^2} = \dfrac{u\left(1-u^2\rho^2\right)}{1+u^2\rho^2} + \dfrac{8u^3}{\left(1+u^2\rho^2\right)^3}\left(\dfrac{d\Phi}{d\phi}\right)^2$.

d. Write the sum $(d\Phi/d\phi)^2 + \Phi^2$ in terms of u, ρ, and a and show that

$$\left(\frac{d\Phi}{d\phi}\right)^2 = a^2 - \rho^2 - \left(\frac{1-u^2\rho^2}{2u}\right)^2$$

e. Show that $f(r) = \dfrac{8a^2\ell^2 r}{\left(r^2 + a^2 - x_c^2 - y_c^2\right)^3}$.

11. Verify that the initial conditions for a circular orbit centered at (x_c, y_c) are

$$\begin{aligned} R_0 &\equiv r(\phi=0) = x_c + \sqrt{a^2 - y_c^2} \\ D_0 &\equiv \left.\frac{dr}{d\phi}\right|_{\phi=0} = y_c \frac{x_c + \sqrt{a^2 - y_c^2}}{\sqrt{a^2 - y_c^2}} \end{aligned}$$

where a is the circle's radius.

12. Use Solve ODE + Numeric to calculate a circular orbit of radius 2 centered at $\left(\frac{1}{2}, 0\right)$. Plot your result, and add the appropriate exact expression to your plot.

13. Use Solve ODE + Numeric to calculate a circular orbit of radius 2 centered at $\left(\frac{1}{2}, \frac{1}{2}\right)$. Plot your result, and add the appropriate exact expression to your plot.

14. Find the radial position as a function of time for an outward Cotes' spiral.

15. Just in case you don't know what a cardioid looks like, make a polar plot of $r(\phi) = 2(1+\cos\phi)$. Be sure to check the Equal Scaling Along Each Axis box.

16. Make polar plots of $r(\phi) = A(1+\cos n\phi)^{-1/n}$ for $n = -\frac{1}{2}, -2, -3$ and any other values you like.

17. Calculate the surface gravity on the Moon. Give your answer in $\mathrm{m/s^2}$ and units of g (the surface gravity on Earth).

18. Consider the even-more-generalized cardioid orbit $r(\phi) = A(1+\epsilon\cos n\phi)^{-1/n}$.

    a. Show that the force law that produces this orbit is
    $$f(r) = \frac{\ell^2}{A^{2n}} r^{2n-3}\left(\left(1-\epsilon^2\right)(n-1) + (2-n)\frac{A^n}{r^n}\right)$$

    b. Verify this reduces to the expected result for generalized and just-plain cardioids.

    c. Analyze the force and orbit when $n = 1$ and $A = \ell^2/\beta$. Look familiar?

19. Let's use Newton's Law of Universal Gravitation to explore the gravitational effects of a rare but possible planetary alignment: when every planet aligns along a straight line.

    a. Calculate the Sun's gravitational pull on the Earth.

    b. Calculate the gravitational force exerted on the Earth by all the planets further from the Sun.

    c. Take the ratio of the two and discuss any effects on the Earth's orbit.

    d. Explain this to any gloom-and-doomers you may encounter.

20. Repeat the previous problem including the effects of the planets closer to the Sun. How does this change the net force exerted on the Earth by the other planets?

21. Show that an elliptical orbit can be written as
    $$r(\phi) = \frac{a\left(1-\varepsilon^2\right)}{1-\varepsilon\cos\phi}$$
    where a is the average distance and ε is the eccentricity.

22. Let's take a look at Eq. (9.27) and its dependence on the eccentricity.

    a. Do a 5-term power series expansion on the integrand in powers of the eccentricity.

    b. Delete the $+O\left(\varepsilon^5\right)$ and integrate the expansion.

    c. Do the same power series expansion on the right side of the equation and compare.

23. There is a geometric way to think about Kepler's 3rd Law that uses the properties of ellipses and Kepler's 2nd Law. Calculate the area enclosed by an elliptical orbit in terms of the eccentricity and average radius. Then use Kepler's 2nd Law to write the areal velocity in terms of the eccentricity and average radius. Given that time equals amount divided by rate, use these results to calculate the orbital period.

24. Use Kepler's Laws to show that the area of an ellipse is $A = \pi a^2\sqrt{1-\varepsilon^2}$.

25. Calculate the radius of the circular orbits and the minimum radial distances for the two effective potentials in Figure 9.5.

26. Suppose evil space monkeys stop the Moon's orbital motion. How much time do we have according to Kepler? How much time do we really have?

27. You may be wondering why the approximate time for the Earth to fall into the Sun is one-half the orbital period of an orbit whose average distance is half an AU.

    a. Plot several elliptical orbits with $R_0 = 1$ with eccentricity $\varepsilon = 0, \frac{1}{2}, \frac{3}{4}$, and $\frac{9}{10}$.

    b. Calculate the ratio of the average distance to the maximum distance as the eccentricity approaches one.

    c. Calculate the minimum distance as the eccentricity approaches one.

28. Consider the potential energy from Example 9.9.

    a. Show that the central force associated with this potential energy is

$$f(r) = \frac{4\pi^2}{r^2}\left(1 - \frac{\mu}{r}\right)e^{\mu/r}$$

    b. Verify that the force and potential energy reduce to familiar results when $\mu \to 0$.

    c. Use the force to verify the radius of the circular orbit.

29. Calculate the orbit for the conditions of the Example 9.9. What is its apsidal angle? What is its period? It's a $\Phi_A = \frac{2}{3}\pi$ periodic orbit.

30. Repeat Example 9.9 with $\mu = -1$.

31. Check if Kepler's 3rd Law works for the orbit in Example 9.12.

32. Let's explore what happens to an orbit under a central force $f(r) = \beta g(r)$ when the angular momentum and force constant change but the ratio ℓ^2/β remains constant.

    a. Show the energy changes such that $E' = \frac{\beta'}{\beta}E$.

    b. Show the orbital period changes such that $T' = \sqrt{\frac{\beta}{\beta'}}\,T$.

    c. Show the orbit's size and shape are unchanged.

33. Use the results from Problem 35 and Example 9.12 to verify the claims in the previous problem.

Calculate and plot these orbits for the central force $f(r) = 4\pi^2 r^n$, each starting at $R_0 = 1$ with $r'(0) = 0$. Estimate the period and verify the apsidal angle for each orbit.

	Angular Momentum	Exponent	Apsidal Angle
34.	4.90781	+6.1	$\frac{1}{3}\pi$
35.	5.74320	−2.735	2π
36.	5.22804	−0.76	$\frac{2}{3}\pi$
37.	5.82446	−2.87	3π
38.	5.16664	−2.53	$\frac{3}{2}\pi$
39.	5.18180	−2.42	$\frac{4}{3}\pi$
40.	5.95155	−2.83	$\frac{5}{2}\pi$
41.	3.77377	−1.60	$\frac{5}{6}\pi$

10 Fluids

A **fluid** is something that can flow and take the shape of its container. In the non-physics world, fluid and liquid are synonymous but here in physics-land both gases and liquids are fluids. Understanding fluids is useful and important. We live on a planet with an atmosphere, so we live at the bottom of a giant pool of a fluid we call air. Our bodies are mostly water so we are made of mostly fluids.

Density and Pressure

Density is an important property of a fluid. Density is mass per unit volume, so it is a measure of how much stuff (mass) occupies how much space (volume). A pound of feathers and a pound of lead have the same weight and mass, but the feathers are less dense so they take up more space than the bricks. A liter of water and a liter of mercury have the same volume, but the mercury has more mass because it's denser.

A substance of mass m that occupies volume V has **density** ρ.

$$\rho \equiv \frac{m}{V} \tag{10.1}$$

The metric unit of density is $\mathrm{kg/\,m^3}$ although $\mathrm{g/\,cm^3}$ is also used.

Density is a physical property of a substance (see Table 10.1 at the end of the chapter). Gases usually have the smallest density, solids have the largest density, and most liquids are in between. The density of water is $\rho_w = 1000\,\mathrm{kg/\,m^3}$. Let's see how much space $1\,\mathrm{kg}$ of water takes up.

$$V = \frac{m}{\rho_w} = \frac{1\,\mathrm{kg}}{1000\,\mathrm{kg/\,m^3}} = \frac{1}{1000}\,\mathrm{m^3}$$

That's one liter! A small bottle holds a kilogram of water.

Example 10.1 *Ugly bags of mostly water*
The average male human has a mass of $76.0\,\mathrm{kg}$ and a density of about $1046\,\mathrm{kg/\,m^3}$. What is his volume? Give your answer in liters.
Solution. Use Eq. (10.1) to create and Evaluate the following expression that gives the volume in liters.

$$V = \frac{m}{\rho} = \frac{76.0\,\mathrm{kg}}{1046\,\mathrm{kg/\,m^3}} \times \frac{1000\,\mathrm{l}}{1\,\mathrm{m^3}} = 72.658\,\mathrm{l}$$

The average male has a volume equivalent to almost 73 small bottles of water. ◄

Doing Physics with Scientific Notebook: A Problem-solving Approach, First Edition. Joseph Gallant.

Pressure is an important process that determines the behavior of a fluid. Pressure is force per unit area so it is a measure of how hard you push (force) and how spread out the push is (area). You can change the pressure by changing the applied force or the area of its application.

Snowshoes don't make you lighter, but they decrease the pressure you exert on the snow by increasing the area. If you stand on one foot, your weight is the same but you've decreased the area by half and doubled the pressure.

A force F applied to an area A exerts a **pressure** P.

$$P \equiv \frac{F}{A} \tag{10.2}$$

The metric unit of pressure is $\mathrm{N/m^2}$ which is called the **pascal** (Pa). A newton is a small unit of force so spreading it out over a square meter means a $\mathrm{N/m^2}$ is a very small unit of pressure. We sometimes use the kilopascal (kPa) so that $1\,\mathrm{kPa} = 1000\,\mathrm{N/m^2}$. The American unit of pressure is the pound per square inch (psi).

Another common and useful unit of pressure is the atmosphere, which is defined as the standard atmospheric pressure at sea level.

$$P_{atm} \equiv 1\,\mathrm{atm} = 1.0133 \times 10^5 \frac{\mathrm{N}}{\mathrm{m^2}} = 101.33\,\mathrm{kPa} = 14.696 \frac{\mathrm{lbf}}{\mathrm{in^2}} \tag{10.3}$$

You can find atm, Pa, and kPa listed under Pressure on *SNB*'s Physical Quantity list of built-in units in the Unit Name Box.

Example 10.2 *Well excuse me!*
You find yourself in a crowded room you want to cross, so you apply a small push (say a $2\,\mathrm{N}$ force) on someone to get around them. Calculate the pressure if you pushed with an open hand or just one finger tip.
Solution. Use Evaluate Numerically and a look at your own hand to approximate the area of an open hand.

$$A \approx 8\,\mathrm{cm} \times 18\,\mathrm{cm} = 0.0144\,\mathrm{m^2}$$

We can use Evaluate and Eq. (10.2) to see this produces a small pressure.

$$P = \frac{F}{A} = \frac{2\,\mathrm{N}}{0.0144\,\mathrm{m^2}} = 139 \frac{\mathrm{N}}{\mathrm{m^2}}$$

The area of one fingertip is about $1\,\mathrm{cm^2} = 0.0001\,\mathrm{m^2}$.

$$P = \frac{F}{A} = \frac{2\,\mathrm{N}}{10^{-4}\,\mathrm{m^2}} = 20000 \frac{\mathrm{N}}{\mathrm{m^2}}$$

Let's use the Standard Method and Eq. (10.3) to convert this into atmospheres.

$$20000 \frac{\mathrm{N}}{\mathrm{m^2}} \times \frac{1\,\mathrm{atm}}{1.0133 \times 10^5\,\mathrm{N/m^2}} = 0.19737\,\mathrm{atm}$$

This significant pressure of about one-fifth of an atmosphere will probably get their attention. The same push produces two very different sensations. ◀

Static Fluids

Let's take a look at the properties of static fluids. The particles of a static fluid are either at rest or moving at a constant velocity (so there is no relative motion among them). In other words, they are not accelerating.

1. Static fluids exert pressure equally in all directions, and the pressure is the same everywhere in the fluid at a given depth. Otherwise, the pressure difference would exert a force and the fluid would accelerate.

2. Pressure always acts perpendicular to any surface. If the pressure acted at any angle other than 90°, there would be a component of the force parallel to the surface. The fluid would accelerate in the direction of the parallel component.

3. An external pressure applied to a confined fluid is transmitted unchanged throughout the fluid. This is called Pascal's Principle and it's the basis of all hydraulics. The brakes in your car wouldn't work without it.

4. The deeper you go in a fluid, the more pressure you feel. This increase in pressure is due to the weight of the fluid above you. For a static fluid with constant density ρ, the pressure at depth h is

$$P = P_0 + \rho g h \tag{10.4}$$

where P_0 is the external pressure applied to the fluid at the surface. This includes Pascal's Principle, since changing the external pressure changes the pressure at any depth.

The odd but standard notation that uses h for depth presumably avoids confusion with the d in derivatives. Depth is measured from the surface down, so it's opposite to the change in height $h = -\Delta z$.

Example 10.3 *Feeling the pressure*
What is the pressure on a diver 10 ft deep in a pool of water? How deep does the diver have to go to experience a pressure that's twice atmospheric pressure?
Solution. For a diver in a pool on Earth, the external pressure is $P_0 = 1\,\mathrm{atm}$. We can use Evaluate and Eq. (10.4) to calculate the pressure at a depth of $h = 10\,\mathrm{ft} = 3.048\,\mathrm{m}$.

$$P = 1.0133 \times 10^5 \frac{\mathrm{N}}{\mathrm{m}^2} + \left(1000 \frac{\mathrm{kg}}{\mathrm{m}^3}\right)\left(9.806\,7 \frac{\mathrm{N}}{\mathrm{kg}}\right)(3.048\,\mathrm{m}) = 1.312\,2 \times 10^5 \frac{\mathrm{N}}{\mathrm{m}^2}$$

Let's use the Standard Method to convert this to atmospheres.

$$1.312\,2 \times 10^5 \frac{\mathrm{N}}{\mathrm{m}^2} \times \frac{1\,\mathrm{atm}}{1.013\,3 \times 10^5\,\mathrm{N}/\mathrm{m}^2} = 1.295\,0\,\mathrm{atm}$$

That's a 29.5% increase in pressure. We can use Eq. (10.4) with $P = 2\,\mathrm{atm}$ and Solve Exact to find the depth.

$$2\,\mathrm{atm} = 1\,\mathrm{atm} + \left(1000\,\mathrm{kg}/\mathrm{m}^3\right)\left(9.806\,7\,\mathrm{m}/\mathrm{s}^2\right) h, \text{ Solution is: } 10.332\,\mathrm{m}$$

The pressure is $2\,\mathrm{atm}$ at a depth of about $34\,\mathrm{ft}$. ◀

Buoyancy

When an object is in a fluid, it either floats or it sinks slowly. Either way, it feels an upward force called the buoyant force. Buoyancy is due to the increase of pressure with depth. The pressure on the bottom of the object is larger than the pressure on the top because the bottom is deeper. The buoyant force is due to this pressure difference.

The upward **buoyant force** on an object in a fluid is

$$F_B = \rho_f \, g V_{\text{sub}} \tag{10.5}$$

where ρ_f is the fluid density and V_{sub} is the submerged volume. This is called Archimedes' Principle, which says the buoyant force on an object immersed in a fluid equals the weight of the fluid displaced by the object.

$$m_{\text{fluid}} \, g = \left(\rho_f \, V_{\text{sub}}\right) g = \rho_f \, g V_{\text{sub}}$$

Only the part of the object submerged in the fluid contributes to the buoyant force.

Example 10.4 *An apparently superfluous name*
What is the minimum force needed to lift a treasure chest off the bottom of the ocean? The chest weighs $2200\,\text{N}$ in open air and is $60\,\text{cm}$ long, $30\,\text{cm}$ wide, and $20\,\text{cm}$ deep.
Solution. There are three forces acting on the chest as it sits on the ocean floor. Its weight pulls it down, and the normal and buoyant forces push it up. The chest sits in equilibrium, so the up and down forces balance.

$$F_B + N = mg$$

The chest is completely submerged, so V_{sub} is the total volume and something we can Evaluate Numerically.

$$V_{\text{sub}} = 60\,\text{cm} \times 30\,\text{cm} \times 20\,\text{cm} = 0.036\ \text{m}^3$$

The buoyant force on the chest is $F_B = \rho_{sw} \, g V_{\text{sub}}$ where the density of seawater is $\rho_{sw} = 1025\,\text{kg/}\,\text{m}^3$. We can use Evaluate and Eq. (10.5) to calculate it.

$$F_B = \rho_{sw} \, g V_{\text{sub}} = \left(1025 \frac{\text{kg}}{\text{m}^3}\right) \left(9.806\,7 \frac{\text{N}}{\text{kg}}\right) \left(0.036\,\text{m}^3\right) = 361.87\,\text{N}$$

The normal force is reduced by the buoyant force.

$$N = mg - F_B = 2200\,\text{N} - 361.87\,\text{N} = 1838.1\,\text{N}$$

This is the minimum force needed to lift the chest off the ocean floor. If the chest was sitting on a scale under water, the scale would read $1838.1\,\text{N}$. For this reason, this is sometimes called the "apparent weight" but it's really not worthy of a new name. It's just the normal force.

Of course, the chest in "open air" is also immersed in a fluid that exerts a buoyant force. The buoyant force is proportional to the fluid density, and air has a very small density.

$$\frac{\rho_{air}}{\rho_w} = \frac{1.29\,\text{kg/}\,\text{m}^3}{1000\,\text{kg/}\,\text{m}^3} = 0.001\,29$$

Compared with water, the buoyant effects of air can be ignored. ◀

Example 10.5 *Last nine months floating...*
How far does a solid ball sink when it floats in water?
Solution. The floating ball is in equilibrium, so its downward weight equals the upward buoyant force. Only the submerged volume contributes to the buoyant force so $F_B = \rho_w g V_{\text{sub}}$. The ball's total volume contributes to its weight, so $mg = \rho_b g V$ where ρ_b is the density of the ball.

Setting the two forces equal gives us a completely general result for the fraction of any floating object that is underwater.

$$\frac{V_{\text{sub}}}{V} = \frac{\rho_{\text{obj}}}{\rho_w} \equiv s$$

The ratio of the object's density to that of water is called its specific gravity (s).

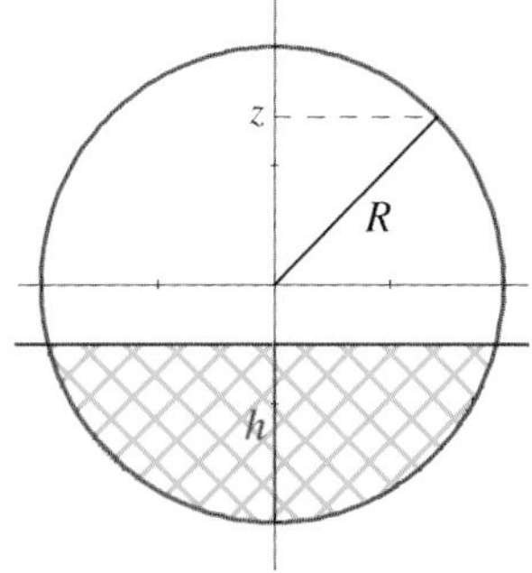

Figure 10.1 A spherical witch

To calculate the depth h, we need to know the submerged volume (see Figure 10.1). A horizontal slice of the sphere (radius R) that's dz thick and a distance z from the center has volume $dV = \pi\left(R^2 - z^2\right) dz$.

Let's use Factor to calculate the sphere's submerged volume.

$$V_{\text{sub}} = \pi \int_{R-h}^{R} \left(R^2 - z^2\right) dz = \tfrac{1}{3}\pi h^2 (3R - h)$$

The balls is a sphere so its total volume is $V = \frac{4}{3}\pi R^3$. Let's Expand the ratio of the submerged volume to the total volume.

$$s = \frac{V_{\text{sub}}}{V} = \frac{\frac{1}{3}\pi h^2 (3R - h)}{\frac{4}{3}\pi R^3} = \frac{3}{4R^2}h^2 - \frac{1}{4R^3}h^3$$

A little editing by-hand produces an equation for a floating sphere's depth.

$$0 = h^3 - 3Rh^2 + 4sR^3$$

This cubic equation has an exact general solution, but it's too cumbersome to be useful. Instead, let's specify that our ball is the size of a basketball ($R = 11.4\,\text{cm}$) and made of light wood $\left(\rho_b = \frac{1}{3}\rho_w\right)$. Tell *SNB* to assume the depth is positive and less than the ball's diameter.

$$\text{assume}\,(h > 0) = (0, \infty)$$

$$\text{additionally}\,(h < 22.8) = (0, 22.8)$$

Then use Solve Exact to find the depth h when $R = 11.4$ and $s = \frac{1}{3}$.

$$0 = \left[h^3 - 3Rh^2 + 4sR^3\right]_{R=11.4, s=1/3}, \text{ Solution is: } 8.8228$$

The ball sinks $8.822\,8\,\text{cm}$ into the water. Less than half the ball is submerged since the depth is less than the ball's radius. ◄

Fluids in Motion

It's time to go with the flow and let our fluids be in motion. Suppose we have a fluid of density ρ moving at velocity v through a pipe or channel with cross-sectional area A. We can quantify this flow with the **mass flow rate**.

$$\frac{\Delta m}{\Delta t} = \rho A v \tag{10.6}$$

This tells us the mass of fluid that flows past a given point per unit time. The metric unit of mass flow rate is $\mathrm{kg/\,s}$. We can also use the **volume flow rate**.

$$\frac{\Delta V}{\Delta t} = A v \tag{10.7}$$

This tells us the volume of fluid that flows past a given point per unit time. The metric unit of volume flow rate is $\mathrm{m^3/\,s}$.

Caveo reputo! Eq. (10.7) is a letter-thinker's nightmare. Be sure to distinguish between the volume V and the velocity v. In this chapter, an uppercase V always indicates volume and a lowercase v always refers to velocity.

Example 10.6 *A typically typical type of problem*
A typical above-ground backyard pool is circular with a $20\,\mathrm{ft}$ diameter and a $4\,\mathrm{ft}$ depth. A typical garden hose has a $\frac{3}{4}$ in inner diameter through which water flows at $1.0\,\mathrm{m/\,s}$. How long does it typically take to fill the pool?
Solution. Instead of converting everything, remember Evaluate Numerically converts all units to metric. Let's use it to calculate the volume flow rate through the hose.

$$Av = \pi\left(\tfrac{3}{8}\,\mathrm{in}\right)^2\left(1.0\frac{\mathrm{m}}{\mathrm{s}}\right) = 2.850\,2\times 10^{-4}\frac{\mathrm{m}^3}{\mathrm{s}}$$

Let's use it again to calculate the volume of the pool.

$$\Delta V = \pi\,(10\,\mathrm{ft})^2\,(4\,\mathrm{ft}) = 35.584\,\mathrm{m}^3$$

Use Eq. (10.7) to create and Evaluate an expression for the time.

$$\Delta t = \frac{\Delta V}{Av} = \frac{35.584\,\mathrm{m}^3}{2.850\,2\times 10^{-4}\,\mathrm{m}^3/\,\mathrm{s}} = 1.248\,5\times 10^5\,\mathrm{s}$$

Let's convert this into days with the Standard Method.

$$\Delta t = 1.248\,5\times 10^5\,\mathrm{s}\times\frac{1\,\mathrm{d}}{86400\,\mathrm{s}} = 1.445\,\mathrm{d}$$

It typically takes about a day and a half to fill the pool. Let's check our value for the volume flow rate by using the Solve Method to convert it into gallons-per-minute.

$$2.850\,2\times 10^{-4}\frac{\mathrm{m}^3}{\mathrm{s}} = R\frac{\mathrm{gal}}{\mathrm{min}},\ \text{Solution is: } 4.517\,7$$

That seems reasonable, but you should grab a bucket, find a hose, and test it! ◀

Now we apply the little-known but important Beakman's Rule: Everything has to go somewhere. When fluid flows along a closed path, all the mass that passes one point must also pass every other point. This leads us to the **Continuity Equation**.

$$\rho_1 A_1 v_1 = \rho_2 A_2 v_2 \tag{10.8}$$

If the fluid is incompressible, we can't squeeze its molecules closer and its density is constant. This leads us to the Continuity Equation for incompressible fluids.

$$A_1 v_1 = A_2 v_2 \tag{10.9}$$

This is the "garden hose" equation! When you put your thumb over the end of a hose, the area of the opening decreases so the water's speed increases as it leaves the hose. The speed and area can change, but the mass flow rate is always constant and the volume flow rate is constant for incompressible fluids.

Example 10.7 *You can water the flowers*
After you fill the pool, you put an adjustable nozzle on the end of the hose. How fast does the water leave the nozzle if you adjust the radius to $\frac{3}{16}$ in? What nozzle-radius produces a water speed of $4\,\mathrm{m/s}$?
Solution. Water is an incompressible fluid, so we can use Eq. (10.9). The hose has a circular cross-section so $A = \pi r^2$ for a given radius. Create and Evaluate an expression for the water's velocity as it leaves the nozzle.

$$v_2 = \frac{A_1}{A_2} v_1 = \frac{\pi \left(\frac{3}{4}\,\mathrm{in}\right)^2}{\pi \left(\frac{3}{16}\,\mathrm{in}\right)^2} \left(1 \frac{\mathrm{m}}{\mathrm{s}}\right) = 16 \frac{\mathrm{m}}{\mathrm{s}}$$

Reducing the radius by a factor of 4 increases the water's velocity by a factor of 16! Let's use Solve Exact on the Continuity Equation for an expression for the nozzle radius.

$$\pi r_1^2 v_1 = \pi r_2^2 v_2, \text{ Solution is: } r_1 \frac{\sqrt{v_1}}{\sqrt{v_2}}$$

Now we can Evaluate it with the appropriate numerical values.

$$r_2 = \left(\tfrac{3}{16}\,\mathrm{in}\right) \sqrt{\frac{16\,\mathrm{m/s}}{4\,\mathrm{m/s}}} = \tfrac{3}{8}\,\mathrm{in}$$

To quadruple the water's speed you need to reduce the radius by a factor of two so the area is decreased by a factor of 4. We could also use Solve Exact with the numerical values from the previous example.

$$\pi \left(\tfrac{3}{4}\,\mathrm{in}\right)^2 \left(1 \frac{\mathrm{m}}{\mathrm{s}}\right) = \pi r_2^2 \left(4 \frac{\mathrm{m}}{\mathrm{s}}\right), \text{ Solution is: } 9.525 \times 10^{-3}\,\mathrm{m}$$

Let's use the Standard Method to verify this is the same answer.

$$r_2 = 9.525 \times 10^{-3}\,\mathrm{m} \times \frac{100\,\mathrm{in}}{2.54\,\mathrm{m}} = 0.375\,\mathrm{in}$$

Yes it is! ◄

Bernoulli's Equation

Fluid Mechanics is the study of the behavior of flowing fluids. Many problems in Fluid Mechanics are complicated enough to require some simplifying assumptions. Fortunately this leads to an important and useful exact solution known as Bernoulli's Equation. This result applies to the laminar flow of incompressible fluids with zero viscosity.

Laminar flow is smooth and steady flow along streamlines that never cross. This flow has no turbulence or whirlpools. Incompressible fluids have constant density. Many common liquids such as water are nearly incompressible. Viscosity is a kind of internal friction or stickiness that impedes a fluid's motion. Honey has a large viscosity and oil has a small viscosity. Water has almost no viscosity. Essentially our fluids always flow smoothly and never stick to the pipe.

Consider the case of water flowing though a smooth pipe (see Figure 10.2) with varying height and cross-sectional area.

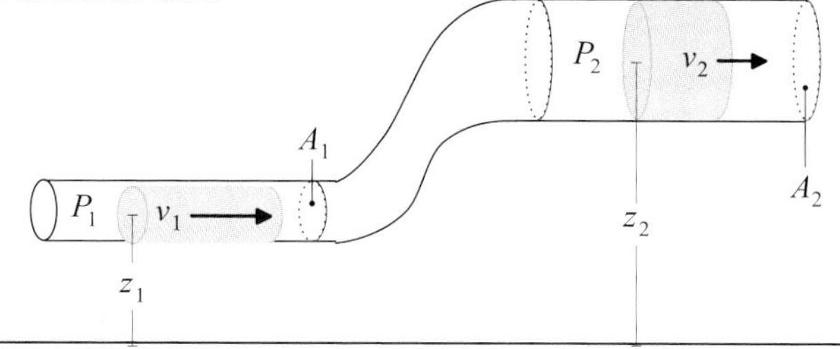

Figure 10.2 Go with the flow

Bernoulli's Equation relates the work done on the fluid by gravity and pressure to the total work done per the Work-Energy Theorem.

$$P + \tfrac{1}{2}\rho v^2 + \rho g z = constant \tag{10.10}$$

The constant has units $\mathrm{J/\,m^3}$. Bernoulli's Equation relates the fluid's pressure P, speed v, and height z at any two points along the flow.

$$P_1 + \tfrac{1}{2}\rho v_1^2 + \rho g z_1 = P_2 + \tfrac{1}{2}\rho v_2^2 + \rho g z_2 \tag{10.11}$$

As the units suggest, this complicated equation is just energy conservation per-unit-volume for a fluid.

Example 10.8 *Can you also water the upstairs window boxes?*
A garden hose laying on the ground has a $2\,\mathrm{cm}$ diameter and water flowing in it at $1\,\mathrm{m/\,s}$ and a pressure of $1.6\,\mathrm{atm}$. At the end of the hose is a nozzle with a $\frac{3}{4}\,\mathrm{cm}$ diameter. You pick up the end with the nozzle and hold it $125\,\mathrm{cm}$ off the ground to water the flowers. Find the water's speed and pressure in the nozzle, and the water's maximum height once it leaves the nozzle.

Solution. We have enough information to calculate the volume flow rate and the constant in Eq. (10.10). Let's use Evaluate Numerically and Eq. (10.7) to calculate the volume flow rate.

$$\pi r_1^2 v_1 = \pi\,(1\,\mathrm{cm})^2\left(1\frac{\mathrm{m}}{\mathrm{s}}\right) = 3.141\,6 \times 10^{-4}\frac{\mathrm{m}^3}{\mathrm{s}}$$

Let's use it again to calculate the total energy (per unit volume) of the water in the hose on the ground.

$$P_1 + \frac{1}{2}\rho v_1^2 + \rho g y_1 = 1.6\,\text{atm} + \frac{1}{2}\left(1000\frac{\text{kg}}{\text{m}^3}\right)\left(1\frac{\text{m}}{\text{s}}\right)^2 = 1.626\,2 \times 10^5 \frac{\text{kg}}{(\text{m})\,\text{s}^2}$$

Notice the units are equal to $\text{J}/\,\text{m}^3$.

The Continuity Equation and Solve Exact tell us the water's speed in the nozzle.

$$3.141\,6 \times 10^{-4}\frac{\text{m}^3}{\text{s}} = \pi\left(\tfrac{1}{2}0.75\,\text{cm}\right)^2 v_2, \text{ Solution is: } 7.111\,1\frac{\text{m}}{\text{s}}$$

Bernoulli's Equation and Solve Exact tell us the water pressure in the nozzle.

$$1.6262 \times 10^5 \tfrac{\text{J}}{\text{m}^3} = P_2 + \tfrac{1}{2}\left(1000\tfrac{\text{kg}}{\text{m}^3}\right)\left(7.1111\tfrac{\text{m}}{\text{s}}\right)^2 + \left(1000\tfrac{\text{kg}}{\text{m}^3}\right)\left(9.8067\tfrac{\text{m}}{\text{s}^2}\right)(1.25\,\text{m}),$$

$$\text{Solution is: } 1.250\,8 \times 10^5 \frac{\text{kg}}{(\text{m})\,\text{s}^2}$$

Notice the units are equal to $\text{N}/\,\text{m}^2$. Let's convert this pressure to atmospheres with the Standard Method.

$$P_2 = 1.250\,8 \times 10^5 \frac{\text{N}}{\text{m}^2} \times \frac{1\,\text{atm}}{1.013\,3 \times 10^5\,\text{N}/\,\text{m}^2} = 1.234\,4\,\text{atm}$$

The water in the nozzle moves at $7.111\,1\,\text{m}/\,\text{s}$ under a pressure of $1.234\,4\,\text{atm}$.

The water reaches its maximum height when you project it straight up. Once it leaves the nozzle, the pressure on the water is $1\,\text{atm}$ and the water stops at its maximum height.

$$1.626\,2 \times 10^5 = 1.013\,3 \times 10^5 + (1000)(9.806\,7)\,z, \text{ Solution is: } 6.249\,8$$

The water reaches a maximum height of almost $6.25\,\text{m}$. If the upstairs window boxes are on the second floor, you can probably reach them.

We could have found these answers all at once with Solve Exact (without the units so the expression fits on the printed page).

$$\begin{bmatrix} 1.626\,2 \times 10^5 = P_2 + \frac{1}{2}(1000)\,v_2^2 + (1000)(9.8067)(1.25) \\ 1.626\,2 \times 10^5 = 1.013\,3 \times 10^5 + (1000)(9.8067)\,z \\ \pi\,(0.01)^2\,(1) = \pi\,(0.003\,75)^2\,v_2 \end{bmatrix},$$

$$\text{Solution is: } \left[z = 6.249\,8, P_2 = 1.250\,8 \times 10^5, v_2 = 7.111\,1\right]$$

Different way, same solutions. ◄

Remember that Bernoulli's Equation has limitations. It applies to incompressible fluids, so it only works for gases when the pressure differences are small. It applies to laminar flow, so it only works for low-velocity flow. It applies to fluids with no viscosity, so it only works for thin, runny fluids.

Applications of Bernoulli's Equation

In spite of its limitations, Bernoulli's Equation is useful. It applies to many physically relevant situations, including the following interesting and enlightening special cases.

Constant Height: $z_1 = z_2$
When a fluids flows over a surface with no significant change in height, the potential energy terms in Bernoulli's Equation cancel. The result is a simple relation between the pressure and fluid speed above and below the surface.

$$P_b - P_a = \tfrac{1}{2}\rho\left(v_a^2 - v_b^2\right) \tag{10.12}$$

This applies to wind blowing over a roof or air flowing around an airplane wing, and it also applies to the flow inside a horizontal pipe.

The pressure below the surface is larger when the fluid speed above is faster, so there is a net upward force on the surface.

$$F = \left(P_b - P_a\right)A = \tfrac{1}{2}\rho\left(v_a^2 - v_b^2\right)A \tag{10.13}$$

This is good for airplanes but bad for roofs.

In practice, pilots rarely know the air speed around their wings but they do know how fast they're moving. When an object is moving through the fluid, it's more useful to relate the lift force to the object's speed v.

$$F \equiv \tfrac{1}{2}C_L\rho v^2 A \tag{10.14}$$

The dimensionless lift coefficient C_L is usually determined with wind-tunnel experiments. Remember ρ is the density of the fluid.

Example 10.9 *Where'd that extra two come from?*
One model of the Boeing 747 has a maximum takeoff weight of about $910\,000$ pounds and a wing area of 5650 square feet. At liftoff, its lift coefficient is about 2. What is the minimum pressure difference needed to lift this plane off the ground? Estimate the minimum speed needed for takeoff.
Solution. Since we're given the weight in pounds, let's use the Standard Method to convert the area into square inches.

$$A = 5650\,\text{ft}^2 \times \left(\frac{12\,\text{in}}{1\,\text{ft}}\right)^2 = 813\,600\,\text{in}^2$$

The upward pressure on the wings must support the plane's weight, so use Eq. (10.2) to create and Evaluate an expression for the minimum pressure difference in psi.

$$\Delta P = \frac{F}{A} = \frac{910\,000.\,\text{lbf}}{813\,600\,\text{in}^2} = 1.1185\frac{\text{lbf}}{\text{in}^2}$$

This is only a $7.6\,\%$ increase over atmospheric pressure! Use this pressure difference and Eq. (10.14) to create and Simplify an expression for the minimum speed.

$$v_{\text{min}} \approx \sqrt{\frac{2\Delta P}{C_L\rho}} = \sqrt{\frac{2\left(1.1185\,\text{lbf}/\,\text{in}^2\right)}{2\left(1.29\,\text{kg}/\,\text{m}^3\right)}} = 77.318\frac{\text{m}}{\text{s}}$$

Let's use the Solve Method to convert this into miles-per-hour.

$$77.318\frac{\text{m}}{\text{s}} = v\frac{\text{mi}}{\text{h}}, \text{ Solution is: } 172.96$$

This is the slowest speed the plane can have and generate enough lift to get off the ground. Safety margins typically require the liftoff speed to be at least 10% larger.

Experiments show that the wind speed below the wing can be as small as half the speed above the wing. If we take the plane's speed as the wind speed above, there isn't enough pressure to lift the plane.

$$\frac{\frac{1}{2}\rho\left(v_a^2 - v_b^2\right)}{\Delta P} = \frac{\frac{1}{2}\left(1.29\frac{\text{kg}}{\text{m}^3}\right)\left(1 - \frac{1}{2^2}\right)\left(77.318\frac{\text{m}}{\text{s}}\right)^2}{1.1185\frac{\text{lbf}}{\text{in}^2}} = 0.37500$$

The "extra" lift during takeoff is generated by extending the flaps on the wings. ◄

Constant Pressure: $P_1 = P_2$

When a fluid flows between two points at the same pressure, the pressure terms in Bernoulli's Equation cancel. The result is a simple relation between the height and speed of the fluid that is independent of the fluid's density.

$$\tfrac{1}{2}v^2 + gz = \tfrac{1}{2}v_0^2 + gz_0 \tag{10.15}$$

For fluid flow from a large open reservoir through an open valve, the pressure is $1\ \text{atm}$ at both points and the surface fluid barely moves so $v_0 \approx 0$. The fluid flows through the valve at speed

$$v = \sqrt{2g\left(z_0 - z\right)} \tag{10.16a}$$

$$= \sqrt{2gh} \tag{10.16b}$$

where the valve's depth $h = z_0 - z$ is the difference between the valve and surface heights. This is known as Torricelli's Law.

The fluid comes out at the same speed that a ball dropped a distance h impacts the ground. This reminds us that fluids obey the same Newtonian rules as balls, and that Bernoulli's Equation is just energy conservation

Constant Speed: $v_1 = v_2$

When a fluids flows at constant speed, the kinetic energy terms in Bernoulli's Equation cancel. The result is a simple relation between the pressure and fluid depth.

$$P + \rho gz = P_0 + \rho gz_0 \tag{10.17}$$

This should seem familiar. It's the pressure-depth relation for static fluids!

$$P = P_0 + \rho g\left(z_0 - z\right) \tag{10.18a}$$

$$= P_0 + \rho gh \tag{10.18b}$$

This relation holds when the fluid speed is constant (but not necessarily zero). Newton's Laws are the same for $v = 0$ and any other constant velocity. Zero acceleration is zero acceleration!

A More Realistic Approach

We have only considered ideal fluids with no viscosity that move without friction. Real fluids have viscosity so they lose energy due to friction and that causes a pressure loss. A fluid with no viscosity acts like a hockey puck without friction: once set in motion it moves forever. A fluid with viscosity is like a hockey puck with friction: it requires a constant push to keep it moving at constant speed.

When an ideal fluid that obeys Bernoulli's Equation flows through a pipe, its velocity at any given point is uniform across the pipe's cross-section. That's what "uniform flow" means in Fluid Mechanics. When a viscous fluid flows through a pipe, its viscosity creates a velocity gradient. The closer the fluid is to the pipe walls, the slower it moves.

Newton realized the pressure needed to maintain this was proportional to the velocity gradient. The proportionality constant is the fluid's viscosity and its unit is $\mathrm{N\,s/\,m^2}$. It's a unit of pressure per unit velocity-change per-unit-distance.

$$\frac{\mathrm{N\,s}}{\mathrm{m}^2} = \frac{\mathrm{N}}{\mathrm{m}^2} \div \left(\frac{\mathrm{m}}{\mathrm{s}} \div \mathrm{m}\right)$$

Fluids with constant viscosity at a given temperature are called Newtonian fluids.

Like density, viscosity is a physical property that varies with temperature. The viscosity of liquids tends to decrease with increasing temperature, so honey and motor oil flow more readily as they warm up. Water's viscosity drops from about $10^{-3}\,\mathrm{N\,s/\,m^2}$ at $20\,°\mathrm{C}$ to half that value at $55\,°\mathrm{C}$. Viscosity is usually independent of pressure.

Bernoulli's Equation applies to the 1-dimensional steady laminar flow of incompressible fluids with zero viscosity. All real fluids except superfluids have some viscosity, and the **Navier–Stokes Equation** describes their flow. The complete Navier–Stokes Equation is a complicated 3-d mess that applies to the unsteady, viscous flow of compressible fluids. Unsteady flow includes time-dependent changes in velocity and density. Exact solutions exist for some simple laminar flows, but not for any turbulent flows.

For the laminar flow of an incompressible fluid with constant viscosity, the Navier–Stokes Equation simplifies somewhat to

$$\rho\frac{\partial\vec{v}}{\partial t} + \rho\left(\vec{v}\cdot\vec{\nabla}\right)\vec{v} = -\vec{\nabla}P + \mu\nabla^2\vec{v} + \rho\,\vec{g} \tag{10.19}$$

where μ is the fluid's viscosity. Even for laminar flow along a 1-d streamline, this equation is complicated.

$$\rho\frac{\partial v}{\partial t} + \rho v\frac{\partial v}{\partial s} = -\frac{\partial P}{\partial s} + \mu\frac{\partial^2 v}{\partial s^2} - \rho g\frac{\partial z}{\partial s} \tag{10.20}$$

For steady non-viscous flow, the viscosity and the time derivative of the velocity are both zero and this reduces to Bernoulli's Equation.

All flows obey Beakman's Rule, and the differential continuity equation is a general statement of mass conservation.

$$\frac{\partial\rho}{\partial t} + \vec{\nabla}\cdot(\rho\,\vec{v}) = 0 \tag{10.21}$$

For steady flow, the time derivative of the density is zero and this reduces to Eq. (10.8).

Flow in a Pipe

The analysis of fluid flow depends on the physical details of the system (such as a closed pipe or an open channel) and type of flow (laminar or turbulent). Typically, students use Bernoulli's Equation to analyze the laminar flow of water through pipes. There are many common everyday examples, such as water flow through home plumbing and drinking straws or blood flow through blood vessels, that suggest this is a worthy endeavor.

Now we'll extend this analysis and look at the flow of a viscous fluid through a pipe. This flow is similar to those typical Bernoulli problems. The fluid's velocity and pressure change as the pipe's height and size change, but we'll also account for the pressure loss due to viscosity. First we need a way to determine whether a flow is laminar or turbulent.

Reynolds Number

The **Reynolds number** is a dimensionless number used to characterize flows. It depends on two properties of the fluid (density ρ and viscosity μ), the fluid's average speed, and a characteristic length of the system.

$$\mathrm{Re} \equiv \frac{\rho D v_{\mathrm{ave}}}{\mu} \tag{10.22}$$

The characteristic length for flow in a pipe is the diameter D.

Physically, the Reynolds number is the ratio of the dynamic effects due to the fluid's mass and motion to the viscous effects slowing the fluid. Mathematically, it's the ratio of two terms of the Navier–Stokes Equation.

$$\frac{\rho\left(\vec{v}\cdot\vec{\nabla}\right)\vec{v}}{\mu\nabla^2\vec{v}} \sim \frac{\rho v^2/L}{\mu v/L^2} = \frac{\rho L v}{\mu}$$

A large Reynolds number means the dynamic effects dominate over the viscous effects so the fluid flows faster and the flow is turbulent. A small Re means the viscous effects dominate so the fluid flows slower and the flow is laminar. For fluid flow in a pipe, a Reynolds number less than 2000 indicates the flow is laminar, if it's greater than 4000 the flow is turbulent and in between is a transition zone.

Example 10.10 *The usual state...*
Calculate the Reynolds number for the flow in the nozzle from Example 10.8.
Solution. Apply Evaluate Numerically to Eq. (10.22) with the appropriate numerical values.

$$\mathrm{Re} = \frac{\left(1000\,\mathrm{kg/\,m^3}\right)\left(0.007\,5\,\mathrm{m}\right)\left(7.111\,1\,\mathrm{m/\,s}\right)}{10^{-3}\,\mathrm{N\,s/\,m^2}} = 53333$$

Even the slower flow on the ground has a high Reynolds number.

$$\mathrm{Re} = \frac{\left(1000\,\mathrm{kg/\,m^3}\right)\left(0.02\,\mathrm{m}\right)\left(0.75\,\mathrm{m/\,s}\right)}{10^{-3}\,\mathrm{N\,s/\,m^2}} = 15000$$

Both numbers are much larger than 2000, so both flows are probably turbulent. Laminar flow in a pipe is possible for large Reynolds numbers if the pipe has smooth walls and the fluid enters the pipe without any disturbances. ◀

Laminar Flow

Laminar flow is the smooth, orderly flow of slow-moving fluids. Experience tells us fluids with high viscosity tend to flow slower and are more likely to undergo laminar flow. Here we look at the effects of viscosity on the laminar flow through a pipe. Keep in mind that the results in this section *only* apply to the laminar flow of a viscous fluid through a pipe.

We have to push the fluid through the pipe. We can generate this push with pressure and gravity by letting them vary along the length of the pipe. Let's incorporate both gradients in a single parameter.

$$\kappa \equiv -\left(\frac{dP}{dx} + \rho g \frac{dz}{dx}\right) \tag{10.23a}$$

For a straight pipe of length L with a constant pressure gradient, κ depends on simple differences in pressure and height.

$$\kappa = -\left(\frac{\Delta P}{L} + \rho g \frac{\Delta z}{L}\right) \tag{10.23b}$$

A negative pressure difference means the pressure is lower to the right and a negative height difference means the right end of the pipe is lower. When κ is positive the net push on the fluid is to the right.

There is an exact solution to the Navier–Stokes Equation for laminar flow of a viscous fluid through a pipe. It gives the fluid's velocity as a function of radial distance from the center of the pipe.

$$v(r) = \kappa \frac{R^2 - r^2}{4\mu} \tag{10.24}$$

This quadratic velocity profile depends on the fluid's viscosity μ, the pipe's radius R, and the gradients (κ).

The flow is not uniform! Fluid nearer the pipe wall moves slower than the fluid at the center. The long dashed line is the center of the pipe ($r = 0$) where the fluid moves at its maximum velocity. The two short dashed lines indicate the fluid's speed at $r = \pm\frac{1}{2}R$. The fluid's velocity is zero at the pipe wall ($r = R$) where the fluid feels the maximum frictional force.

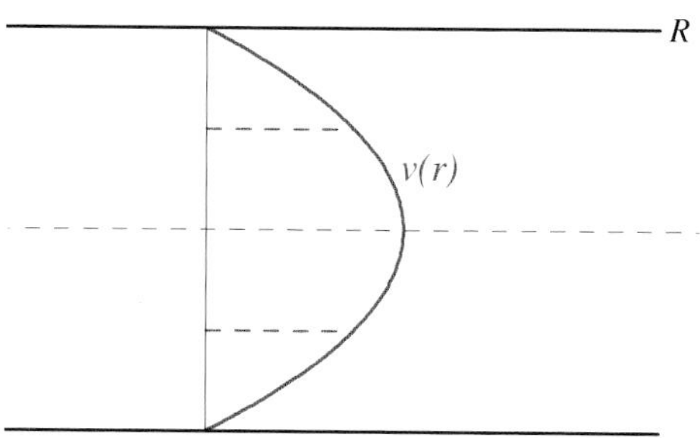

Figure 10.3 Laminar velocity profile

The maximum fluid velocity occurs at the pipe center ($r = 0$).

$$v_{\text{max}} = \frac{1}{4}\frac{\kappa}{\mu}R^2 \tag{10.25}$$

Here is the velocity profile in terms of the maximum velocity.

$$v(r) = v_{\text{max}}\left(1 - \frac{r^2}{R^2}\right) \tag{10.26}$$

Notice this has the correct values at the pipe center and wall.

The flow is not uniform, so let's Evaluate an integral across the pipe's cross-sectional area to calculate the fluid's average velocity.

$$\overline{v} = \frac{1}{A}\int v dA = \frac{1}{\pi R^2}\frac{\kappa}{4\mu}\int_0^R \left(R^2 - r^2\right) 2\pi r dr = \tfrac{1}{8}R^2\frac{\kappa}{\mu}$$

This is the fluid's average velocity at one point in the flow. It is *not* the average between two different points in the flow. We can use Eq. (10.25) to relate the average velocity to the maximum velocity.

$$\frac{\overline{v}}{v_{\text{max}}} = \frac{\tfrac{1}{8}R^2\kappa/\mu}{\tfrac{1}{4}R^2\kappa/\mu} = \frac{1}{2}$$

At any given point in the flow, the fluid's average speed is half its maximum speed. For pipes with constant radius, the average and maximum speeds are constant.

That same integral shows up in calculating the volume flow rate.

$$\frac{\Delta V}{\Delta t} = \int v dA = A\overline{v} = \pi R^2 \times \tfrac{1}{8}R^2\frac{\kappa}{\mu} = \tfrac{1}{8}\pi R^4\frac{\kappa}{\mu}$$

This shows the volume flow rate is related to the average velocity.

$$\frac{\Delta V}{\Delta t} = A\overline{v} \tag{10.27}$$

We can invoke Beakman's Rule to produce a modified Continuity Equation.

$$A_1\overline{v_1} = A_2\overline{v_2} \tag{10.28}$$

These two expressions are identical to their no-viscosity counterparts with the simple modification of using the average velocity.

Here's the full expression for the volume flow rate for laminar flow in a pipe.

$$\frac{\Delta V}{\Delta t} = -\frac{\pi R^4}{8\mu}\left(\frac{\Delta P}{L} + \rho g\frac{\Delta z}{L}\right) \tag{10.29}$$

The flow rate depends on the fourth power of the pipe's radius, so small changes in radius produce significant changes in flow. Combining Eqs. (10.27) and (10.29) gives the pressure loss due to viscous effects. This is called **Poiseuille's Equation**.

$$\Delta P_{\text{vis}} = 8\mu L\frac{\overline{v}}{R^2} \tag{10.30}$$

The lost pressure is proportional to the average velocity and the length of the pipe.

The friction factor f is a dimensionless measure of the frictional force on the fluid. For laminar flow through circular pipes, it depends only on the Reynolds number.

$$f = \frac{64}{\text{Re}} \tag{10.31}$$

The pressure loss can be written in terms of f but don't let this mislead you.

$$\Delta P_{\text{vis}} = \tfrac{1}{4}f\frac{L}{R}\rho\overline{v}^2 \tag{10.32}$$

The friction factor (via the Reynolds number) is inversely proportional to the fluid's average velocity so ΔP_{vis} is proportional to $\overline{v}$ for laminar flow.

The **Bernoulli–Poiseuille Equation** includes the effects of viscosity for any steady flow along the pipe's centerline. [32]

$$P_1 + \tfrac{1}{2}\rho v_1^2 + \rho g z_1 - \Delta P_{\text{vis}} = P_2 + \tfrac{1}{2}\rho v_2^2 + \rho g z_2 \tag{10.33}$$

The pressure loss due to the effects of the fluid's viscosity happens as the fluid flows from point 1 to point 2. For pipes with constant radius, the fluid flows at constant speed and the pressure loss due to viscosity is given by Eq. (10.30). For a pipe with varying radius, that pressure loss is

$$\Delta P_{\text{vis}} = 8\mu \int_{s_1}^{s_2} \frac{v}{r^2}\,ds \approx 8\mu L \left(\frac{v}{R^2}\right)_{\text{ave}} \tag{10.34}$$

where the average is over the pipe's length along the centerline.

Example 10.11 *Draining the tank is not*
A viscous liquid flows out of a large open tank through a horizontal pipe that's $3\,\mathrm{m}$ below the liquid's surface. The pipe is $7\,\mathrm{m}$ long, has a $2\,\mathrm{cm}$ diameter, and is open at the other end. The liquid's density is $910\,\mathrm{kg/\,m^3}$ and its viscosity is $0.021\,\mathrm{N\,s/\,m^2}$. Calculate the liquid's speed and flow rate, and the pressure loss in the pipe. Is this flow laminar?
Solution. The Bernoulli–Poiseuille Equation relates the liquid at the surface (which is at rest for a large tank) to that flowing out the open end of the pipe. Let's apply Solve Exact to Eqs. (10.30) and (10.33) simultaneously.

$$\begin{bmatrix} 1\,\mathrm{atm} + \left(910\,\mathrm{kg/\,m^3}\right)\left(9.8067\,\mathrm{m/\,s^2}\right)(3\,\mathrm{m}) - P_L = 1\,\mathrm{atm} + \frac{1}{2}\left(910\,\mathrm{kg/\,m^3}\right)v^2 \\ P_L = 8\left(0.021\,\mathrm{N\,s/\,m^2}\right)\frac{7\,\mathrm{m}}{(0.01\,\mathrm{m})^2}v \end{bmatrix},$$

Solution is: $\left[v = 2.105\,1\frac{\mathrm{m}}{\mathrm{s}^{1.0}},\ P_L = 24756.\frac{\mathrm{kg}}{\mathrm{m}^{1.0}\,\mathrm{s}^{2.0}}\right]$

Let's use Evaluate Numerically to calculate the volume flow rate.

$$\frac{\Delta V}{\Delta t} = \pi\,(0.01\,\mathrm{m})^2\left(2.105\,1\tfrac{\mathrm{m}}{\mathrm{s}}\right) = 6.613\,4 \times 10^{-4}\,\tfrac{\mathrm{m}^3}{\mathrm{s}}$$

We know the liquid's speed, so the Reynolds number (using Evaluate Numerically again) will tell us if the flow is laminar.

$$\mathrm{Re} = \frac{\left(910\,\mathrm{kg/\,m^3}\right)(2.105\,1\,\mathrm{m/\,s})(0.02\,\mathrm{m})}{0.021\,\mathrm{N\,s/\,m^2}} = 1824.4$$

The Reynolds number is less than 2000 so this flow is probably laminar. Let's see what Bernoulli says if we turn off the pressure-loss due to viscosity (with the flow rate called Q because *SNB* and Solve Exact do not treat $\Delta V/\Delta t$ as a single variable).

$$\begin{bmatrix} \left(910\,\mathrm{kg/\,m^3}\right)\left(9.8067\,\mathrm{m/\,s^2}\right)(3\,\mathrm{m}) = \frac{1}{2}\left(910\,\mathrm{kg/\,m^3}\right)v^2 \\ Q = \pi\,(0.01\,\mathrm{m})^2\,v \end{bmatrix},$$

Solution is: $\left[Q = 2.409\,8 \times 10^{-3}\frac{\mathrm{m}^{3.0}}{\mathrm{s}^{1.0}},\ v = 7.670\,7\frac{\mathrm{m}}{\mathrm{s}^{1.0}}\right]$

The flow rate and speed are over 3.6 times larger if we don't include the effects of viscosity. This liquid has a significant viscosity that we can't ignore. ◀

Example 10.12 *Taking the plunge sometimes is*
A hypodermic needle consists of a needle, a syringe, and a plunger. The needle has a length of 2.5 cm and a diameter of 0.3 mm. The syringe has a length of 5 cm and a diameter of 1 cm. The medicine's density is 800 kg/ m³ and its viscosity is 0.003 N s/ m². How much force is needed on the plunger so the medicine leaves the needle at $\frac{1}{2}$ ml/ s?
Solution. The Bernoulli–Poiseuille Equation relates the flows and pressure losses in the syringe and needle.

$$P_s + \tfrac{1}{2}\rho v_s^2 - \Delta P_s - \Delta P_n = P_n + \tfrac{1}{2}\rho v_n^2$$

Let's apply Solve Exact to the volume flow rate and Continuity Equation to find the fluid's velocity in the syringe and needle.

$$\begin{bmatrix} \frac{1}{2}\frac{\text{ml}}{\text{s}} = \pi\,(0.5\,\text{cm})^2\,v_s \\ \pi\,(0.5\,\text{cm})^2\,v_s = \pi\,(0.15\,\text{mm})^2\,v_n \end{bmatrix},$$

Solution is: $\left[v_n = 7.0736\frac{\text{m}}{\text{s}^{1.0}}, v_s = 6.3662\times10^{-3}\frac{\text{m}}{\text{s}^{1.0}}\right]$

The Reynolds number (and Evaluate Numerically) will tell us if the flow is laminar.

$$\text{Re} = \frac{\rho v_n D_n}{\mu} = \frac{\left(800\,\text{kg/m}^3\right)(7.0736\,\text{m/s})(0.3\,\text{mm})}{0.003\,\text{N s/m}^2} = 565.89$$

The flow is laminar so we can use Eq. (10.30) to Evaluate the pressure lost in the syringe and needle.

$$\Delta P_s = 8\mu\frac{L_s}{R_s^2}v_s = \frac{8\left(0.003\,\text{N s/m}^2\right)(0.05\,\text{m})(0.0063662\,\text{m/s})}{(0.005\,\text{m})^2} = 0.30558\frac{\text{N}}{\text{m}^2}$$

$$\Delta P_n = 8\mu\frac{L_n}{R_n^2}v_n = \frac{8\left(0.003\,\text{N s/m}^2\right)(0.025\,\text{m})(7.0736\,\text{m/s})}{(0.00015\,\text{m})^2} = 1.8863\times10^5\frac{\text{N}}{\text{m}^2}$$

Notice how much smaller the speed and pressure losses are in the syringe. The pressure difference between the plunger and the output end of the needle, which we calculate with the Bernoulli–Poiseuille Equation, is due almost entirely to the fluid in the needle.

$$P_s - P_n = \tfrac{1}{2}\left(800\frac{\text{kg}}{\text{m}^3}\right)\left(7.0736\frac{\text{m}}{\text{s}}\right)^2 + 1.8863\times10^5\frac{\text{N}}{\text{m}^2} = 2.0864\times10^5\frac{\text{kg}}{(\text{m})\,\text{s}^2}$$

The required force is the syringe's cross-sectional area times the pressure difference.

$$F = \pi R_s^2\,(P_s - P_n) = \pi\,(0.5\,\text{cm})^2\left(2.0864\times10^5\frac{\text{N}}{\text{m}^2}\right) = 16.387\,\text{N}$$

This force provides enough pressure to overcome the losses and give the fluid some kinetic energy. The fluid in the syringe is very slow, so most of the pressure loss occurs in the needle. The pressure needed to move the plunger is significantly larger than the Bernoulli's Equation prediction $\Delta P_{\text{Bernoulli}} \approx \frac{1}{2}\rho v_n^2 = 20014\,\text{Pa}$. ◀

Turbulent Flow

Recall Example 10.8 where the water flow through a hose and nozzle at low, typical everyday speeds was turbulent. Let's use Eq. (10.22) with $\mathrm{Re} = 2000$ and Solve Exact to find the maximum water speed through the nozzle for laminar flow.

$$2000 = \frac{(1000\,\mathrm{kg/\,m^3})\,(0.007\,5\,\mathrm{m})\,v}{10^{-3}\,\mathrm{N\,s/\,m^2}}, \text{ Solution is: } 0.266\,67\frac{\mathrm{m}}{\mathrm{s}}$$

That's pretty slow! Turbulent flow is more common than laminar flow. There aren't many interesting real-life scenarios with laminar water flow. Most physically interesting flows are turbulent unless the fluid has a large viscosity or the pipe is very small.

One way to model turbulent flow is the power-law profile

$$v(r) = v_{\max}\left(1 - \frac{r}{R}\right)^n \tag{10.35}$$

where $v(r)$ is the time-averaged velocity in the direction of the flow. The exponent depends on the Reynolds number and it's related to the friction factor $f \approx n^2$.

Figure 10.4 shows this profile for $n = \frac{1}{7}$, which is the exponent for turbulent flows with $\mathrm{Re} \approx 10^5$. It also shows the velocity profile for laminar flow through the same pipe. More turbulent flows have a larger Reynolds number, a smaller power-law exponent, and a flatter velocity profile. The uniform profile provides a good approximation to many turbulent flows.

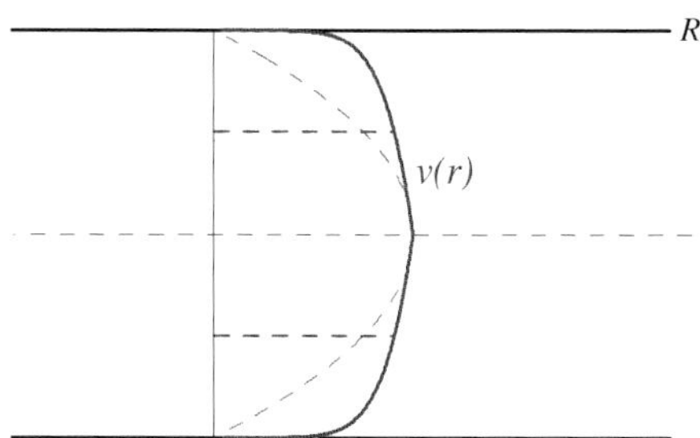

Figure 10.4 Turbulent velocity profile

The expressions for volume flow rate and continuity are the same for all flows. The pressure loss during turbulent flow due to the effects of the fluid's viscosity is

$$\Delta P_{\mathrm{vis}} = \tfrac{1}{4} f \frac{L}{R} \rho \overline{v}^2 \tag{10.36}$$

where R is the pipe radius, L is the pipe length, and $\overline{v}$ is the fluid's average velocity. The pressure loss during turbulent flow in a pipe is proportional to the average-velocity squared.

This appears identical to Eq. (10.32) but there is an important difference. For turbulent flow, there is no simple expression analogous to Eq. (10.31) relating the friction factor to the Reynolds number. You find f from tables or graphs (if you like using tables and graphs to obtain numerical values then Fluid Mechanics is the field for you), or you can estimate it with the Colebrook Equation.

$$\frac{1}{\sqrt{f}} = -2\log_{10}\left(\frac{\delta}{3.70} + \frac{2.51}{\mathrm{Re}\sqrt{f}}\right) \tag{10.37}$$

The relative roughness δ is the ratio of the height of the imperfections on the pipe wall to the pipe's diameter. It's another number you can get from tables and graphs. For smooth pipes $\delta \approx 10^{-6} \to 10^{-4}$ is small, for rough pipes $\delta \approx 0.01 \to 0.05$ is large, and for medium pipes $\delta \approx 10^{-3}$ takes intermediate values.

The Moody diagram is perhaps the most famous of those graphs. [27] It is a 2-dimensional log-log graph with 3 dimensionless axes that relates the Reynolds number, the friction factor, and the relative roughness for any flow through a circular pipe. Once you know any two, you can use the Moody diagram to estimate the third. It tells you graphically what the Colebrook Equation tells you mathematically. Usually you know the Reynolds number (a property of the flow) and the relative roughness (a property of the pipe) and you read the friction factor off the graph.

The subscript in $\log_{10}$ tells *SNB* to interpret the logarithm as base-10. *SNB*'s default interpretation treats both $\ln x$ and $\log x$ as the natural logarithm (base-e). You can tell *SNB* to interpret $\log x$ as base-10 by checking the Change from e to 10 box listed under Base for log function on the General tab of the Computation Setup... menu choice. This can lead to mistakes if you use many computers, so it's safer to use the explicit $\log_{10}$ notation when you want base-10 logarithms.

The Bernoulli–Poiseuille Equation includes the effects of viscosity for any steady flow along the pipe's centerline.

$$P_1 + \tfrac{1}{2}\rho v_1^2 + \rho g z_1 - \Delta P_{\text{vis}} = P_2 + \tfrac{1}{2}\rho v_2^2 + \rho g z_2 \tag{10.38}$$

The pressure loss due to the effects of the fluid's viscosity happens as the fluid flows from point 1 to point 2. For pipes with constant radius, the fluid flows at constant speed and the pressure loss due to viscosity is given by Eq. (10.36). For a pipe with varying radius, that pressure loss is

$$\Delta P_{\text{vis}} = \tfrac{1}{4}\rho \int_{s_1}^{s_2} \frac{f\,v^2}{r} ds \approx \tfrac{1}{4}\rho L \left(\frac{f\,v^2}{R}\right)_{\text{ave}} \tag{10.39}$$

where the average is over the pipe's length along the centerline.

Example 10.13 *A pipe dream*

Water flows through a 12 m-long pipe, entering at the left end (5 cm radius) that is 3 m higher than the right end (radius 2.5 cm). The water's speed at the left end is 2 m/ s and the pressure there is 200 kPa. The pressure loss due to viscosity is 80 kPa. Calculate the water's pressure and speed at the right end of the pipe. Is the flow turbulent? Does the pipe have smooth walls?

Solution. A look at the Reynolds number with Evaluate Numerically tells us this flow is turbulent.

$$\text{Re}_1 = \frac{(1000\,\text{kg/ m}^3)\,(10\,\text{cm})\,(2\,\text{m/ s})}{10^{-3}\,\text{N s/ m}^2} = 2.0 \times 10^5$$

Since the pressures are given in kPa, let's use the Solve Method to convert the initial kinetic and potential energy terms.

$$\tfrac{1}{2}\left(1000\frac{\text{kg}}{\text{m}^3}\right)\left(2\frac{\text{m}}{\text{s}}\right)^2 + \left(1000\frac{\text{kg}}{\text{m}^3}\right)\left(9.806\,7\frac{\text{N}}{\text{kg}}\right)(3\,\text{m}) = P\,\text{kPa},$$

Solution is: 31.42

The Bernoulli–Poiseuille Equation, the Continuity Equation, and Solve Exact tell us the water's pressure and speed at the right end of the pipe.

$$\begin{bmatrix} 200\,\mathrm{kPa} + 31.42\,\mathrm{kPa} - 80\,\mathrm{kPa} = P_2 + \frac{1}{2}\left(1000\,\mathrm{kg/\,m^3}\right) v_2^2 \\ \pi\,(5\,\mathrm{cm})^2\,(2\,\mathrm{m/\,s}) = \pi\,(2.5\,\mathrm{cm})^2\, v_2 \end{bmatrix},$$

Solution is: $\left[P_2 = \frac{1.194\,2\times10^5}{\mathrm{m}^{1.0}\,\mathrm{s}^{2.0}}\,\mathrm{kg},\ v_2 = 8.0\frac{\mathrm{m}}{\mathrm{s}^{1.0}}\right]$

The $119.42\,\mathrm{kPa}$ pressure at the right end is significantly less than the $200\,\mathrm{kPa}$ pressure on the left, even though water has a "small" viscosity.

The water's increased speed makes the flow more turbulent.

$$\mathrm{Re}_2 = \frac{\left(1000\,\mathrm{kg/\,m^3}\right)(5\,\mathrm{cm})\,(8.0\,\mathrm{m/\,s})}{10^{-3}\,\mathrm{N\,s/\,m^2}} = 4.0\times10^5$$

Let's look at the results if we ignore viscosity with Bernoulli's Equation and the Continuity Equation.

$$\begin{bmatrix} 200\,\mathrm{kPa} + 31.42\,\mathrm{kPa} = P_2 + \frac{1}{2}\left(1000\,\mathrm{kg/\,m^3}\right) v_2^2 \\ \pi\,(5\,\mathrm{cm})^2\,(2\,\mathrm{m/\,s}) = \pi\,(2.5\,\mathrm{cm})^2\, v_2 \end{bmatrix},$$

Solution is: $\left[P_2 = \frac{1.994\,2\times10^5}{\mathrm{m}^{1.0}\,\mathrm{s}^{2.0}}\,\mathrm{kg},\ v_2 = 8.0\frac{\mathrm{m}}{\mathrm{s}^{1.0}}\right]$

The fluid speed is the same, but without the losses due to viscosity the pressure at the right end is only slightly less than the pressure on the left.

Let's use Solve Exact and Eq. (10.39) to estimate the average friction factor.

$$80\,\mathrm{kPa} = \frac{1}{4}\left(1000\frac{\mathrm{kg}}{\mathrm{m}^3}\right)(12\,\mathrm{m})\,\frac{1}{2}\left(\frac{\left(8\frac{\mathrm{m}}{\mathrm{s}}\right)^2}{2.5\,\mathrm{cm}} + \frac{\left(2\frac{\mathrm{m}}{\mathrm{s}}\right)^2}{5\,\mathrm{cm}}\right) f,$$

Solution is: $2.020\,2\times10^{-2}$

Now we can use Solve Numeric and the Colebrook Equation to estimate the relative roughness of the pipe.

$$\frac{1}{\sqrt{0.020\,202}} = -2\log_{10}\left(\frac{\delta}{3.70} + \frac{2.51}{3\times10^5\sqrt{0.020\,202}}\right),$$

Solution is: $\{[\delta = 9.052\,2\times10^{-4}]\}$

This pipe has $\delta \approx 10^{-3}$ so its walls are moderately smooth. ◀

Example 10.14 *Not exactly a Roman aqueduct*
Water flows into a pipe 950 m long at a point that's 16 m higher and has 6 kPa more pressure than the discharge point. What size pipe is needed for a flow rate of 50 000 gal/ h? Use $f \approx \frac{1}{40}$ for the friction factor.
Solution. Let's use Evaluate Numerically to convert the flow rate into the standard metric unit.

$$5 \times 10^4 \frac{\text{gal}}{\text{h}} = 5.2575 \times 10^{-2} \frac{\text{m}^3}{\text{s}}$$

Let's use the Bernoulli–Poiseuille Equation with Eq. (10.36), the definition of volume flow rate, and Solve Exact to find the pipe's radius and the water's speed. The pipe has a constant radius so the water flows at a constant speed and the kinetic energy terms cancel.

$$\begin{bmatrix} 6000 + (1000)(9.8067)(16) - \left(\frac{1}{40}\right)\frac{950}{R}\frac{1}{4}(1000)v^2 = 0 \\ 5.2575 \times 10^{-2} = \pi R^2 v \end{bmatrix},$$

Solution is: $[R = 0.10041, v = 1.6598]$

Water flowing through a 20.082 cm-diameter pipe at 1.6598 m/ s will produce the desired flow rate. This flow is turbulent.

$$\text{Re} = \frac{\rho v d}{\mu} = \frac{\left(1000\frac{\text{kg}}{\text{m}^3}\right)\left(1.6598\frac{\text{m}}{\text{s}}\right)(20.082\,\text{cm})}{1 \times 10^{-3}\frac{\text{N s}}{\text{m}^2}} = 3.3332 \times 10^5$$

It does not require particularly smooth pipe walls.

$$\frac{1}{\sqrt{\frac{1}{40}}} = -2\log_{10}\left(\frac{\delta}{3.70} + \frac{2.51}{3.3332 \times 10^5 \sqrt{\frac{1}{40}}}\right),$$

Solution is: $\{[\delta = 2.3702 \times 10^{-3}]\}$

Let's see what Bernoulli's Equation says about the required pipe size for water entering the pipe under the same conditions.

$$\begin{bmatrix} 6000 + \frac{1}{2}(1000)(1.6598)^2 + (1000)(9.8067)(16) = \frac{1}{2}(1000)v^2 \\ 5.2575 \times 10^{-2} = \pi r^2 v \end{bmatrix},$$

Solution is: $[r = 3.0385 \times 10^{-2}, v = 18.126]$

This pipe radius at the discharge end has to be 3.3 times smaller, causing the water to move $3.3^2 \approx 10.9$ times faster. ◀

Stokes' Law

Viscosity is the internal friction that impedes a fluid's motion. It also affects the motion of objects moving through the fluid. **Stokes' Law** gives the drag force on a small, slow-moving sphere in a viscous fluid. The drag force on a sphere moving at velocity v is

$$F_d = 6\pi\mu R v \tag{10.40}$$

where R is the sphere's radius and μ is the fluid's viscosity.

According to Newton's 2nd Law, the net force on a sphere (mass m) descending through the fluid depends on the buoyant force F_B of Eq. (10.5).

$$ma = mg - F_B - F_d$$

When the sphere is completely submerged, the buoyant force on it is

$$F_B = \tfrac{4}{3}\pi R^3 \rho_f\, g$$

and the sphere's mass is $m = \frac{4}{3}\pi R^3 \rho_s$.

When the sphere reaches its terminal velocity, the net force on it is zero. Let's put this all together and use Solve Exact and Factor in-place to find an expression for the sphere's terminal velocity.

$$0 = \tfrac{4}{3}\pi R^3 \rho_s g - \tfrac{4}{3}\pi R^3 \rho_f g - 6\pi\mu R v_{\mathrm{T}}, \text{ Solution is: } -\tfrac{2}{9}R^2 g\frac{\rho_f - \rho_s}{\mu}$$

The sphere's terminal velocity is

$$v_{\mathrm{T}} = \tfrac{2}{9}\frac{\rho_s - \rho_f}{\mu}R^2 g \tag{10.41}$$

where ρ_f is the fluid density and ρ_s is the density of the sphere. The terminal velocity is proportional to the density difference and inversely proportional to the fluid's viscosity.

Example 10.15 *A lot of clean*
A commercial from ancient times shows a pearl falling through shampoo at about $1\,\mathrm{cm/s}$. Estimate the viscosity of the shampoo.
Solution. A typical pearl is a sphere approximately $8\,\mathrm{mm}$ in diameter with a mass of $0.6\,\mathrm{g}$. Let's use Evaluate Numerically to calculate its density.

$$\rho_s = \frac{0.6\,\mathrm{g}}{\frac{4}{3}\pi\,(4\,\mathrm{mm})^3} = 2238.1\frac{\mathrm{kg}}{\mathrm{m}^3}$$

A typical shampoo has a density $\rho_f = 1020\,\mathrm{kg/m^3}$ that's about 2% larger than water. Now we can use Solve Exact and Eq. (10.41) to find the shampoo's viscosity.

$$1\frac{\mathrm{cm}}{\mathrm{s}} = \tfrac{2}{9}\frac{\left(2238.1\frac{\mathrm{kg}}{\mathrm{m}^3} - 1020\frac{\mathrm{kg}}{\mathrm{m}^3}\right)}{\mu}(3.75\,\mathrm{mm})^2\left(9.8067\frac{\mathrm{m}}{\mathrm{s}^2}\right),$$

$$\text{Solution is: } \frac{3.7330}{\mathrm{m}^{1.0}\,\mathrm{s}^{1.0}}\,\mathrm{kg}$$

According to our estimate, the shampoo has about the same viscosity $\mu \approx 3.7\,\mathrm{N\,s/m^2}$ as maple syrup. ◀

Problems

1. Use the Standard Method to convert $1\,\mathrm{g/cm^3}$ into $\mathrm{kg/m^3}$. Use Evaluate Numerically to verify your answer.

2. Prove that one megapascal (MPa) equals a $\mathrm{N/mm^2}$.

3. Olympic swimming pools are $25\,\mathrm{m}$ wide, $50\,\mathrm{m}$ long, and $2\,\mathrm{m}$ deep. How much water do they hold? Give your answer in cubic meters and gallons.

4. How much water does a typical backyard pool with a $20\,\mathrm{ft}$ diameter and a $4\,\mathrm{ft}$ depth hold? Give your answer in cubic meters and gallons.

5. What are the pressure and downward force on the bottom of a filled Olympic pool?

6. Some above-ground backyard pools have a $24\,\mathrm{ft}$ diameter and a $54\,\mathrm{in}$ depth. How long does it take to fill these pools with a typical garden hose?

7. Home fire-fighting equipment can have volume flow rates of $50\,\mathrm{gal/min}$. Using this equipment, how long would it take you to fill

   a. a typical above-ground backyard pool, and

   b. an Olympic-sized pool?

8. Fire trucks can have volume flow rates of $300\,\mathrm{gal/min}$. Using a fire truck, how long would it take you to fill

   a. a typical above-ground backyard pool, and

   b. an Olympic-sized pool?

9. Satellites experience a pressure of 2 million atmospheres when they're 12 000 miles deep in Jupiter's atmosphere. Assuming it's constant, what is the density of Jupiter's atmosphere at that depth?

10. How deep in the ocean is the pressure 100 atmospheres?

11. The bathyscaphe Trieste set a world record in 1960 by diving $35\,810\,\mathrm{ft}$ to the bottom of the Mariana Trench in the Pacific Ocean. What is the pressure at that depth? Give your answer in atmospheres.

12. The current record for the deepest free dive by a human is $209.6\,\mathrm{m}$. What is the pressure at that depth?

13. What fraction of a piece of wood is submerged when it floats in water?

14. What fraction of a piece of aluminum is submerged when it floats in bromine?

15. What fraction of a piece of lead is submerged when it floats in mercury?

16. How much lift is generated by $1\,\mathrm{m^3}$ of helium?

17. This one has nothing much to do with physics, it's just cool. Suppose you have a cone, a sphere, and a cylinder with the same radius. Show that if they also have the same height, the ratio of their volumes is 1:2:3.

18. Suppose a cylinder of radius R and height H floats in a fluid with a flat side down (like a hockey puck). Show that it sinks to a depth $h = sH$ where s is the specific gravity of the puck's material.

19. Suppose a cylinder of radius R and height H floats in a fluid with its curved surface down (like a log). Show that it sinks to a depth h such that

$$\frac{h-R}{R^2}\sqrt{h\,(2R-h)} + \tan^{-1}\frac{h-R}{\sqrt{h\,(2R-h)}} = \left(s - \tfrac{1}{2}\right)\pi$$

where s is the specific gravity of the log's wood. Find the depth numerically for a $1.2\,\mathrm{m}$-diameter log when $s = \frac{1}{3}, \frac{1}{2}$, and $\frac{2}{3}$.

20. A baseball has a mass of $0.142\,\mathrm{kg}$ and radius $3.68\,\mathrm{cm}$. Assuming the baseball is a uniform sphere, does it float in water? If so, how deep does it sink?

21. A golf ball has mass $46\,\mathrm{g}$ and radius $2.15\,\mathrm{cm}$. Assuming the golf ball is a uniform sphere, does it float in water? Does it float in bromine? If so, how deep does it sink?

22. A bowling ball weighs $16\,\mathrm{lbf}$ and has a $4.25\,\mathrm{in}$ radius. Assuming the ball is a uniform sphere, does it float in water? Does it float in bromine? If so, how deep does it sink?

23. A shot put weighs $16\,\mathrm{lbf}$ and has a $6\,\mathrm{cm}$ radius. Assuming the shot put is a uniform sphere, does it float in bromine? In mercury? If so, how deep does it sink?

24. Show that a sphere will be exactly half-submerged if it is made of a material whose density is half that of the fluid.

25. Show that a sphere floating in a fluid will be completely submerged when it has the same density the fluid.

26. Check my expression for the submerged volume of a sphere by evaluating it with the depths of a half-submerged and fully-submerged sphere.

27. Calculate the volume and mass flow rates when water flows through a $12\,\mathrm{cm}$-diameter pipe at $3.5\,\mathrm{m/\,s}$.

28. What fluid velocity produces a flow of 1 million gallons of water per day through a pipe with a $5\,\mathrm{cm}$ radius?

29. What size pipe can carry $2500\,\mathrm{kg/\,s}$ of water at $4.95\,\mathrm{m/\,s}$?

30. Water flows through a horizontal $15\,\mathrm{cm}$-diameter pipe under a pressure of $5\,\mathrm{atm}$. The pipe narrows to a $5\,\mathrm{cm}$ radius and the pressure drops to $2\,\mathrm{atm}$. What is the volume flow rate?

31. The flow rate through a 15 cm water pipe is $0.1\,\mathrm{m}^3/\,\mathrm{s}$. The pipe branches into two pipes, one 5 cm in diameter and the other 10 cm in diameter. The water's velocity in the smaller branch is $13\,\mathrm{m}/\,\mathrm{s}$.

    a. What is the water's velocity before the branch?

    b. What is the water's velocity in the larger branch?

32. Show that the friction factor for laminar flow in a pipe is $f = \dfrac{32\mu}{\rho R v_{\text{ave}}}$.

33. Consider fluid flow in a channel between two large horizontal plates, one at $z = 0$ and one at $z = z_0$. The velocity profile for the laminar flow of a viscous fluid between these plates is

$$v(r) = \frac{P}{2\mu L}(z_0 - z)\,z$$

    where P is the pressure loss and L is the length of the channel. What are the fluid's maximum and average speeds? Calculate the volume flow rate and the pressure drop.

34. Show that the full expression for the volume flow rate for laminar flow in a pipe is just a special case of the Bernoulli–Poiseuille Equation.

35. Show that Poiseuille's Equation can also be written as

$$\Delta P_{\text{vis}} = 8\mu \frac{L}{\pi R^4} Q$$

    where Q is the volume flow rate.

36. Repeat Example 10.11 using glycerin as the viscous liquid. Glycerin has a density of $1260\,\mathrm{kg}/\,\mathrm{m}^3$ and a viscosity of $1.42\,\mathrm{N\,s}/\,\mathrm{m}^2$. How did the results change?

37. Repeat Example 10.12 using cortisone as the medicine. Cortisone's viscosity is $0.0012\,\mathrm{N\,s}/\,\mathrm{m}^2$ and its density is $790\,\mathrm{kg}/\,\mathrm{m}^3$.

38. A cylinder with a 40 cm diameter contains 34.4 cm of water that flows through a 30 cm-long tube with a 1 mm diameter that extends horizontally from its base. How much time does it take for the cylinder to empty? Give your answer in hours, with and without viscosity. Assume the flow is laminar, and verify that it is, but do not assume the fluid speed at the top of the cylinder is zero.

39. Repeat the previous problem with the tube extending downward from the center of the cylinder's base.

40. Show that the average velocity for turbulent flow with a power-law profile is

$$\overline{v} = \frac{2}{(n+2)(n+1)} v_{\max}\,.$$

41. Show that the pressure loss due to viscosity in pipes of varying radius can be written

$$\Delta P_{\text{vis}} = \rho \frac{Q^2}{4\pi^2} \int_{s_1}^{s_2} \frac{f}{r^5}\,ds$$

    for both laminar and turbulent flows.

42. Use the result from the previous problem to verify the result in Example 10.13 for the average friction factor.

43. Do a little research and find a Moody diagram online or in a textbook. Use it to check the results of Example 10.13 for the friction factor and relative roughness.

44. Repeat Example 10.14 for gasoline, which has a density of $719\,\mathrm{kg/m^3}$ and a viscosity of $2.92 \times 10^{-4}\,\mathrm{N\,s/m^2}$. Use 0.0235 for the friction factor.

45. Repeat Example 10.14 for a flow rate of 9 million gallons per hour. This is one modern estimate for the flow rate of the Roman aqueducts.

46. Show that for slow turbulent flow along a pipe with very smooth walls, the Colebrook Equation reduces to

$$\frac{1}{\sqrt{f}} = 2\log_{10}\left(\frac{\mathrm{Re}\sqrt{f}}{2.51}\right).$$

47. What is the terminal velocity of a golf ball in water?

48. A human skydiver will reach a terminal velocity of $120\,\mathrm{mi/h}$ before opening his parachute. Use the information in Example 10.1 to treat the skydiver as an appropriately sized sphere. Calculate the terminal velocity as predicted by Stokes' Law. Is the skydiver a slow-moving sphere?

You may find this table useful...

Material	Density ($\mathrm{kg/m^3}$)	Material	Density ($\mathrm{kg/m^3}$)
Air	1.292	Mercury	13 530
Aluminum	2700	Water (Pure)	1000
Bromine	3103	Water (Ocean)	1025
Helium	0.178 6	Water (Dead Sea)	1240
Lead	11 340	Wood	670

Table 10.1 Density

One more thing...

If you continue your study of fluids, you'll find that engineers and other fluid mechanics often speak in terms of "head" instead of pressure. This is an example of physicists and engineers saying the same thing with different jargon. Head has units of meters or feet, but it's just pressure divided by density-times-gravity.

$$H = \frac{P}{\rho g} \sim \frac{\mathrm{Pa}}{\frac{\mathrm{kg}}{\mathrm{m^3}}\frac{\mathrm{m}}{\mathrm{s^2}}} = \mathrm{m}$$

Physically, a system with head H is under pressure $P = \rho g H$ that can support a column of fluid with height H and density ρ on a planet with surface gravity g.

11 Temperature and Heat

In our study of Mechanics, the calculations involved just three physical quantities: mass, length, and time. Now we enter the world of Thermodynamics and the study of processes involving heat flow. We'll need a fourth physical quantity: **temperature**.

You already understand temperature. You know what "hot", "cold", and "just right" mean. You also understand **heat**. You know that heat flows from a hot stove to cold water. Heat and temperature are related – the water's temperature changes when we add heat to it – but heat and temperature are different. One is a property of an object and the other is a process that happens to an object.

Heat is the energy transferred spontaneously between objects due to a temperature difference. This energy flow is a process so objects do not "have" heat. When two objects with different temperatures are in thermal contact, heat flows from one to the other. The higher-temperature object loses energy and its temperature decreases. The lower-temperature object gains energy and its temperature increases.

This process stops when the objects reach **thermal equilibrium**, a fancy way of saying they have the same temperature. The heat flows until the objects have the same temperature. There is no heat flow between objects in thermal contact at the same temperature.

How do you take a baby's temperature? You put the thermometer in and wait. Wait for what? Thermal equilibrium. You wait until the baby and the thermometer have the same temperature. You wait until the heat flow stops. The only thing a thermometer ever tells you is its *own* temperature!

Temperature Scales

There are at least four temperature scales we can use to quantify temperature. We'll start with the two common scales, Celsius (°C) and Fahrenheit (°F). To access *SNB*'s built-in temperature units, click the Unit Name button and look on the Physical Quantity list under Temperature.

The Celsius scale is based on water's freezing and boiling points. Water freezes at 0 °C and water boils at 100 °C, so there are 100 Celsius degrees between the two. These choices are quite arbitrary. There is nothing fundamental about those particular numerical values or using water as the standard.

The Fahrenheit scale is based on Fahrenheit, apparently a hot guy on a cold day. Water freezes at 32 °F and water boils at 212 °F, so there are 180 Fahrenheit degrees between the two. These choices are also arbitrary and different from the Celsius scale.

Doing Physics with Scientific Notebook: A Problem-solving Approach, First Edition. Joseph Gallant.

We'll need to convert between the two scales. For particular values, we can use the Solve Method. Here is a Fahrenheit-to-Celsius conversion.

$75\,°\text{F} = T_C\,°\text{C}$, Solution is: 23.889

Here is a Celsius-to-Fahrenheit conversion.

$-4.5\,°\text{C} = T_F\,°\text{F}$, Solution is: 23.9

A lovely, warm $75\,°\text{F}$ day is about $24\,°\text{C}$, and a very cold $-4.5\,°\text{C}$ day is about $24\,°\text{F}$.

It's easy to derive a general conversion formula. It's a linear relationship so let's guess $T_F = aT_C + b$ and use Solve Exact and the two values we know.

$$\begin{bmatrix} 212 = 100a + b \\ 32 = 0a + b \end{bmatrix}, \text{ Solution is: } \left[a = \frac{9}{5}, b = 32\right]$$

This gives us the Celsius-to-Fahrenheit conversion.

$$T_F = \frac{9}{5}\,T_C + 32 \tag{11.1}$$

Solving for T_C gives us the Fahrenheit-to-Celsius conversion.

$$T_C = \frac{5}{9}\,(T_F - 32) \tag{11.2}$$

Notice that $5/9 = 100/180$ is the ratio of the number of degrees between the two points we used. Let's use Substitute to check these conversion formulas against our previous results from the Solve Method.

$$T_C = \left[\tfrac{5}{9}\,(T_F - 32)\right]_{75} = 23.889$$
$$T_F = \left[\tfrac{9}{5}T_C + 32\right]_{-4.5} = 23.9$$

They agree! If it's $24°$ *something* out, you better ask $24°$ *what*? There is a temperature that's the same for both scales. Let's set the temperatures equal and use Solve Exact.

$T_F = \frac{5}{9}\,(T_F - 32)$, Solution is: -40

When the temperature is a bone-chilling -40, you don't need to ask -40 *what* unless you're from another planet.

Example 11.1 *Oh, X is for XO*
Denizens of an exoplanet use bromine to define their X temperature scale. They call its melting point $12°X$ and its boiling point $144°X$. Here on Earth, its melting point is $-7.20\,°\text{C}$ and its boiling point is $+58.8\,°\text{C}$. How should we dress on a $24°X$ day?
Solution. Let's apply Solve Exact to the linear relationship $T_C = aT_X + b$ and the two values we know.

$$\begin{bmatrix} +58.8 = 144a + b \\ -7.20 = 12a + b \end{bmatrix}, \text{ Solution is: } [a = 0.5, b = -13.2]$$

The $°X \rightarrow °\text{C}$ conversion is $T_C = 0.5T_X - 13.2$.

Let's solve for T_X to get the $^\circ\mathrm{C} \to {}^\circ X$ conversion.

$T_C = 0.5T_X - 13.2$, Solution is: $2.0T_C + 26.4$

Here are the $^\circ X \leftrightarrow {}^\circ\mathrm{C}$ conversion formulas.

$$\begin{aligned} T_C &= 0.5T_X - 13.2 \\ T_X &= 2.0T_C + 26.4 \end{aligned}$$

Now we can use Substitute (with Evaluate) to convert $24^\circ X$ to Celsius.

$T_C = [0.5T_X - 13.2]_{24} = -1.2$

We should dress warmly on this very cold (for humans anyway) $-1.2\,^\circ\mathrm{C}$ day. ◄

Absolute Temperature

The universe has a minimum temperature. That's not obvious without a better physical interpretation of temperature, but experiments show it.

Figure 11.1 shows volume-temperature plots for three different gases maintained at constant pressure and constant mass.

The V-T plots for the three gases have three different slopes but when extrapolated they cross the temperature axis at the same point!

Physically, this T-intercept is interpreted as the temperature for zero volume ($V = 0$).

The experimental value for the T-intercept is $-273.15\,^\circ\mathrm{C}$ (a rather chilly $-459.67\,^\circ\mathrm{F}$).

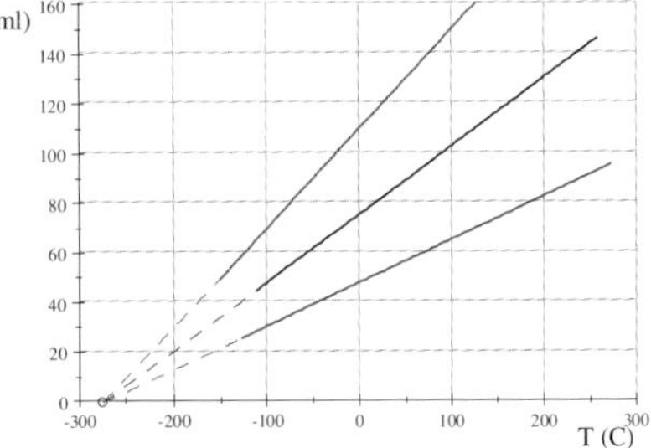

Figure 11.1 Volume vs. temperature

Experiments that measure pressure versus temperature at constant volume and mass produce the same result. The P-T graphs for different gases have different slopes but they all extrapolate to the same temperature for zero pressure as the V-T plots did for zero volume. Since the smallest possible pressure is zero and the smallest possible volume is zero, that T-intercept is the minimum temperature possible. We call this temperature **absolute zero**.

The Kelvin scale is based on these experiments. It is not arbitrary, but it is designed so its zero point is absolute zero.

$$0\,\mathrm{K} = -273.15\,^\circ\mathrm{C} = -459.67\,^\circ\mathrm{F} \tag{11.3}$$

The **kelvin** (K) is the official metric unit of temperature. Notice there is no "degree" with the kelvin. You read $3\,\mathrm{K}$ as "three kelvins".

Here is the Celsius-to-Kelvin conversion.

$$T = T_C + 273.15 \tag{11.4}$$

Here is the Kelvin-to-Celsius conversion.

$$T_C = T - 273.15 \tag{11.5}$$

The Kelvin scale is designed so one Celsius degree and one kelvin are exactly the same.

Table 11.1 gives some common temperatures in the three scales we will use.

	Cold	Freeze	Room	Body	Boil
Fahrenheit	$0\,°\mathrm{F}$	$32\,°\mathrm{F}$	$70\,°\mathrm{F}$	$98.6\,°\mathrm{F}$	$212\,°\mathrm{F}$
Celsius	$-18\,°\mathrm{C}$	$0\,°\mathrm{C}$	$21\,°\mathrm{C}$	$37\,°\mathrm{C}$	$100\,°\mathrm{C}$
Kelvin	$255\,\mathrm{K}$	$273\,\mathrm{K}$	$294\,\mathrm{K}$	$310\,\mathrm{K}$	$373\,\mathrm{K}$

Table 11.1

Note the difference between water's boiling and freezing temperatures. There are 100 Celsius degrees and $373\,\mathrm{K} - 273\,\mathrm{K} = 100\,\mathrm{K}$ between the two, confirming one Celsius degree equals one kelvin. There is a similar relationship between the Fahrenheit and Rankine scales. The Rankine scale is rarely used, and *SNB* doesn't even include it among the built-in temperature units.

One more thing about absolute zero: you can't get there from here. Absolute zero is an unreachable lower limit. Having a $0\,\mathrm{K}$ temperature is inconsistent with the laws of physics. Zero volume violates the Standard Model – we can't freeze subatomic particles out of existence. All molecular motion stops at absolute zero, and that violates the Uncertainty Principle – we would know the location of the stopped particles exactly.

Heat and Work

Heat is a form of energy, and there is a connection between heat and work. Heat and work are both processes that involve the transfer of energy into or out of a system. Mechanical work can produce heat. That's why you rub your hands together when they're cold. Joule (for whom the metric unit of energy is named) showed that doing work can increase temperature, which is exactly what heat does.

Before Joule discovered the connection, heat and work were measured in different units. Work was measured in foot-pounds (and eventually joules), and heat in calories or Calories. One Calorie is 1000 calories (yes really) so it's a kilocalorie ($1\,\mathrm{kcal} = 1000\,\mathrm{cal}$). The Calorie is the unit used for food energy.

Officially, it takes $1\,\mathrm{kcal}$ of heat to raise the temperature of $1\,\mathrm{kg}$ of water from $14.5\,°\mathrm{C}$ to $15.5\,°\mathrm{C}$. What Joule called the **mechanical equivalent of heat** combines the important recognition that heat is a form of energy with these conversion formulas.

$$1\,\mathrm{cal} = 4.186\,\mathrm{J} \tag{11.6a}$$

$$1\,\mathrm{kcal} = 4186\,\mathrm{J} \tag{11.6b}$$

So $1\,\mathrm{kcal}$ is approximately $4000\,\mathrm{J}$.

The American unit of heat is the British Thermal Unit (yes really). Officially, it takes $1\,\mathrm{Btu}$ of heat to raise the temperature of 1 pound of water from $39\,°\mathrm{F}$ to $40\,°\mathrm{F}$.

$$1\,\mathrm{Btu} = 0.252\,\mathrm{kcal} = 1055\,\mathrm{J} \tag{11.7}$$

So $1\,\mathrm{Btu}$ is approximately $\frac{1}{4}$ kcal.

You can find these heat units by clicking *SNB*'s Unit Name button and looking on the Physical Quantity list under Energy. We also need a symbol for heat. Traditionally textbooks use the letter Q for heat; perhaps it stands for *Q*uantity of heat.

Example 11.2 *Hey LB, turn right for coffee milk shakes*
Showing uncharacteristic restraint, your $94\,\mathrm{kg}$ physics author drinks half of a $770\,\mathrm{kcal}$ double-thick coffee milk shake. How high a hill does he need to hike to work it off?
Solution. Using the mechanical equivalent of heat, the heat released by the shake equals the work done against gravity.

$$Q = mgh$$

Let's convert the kilocalories into joules with Evaluate.

$$Q = 385\,\mathrm{kcal} \times \frac{4186.\ \mathrm{J}}{\mathrm{kcal}} = 1.6116 \times 10^6\ \mathrm{J}$$

Set up the equation and use Solve Exact to find the height.

$$1.6116 \times 10^6\ \mathrm{J} = (94\,\mathrm{kg})\left(9.8067\,\mathrm{m/\,s^2}\right) h, \text{ Solution is: } 1748.3\,\mathrm{m}$$

That hill needs to be almost $1.75\,\mathrm{km}$ high! That's over a mile! The moral of this story: it's easier to lose weight by reducing food input than increasing exercise output. ◀

Heat Flow

When heat flows into or out of an object, several things can happen. The object's temperature, state, volume, density, and color can all change. Here we consider changes in temperature and state only.

Change in Temperature: Specific Heat

The heat units are defined in terms of water. The heat needed for a specific change in water's temperature defines the calories and the Btu. That's a reasonable definition, but again it's arbitrary.

It takes $4186\,\mathrm{J}$ of heat to raise the temperature of $1\,\mathrm{kg}$ of water by $1\,^\circ\mathrm{C}$, but only $128\,\mathrm{J}$ does the same to lead. Something is different! Water and lead are different substances and they have different specific heats.

The **specific heat** is the amount of heat needed to raise the temperature of $1\,\mathrm{kg}$ of a substance by $1\,^\circ\mathrm{C}$. It's a measure of heat per unit mass per unit temperature. Its metric unit is $\mathrm{J/\,kg/\,^\circ C}$, but $\mathrm{kJ/\,kg/\,^\circ C}$ and $\mathrm{kcal/\,kg/\,^\circ C}$ are also used. Specific heat is a property of the substance, and it only depends on what the object is made of.

The amount of heat needed to change an object's temperature is

$$Q = mc\Delta T \tag{11.8}$$

where m is the object's mass, c is the specific heat of the substance the object is made of, and ΔT is the change in the object's temperature.

A positive Q indicates heat flows into the object; a negative Q means heat flows out of the object. Mass and specific heat are always positive, so the heat flow and temperature change always have the same sign. When heat flows in, ΔT and Q are both positive and the temperature increases. When heat flows out, ΔT and Q are both negative and the temperature decreases.

Example 11.3 *Why do coffee cups have handles?*
How much heat must you add to 400 ml of $0\,^\circ\mathrm{C}$ water to raise its temperature to $90\,^\circ\mathrm{C}$?
Solution. Let's use Evaluate Numerically and the definition of density to calculate the water's mass.

$$m = \rho V = \left(1000\,\mathrm{kg/\,m^3}\right)(400\,\mathrm{ml}) = 0.4\,\mathrm{kg}$$

Use Evaluate and Eq. (11.8) to calculate the amount of heat. Notice the specific heat of water is $1\,\mathrm{kcal/\,(\,kg\,^\circ C)}$.

$$Q = mc\Delta T = (0.4\,\mathrm{kg})\left(1\frac{\mathrm{kcal}}{\mathrm{kg}\,^\circ\mathrm{C}}\right)(+90\,^\circ\mathrm{C}) = 36.0\,\mathrm{kcal}$$

Let's repeat the calculation for aluminum, which has a significantly smaller specific heat.

$$Q_{\mathrm{al}} = mc_{\mathrm{al}}\Delta T = (0.4\,\mathrm{kg})\left(0.215\frac{\mathrm{kcal}}{\mathrm{kg}\,^\circ\mathrm{C}}\right)(+90\,^\circ\mathrm{C}) = 7.74\,\mathrm{kcal}$$

Water has a larger specific heat, so it needs more heat for the same temperature change. Water's specific heat is larger than most common substances. ◂

Change in State: Latent Heat

Under ordinary circumstances, there are three states of matter – solid, liquid, gas – and heat must flow to transform matter from one state to another.

The **latent heat** is the amount of heat needed to change the state of 1 kg of a substance that is at its transition temperature. It's a measure of heat per unit mass. Its metric unit is $\mathrm{J/\,kg}$, but $\mathrm{kJ/\,kg}$ and $\mathrm{kcal/\,kg}$ are also used. Latent heat is a property of the substance, and it only depends on what the object is made of.

The amount of heat needed to change an object's state is

$$Q = mL \tag{11.9}$$

where m is the object's mass and L is the latent heat of the substance the object is made of. There is no temperature change during a state transition.

To change an object's state, you first must add or remove enough heat to get it to the appropriate transition temperature. Then you must add or remove more heat. To boil a liquid, you must add heat to increase its temperature to its boiling point. Then you need to add more heat to change its state from liquid to gas. To freeze a liquid, you must remove heat to decrease its temperature to its freezing point. Then you need to remove more heat to change its state from liquid to solid.

The latent heat of fusion L_f involves the transformation between solid and liquid. Adding heat mL_f to a solid at its melting temperature melts it into a liquid. Removing heat mL_f from a liquid at that same temperature freezes it into a solid.

The latent heat of vaporization L_v involves the transformation between liquid and gas. Adding heat mL_v to a liquid at its boiling temperature boils it into a gas. Removing heat mL_v from a gas at that same temperature condenses it into a liquid.

Table 11.2 shows the specific and latent heats related to water. Water's heat of vaporization is 6.74 times larger than its heat of fusion. Vaporizing water requires breaking all bonds between liquid water molecules, but melting ice only needs some bonds between ice molecules to be broken. It takes much more heat to boil water than to melt ice.

Water	Specific	Heat		Latent	Heat	
Ice	2.108	0.5035	Ice ↔ Water	335.0	80.01	0 °C
Water	4.186	1	Water ↔ Steam	2258	539.3	100 °C
Steam	1.996	0.4767				
	$\frac{\text{kJ}}{\text{kg}\,^\circ\text{C}}$	$\frac{\text{kcal}}{\text{kg}\,^\circ\text{C}}$		$\frac{\text{kJ}}{\text{kg}}$	$\frac{\text{kcal}}{\text{kg}}$	

Table 11.2

At atmospheric pressure, water and ice can coexist only when the temperature is 0 °C. If the temperature is any lower the water freezes; any higher, the ice melts. Water and steam can coexist only when the temperature is 100 °C. If the temperature is any lower the steam condenses; any higher, the water boils.

Example 11.4 *Sometimes the iceman goeth*
How much heat must you add to 700 g of ice at −40 °C to completely melt it?
Solution. First you need to raise the ice's temperature to its melting point.

$$Q_1 = mc_i\Delta T_i = (0.7\,\text{kg})\left(0.503\,5\frac{\text{kcal}}{\text{kg}\,^\circ\text{C}}\right)(0\,^\circ\text{C} - (-40\,^\circ\text{C})) = 14.098\,\text{kcal}$$

Then you need to turn the 0 °C ice into 0 °C water.

$$Q_2 = mL_f = (0.7\,\text{kg})\left(80.01\frac{\text{kcal}}{\text{kg}}\right) = 56.007\,\text{kcal}$$

The total heat you need is the sum of the two.

$$Q = Q_1 + Q_2 = 14.098\,\text{kcal} + 56.007\,\text{kcal} = 70.105\,\text{kcal}$$

It takes about 70.1 kcal of heat to melt this ice. ◄

Calorimetry

Heat flows spontaneously between objects in thermal contact with different temperatures until they reach thermal equilibrium. Calorimetry provides a way to relate how much heat flows to the temperature differences and state changes. Calorimetry involves an old friend under new circumstances.

We here at *DPwSNB* aspire to be the best full-service operation we can. In that spirit, I present to you the solution to every calorimetry problem.

$$0 = Q_{\text{gained}} + Q_{\text{lost}} \tag{11.10}$$

This simple calorimetry equation is just a statement of energy conservation. When objects with different temperatures are in thermal contact, heat flows until they reach thermal equilibrium. Energy is conserved so the overall change in energy, which here equals the heat gained plus the heat lost, is zero.

Example 11.5 *Sometimes the iceman stayeth*
Seven-hundred grams of ice at $-40\,^\circ\mathrm{C}$ are added to $400\,\mathrm{ml}$ of hot $90\,^\circ\mathrm{C}$ water. What is the final temperature of this system? How much of the ice melted? Assume no heat is gained by the well-insulated container holding the water.
Solution. Example 11.3 tells us the water needs to lose $36.0\,\mathrm{kcal}$ to reach $0\,^\circ\mathrm{C}$. Example 11.4 tells us the ice needs to gain $70.105\,\mathrm{kcal}$ to completely melt. The water can't supply enough heat to melt all the ice so the final temperature of the system is $0\,^\circ\mathrm{C}$.

To melt any ice, we must heat it up to $0\,^\circ\mathrm{C}$ and then add more heat to change it to water. From Example 11.4 we know this ice needs to gain $14.098\,\mathrm{kcal}$ to reach $0\,^\circ\mathrm{C}$, so the ice gains heat

$$Q_{\text{gained}} = 14.098\,\mathrm{kcal} + m\left(80.01\frac{\mathrm{kcal}}{\mathrm{kg}}\right)$$

where m is the mass of the ice that melts. The heat gained by the ice is lost by the water.

$$Q_{\text{lost}} = -36.0\,\mathrm{kcal}$$

We can use Eq. (11.10) and Solve Exact to find the mass of the melted ice.

$$0 = 14.098\,\mathrm{kcal} + m\left(80.01\frac{\mathrm{kcal}}{\mathrm{kg}}\right) - 36.0\,\mathrm{kcal}, \text{ Solution is: } 0.273\,74\,\mathrm{kg}$$

About 40% of the ice melts. ◄

Example 11.6 *Sometimes the steamman cometh too*
Seven-hundred grams of $-40\,^\circ\mathrm{C}$ ice and 167.23 liters of $110\,^\circ\mathrm{C}$ steam are added to an insulated container holding $400\,\mathrm{ml}$ of $90\,^\circ\mathrm{C}$ water. What is the final temperature of this system? Assume no heat is gained by the container. The density of steam under these conditions is $0.598\,\mathrm{kg/\,m^3}$.
Solution. We need to figure out if the ice and steam completely transform into water. Mathematically, it doesn't matter if the $90\,^\circ\mathrm{C}$ water gains or loses heat – Eq. (11.10) will be unaffected – but we'd like to know physically.

Let's look at what happens if we mix just the $-40\,^\circ\mathrm{C}$ ice with the $110\,^\circ\mathrm{C}$ steam. We'll need the steam's mass, so use the definition of density and Evaluate Numerically.

$$m_s = V_s\rho_s = (167.231)\left(0.598\,\mathrm{kg/\,m^3}\right) = 0.1\,\mathrm{kg}$$

The ice must gain $70.105\,\mathrm{kcal}$ of heat to melt into $0\,^\circ\mathrm{C}$ water.

$$Q_{\text{ice}} = (0.7\,\mathrm{kg})\left(0.503\,5\frac{\mathrm{kcal}}{\mathrm{kg}\,^\circ\mathrm{C}}\right)(+40\,^\circ\mathrm{C}) + (0.7\,\mathrm{kg})\left(80.01\frac{\mathrm{kcal}}{\mathrm{kg}}\right) = 70.105\,\mathrm{kcal}$$

The steam must lose $54.407\,\mathrm{kcal}$ of heat to condense into $100\,^\circ\mathrm{C}$ water.

$$Q_{\text{steam}} = (0.1\,\mathrm{kg})\left(0.4767\frac{\mathrm{kcal}}{\mathrm{kg}\,^\circ\mathrm{C}}\right)(-10\,^\circ\mathrm{C}) - (0.1\,\mathrm{kg})\left(539.3\frac{\mathrm{kcal}}{\mathrm{kg}}\right) = -54.407\,\mathrm{kcal}$$

It must lose an additional $10.0\,\mathrm{kcal}$ of heat to cool to $0\,^\circ\mathrm{C}$ water.

$$Q_{\text{cool}} = (0.1\,\mathrm{kg})\left(1\frac{\mathrm{kcal}}{\mathrm{kg}\,^\circ\mathrm{C}}\right)(-100\,^\circ\mathrm{C}) = -10.0\,\mathrm{kcal}$$

A total heat loss of $64.407\,\mathrm{kcal}$ turns the $110\,^\circ\mathrm{C}$ steam into $0\,^\circ\mathrm{C}$ water.

The steam can't supply enough heat to melt all the ice so if we mixed just the ice and steam, most but not all of the ice would melt.

$$0 = 14.098\,\text{kcal} + m\left(80.01\frac{\text{kcal}}{\text{kg}}\right) - 64.407\,\text{kcal}, \text{ Solution is: } 0.628\,78\,\text{kg}$$

We'd end up with $0.7\,\text{kg} - 0.628\,78\,\text{kg} = 0.071\,22\,\text{kg}$ of ice in $0\,°\text{C}$ water.

Now let's include the $90\,°\text{C}$ water. It provides more than enough heat to melt the remaining ice (see Example 11.5), so all the ice melts, the $90\,°\text{C}$ water loses heat, and the final temperature is less than $90\,°\text{C}$.

The ice gains $70.105\,\text{kcal}$ of heat to become $0\,°\text{C}$ water, and then it gains more heat to reach the final temperature T.

$$Q_{\text{gained}} = 70.105\,\text{kcal} + (0.7\,\text{kg})\left(1\frac{\text{kcal}}{\text{kg}\,°\text{C}}\right)(T - 0\,°\text{C})$$

The steam loses $54.407\,\text{kcal}$ of heat to become $100\,°\text{C}$ water, then it and the $90\,°\text{C}$ water lose more heat to reach the final temperature.

$$\begin{aligned} Q_{\text{lost}} &= -54.407\,\text{kcal} + (0.1\,\text{kg})\left(1\frac{\text{kcal}}{\text{kg}\,°\text{C}}\right)(T - 100\,°\text{C}) \\ &\quad + (0.4\,\text{kg})\left(1\frac{\text{kcal}}{\text{kg}\,°\text{C}}\right)(T - 90\,°\text{C}) \end{aligned}$$

The final temperature is less than $90\,°\text{C}$ so the two ΔT terms are negative.

Use Eq. (11.10) to create the calorimetry equation and use Solve Exact to find the final temperature.

$$0 = 70.105\,\text{kcal} + (0.7\,\text{kg})\left(1\frac{\text{kcal}}{\text{kg}\,°\text{C}}\right)(T - 0\,°\text{C}) + (0.4\,\text{kg})\left(1\frac{\text{kcal}}{\text{kg}\,°\text{C}}\right)(T - 90\,°\text{C})$$
$$- 54.407\,\text{kcal} + (0.1\,\text{kg})\left(1\frac{\text{kcal}}{\text{kg}\,°\text{C}}\right)(T - 100\,°\text{C}), \text{ Solution is: } 25.252\,\text{tmpK}$$

This shows a bug in *SNB*. Solve Exact changes units to the metric system. It changed the $°\text{C}$ to tmpK, but it did not convert the numerical value. The correct answer is $25.252\,°\text{C}$. *SNB* has two versions of the kelvin, the K you'll find in the built-in units and this tmpK which shows up when you use Solve Exact or Evaluate Numerically.

$$T = 25.252\,°\text{C} = 298.4\,\text{tmpK}$$

It's best to avoid this issue by writing the heat flows carefully without units. Use Evaluate in-place on the temperature-dependent terms to avoid a small rounding error.

$$Q_{\text{gained}} = 70.105 + (0.7)(1)(T - 0) = 0.7T + 70.105$$
$$Q_{\text{lost}} = -54.407 + (0.1)(1)(T - 100) + (0.4)(1)(T - 90) = -54.407 + 0.5T - 46.0$$

Then apply Solve Exact to the calorimetry Eq. (11.10).

$$0 = 0.7T + 70.105 - 54.407 + 0.5T - 46.0, \text{ Solution is: } 25.252$$

We could've started with the calorimetry equation. If the mathematical solution has $T < 0$ then all the ice didn't melt and $T = 0\,°\text{C}$ is the physical final temperature. If the solution has $T > 100$ then all the steam didn't condense and $T = 100\,°\text{C}$. ◄

Varying Specific Heat

In our previous discussions, we treated the specific and latent heats as constant properties of matter. The specific heat of most substances varies with temperature. Over most common temperature ranges, the changes are often small. Many tables list values at one particular temperature and pressure, or give averages over a small range of conditions.

When the specific heat is a function of temperature, we can generalize Eq. (11.8) so the heat flow needed to change an object's temperature is

$$Q = m\int_{T_1}^{T_2} c(T)\ dT \tag{11.11}$$

where $c(T)$ is the temperature-dependent specific heat. This heat flow is also

$$Q = m\overline{c}\Delta T \tag{11.12}$$

where $\overline{c}$ is the average specific heat in the temperature interval ΔT.

Figure 11.2 shows the specific heat of water as a function of temperature at a pressure of one atmosphere from $0\,°\mathrm{C}$ to $100\,°\mathrm{C}$.

Don't let the shape of the graph mislead you.

Look carefully at the vertical-axis scale.

The difference between the extreme values of water's specific heat is less than 1%.

Treating water's specific heat as a constant is a reasonable and valid approximation.

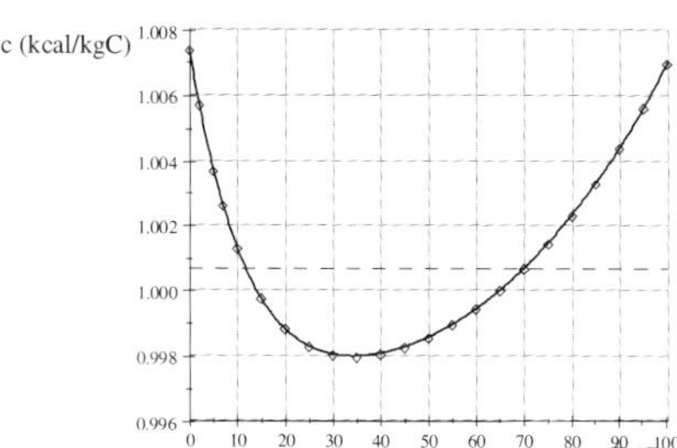

Figure 11.2 Water's c vs. T

The points on the graph are data and the curve comes from fitting a Polynomial of Degree 4 using Statistics + Fit Curve to Data.

Polynomial fit: $c = 8.1575 \times 10^{-10}T^4 - 1.9764 \times 10^{-7}T^3 + 1.8762 \times 10^{-5}T^2 - 7.1367 \times 10^{-4}T + 1.007$

That's a cumbersome expression, so let's use New Definition to define a function $c(T)$.

$c(T) = 8.1575 \times 10^{-10}T^4 - 1.9764 \times 10^{-7}T^3 + 1.8762 \times 10^{-5}T^2 - 7.1367 \times 10^{-4}T + 1.007$

Let's use the fit and Evaluate Numerically to check the definition of a kilocalorie.

$$\overline{c} = \frac{1}{15.5 - 14.5}\int_{14.5}^{15.5} c(T)\,dT = 0.99989$$

The answer is only 0.011% off the exact answer of $1\,\mathrm{kcal}$. Let's calculate the heat needed to raise the temperature of $1\,\mathrm{kg}$ of water from $0\,°\mathrm{C}$ to $100\,°\mathrm{C}$.

$$Q = (1)\int_0^{100} c(T)\,dT = 100.08$$

The answer is only 0.08% off the exact answer of $100\,\mathrm{kcal}$. The fit does a good job!

The Specific Heat of Solids

The specific heat of solids are functions of temperature, but in 1819 Dulong and Petit discovered experimentally that many solids have approximately the same specific heat. In modern lingo, the Dulong–Petit law says the specific heat per mole of any solid *at room temperature* is

$$C_V \approx 3R = 24.944 \text{ J}/(\text{mol K}) \tag{11.13}$$

where $R = 8.3145\,\text{J}/\text{mol}/\text{K}$ is the universal gas constant. A gas constant that shows up in a discussion about solids truly is universal.

Dulong and Petit discovered the classical limit of the specific heat of solids for temperatures where the energy due to the atomic oscillations is the dominant effect. The atoms in a solid are not free to translate or rotate; they can only oscillate. Each atom has energy $3kT$ where k is Boltzmann's constant, so the solid has $3RT$ of energy per mole. The specific heat relates changes in energy and temperature at constant volume.

$$C_V = \left(\frac{\partial E}{\partial T}\right)_V = \frac{\partial\,(3RT)}{\partial T} = 3R$$

The Dulong–Petit law fails at low temperatures where a solid's specific heat is not constant due to quantum effects. It provides good predictions for some simple crystals at high temperatures but it fails at moderate temperature for others with light, tightly-bound atoms. We need a better model.

In what follows, each model has a characteristic temperature that defines high and low. As a rule of thumb, "low" typically refers to temperatures between absolute zero and $50\,\text{K}$ or so, "moderate" means room temperature plus-or-minus a hundred kelvins or two, and "high" is $500\,\text{K}$ and above. Of course, your mileage may vary...

The Einstein Model

Much of Albert Einstein's early work was in thermodynamics, including his doctoral dissertation on molecular dimensions and a paper on Brownian motion (both published in his miracle-year). In 1907, he published a paper proposing a model based on Planck's quantum hypothesis which gave an improved explanation for the specific heat of solids.

Einstein's solid is a 3-dimensional lattice of evenly-spaced, independent atoms. Einstein treated the atoms as identical 3-d quantum harmonic oscillators that vibrate independently at the same frequency. Imagine each atom is at the center of 6 identical perpendicular springs, one for each direction, in a kind of microscopic bedspring.

The energy gap between states is constant ϵ (those of you with quantum experience know $\epsilon = hf$ where h is Planck's constant and f is the oscillator's frequency). At the Einstein temperature T_E, each atom's thermal energy kT_E equals this gap.

$$T_E = \frac{\epsilon}{k} \tag{11.14}$$

For temperatures near and above T_E, the Dulong–Petit law gives a good approximation. But for temperatures less than T_E, the atoms don't have enough energy to occupy excited states so they stay in the ground state. The solid can't absorb much energy and the specific heat is significantly less than the classical limit.

Using Planck's result for the average energy of a quantized oscillator, we can calculate the energy per mole of all N atoms in the solid.

$$E = 3N\frac{\epsilon}{e^{T_E/T}-1} \tag{11.15a}$$

$$= \frac{3RT_E}{e^{T_E/T}-1} \tag{11.15b}$$

The "extra" factor of 3 is there because Einstein used 3 oscillators to model each atom. Now we can Evaluate the expression for the heat capacity per mole.

$$C_V = \left(\frac{\partial E}{\partial T}\right)_V = 3RT_E\frac{\partial}{\partial T}\left(e^{T_E/T}-1\right)^{-1} = 3\frac{R}{T^2}T_E^2\frac{e^{\frac{1}{T}T_E}}{\left(e^{\frac{1}{T}T_E}-1\right)^2}$$

A little editing by-hand tells us the heat capacity per mole of an Einstein solid.

$$C_V = 3R\left(\frac{T_E}{T}\right)^2\frac{e^{T_E/T}}{\left(e^{T_E/T}-1\right)^2} \tag{11.16}$$

Each substance has its own Einstein temperature whose value is picked so the model fits the experimental data, and typically it's around 300 K.

Let's look at the high- and low-temperature behavior of Einstein's result. In Einstein's model, "high temperature" means $T \gg T_E$ so T_E/T is very small and we can expand the exponentials. *SNB* won't expand in terms of a variable in the denominator of an exponent, so let's do a 3-term Power Series and Expand in Powers of T_E.

$$C_V \to 3R\left(\frac{T_E}{T}\right)^2\frac{e^{T_E/T}}{\left(e^{T_E/T}-1\right)^2} = 3R - \tfrac{1}{4}\frac{R}{T^2}T_E^2 + O\left(T_E^3\right)$$

In Einstein's model, "low temperature" means $T \ll T_E$ so T_E/T is very large and the exponential in the denominator is much larger than 1. Let's impose that condition by-hand and Simplify the resulting expression.

$$C_V \to 3R\left(\frac{T_E}{T}\right)^2\frac{e^{T_E/T}}{\left(e^{T_E/T}\right)^2} = 3\frac{R}{T^2}T_E^2\,e^{-\frac{T_E}{T}}$$

A little editing by-hand gives us the two limiting expressions for Einstein's result.

$$C_V \approx 3R\begin{cases}\dfrac{T_E^2}{T^2}e^{-T_E/T} & \text{low-}T\\[2ex] 1-\tfrac{1}{12}\dfrac{T_E^2}{T^2} & \text{high-}T\end{cases} \tag{11.17}$$

Near absolute zero, Einstein predicts the specific heat of a solid approaches zero exponentially. For high temperatures, Einstein agrees with the Dulong–Petit law. In between, this model works well at moderate temperatures where the vibrational energy of the atoms is the dominant contributor to the specific heat.

Example 11.7 *How low is low?*
Calculate the specific heat of an Einstein solid at $T = 2T_E$ and $T = \frac{1}{10}T_E$. Compare each result with the appropriate limiting expression.
Solution. We can use Substitute (with Evaluate Numerically) on Einstein's result for the specific heat of a solid at various temperatures. Let's start with the specific heat at the Einstein temperature.

$$\frac{C_V}{3R} = \left[\left(\frac{T_E}{T}\right)^2 \frac{e^{T_E/T}}{\left(e^{T_E/T} - 1\right)^2}\right]_{T=T_E} = 0.920\,67$$

At $T = T_E$ the Einstein solid is within 8% of the classical limit. Now let's look at $T = 2T_E$ using Eqs. (11.16) and (11.17).

$$\frac{C_V}{3R} = \left[\left(\frac{T_E}{T}\right)^2 \frac{e^{T_E/T}}{\left(e^{T_E/T} - 1\right)^2}\right]_{T=2T_E} = 0.979\,42$$

$$\frac{C_V}{3R} \approx \left[1 - \frac{1}{12}\frac{T_E^2}{T^2}\right]_{T=2T_E} = 0.979\,17$$

At $T = 2T_E$ the high-temperature approximation is valid and the Einstein solid is within 2.1% of the classical limit. Let's look at $T = \frac{1}{10}T_E$ next.

$$\frac{C_V}{3R} = \left[\left(\frac{T_E}{T}\right)^2 \frac{e^{T_E/T}}{\left(e^{T_E/T} - 1\right)^2}\right]_{T=T_E/10} = 4.540\,4 \times 10^{-3}$$

$$\frac{C_V}{3R} \approx \left[\frac{T_E^2}{T^2} e^{-T_E/T}\right]_{T=T_E/10} = 4.540\,0 \times 10^{-3}$$

At $T = \frac{1}{10}T_E$ the low-temperature approximation is valid. ◀

The Debye Model

The Einstein model gives accurate predictions at moderate and high temperatures, but its low-temperature prediction that $C_V \to 0$ exponentially does not agree with experiments that show $C_V \sim T^3$ near absolute zero. The problem was Einstein's assumption that the atoms vibrate independently at the same frequency.

In 1912, Peter Debye published a better but more complicated model. Debye realized that sound waves couldn't propagate through a solid without some interaction between atoms. He modeled the interaction between the vibrating atoms with sound waves that he quantized in the same way Planck quantized light waves.

The interacting atoms vibrate over a range of frequencies with wavelengths $\lambda_n = 2L/n$ where n is a positive integer that counts the allowed oscillations. The wavelength limits the allowed oscillations: the longest possible wavelength is twice the solid's length (so $\lambda = 2L$ when $n = 1$) and the shortest is twice the spacing between atoms (for the maximum n we don't yet know).

The energy gap between states is smaller for low-frequency oscillations.

$$\epsilon_n = hf_n = \frac{hc}{\lambda_n} = \frac{hc}{2L}n \tag{11.18}$$

Here c is the speed of sound through the solid. Using Planck's result, the average energy of the atoms oscillating at frequency f_n is

$$E_n = \frac{\epsilon_n}{e^{\epsilon_n/kT} - 1} = \frac{kKn}{e^{nK/T} - 1} \tag{11.19}$$

where $K \equiv hc/(2Lk)$ is a collection of constants that has a unit of temperature.

Each allowed oscillation contributes to the solid's total energy which is the sum of this average over every value of n. A clever math trick changes a sum we can't do into an integral we can't do.

$$\sum_{\text{all } n}^{n_{\text{max}}} E_n \rightarrow \frac{\pi}{2}\int_0^{n_{\text{max}}} E_n\, n^2\, dn \tag{11.20}$$

Physically, we're now treating a solid composed of discrete closely-packed atoms as a continuous flexible blob. Mathematically, we're using part of a make-believe sphere of radius n_{max} to count the allowed oscillations. Pragmatically, this is progress because it's easy to approximate integrals and now we can determine n_{max}.

Brief physics time-out: Now you can see the differences between Debye and Planck. For electromagnetic waves c is the speed of light (way faster than the speed of sound) and $n_{\text{max}} = \infty$ since there is no lower limit on the wavelength.

The allowed oscillations have positive n, so only one octant of our sphere encloses physically meaningful values (that explains the $\pi/2$). The total number of allowed oscillations enclosed by our octi-sphere must equal the number of atoms in the solid. We have a strict one-frequency-per-atom policy.

We can use Solve Exact to find n_{max} (which we'll temporarily call n_m to avoid computational conflict with *SNB*'s built-in $\max$ function).

$N = \frac{1}{8}\frac{4}{3}\pi n_m^3$, Solution is: $\sqrt[3]{\frac{6}{\pi}N}$

Now we can calculate the total energy of all N atoms in the solid.

$$E = \frac{3\pi}{2}kK\int_0^{(6N/\pi)^{1/3}} \frac{n^3}{e^{nK/T} - 1}\, dn \tag{11.21}$$

The "extra" factor of 3 is there because the atoms can wiggle in all three dimensions.

The specific heat is the derivative of the energy with respect to temperature at constant volume. The integrand depends on temperature but the integration does not, so we can pass the derivative through the integral sign. Let's Evaluate just the derivative of the integrand.

$$C_V = \left(\frac{\partial E}{\partial T}\right)_V \sim \frac{\partial}{\partial T}\left(\frac{n^3}{e^{nK/T} - 1}\right) = \frac{K}{T^2}n^4 \frac{e^{\frac{K}{T}n}}{\left(e^{\frac{K}{T}n} - 1\right)^2}$$

Let's combine this result and the rest of Eq. (11.21) and see what we have.

$$C_V = \frac{3\pi}{2}k\frac{K^2}{T^2}\int_0^{(6N/\pi)^{1/3}} \frac{n^4\, e^{nK/T}}{\left(e^{nK/T}-1\right)^2}\,dn$$

That's another integral we can't do, but life will be simpler if we use the dimensionless variable $x \equiv nK/T$. The upper integration limit becomes $x_{\text{max}} = T_D/T$ where we've defined the Debye temperature.

$$T_D \equiv n_{\text{max}}K = \left(\tfrac{6N}{\pi}\right)^{1/3}\frac{hc}{2Lk} \tag{11.22}$$

At the Debye temperature T_D, each atom's thermal energy kT_D equals the largest energy gap (due to the maximum frequency) so all excited states are available. For temperatures near and above T_D, the Dulong–Petit law gives a good approximation. But for $T < T_D$, some low-frequency oscillators have enough energy to occupy excited states so C_V is significantly less than the classical limit but larger than Einstein's prediction.

A little editing by-hand tells us the heat capacity per mole of a Debye solid.

$$C_V = 9R\frac{T^3}{T_D^3}\int_0^{T_D/T}\frac{x^4 e^x}{\left(e^x-1\right)^2}dx \tag{11.23}$$

Each substance has its own Debye temperature which is usually picked so the model fits the experimental data.

Let's look at the high and low-temperature behavior of Debye's result. In Debye's model, "high temperature" means $T \gg T_D$ so x is very small and we can expand the integrand. Let's do a 3-term Power Series and Expand in Powers of x.

$$\frac{x^4 e^x}{\left(e^x-1\right)^2} = x^2 - \tfrac{1}{12}x^4 + O\left(x^5\right)$$

Let's use this to Expand our high-temperature approximation.

$$C_V \to 9R\frac{T^3}{T_D^3}\int_0^{T_D/T}\left(x^2 - \tfrac{1}{12}x^4\right)dx = 3R - \tfrac{3}{20}\frac{R}{T^2}T_D^2$$

In Debye's model, "low temperature" means $T \ll T_D$ so $T_D/T \to \infty$ and the integral becomes doable. Let's impose that condition by-hand and use Evaluate Numerically in-place on the integral in the resulting expression.

$$C_V \to 3R\frac{T^3}{T_D^3}\int_0^{\infty}\frac{3x^4 e^x}{\left(e^x-1\right)^2}dx = 3R\frac{T^3}{T_D^3}(77.927)$$

A little editing by-hand gives us the two limiting expressions for Debye's result.

$$C_V = 3R\begin{cases} \tfrac{12}{15}\pi^4\dfrac{T^3}{T_D^3} & \text{low-}T \\ 1 - \tfrac{1}{20}\dfrac{T_D^2}{T^2} & \text{high-}T \end{cases} \tag{11.24}$$

Near absolute zero, Debye predicts the correct low-temperature $C_V \sim T^3$ behavior. For high temperatures, Debye agrees with the Dulong–Petit law.

Example 11.8 *How low is not-so-low?*

Calculate the specific heat of a Debye solid at $T = 2T_D$ and $T = \frac{1}{10}T_D$. Compare each result with the appropriate limiting expression.

Solution. Let's use Evaluate Numerically on Debye's result for the specific heat of a solid at various temperatures. Let's start by using Eq. (11.23) to calculate the specific heat at the Debye temperature (so $T/T_D = 1$).

$$\frac{C_V}{3R} = 3\left(1^3\right)\int_0^1 \frac{x^4 e^x}{\left(e^x - 1\right)^2}dx = 0.951\,73$$

At $T = T_D$ the Debye solid is within 5% of the classical limit. Now let's look at $T = 2T_D$ using Eqs. (11.23) and (11.24). The upper limit on the integration is T_D/T.

$$\frac{C_V}{3R} = 3\left(2^3\right)\int_0^{1/2} \frac{x^4 e^x}{\left(e^x - 1\right)^2}dx = 0.987\,61$$

$$\frac{C_V}{3R} \approx \left[1 - \frac{1}{20}\frac{T_D^2}{T^2}\right]_{T=2T_D} = 0.987\,5$$

At $T = 2T_D$ the high-temperature approximation is valid and the Debye solid is within 1.3% of the classical limit. Let's look at $T = \frac{1}{10}T_D$ next.

$$\frac{C_V}{3R} = 3\left(\tfrac{1}{10}\right)^3\int_0^{10} \frac{x^4 e^x}{\left(e^x - 1\right)^2}dx = 7.582\,1 \times 10^{-2}$$

$$\frac{C_V}{3R} \approx \left[\frac{12}{15}\pi^4\frac{T^3}{T_D^3}\right]_{T=T_D/10} = 7.792\,7 \times 10^{-2}$$

At $T = \frac{1}{10}T_D$ the complete expression and low-temperature approximation differ by about 2.8%. This percent difference is much larger than we found for Einstein's model. Apparently a temperature of $\frac{1}{10}T_E$ is lower than a temperature of $\frac{1}{10}T_D$. ◀

Let's compare the two models directly. That's easy to do with *SNB*'s New Definition, which lets us define generic specific-heat functions for the two models.

$$\text{Debye: } C_D(t) = 3x^3\int_0^{1/t} \frac{y^4 e^y}{\left(e^y - 1\right)^2}dy$$

$$\text{Einstein: } C_E(t) = \frac{1}{t^2}\frac{e^{1/t}}{\left(e^{1/t} - 1\right)^2}$$

These dimensionless functions are actually $C_V/(3R)$ so they should approach 1 in the classical high-temperature limit. Both models depend only on a ratio of temperatures so the variable t is a dimensionless temperature. For the Debye model, it's $t \equiv T/T_D$ and for Einstein it's $t \equiv T/T_E$.

The Debye function contains a definite integral, but *SNB* can graph it. Figure 11.3 shows a plot of the dimensionless specific heat $C^* = C_V / (3R)$ as a function of the dimensionless temperature t for the two models.

These are the theoretical predictions of the Debye model (larger blue curve) and the Einstein model (smaller red curve).

Both approach 1 for high temperatures.

As $T \to 0$, the Einstein curve approaches zero much faster than the Debye curve.

In Debye's model, the energy gap between states is smaller for the low-frequency oscillators, so they can contribute to the specific heat at lower temperatures.

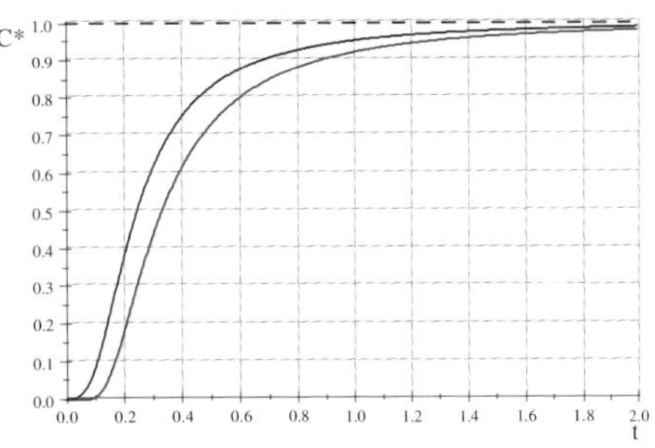

Figure 11.3 Einstein vs. Debye

Now it's time to apply the general models to a particular problem.

Example 11.9 *That's aluminium to Sarah and Zoë*

How much heat is needed to raise the temperature of a $20\,\mathrm{g}$ slug of aluminum from $200\,\mathrm{K}$ to $300\,\mathrm{K}$? The reported experimental value is $1634.6\,\mathrm{J}$.

Solution. The specific heat of aluminum is approximately $0.897\,\mathrm{kJ}/(\mathrm{kg\,K})$ at room temperature, but it varies by about 10% between $200\,\mathrm{K}$ and $300\,\mathrm{K}$. We can use the Debye and Einstein models to account for this, but we'll need the constant R in the appropriate units and the two characteristic temperatures.

Let's convert the constant R from its universal "per mole" to "per gram of aluminum" with Evaluate and aluminum's atomic weight.

$$3R = 3\left(8.3145\frac{\mathrm{J}}{\mathrm{mol\,K}}\right) \div \left(26.982\frac{\mathrm{g}}{\mathrm{mol}}\right) = 0.924\,45\frac{\mathrm{J}}{\mathrm{K\,g}}$$

Here's an interesting ratio you can Evaluate Numerically.

$$\frac{C}{3R} = \frac{0.897\,\mathrm{kJ}/(\mathrm{kg\,K})}{0.924\,45\,\mathrm{J}/(\mathrm{K\,g})} = 0.970\,31$$

Aluminum's room-temperature specific heat is within 3% of the classical limit.

Quoted values for aluminum's Debye temperature range from $240\,\mathrm{K}$ to $428\,\mathrm{K}$, with values typically near $390\,\mathrm{K}$ (which fits these data best). Let's use New Definition to define aluminum-specific specific-heat functions from the two models.

$$\text{Debye: } C_D(T,\Theta) = 3\,(0.924\,45)\left(\frac{T}{\Theta}\right)^3 \int_0^{\Theta/T} \frac{y^4 e^y}{(e^y - 1)^2}\,dy$$

$$\text{Einstein: } C_E(T,\Theta) = 0.924\,45\left(\frac{\Theta}{T}\right)^2 \frac{e^{\Theta/T}}{\left(e^{\Theta/T} - 1\right)^2}$$

These 2-variable functions let us easily adjust the two temperature parameters to fit the data. Figure 11.4 shows a graph of the results.

The points are experimental data [19] for the specific heat of aluminum and the curves are the predictions of the Debye model with $T_D = 390\,\mathrm{K}$ and the Einstein model with $T_E = 300\,\mathrm{K}$.

That's close to the value Einstein proposed in his 1907 paper ($284\,\mathrm{K}$).

The two curves are almost identical and fit the experimental data well.

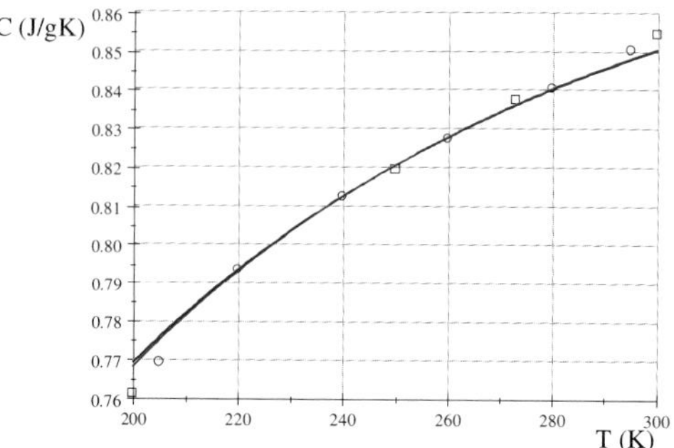

Figure 11.4 Einstein vs. Debye

These curves are small parts (up to a factor of $3R$) of the curves shown in Figure 11.3. At $T = 240\,\mathrm{K}$ the dimensionless temperature is $\frac{4}{5}$ for Einstein and about $\frac{3}{5}$ for Debye.

Let's see how the predictions of Einstein and Debye compare with the experimental value of the heat. Let's use Eq. (11.11) with our new specific-heat functions and apply Evaluate Numerically.

$$Q_D = 20\int_{200}^{300} C_D\,(T, 390)\,dT = 1635.2$$

$$Q_E = 20\int_{200}^{300} C_E\,(T, 300)\,dT = 1634.9$$

We could have taken an alternative approach and calculated the average specific heat over the given temperature range and then use Eq. (11.12) to calculate the heat. First use Evaluate Numerically to calculate the averages.

$$\overline{c}_D = \frac{1}{300-200}\int_{200}^{300} C_D\,(T, 390)\,dT = 0.817\,58$$

$$\overline{c}_E = \frac{1}{300-200}\int_{200}^{300} C_E\,(T, 300)\,dT = 0.817\,47$$

Then apply Evaluate to Eq. (11.12) to calculate the heat.

$$Q_D = m\overline{c}_D\,\Delta T = (20\,\mathrm{g})\left(0.817\,58\frac{\mathrm{J}}{\mathrm{g\,K}}\right)(+100\,\mathrm{K}) = 1635.2\,\mathrm{J}$$

$$Q_E = m\overline{c}_E\,\Delta T = (20\,\mathrm{g})\left(0.817\,47\frac{\mathrm{J}}{\mathrm{g\,K}}\right)(+100\,\mathrm{K}) = 1634.9\,\mathrm{J}$$

Both models fit the aluminum data and agree with the given heat value. This shows us that both the Einstein and Debye models work in this moderate-temperature range. ◀

Problems

1. Find the conversion formulas between the Fahrenheit and X temperature scales.

2. At what temperature do the Celsius and X scales agree?

3. At what temperature do the Fahrenheit and X scales agree?

4. Use the Solve Method to convert absolute zero into Celsius and Fahrenheit.

5. Show the Fahrenheit-to-Kelvin conversion formula is $T = \frac{5}{9}(T_F + 459.67)$.

6. Show the Rankine-to-Kelvin conversion formula is $T = \frac{5}{9}T_R$.

7. Construct the temperature conversion formulas between Celsius and the $^\circ H$ scale based on mercury where mercury freezes at $-99\,^\circ H$ and boils at $+198\,^\circ H$. Calculate water's melting and boiling temperatures in $^\circ H$. Mercury's melting point is $-39\,^\circ\mathrm{C}$ and its boiling point is $357\,^\circ\mathrm{C}$.

8. Construct the temperature conversion formulas between Celsius and the $^\circ E$ scale based on ethanol where ethanol freezes at $0\,^\circ E$ and boils at $+193\,^\circ E$. Calculate water's melting and boiling temperatures in $^\circ E$. Ethanol's melting point is $-115\,^\circ\mathrm{C}$ and its boiling point is $78\,^\circ\mathrm{C}$.

9. Construct the temperature conversion formulas between Celsius and the $^\circ M$ scale based on methanol where methanol melts at $-100\,^\circ M$ and boils at $+100\,^\circ M$. Calculate water's melting and boiling temperatures in $^\circ M$. Methanol's melting point is $-97.5\,^\circ\mathrm{C}$ and its boiling point is $65\,^\circ\mathrm{C}$.

10. Consider the generic conversion $T_B = aT_A + b$ between temperature scales T_A and T_B. Show that $T_B = T_A$ when they both equal $b/(1-a)$. What happens if $a = 1$?

11. Use the Solve Method and the Default Method to verify that $1\,\mathrm{kcal} = 4186.8\,\mathrm{J}$ according to *SNB*.

Treat all specific and latent heats as constants for the following problems.

12. Why does it take more heat to boil a pot of water to make spaghetti than it does to boil a cup a water to make coffee?

13. You pour $500\,\mathrm{ml}$ of hot $95\,^\circ\mathrm{C}$ into a $20\,^\circ\mathrm{C}$ glass beaker. The beaker's mass is $150\,\mathrm{g}$ and the specific heat of glass is $0.84\,\mathrm{kJ}/(\mathrm{kg}\,^\circ\mathrm{C})$. What is the final temperature of this system? That's why coffee cups have handles.

14. Three hundred grams of ice at $-20\,^\circ\mathrm{C}$ and $750\,\mathrm{ml}$ of hot $95\,^\circ\mathrm{C}$ water are mixed in a well-insulated container. What is the final temperature of this system?

15. Three hundred grams of ice at $-20\,^\circ\mathrm{C}$ and $750\,\mathrm{ml}$ of hot $95\,^\circ\mathrm{C}$ water are mixed in a $20\,^\circ\mathrm{C}$ glass beaker. The beaker's mass is $150\,\mathrm{g}$ and the specific heat of glass is $0.84\,\mathrm{kJ}/(\mathrm{kg}\,^\circ\mathrm{C})$. What is the final temperature of this system?

16. Eight hundred grams of $-50\,^{\circ}\mathrm{C}$ ice are added to one liter of hot $95\,^{\circ}\mathrm{C}$ water. What is the final temperature of this system? How much of the ice melted? Assume no heat is gained by the container holding the water.

17. Captain Picard decides he'd like his tea, Earl Gray, iced. Being the dutiful yeoman, you've already made one liter of his tea, Earl Gray, hot. How much $-10\,^{\circ}\mathrm{C}$ ice should you add to the $95\,^{\circ}\mathrm{C}$ hot tea to lower its temperature to $7\,^{\circ}\mathrm{C}$?

18. Sublimation bypasses the liquid phase and the solid transforms directly to a gas, but it requires the same amount of heat as melting, heating to the boiling point, and vaporizing. Show that $719.31\,\mathrm{kcal}$ of heat are needed to sublimate $1\,\mathrm{kg}$ of $0\,^{\circ}\mathrm{C}$ snow.

19. How much heat is required to change $0.25\,\mathrm{kg}$ of solid mercury at its freezing point into mercury vapor at its boiling point? You'll find a table of the specific and latent heats for mercury at the end of the chapter.

20. How much heat must be released to change 6 liters of mercury vapor at $446\,^{\circ}\mathrm{C}$ to liquid mercury at $100\,^{\circ}\mathrm{C}$? The density of mercury vapor at $446\,^{\circ}\mathrm{C}$ is about $3.45\,\mathrm{kg/\,m^3}$.

The next 5 questions follow the scenario of Example 11.6.

21. How much ice should you add to the hot water and steam so the end product is all $100\,^{\circ}\mathrm{C}$ water?

22. How much ice should you add to the hot water and steam so the end product is all $0\,^{\circ}\mathrm{C}$ water?

23. How much steam should you add to the hot water and ice so the end product is all $100\,^{\circ}\mathrm{C}$ water?

24. How much steam should you add to the hot water and ice so the end product is all $0\,^{\circ}\mathrm{C}$ water?

25. How much hot water should you add to the steam and ice so the end product is $50\,^{\circ}\mathrm{C}$ water?

Treat specific and latent heats as functions of temperature for the remaining problems.

26. Use the degree-4 fit from the text to calculate water's average specific heat over the entire temperature range. Compare your result with the dashed line in Figure 11.2.

27. Water can evaporate at any temperature, not just its boiling point. Here are six data points for water's latent heat of vaporization (in $\mathrm{kJ/\,kg}$) versus temperature.

$$\begin{bmatrix} T & 0 & 20 & 40 & 60 & 80 & 100 \\ L_v & 2500.9 & 2454.2 & 2406.3 & 2358.1 & 2308.3 & 2256.8 \end{bmatrix}$$

Plot the data as points, calculate the linear fit to the data, and add it to the plot. What can you say about the dependence of water's L_v on temperature?

28. The latent heat of vaporization for water as a function of temperature between $-40\,^\circ\mathrm{C}$ and $+40\,^\circ\mathrm{C}$ can be approximated by the cubic fit

$$L_v(T) = 2500.8 - 2.3642T + 1.5893 \times 10^{-3}T^2 - 6.1434 \times 10^{-5}T^3$$

where T is in $^\circ\mathrm{C}$ and L_v is in $\mathrm{kJ/\,kg}$.

a. Plot this cubic fit between $-40\,^\circ\mathrm{C}$ and $+100\,^\circ\mathrm{C}$.

b. Plot the linear fit from the previous problem in the same range.

c. At what (positive) temperature do the two fits differ by 1%?

d. Use both fits to calculate the average latent heat between $-40\,^\circ\mathrm{C}$ and the result from part c.

29. Here is an old and odd empirical fit that describes the specific heat of water as a function of temperature. [29]

$$C = 0.996185 + 0.0002874\left(1 + \tfrac{1}{100}T\right)^{5.26} + 0.011160 \times 10^{-0.036T}$$

The temperature is in $^\circ\mathrm{C}$ and the specific heat is in $\mathrm{kcal/\,(\,kg\,^\circ C)}$.

a. Plot the fit between $0\,^\circ\mathrm{C}$ and $100\,^\circ\mathrm{C}$.

b. At what temperature does the specific heat have its minimum value?

c. What is the minimum value?

d. What is the average value over the whole temperature range?

e. Verify the fit reproduces the definition of a kcal.

30. How would you modify the fit from the previous problem so it gives the specific heat in $\mathrm{kJ/\,(\,kg\,^\circ C)}$ and the temperature in Fahrenheit? Make it so.

31. Liquid carbon dioxide cannot exist at pressures less than $5.12\,\mathrm{atm}$. At atmospheric pressure, solid CO_2 (dry ice) sublimates at $-78.5\,^\circ\mathrm{C}$ and its sublimation heat is $571.1\,\mathrm{kJ/\,kg}$. This quadratic polynomial models the specific heat of dry ice (in $\mathrm{J/\,kg/\,K}$) between $10\,\mathrm{K}$ and $195\,\mathrm{K}$.

$$C\,(T) = -0.02197T^2 + 11.293T - 112.40$$

How much heat is needed to change a $1.5\,\mathrm{m}^3$ chunk of $10\,\mathrm{K}$ dry ice into carbon dioxide gas? The density of dry ice is about $1560\,\mathrm{kg/\,m^3}$.

32. The Dulong–Petit law of Eq. (11.13) applies to many monatomic solids. It works for some multi-atomic substances if you apply it per mole of atoms instead of per mole of molecules. Are the following substances consistent with Dulong–Petit?

Substance	Water	Salt	KCl	Diamond	Beryllium
$C\left(\frac{\mathrm{J}}{\mathrm{kg\,^\circ C}}\right)$	4186	850	690	510	1825

Look up the Debye temperatures for these substances, and explain your answers.

33. Calculate the average specific heat for aluminum over the temperature range at atmospheric pressure where nitrogen is liquid. Use both the Einstein and Debye models. At $1\,\mathrm{atm}$, liquid nitrogen boils at $-196\,^\circ\mathrm{C}$ and it freezes at $-210\,^\circ\mathrm{C}$.

34. Show that Einstein's result for the specific heat of a solid can be written as

$$C_V = 3R\left(\frac{T_E}{2T}\right)^2 \sinh^{-2}\frac{T_E}{2T}\,.$$

35. Let's verify my claim that Einstein's model fits aluminum for $T_E = 300\,\mathrm{K}$.

a. Plot the following data for aluminum's specific heat (in $\mathrm{kJ/(\,kg\,K)}$) as points.

$$\begin{bmatrix} T & 205 & 220 & 240 & 250 & 260 & 273 & 280 & 295 & 300 \\ C & 0.770 & 0.794 & 0.813 & 0.820 & 0.828 & 0.838 & 0.841 & 0.851 & 0.855 \end{bmatrix}$$

b. Plot the Einstein model with $T_E = 300\,\mathrm{K}$ between $200\,\mathrm{K}$ and $300\,\mathrm{K}$.

c. Calculate aluminum's average specific heat over this temperature range.

36. In his 1907 paper, Einstein estimated what we now call the Einstein temperature for various solids.

	Silver	Aluminum	Diamond
T_E	$151\,\mathrm{K}$	$284\,\mathrm{K}$	$1325\,\mathrm{K}$
T_{melt}	$1234\,\mathrm{K}$	$933\,\mathrm{K}$	none @ $1\,\mathrm{atm}$

Use these values to plot Einstein's specific heat for all three solids between $0\,\mathrm{K}$ and $1500\,\mathrm{K}$ (stopping at the appropriate melting points of course) on the same graph. Comment on the relationship between T_E and C_V.

37. Verify this definite integral numerically: $\displaystyle\int_0^\infty \frac{x^4 e^x}{(e^x-1)^2}dx = \frac{4}{15}\pi^4$

38. The math trick changes a 3-d sum over (n_x, n_y, n_z) to a 3-d integral over the spherical coordinates (n, θ, ϕ). Choose the correct limits and Evaluate the following integrals over the positive octant of an n-sphere. Verify that the total integral equals the number of atoms N.

a. The angular part: $\displaystyle\iint \sin\theta\, d\theta\, d\phi$

b. The radial part: $\displaystyle\int n^2 dn$

39. Verify that the collection of constants $K \equiv \frac{hc}{2Lk}$ introduced in Eq. (11.19) has a unit of temperature.

40. Show that the energy of a Debye solid of Eq. (11.21) can be written

$$E = 9R\frac{T^4}{T_D^3}\int_0^{T_D/T}\frac{x^3}{e^x-1}dx$$

where x is the same dimensionless variable defined in the text. Verify numerically that the integral equals $\pi^4/15$ in the zero-temperature limit.

41. Verify the Einstein and Debye models agree for aluminum at $T = 240\,\mathrm{K}$. Show that Debye's prediction is 31 times larger than Einstein's at $24\,\mathrm{K}$.

42. In 1915, Arthur Holly Compton published his model for the specific heat of a solid based on agglomeration, a proposed alternative to Planck's quantum hypothesis. [21] The agglomeration hypothesis describes a low-temperature effect where the interaction between atoms suppresses some oscillations when the energy-per-atom falls below a critical value. Compton proposed that the specific heat is

$$\frac{C_V}{3R} = e^{-T_C/T}\left(1 + \frac{T_C}{T}\right)$$

where T_C is a parameter related to the critical energy.

a. Use $T_C = 143\,\mathrm{K}$ and calculate the heat needed to raise the temperature of 20 g of aluminum from 200 K to 300 K. Compare your answer with that of Example 11.9.

b. Plot Compton's specific heat with my $T_C = 143\,\mathrm{K}$ between 0 K and 400 K.

c. Add Compton's specific heat with his $T_C = 167\,\mathrm{K}$ for aluminum to your plot.

d. Add Einstein's and Debye's results for aluminum to your plot.

e. Notice neither Compton plot matches Debye for both low and high temperatures. What does this suggest about the agglomeration hypothesis?

(Hint: discuss this with your Agglomeration Mechanics professor.)

43. What are the high- and low-temperature predictions of Compton's agglomeration model? Compare these predictions to those of Einstein and Debye, and to the Dulong–Petit law where appropriate.

Here at *DPwSNB*, we're gruntled and chalant about the advertent and ept invention of words to combobulate students. We are also heveled, ruly, and appropriately whelmed.

Mercury	**Specific**	**Heat**		**Latent**	**Heat**	
Solid	141	33.7	Solid ↔ Liquid	11.4	2.72	−38.9 °C
Liquid	139	33.2	Liquid ↔ Vapor	295	70.5	357 °C
Vapor	104	24.8				
Note →	$\frac{\mathrm{J}}{\mathrm{kg\,°C}}$	$\frac{\mathrm{cal}}{\mathrm{kg\,°C}}$		$\frac{\mathrm{kJ}}{\mathrm{kg}}$	$\frac{\mathrm{kcal}}{\mathrm{kg}}$	

Table 11.3 Mercury

12 Special Relativity

At the start of the 20th century, the two dominant theories in physics were Newton's Mechanics and Maxwell's Electrodynamics. Each was hugely successful and highly regarded. Newton's Laws of Motion and Universal Gravitation explained the motion of everything from tennis balls to planets, the Earth's surface gravity, and it was used to predict the location of a previously unknown planet. Maxwell's theory unified Electricity and Magnetism, established light as a traveling electromagnetic wave, and predicted the value of the speed of light.

Considered separately, the two theories worked. But problems arose when they were brought together. According to Newton, it's possible for objects to travel at speeds equal to or greater than the speed of light. Starting from rest, an object experiencing a constant acceleration of $1g$ will reach the speed of light (c) in slightly less than one year. According to Maxwell, light is a travelling wave of oscillating electric and magnetic fields. An observer moving at the speed of light would see a standing electromagnetic wave, and such a wave is not a solution to Maxwell's equations.

This brings us to an important point: Special Relativity (SR) relates the descriptions made by observers with different velocities with respect to the same events. Often one observer is at rest with respect to the events, and the other is not. Each observer measures or calculates some physical quantity that occurs during the events.

If all observers measure the same value, then that quantity is **absolute**. Absolute means the measured value does not depend on the measurer's relative velocity, so all observers agree on the value. If any two observers measure different values, then that quantity is **relative**. Relative means the measured value does depend on the measurer's relative velocity, so all observers do not agree on the value.

The "special" in SR indicates it applies to observers in **inertial reference frames** (IRF). For our purposes, an IRF is the point of view of an observer moving at constant velocity. Typically, our observers occupy IRFs labelled S and S'. The S' frame is the one you would've called "moving" before you started thinking about relativity. If I'm standing at a bus stop and you're driving the bus, then I'm in frame S and you're in frame S'.

An **event** is anything with a location in time and space. Think of it as a time and a place. Every event is 4-dimensional so it has 4 coordinates, one for time and three for space. In frame S the event has coordinates $s = (ct, x, y, z)$. The time component is always first and it's ct so all four components have the same units. To simplify matters, we'll often consider straight-line motion in one spatial dimension. Then each event is 2-d and has 2 coordinates, one for time and one for space.

Doing Physics with Scientific Notebook: A Problem-solving Approach, First Edition. Joseph Gallant.

The Two Postulates

Einstein realized that the standing-wave problem and other inconsistencies between Newton and Maxwell were all related to the speed of light. He had two choices, either their theories were invalid or there's something special about the speed of light. He chose wisely.

Special Relativity (SR) is Einstein's solution to those problems. In 1905, Einstein realized these two simple sentences fix all the problems.

1. The Laws of Physics are absolute.
2. The Speed of Light is absolute.

That's genius at work. That's why we call him "Big Al"!

The 1st Postulate is shorthand for a reasonable extension of Galileo's work, with one important improvement. Galileo said the laws of mechanics were such that no mechanical experiment could detect constant-velocity motion. Einstein's 1st Postulate is an upgrade that includes the physics discovered since Galileo and any new physics to come. No experiment of any kind can detect constant-velocity motion. For any experiment, the result at rest is the same as the result at any constant velocity. The only way to tell you're moving at a constant velocity is to look out the window. All inertial reference frames, at rest or not, are equivalent!

The more you think about the 2nd Postulate, the stranger it seems. No matter what the relative velocity is between a light source and an inertial observer, the observer always measures Maxwell's predicted value for the speed of light. It doesn't matter if the source and observer are moving towards or away from each other, or not moving at all. As long as each has a constant velocity (including zero) all observers always measure the same value. Baseballs and bullets don't act this way, but light is special and the speed of light is absolute.

The 2nd Postulate suggests the speed of light plays an important role in SR. You can find the speed of light in a vacuum on the popup list on *SNB*'s Fragments toolbar.

$$c = 2.9979 \times 10^{8} \frac{\mathrm{m}}{\mathrm{s}} = 1.8628 \times 10^{5} \frac{\mathrm{mi}}{\mathrm{s}}$$

For many applications, we'll use $c \approx 3 \times 10^{8}\ \mathrm{m/s}$.

An important unit of distance is the light-year, which is defined at the distance light travels in one year. Even though it contains the word "year" a light-year is a unit of distance! Here is a light-year $(c\mathrm{y})$ in meters.

$$1c\mathrm{y} = \left(2.9979 \times 10^{8} \frac{\mathrm{m}}{\mathrm{s}}\right)\left(3.1557 \times 10^{7}\ \mathrm{s}\right) = 9.4605 \times 10^{15}\ \mathrm{m}$$

Let's use the Solve Method to convert the speed of light into feet-per-nanosecond.

$$2.9979 \times 10^{8} \frac{\mathrm{m}}{\mathrm{s}} = c \frac{\mathrm{ft}}{\mathrm{ns}}, \text{ Solution is: } 0.98356$$

Light travels almost one foot every billionth of a second.

The Consequences

In the classical physics of Newton and Maxwell, there was a tacit assumption that time and space are absolute. The measured values for the time interval and the space interval between two events are the same for all inertial observers. All observers measure the same intervals.

Big Al made different assumptions – his postulates – that solved the compatibility problems but had other consequences. The rest of this chapter is about those consequences, which lead us to new ideas about space and time, momentum and energy, and the addition of velocities. We'll also have to explain how no one noticed any relativistic effects in the more than 200 years between Newton and Einstein.

Before we explore the consequences in detail, let's introduce the **relativistic gamma factor** γ that appears in many SR results.

$$\gamma = \frac{1}{\sqrt{1 - \dfrac{v^2}{c^2}}} \tag{SR.1}$$

Gamma depends only on the relative velocity v between the two frames of reference. Gamma is a positive, dimensionless number that is never less than 1, and it equals 1 only when the frames are at rest relative to each other ($v = 0$).

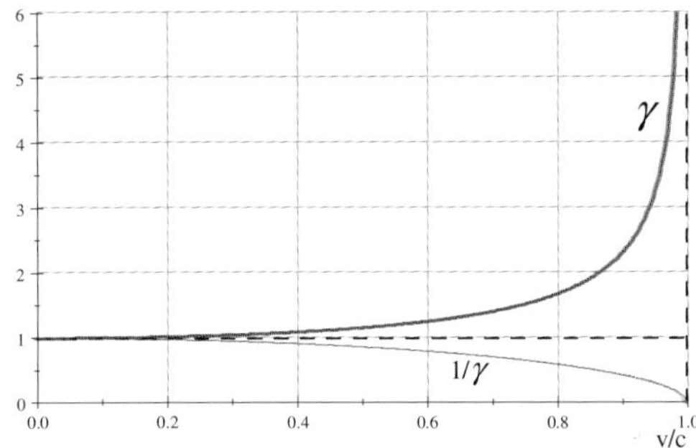

Figure SR.1 SR's γ and $1/\gamma$

The larger the value of γ the more significant are relativistic effects. For speeds much less than the speed of light, γ varies very slowly. But for speeds beyond $v \approx 0.9c$, small increases in speed produce large increases in γ.

Example SR.1 *Perfect speed is being there*
What is the value for gamma for an object moving at 99% the speed of light?
Solution. Use Eq. (SR.1) to create an expression for γ, and then Substitute the value for v with Evaluate.

$$\gamma = \left[\frac{1}{\sqrt{1 - \frac{v^2}{c^2}}}\right]_{v=0.99c} = 7.0888$$

At this speed, relativistic effects are significant. ◀

The current human speed record is approximately $25\,000\ \mathrm{mi}/\mathrm{h}$, reached by the Apollo astronauts on their trips to the moon. At that speed, which is about $c\,/\,27\,000$, relativistic effects are negligible. Let's use Evaluate Numerically on the following expression to verify this.

$$\gamma - 1 = \left[\frac{1}{\sqrt{1-\frac{v^2}{c^2}}} - 1\right]_{v=c/27\,000} = 6.858\,7 \times 10^{-10}$$

For the fastest human travel ever, gamma differs from its rest value of 1 by less than one part per billion!

This explains why we don't notice relativistic effects in our everyday lives. At typical human speeds, which are extremely small compared with the speed of light, these effects are too small for us to notice. Relativistic effects often seem counter-intuitive (a nice way of saying "goofy") to us because they are beyond our experience.

Let's Evaluate gamma for $v = \frac{1}{10}c$.

$$\gamma = \left[\frac{1}{\sqrt{1-\frac{v^2}{c^2}}}\right]_{v=0.10c} = 1.005$$

Even at one-tenth the speed of light, a speed 2700 times larger than the current human record, gamma is only 0.5% larger than 1.

Newton's classical mechanics (CM) is valid for slow-moving objects. It's a low-speed theory. For small speeds $(v \ll c)$, CM gives the right answer and SR gives the same answer. For speeds significant compared with light-speed, SR gives the right answer and Newton does not. CM is valid only for small speeds but SR is valid for all speeds. Special Relativity is not a high-speed theory, it's an all-speed theory.

Example SR.2 *How fast is half-fast?*
How fast must an object be going so that $\gamma = 2$?
Solution. Use Eq. (SR.1) to create an equation for γ, setting the left-hand side equal to 2. Leave the insertion point anywhere in the equation, click the Solve Exact button, and enter v in the Variable(s) to Solve for box.

$$2 = \frac{1}{\sqrt{1-\frac{v^2}{c^2}}}, \text{ Solution is: } -\tfrac{1}{2}\sqrt{3}c,\ \tfrac{1}{2}\sqrt{3}c$$

As we are about to see, to an observer moving at speed $v = \frac{\sqrt{3}}{2}c \approx 0.866\,c$, Newton's value for the time interval between two events is one-half Big Al's prediction. ◀

Time Dilation

Consider two events that happen at the same place but at different times in some reference frame. An observer at rest in that frame measures the time interval Δt_0 between the events where the subscript indicates this observer has a zero velocity relative to the events. For an observer that has velocity v relative to the events, the time interval between the two events is

$$\Delta t = \gamma \, \Delta t_0 \tag{SR.2}$$

where γ is the relativistic gamma factor of Eq. (SR.1).

This effect is called **time dilation** because the time interval Δt is always larger than Δt_0. The time interval Δt_0 between events that happen at the same place is called the **proper time interval**. The proper time is the *shortest* time interval any observer will measure between the two events.

Time dilation is a property of time. Clocks simply measure time intervals. Moving clocks run slower because the time they're measuring elapses slower.

All clocks at rest relative to the events, including biological clocks, measure the same proper time interval. All clocks with the same velocity relative to the events, including biological clocks, measure the same dilated time interval. Any discrepancy between clocks would violate the 1st Postulate!

Example SR.3 *It's about time...*
The starship Enterprise's crew installs external lights that flash 30 times per minute. How often do they flash as seen from the star base she moves past at $0.60\,c$?
Solution. The ship's crew is at rest relative to the lights, so they observe the proper time interval $\Delta t_0 = 2\,\text{s}$ between flashes. We need to Evaluate gamma for a relative velocity of $v = 0.60c$.

$$\gamma = \left[\left(1 - \frac{v^2}{c^2}\right)^{-1/2}\right]_{v=0.60c} = 1.25$$

The observers on the star base see the dilated time interval between flashes.

$$\Delta t = \gamma \, \Delta t_0 = (1.25)\,(2\,\text{s}) = 2.5\,\text{s}$$

Evaluate the following expression to get the flash rate as measured on the star base.

$$\frac{60\,\text{s}/\min}{2.5\,\text{s}} = \frac{24.0}{\min}$$

Seen from the star base, the lights are moving clocks that flash 24 times per minute. ◀

Who is the "moving" observer? All observers see the other frame move, but any reasonable person would agree that the starship is moving (frame S') and the star base is not (frame S). The issue is which observers have a zero velocity relative to the events under consideration. Even though the starship crew is physically moving through space, so their clocks run slow, they are at rest relative to the flashing-light events and they measure the proper time.

Length Contraction

Consider two events that happen at different places but at the same time in some reference frame. An observer at rest in that frame measures the space interval x_0 between the events where the subscript indicates this observer has a zero velocity relative to the events. For an observer that has velocity v relative to the events, the space interval between the two events is

$$\Delta x = \frac{\Delta x_0}{\gamma} \tag{SR.3}$$

where γ is the relativistic gamma factor of Eq. (SR.1).

This effect is called **length contraction** because the space interval Δx is always smaller than Δx_0. The space interval Δx_0 between events that happen at the same time is called the **proper length**. The proper length is the *longest* space interval any observer will measure between the two events.

In the lingo of Special Relativity, "space" refers to the usual (x, y, z) coordinates that indicate the dimensions of length, width, and height. Length is the 1-dimensional space interval between the x-coordinates of the object's ends.

Length contraction is a property of space. Meter sticks simply measure space intervals. Moving meter sticks are shorter because the space they occupy is contracted.

Length contraction applies only to space intervals along the direction of motion. The length of an object as seen from a moving frame is contracted in the direction of motion, but the other two spatial dimensions are not affected.

Example SR.4 *It's about space...*
The starship Enterprise is 289 m long and 72.6 m high when measured by her crew. How long and how tall is she as seen from the star base she moves past at $0.75\,c$?
Solution. The crew is at rest relative to the starship, so they measure its proper length $\Delta x_0 = 289\,\text{m}$. Let's Evaluate gamma for a relative velocity of $v = 0.75c$.

$$\gamma = \left[\left(1 - \frac{v^2}{c^2}\right)^{-1/2}\right]_{v=0.75c} = 1.5119$$

The observers on the star base see the ship's contracted length.

$$\Delta x = \frac{\Delta x_0}{\gamma} = \frac{289\,\text{m}}{1.5119} = 191.15\,\text{m}$$

The ship's height is perpendicular to its motion and is unaffected. The star base crew sees a ship that's $191.15\,\text{m}$ long and $72.6\,\text{m}$ high pass by. ◀

Space intervals and time intervals are relative. Moving clocks run slower and moving sticks are shorter. The faster you move through space, the slower you move through time and the less space you occupy along your direction of motion.

Addition of Velocities

The 2nd Postulate tells us the speed of light is absolute. Time dilation and length contraction are both consequences of that, but something strange happens at the speed of light. As seen by an observer moving at light speed, the time interval is infinite and the space interval is zero. At the speed of light, time stops and space vanishes! This suggests the speed of light is the cosmic speed limit.

Suppose we strap a cannon to a truck. The truck moves at velocity v with respect to the target and the cannon has a muzzle velocity u' with respect to the truck. According to Newton, the cannon ball hits the target with velocity $u = u' + v$. If we replace the cannon with a laser, Newton says the laser beam hits the target with velocity $u = c + v$.

This violates the 2nd Postulate! It doesn't include the relative nature of space and time. An object's velocity in different frames depends on different time and space intervals.

$$u = \frac{\Delta x}{\Delta t} \qquad u' = \frac{\Delta x'}{\Delta t'} \tag{SR.4}$$

We must account for this when we transform an object's velocity from one inertial reference frame to another.

Here is the SR **velocity transformation** when all the velocities are along the x and x' axes and the frame S' is moving to the right at speed v with respect to frame S.

$$u = \frac{u' + v}{1 + \dfrac{u'v}{c^2}} \tag{SR.5a}$$

$$u' = \frac{u - v}{1 - \dfrac{uv}{c^2}} \tag{SR.5b}$$

Observers in frame S see the object move with velocity u; observers in frame S' see it move with velocity u'. Both expressions have Newton's result in the numerator with the SR–IRF correction in the denominator.

Notice the symmetry. Swapping $u \leftrightarrow u'$ and letting $v \rightarrow -v$ produces the same result as solving the u' equation for u (and vice versa). This mathematical symmetry reflects the physical symmetry of relative motion: Observers in frame S see frame S' moving to the right at velocity v and observers in frame S' see frame S moving to the left at velocity $-v$. All inertial reference frames are equivalent so each observer makes physically relevant statements about the other.

What about that cosmic speed limit? Let's use Substitute (with Evaluate) and see what SR says about the speed of our laser beam with respect to the target.

$$u = \left[\frac{u' + v}{1 + \frac{u'v}{c^2}}\right]_{u'=c} = c$$

The laser beam hits the target with velocity c for any and all values of the truck's velocity. That's the cosmic speed limit. No matter how fast IRFs move relative to each other, no velocity ever exceeds the speed of light. That's also the 2nd Postulate. When observing the speed of a light beam, all observers see c. See? c? *Si*!

Example SR.5 *Some starships bring probes wherever they go...*
As the starship Boobyprize approaches an uncharted nebula at $0.50c$, it launches a probe toward the nebula. Its on-board probe-launcher is rated at $0.75c$. Before launch, the crew measured the probe's length to be $15\,\mathrm{m}$ and installed a beacon that flashes every $5\,\mathrm{s}$. As seen from the nebula, how fast is the probe moving, how long is the probe, and how often does its beacon flash?
Solution. Let S be the nebula frame and S' be the starship frame. The starship moves at $v = 0.50c$ relative to the nebula, and the probe launches at $u' = 0.75c$ relative to the ship. We want to calculate u, the velocity of the probe relative to the nebula. Substitute (with Evaluate) the appropriate values for u' and v into Eq. (SR.5a).

$$u = \left[\frac{u'+v}{1+\frac{u'v}{c^2}}\right]_{u'=0.75c,v=0.50c} = 0.909\,09c$$

This is the velocity we use to calculate the effects of length contraction and time dilation on the probe as seen from the nebula. Let's Evaluate the following expression for the probe's gamma factor.

$$\gamma_u = \frac{1}{\sqrt{1-0.909\,09^2}} = 2.400\,4$$

The proper time interval between flashes is $\Delta t_0 = 5\,\mathrm{s}$, and an observer in the nebula would see the dilated time Δt, which you can calculate with Eq. (SR.2) and Evaluate.

$$\Delta t = (2.400\,4)\,(5\,\mathrm{s}) = 12.002\,\mathrm{s}$$

The rest length of the probe is $\Delta x_0 = 15\,\mathrm{m}$, and an observer in the nebula sees the contracted length Δx, which you can calculate with Eq. (SR.3) and Evaluate.

$$\Delta x = \frac{15\,\mathrm{m}}{2.400\,4} = 6.249\,0\,\mathrm{m}$$

To an observer in the nebula's frame, a $6.249\,0\,\mathrm{m}$ probe with a beacon flashing every $12.002\,\mathrm{s}$ approaches at $0.909\,09\,c$. ◀

Example SR.6 *...some starships bring probes whenever they go*
How fast must the ship be moving so the probe launched toward the nebula at $0.75c$ relative to the ship approaches the nebula at $0.4\,c$?
Solution. The probe's velocity is still $u' = 0.75c$ relative to the ship but now $u = 0.4\,c$ relative to the nebula. Use Eq. (SR.5a) to create the following expression and Solve Exact to find v.

$$0.4c = \frac{0.75c+v}{1+\frac{(0.75c)v}{c^2}}, \text{ Solution is: } -0.5c$$

We took toward the nebula as the positive direction, so the starship is moving *away* from the nebula at $v = -0.50c$ as it launches the probe and this reduces the probe's velocity as seen from the nebula. ◀

Simultaneity

Simultaneous events occur at the same time, so they have the same time coordinate and the time interval between them is zero. Simultaneity is relative. Events that are simultaneous in one IRF are not necessarily simultaneous in another. In fact, if the two events are simultaneous but spatially separated in S (so $\Delta t = 0$ but $\Delta x \neq 0$), then they are not simultaneous in S' ($\Delta t' \neq 0$). Only events that occur at the same time and place are simultaneous in all IRFs. Such events are sometimes called space-time coincidences.

The Lorentz Transformation

SR relates the descriptions of an event made by observers in different IRFs. To observers in S the event happens at coordinates $s = (ct, x, y, z)$. What are its coordinates in S'?

Here is the physical situation: IRF S' moves to the right at velocity v relative to IRF S so that the x-axis coincides with the x'-axis. Clocks in both frames read zero when the two origins coincide. An event happens.

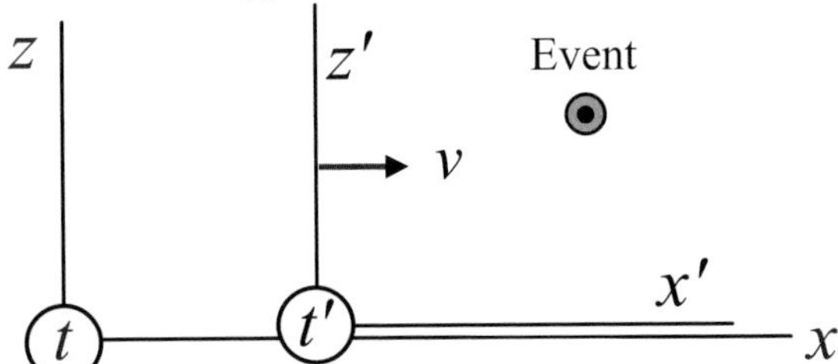

Figure SR.2 IRFs S and S'

The **Lorentz Transformation** (LT) is the set of equations that gives the event's space and time coordinates in S' in terms of its coordinates in S.

$$ct' = \gamma\left(ct - \frac{v}{c}\,x\right) \quad \text{(SR.6a)}$$

$$x' = \gamma\left(x - \frac{v}{c}\,ct\right) \quad \text{(SR.6b)}$$

$$y' = y \quad \text{(SR.6c)}$$

$$z' = z \quad \text{(SR.6d)}$$

The inverse Lorentz Transformation gives the event's space and time coordinates in S in terms of its coordinates in S'. We can again use symmetry to swap primed and unprimed coordinates and let $v \rightarrow -v$.

$$ct = \gamma\left(ct' + \frac{v}{c}\,x'\right) \quad \text{(SR.7a)}$$

$$x = \gamma\left(x' + \frac{v}{c}\,ct'\right) \quad \text{(SR.7b)}$$

$$y = y' \quad \text{(SR.7c)}$$

$$z = z' \quad \text{(SR.7d)}$$

The transverse spatial components are the same in both frames, but the time and space component along the direction of motion are not. Their values in one frame each depends on the value of *both* in the other frame. This mixing is our first hint that space and time are not separate entities.

Example SR.7 *What ever happened to Optimus Prime?*

Two inertial frames move with a speed $\frac{1}{2}c$ relative to each other. Observers in S' are at rest relative to an event they see occur at $(2\,\text{m}, 1\,\text{m})$. What are the coordinates of the event in the other frame?

Solution. The given coordinates are $ct = 2\,\text{m}$ and $x = 1\,\text{m}$ in frame S. We're given the relative velocity between the two frames, so we can Evaluate the gamma factor.

$$\gamma = \frac{1}{\sqrt{1 - \left(\frac{1}{2}\right)^2}} = \tfrac{2}{3}\sqrt{3} = 1.1547$$

Use Eq. (SR.6a) and Evaluate to calculate the time coordinate of the event in S'.

$$ct' = \tfrac{2}{\sqrt{3}}\left(2\,\text{m} - \tfrac{1}{2}\,(1\,\text{m})\right) = \sqrt{3}\,\text{m}$$

Use Eq. (SR.6b) and Evaluate to calculate the space coordinate of the event in S'.

$$x' = \tfrac{2}{\sqrt{3}}\left(1\,\text{m} - \tfrac{1}{2}\,(2\,\text{m})\right) = 0$$

To the observer in S', the event occurred at $ct' = \sqrt{3}\,\text{m}$ and $x' = 0\,\text{m}$. Perhaps you'd prefer a more traditional unit of time.

$$t' = \frac{ct'}{c} = \frac{\sqrt{3}\,\text{m}}{0.3\,\text{m/ ns}} = 5.773\,4\,\text{ns}$$

The event occurred at $t' = 5.773\,4\,\text{ns}$ in S'. ◀

The Lorentz Transformation is linear, so the transformation of the difference is the difference of the transformations. Here is the $S \rightarrow S'$ LT for the space and time intervals between two events.

$$c\Delta t' = \gamma\left(c\Delta t - \frac{v}{c}\,\Delta x\right) \quad \text{(SR.8a)}$$

$$\Delta x' = \gamma\left(\Delta x - \frac{v}{c}\,c\Delta t\right) \quad \text{(SR.8b)}$$

$$\Delta y' = \Delta y \quad \text{(SR.8c)}$$

$$\Delta z' = \Delta z \quad \text{(SR.8d)}$$

Symmetry and swapping the primed and unprimed coordinates gives us the $S' \rightarrow S$ inverse LT.

$$c\Delta t = \gamma\left(c\Delta t' + \frac{v}{c}\,\Delta x'\right) \quad \text{(SR.9a)}$$

$$\Delta x = \gamma\left(\Delta x' + \frac{v}{c}\,c\Delta t'\right) \quad \text{(SR.9b)}$$

$$\Delta y = \Delta y' \quad \text{(SR.9c)}$$

$$\Delta z = \Delta z' \quad \text{(SR.9d)}$$

Intervals transform the same way coordinates do.

Example SR.8 *After you. No, after you!*
Spock, standing $9\,\mathrm{m}$ to the right of Kirk, fires a phaser $20\,\mathrm{ns}$ after Kirk fires his. A passing space monkey observes that the phasers fire simultaneously. How fast is the space monkey moving? Where is Spock in the monkey's frame?
Solution. The two events are Kirk and Spock each firing his own phaser. Let's define an interval as "Spock minus Kirk". In the Kirk-Spock frame (S), Spock is to the right of Kirk, so the space interval between them is $\Delta x = +9\,\mathrm{m}$. The time interval between the two events is $\Delta t = +20\,\mathrm{ns}$ since Spock fires $20\,\mathrm{ns}$ later than Kirk. In the space monkey's frame (S') the shots are simultaneous so $\Delta t' = 0$.

Use Eq. (SR.8a) and Solve Exact to find the monkey's speed. Solve Exact handles *SNB*'s built-in metric prefixes like "nano" numerically.

$$0 = \left(3 \times 10^{8} \frac{\mathrm{m}}{\mathrm{s}}\right)\left(+20 \times 10^{-9}\,\mathrm{s}\right) - \frac{v}{c}\left(+9\,\mathrm{m}\right), \text{ Solution is: } \tfrac{2}{3}c$$

$$0 = \left(\tfrac{3}{10} \frac{\mathrm{m}}{\mathrm{ns}}\right)\left(+20\,\mathrm{ns}\right) - \frac{v}{c}\left(+9\,\mathrm{m}\right), \text{ Solution is: } 0.666\,67c$$

The space monkey is moving to the right at $v = \frac{2}{3}c$. To calculate the space interval between Spock and Kirk in the monkey's frame, use Eq. (SR.8b) and apply Evaluate and Evaluate Numerically to the following expression.

$$\Delta x' = \frac{1}{\sqrt{1-\left(\frac{2}{3}\right)^2}}\left(+9\,\mathrm{m} - \tfrac{2}{3}\left(\tfrac{3}{10}\frac{\mathrm{m}}{\mathrm{ns}}\right)\left(+20\,\mathrm{ns}\right)\right) = 3\sqrt{5}\,\mathrm{m} = 6.708\,2\,\mathrm{m}$$

The space monkey sees Spock standing $6.708\,2\,\mathrm{m}$ to the right of Kirk. We can also use Length Contraction because Δx is the at-rest space interval in S between events that are simultaneous in the monkey's frame S'.

$$\Delta x' = \frac{\Delta x}{\gamma} = \sqrt{1-\left(\tfrac{2}{3}\right)^2}\,(9\,\mathrm{m}) = 6.708\,2\,\mathrm{m}$$

Suppose the space monkey was moving at a larger speed, say $\frac{3}{4}c$. We can Evaluate the time and space intervals between the phaser shots in this new frame with Eqs. (SR.8).

$$\Delta t' = \frac{1}{\sqrt{1-0.75^2}}\left(+20\,\mathrm{ns} - \tfrac{3}{4}\left(\frac{+9\,\mathrm{m}}{0.3\,\mathrm{m/\,ns}}\right)\right) = -3.779\,6\,\mathrm{ns}$$

$$\Delta x' = \frac{1}{\sqrt{1-0.75^2}}\left(+9\,\mathrm{m} - \tfrac{3}{4}\left(0.3\frac{\mathrm{m}}{\mathrm{ns}}\right)\left(+20\,\mathrm{ns}\right)\right) = 6.803\,4\,\mathrm{m}$$

The negative time interval tells us Spock fires his phaser *before* Kirk in this frame! The space interval is larger than it was in the $v = \frac{2}{3}c$ frame. The smallest space interval between two events is in the frame where the events are simultaneous. ◀

Three different inertial reference frames have three different time intervals and three different space intervals between the same two events. The events are simultaneous in one frame, and happen in different order in the other two. Time and space and simultaneity are all relative!

As the previous example suggests, the Lorentz Transformation includes all the SR consequences of the Two Postulates we discussed above.

Time Dilation: When events happen at the same place in S, $\Delta x = 0$ and Eq. (SR.8a) reduces to Eq. (SR.2).

$$c\Delta t' = \gamma\left(c\Delta t - \frac{v}{c}\,\Delta x\right) \rightarrow \boxed{\Delta t' = \gamma\,\Delta t}$$

The interval Δt is the proper time interval. When you (in frame S) look at your watch to measure the time elapsed since the start of class, you're measuring the proper time interval. An observer on a passing starship (in frame S') would measure the dilated time.

Length Contraction: When events happen at the same time in S, $\Delta t = 0$ and Eq. (SR.8b) reduces to Eq. (SR.3).

$$\Delta x' = \gamma\left(\Delta x - \frac{v}{c}\,c\Delta t\right) \rightarrow \boxed{\Delta x' = \gamma\,\Delta x}$$

The interval $\Delta x'$ is the proper length. You measure the passing starship's contracted length while her crew measures its proper length.

Addition of Velocities: It takes very little math for the ratio of Eq. (SR.8a) to Eq. (SR.8b) to reduce to Eq. (SR.5b).

$$\frac{\Delta x'}{\Delta t'} = \frac{\gamma\left(\Delta x - \frac{v}{c}\,c\Delta t\right)}{\gamma\left(\Delta t - \frac{v}{c^2}\,\Delta x\right)} \times \frac{\frac{1}{\Delta t}}{\frac{1}{\Delta t}} = \frac{\frac{\Delta x}{\Delta t} - v}{1 - \frac{v}{c^2}\frac{\Delta x}{\Delta t}}$$

The ratios of space intervals to time intervals are u' and u per Eq. (SR.4). The Lorentz Transformation accounts for the different time and space intervals measured by you and the starship's crew.

Simultaneity: When events are simultaneous in S, $\Delta t = 0$ and Eq. (SR.8a) relates the time interval in S' to the space interval in S.

$$c\Delta t' = \gamma\left(c\Delta t - \frac{v}{c}\,\Delta x\right) \rightarrow \boxed{c\Delta t' = -\gamma\frac{v}{c}\,\Delta x}$$

The events are simultaneous in S' only if they happen at the same place in S ($\Delta x = 0$). You sat down just as your professor started talking, but those events are not simultaneous in the starship's frame.

The boxed expressions for time dilation, length contraction, and simultaneity are only valid under particular conditions. The Lorentz Transformation is more general and useful. It's a direct consequence of the 2nd Postulate, and it includes all the effects of the relative nature of space and time.

Space-Time

Special Relativity shows the space and time coordinates of any event are mixed according to the Lorentz Transformation. Since space consists of 3 dimensions and time is 1-dimensional, together they comprise an entity we cleverly call the 4-dimensional space-time continuum. The "continuum" simply means there are no missing points in space or instants in time. Space-time is continuous.

Contrary to a popular misconception, the phrase "everything is relative" is not a part of Special Relativity. The very basis of the theory – the postulates – is given in terms of absolute quantities. Considered separately, the space and time intervals between two events are relative, but there is an absolute quantity involving those intervals together.

The coordinates of a light beam emanating from the origin in frame S obey the relation $c^2t^2 = x^2 + y^2 + z^2$. We can generalize this and define the 4-d space-time separation between any event and the origin as the difference between the time and space parts of that relation. For now this appears to be nothing more than an inspired guess, but it turns out to be a very good idea.

In frame S, where the event has coordinates (ct, x, y, z), the space-time separation is

$$s^2 = c^2t^2 - x^2 - y^2 - z^2 \qquad \text{(SR.10)}$$

In frame S' the same event has coordinates (ct', x', y', z') and the separation is

$$(s')^2 = c^2 (t')^2 - (x')^2 - (y')^2 - (z')^2 \qquad \text{(SR.11)}$$

Let's use the Lorentz Transformation from Eqs. (SR.6) to relate the space-time separation in S' to the event's coordinates in S, and Simplify the resulting equation.

$$(s')^2 = \left(\frac{ct - \frac{v}{c}x}{\sqrt{1 - \frac{v^2}{c^2}}}\right)^2 - \left(\frac{x - \frac{v}{c}ct}{\sqrt{1 - \frac{v^2}{c^2}}}\right)^2 - y^2 - z^2 = c^2t^2 - x^2 - y^2 - z^2$$

The space-time separation in S' equals the space-time separation in S. The quantity s^2 is the same for observers in all IRFs. It's absolute!

Space and time intervals transform the same way as coordinates, so the 4-dimensional **space-time interval** between any two events is also absolute.

$$\Delta s^2 = c^2\Delta t^2 - \Delta x^2 - \Delta y^2 - \Delta z^2 \qquad \text{(SR.12)}$$

Changing from one frame to another mixes the time and space intervals, but that mixing leaves the space-time interval unchanged.

This may be your first encounter with traditional but sloppy relativistic notation. Expressions like Δs^2 and Δt^2 would be more correctly written as $(\Delta s)^2$ and $(\Delta t)^2$, meaning the square of the difference and not the difference of the squares. Relativistic equations involving finite intervals or infinitesimal differentials can be complicated, and parentheses tend to make that worse, so we just omit them. You'll have to remind yourself that Δx^2 and $(\Delta x)^2$ mean the same thing in the context of relativity.

The space-time interval is a good way to describe events. When the space-time interval between two events is positive, the interval is called time-like and there can be a cause-and-effect relationship between the two events. Space-like intervals are negative, and there cannot be a causal relationship between the two events. When the space-time interval is zero, the events are separated by a light-like interval. There can be a causal relationship between the events, which are separated by the propagation of a light signal.

While Δs has units of distance, it's not a distance between points in 3-d space. It's a separation between events in 4-d space-time, a space-time that does not follow the rules of Euclidean geometry. There are no minus signs in the Pythagorean Theorem.

Example SR.9 *Monkey see, monkey do*
Calculate the space-time interval between Kirk and Spock firing their phasers in both their frame and the space monkey's frame.
Solution. In the Kirk-Spock frame, the time interval between the events is $\Delta t = +20\,\text{ns}$ and the space interval in $\Delta x = +9\,\text{m}$. We can use Eq. (SR.12) to Evaluate the space-time interval.

$$\Delta s^2 = c^2 \Delta t^2 - \Delta x^2 = \left(3 \times 10^8\,\text{m/s}\right)^2 \left(20 \times 10^{-9}\,\text{s}\right)^2 - (9\,\text{m})^2 = -45\,\text{m}^2$$

In the space monkey's frame, the events are simultaneous so $\Delta t' = 0$, and they are $\Delta x' = +3\sqrt{5}\,\text{m}$ apart in space.

$$(\Delta s')^2 = c^2 (\Delta t')^2 - (\Delta x')^2 = \left(3 \times 10^8\,\text{m/s}\right)^2 (0)^2 - \left(3\sqrt{5}\,\text{m}\right)^2 = -45\,\text{m}^2$$

The interval is the same in both frames. Space-time intervals are absolute! ◄

A **space-time diagram** gives a visual representation of the effects of inertial motion on space and time. Each diagram is a graph of events showing an object's time and space coordinates in two different frames. Unlike traditional physics graphs, here the vertical axis is time and the horizontal axis is space. We can plot single events or a series of events. The resulting trajectory through space-time is called the object's **worldline**.

Figure SR.3 shows the space-time diagram of an event with coordinates (ct, x) in frame S and coordinates (ct', x') in frame S'. The Lorentz Transformation tells us where to put the axes for the primed frame on the diagram. Along the ct'-axis, the space coordinate $x' = 0$ so Eq. (SR.6b) gives the equation of the ct'-axis

$$ct = \frac{1}{\beta}\,x \qquad \text{(SR.13}t'\text{)}$$

where $\beta \equiv v/c$. Along the x'-axis, the time coordinate $ct' = 0$ so Eq. (SR.6a) tells us

$$ct = \beta\, x \qquad \text{(SR.13}x'\text{)}$$

is the equation of the x'-axis. The primed axes are not perpendicular but skewed symmetrically toward the $45°$ $x = ct$ line. Space-time is not Euclidean!

On the diagram, the slope of the ct'-axis is $1/\beta$ and the slope of the x'axis is β.

The line parallel to the ct'-axis connecting the event to the x'-axis has the form

$$ct = \frac{1}{\beta}\,x + b_1$$

The line parallel to the x'-axis connecting the event to the ct'-axis has the form

$$ct = \beta x + b_2$$

All you need to find the intercepts are the space-time coordinates in S of one event.

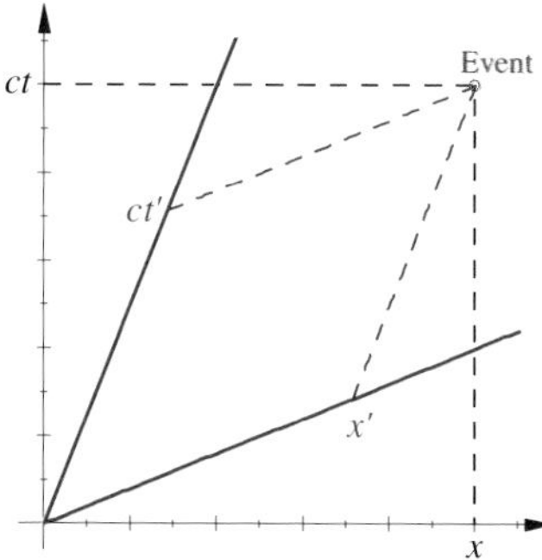

Figure SR.3 A good STD

Lines parallel to the x'-axis have the same ct' coordinate, so any events on such a line are simultaneous in S'. Lines parallel to the ct'-axis have the same x' coordinate, so any events on such a line happen at the same place in S'.

Example SR.10 *It's about space-time*

Observers at rest relative to an event see it happen at $(12\,\text{m}, 10\,\text{m})$. What are the coordinates of this event in a frame moving at $\frac{3}{5}c$? Make the space-time diagram for this event and both frames. Verify that the space-time interval between this event and the origin is the same in both frames.

Solution. The observers are in frame S and frame S' has velocity $v = \frac{3}{5}c$ relative to them. Write the Lorentz Transformation from Eq. (SR.6a) and Eq. (SR.6b) in terms of β and Substitute (with Evaluate) $\beta = \frac{3}{5}$ to calculate the event coordinates in S'.

$$ct' = \left[\frac{1}{\sqrt{1-\beta^2}}\left(12\,\text{m} - 10\,\text{m}\beta\right)\right]_{\beta=3/5} = \frac{15}{2}\,\text{m}$$

$$x' = \left[\frac{1}{\sqrt{1-\beta^2}}\left(10\,\text{m} - 12\,\text{m}\beta\right)\right]_{\beta=3/5} = \frac{7}{2}\,\text{m}$$

The event happens at $\left(ct' = \frac{15}{2}\,\text{m}, x' = \frac{7}{2}\,\text{m}\right)$ in S'. Let's Evaluate expressions for the space-time interval between this event and the origin in both frames.

$$(\Delta s')^2 = \left(\frac{15}{2}\,\text{m}\right)^2 - \left(\frac{7}{2}\,\text{m}\right)^2 = 44\,\text{m}^2$$

$$\Delta s^2 = (12\,\text{m})^2 - (10\,\text{m})^2 = 44\,\text{m}^2$$

The space-time interval is the same in both frames.

Plot the expression $\frac{3}{5}x$ for the x'-axis and $\frac{5}{3}x$ for the ct'- axis, and add $\begin{bmatrix}10 & 12\end{bmatrix}$ as a Point with a Circle for its Point Marker for the event. SR convention has the time coordinate first, but *SNB* always wants the coordinate on the horizontal axis first.

Now we want to create expressions for the projections of the event in S'. Use the event's coordinates in S and Solve Exact to find the intercept of the line connecting the event and the x'-axis.

$$12 = \tfrac{5}{3}(10) + b_1, \text{ Solution is: } -\frac{14}{3}$$

To find out where this line intersects the x'-axis, create an expression equating this line to the line representing the x'-axis and click the Solve Exact button.

$$\tfrac{5}{3}x - \tfrac{14}{3} = \tfrac{3}{5}x, \text{ Solution is: } \frac{35}{8} = 4.375$$

Add the expression $\frac{5}{3}x - \frac{14}{3}$ to your plot and let x run from 4.375 to 10.

Use the event's coordinates in S and Solve Exact to find the intercept of the line connecting the event and the ct'-axis.

$$12 = \tfrac{3}{5}(10) + b_2, \text{ Solution is: } 6$$

To find out where this line intersects the ct'-axis, create an expression equating this line to the line representing the ct'-axis and click the Solve Exact button again.

$\frac{3}{5}x + 6 = \frac{5}{3}x$, Solution is: $\dfrac{45}{8} = 5.625$

Add the expression $\frac{3}{5}x + 6$ to your plot and let x run from 5.625 to 10.

Figure SR.4 shows the space-time diagram for this event and these frames.

Notice the dimensions in the primed frame are stretched out on the graph.

The coordinate $x' = 3\frac{1}{2}$ m on the x'-axis is further from the origin (on the graph) than the point $x = 4$ m on the x-axis.

The coordinate $ct' = 7\frac{1}{2}$ m on the ct'-axis is further from the origin (on the graph) than the point $ct = 9$ m on the ct-axis.

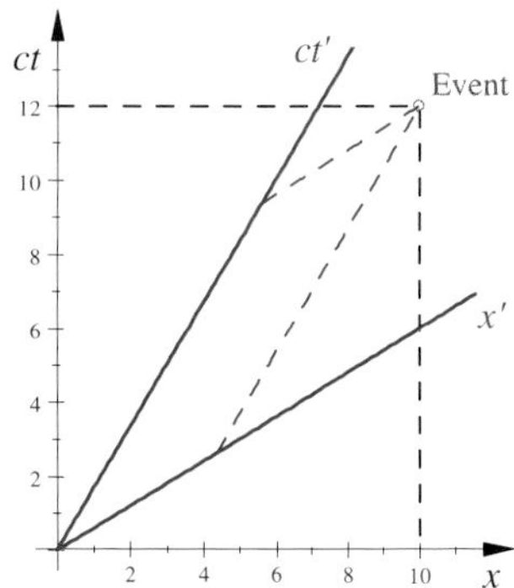

Figure SR.4 An event

An object that is $\Delta x = 4.375$ m long in S is $\Delta x' = 3.5$ m long in S'. The object is physically shorter in the real world but it looks longer on the graph. ◄

Example SR.11 *Monkey see, monkey overdo*

Make the space-time diagram for the events and frames in Example SR.8. Use the diagram to verify the events are simultaneous in a certain frame.

Solution. In frame S, the Kirk event happens when he fires his phaser at $(ct = 0, x = 0)$ and the Spock event happens at $(ct = 6\,\mathrm{m}, x = 9\,\mathrm{m})$. Plot the events as the points $(0, 0)$ and $(9, 6)$ (remember *SNB* always wants the coordinate on the horizontal axis first).

The space-monkey frame S' has velocity

$$v = \frac{2}{3}c$$

relative to the Kirk-Spock frame S, so the x'-axis has slope $\beta = \frac{2}{3}$ and the ct'-axis has slope $\beta^{-1} = \frac{3}{2}$.

Add the lines $\frac{2}{3}x$ and $\frac{3}{2}x$ to your plot.

Add this matrix (as a dashed line)

$$\begin{bmatrix} 0 & 6 \\ 9 & 6 \\ 9 & 0 \end{bmatrix}$$

to the plot to project the coordinates of the Spock event in frame S.

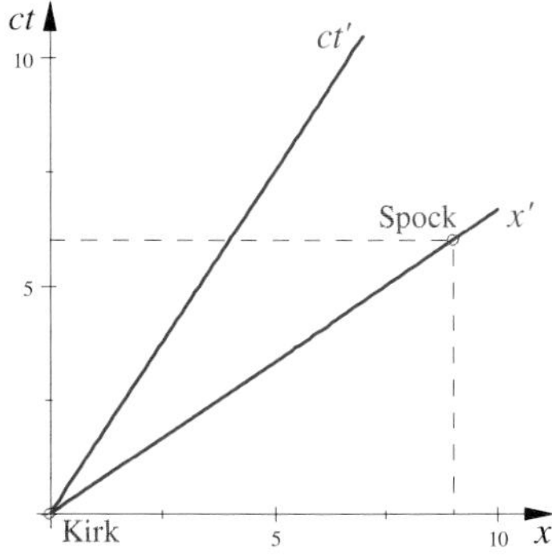

Figure SR.5 More monkeys

There's no need to include coordinate projections in S' because the Kirk and Spock events are both on the x'-axis. Their time coordinates are the same $(ct' = 0)$ so the events are simultaneous in frame S'. ◄

Relativistic Momentum and Energy

The properties of space and time are a direct consequence of the 2nd Postulate and the absolute nature of the speed of light. The relativistic momentum and energy are a direct consequence of the 1st Postulate and the absolute nature of the laws of physics. Energy and momentum must be conserved in all inertial reference frames.

The **relativistic momentum** of an object with mass m moving at velocity $\overrightarrow{u}$ is

$$p = \gamma_u m\overrightarrow{u} = \frac{m\overrightarrow{u}}{\sqrt{1 - \dfrac{u^2}{c^2}}} \qquad \text{(SR.14)}$$

where γ_u is the gamma factor for speed u. The relativistic momentum is gamma times the nonrelativistic momentum $m\overrightarrow{u}$. The relative quantity $\gamma_u m$ is sometimes called the object's **relativistic mass** while the absolute quantity m is called its **rest mass**.

Let's do a simple example with the speed of light set at $c = 3.0\,\mathrm{m/\,s}$ to exaggerate the relativistic effects.

Example SR.12 *On the thin ice of a new day*
Starting from rest, two skaters push-off against each other on smooth level ice with negligible friction. The $45\,\mathrm{kg}$-woman moves away with a velocity of $+2.4\,\mathrm{m/\,s}$ relative to the ice. Let the speed of light be $c = 3.0\,\mathrm{m/\,s}$ and find the recoil velocity of the $75\,\mathrm{kg}$-man relative to the ice.
Solution. After the push-off, the woman has $v = \frac{4}{5}c$ so we have to use the relativistic momentum of Eq. (SR.14). For the sake of comparison, let's first use nonrelativistic momentum conservation

$$0 = m_w v_w + m_m v_m$$

and Solve Exact to find the nonrelativistic prediction for the man's final velocity.

$$0 = (45\,\mathrm{kg})\left(+2.4\frac{\mathrm{m}}{\mathrm{s}}\right) + (75\,\mathrm{kg})\,v_m, \text{ Solution is: } -1.44\frac{\mathrm{m}}{\mathrm{s}}$$

According to Newton, the man moves at speed $1.44\,\mathrm{m/\,s}$ in the opposite direction of the woman. Now let's use relativistic momentum conservation

$$0 = \gamma_w m_w v_w + \gamma_m m_m v_m$$

and Solve Exact to find the relativistic prediction for the man's final velocity.

$$0 = \frac{(45)(+2.4)}{\sqrt{1 - \dfrac{2.4^2}{3.0^2}}} + \frac{(75)\,v_m}{\sqrt{1 - \dfrac{v_m^2}{3.0^2}}}, \text{ Solution is: } -1.8741$$

Big Al agrees the man moves in the opposite direction but at a speed significantly larger than the nonrelativistic result. ◄

An object's **relativistic energy** depends on its speed u and rest mass m.

$$E = \gamma_u mc^2 = \frac{mc^2}{\sqrt{1 - \dfrac{u^2}{c^2}}} \qquad \text{(SR.15)}$$

The relativistic energy is ***not*** gamma times the non-relativistic energy.

Even when the object is at rest (so $\gamma_u = 1$) its energy is not zero. This **rest energy** depends on the object's rest mass.

$$E_{\text{rest}} = mc^2 \tag{SR.16}$$

Mass is a form of energy and this is the conversion formula. If we convert m kilograms of mass completely into energy, we'll get mc^2 joules of energy. Since c^2 is a very big number,

$$c^2 = \left(3 \times 10^8 \,\frac{\text{m}}{\text{s}}\right)^2 = 9.0 \times 10^{16} \,\frac{\text{J}}{\text{kg}}$$

converting a small amount of mass releases a huge amount of energy.

Example SR.13 *Another good use for a baseball*
Calculate the rest energy of a baseball.
Solution. A baseball's mass is only $0.142\,\text{kg}$, but with Eq. (SR.16) and Evaluate we see it has a huge rest energy.

$$E_{\text{rest}} = (0.142\,\text{kg}) \left(9.0 \times 10^{16} \frac{\text{J}}{\text{kg}}\right) = 1.278 \times 10^{16}\,\text{J}$$

That's enough energy to run the entire United States for almost an hour. To give to you an idea just how much energy this is, let's see how long it'll run a $100\,\text{W}$ light bulb.

$$t = \frac{E_{\text{rest}}}{P} = \frac{1.278 \times 10^{16}\,\text{J}}{100\,\text{W}} \times \frac{1\,\text{y}}{3.1557 \times 10^7\,\text{s}} = 4.0498 \times 10^6\,\text{y}$$

That's about 4 million years! Unfortunately, to convert the baseball's mass completely to energy we'd need another baseball made of antimatter. ◄

An object's kinetic energy is just the difference between its total and rest energies.

$$K = (\gamma_u - 1)\, mc^2 = \left(\frac{1}{\sqrt{1 - \dfrac{u^2}{c^2}}} - 1\right) mc^2 \tag{SR.17}$$

The total energy $E = E_{\text{rest}} + K$ has two parts, the rest energy due to its mass and the kinetic energy due to its motion.

Example SR.14 *A wicked fast electron*
Calculate the speed and momentum of an electron whose kinetic energy is $5\,\text{MeV}$.
Solution. We're given the kinetic energy $K = 5\,\text{MeV}$, so let's use Eq. (SR.17), the electron's rest energy and Solve Exact to find the electron's gamma factor.

$5\,\text{MeV} = (\gamma_u - 1)\, 0.511\,\text{MeV}$, Solution is: 10.785

Once we know gamma, we can apply Solve Exact to Eq. (SR.1) to find the speed.

$10.785 = \left(1 - \dfrac{u^2}{c^2}\right)^{-1/2}$, Solution is: $-0.99569c, 0.99569c$

The electron's speed is almost 99.6% of the speed of light.

Create and Evaluate an expression for the momentum with Eq. (SR.14).

$$p = \gamma_u mu = (10.785)\left(0.511\,\mathrm{MeV}/c^2\right)(0.99569c) = 5.4874\frac{\mathrm{MeV}}{c}$$

If you prefer more familiar units, apply Evaluate Numerically to these expressions.

$$p = 5.4874\frac{\mathrm{MeV}}{3\times10^8\,\mathrm{m/s}} = 2.9306\times10^{-21}\frac{\mathrm{m}}{\mathrm{s}}\,\mathrm{kg}$$

$$K = 5\,\mathrm{MeV} = 8.0109\times10^{-13}\frac{\mathrm{m}^2}{\mathrm{s}^2}\,\mathrm{kg}$$

Even very fast electrons don't have much kinetic energy and momentum. ◀

Here is an important relativistic **energy-momentum relation** where p and E are given by Eqs. (SR.14) and (SR.15), respectively.

$$E^2 = c^2p^2 + m^2c^4 \tag{SR.18}$$

This relation has no explicit γ-dependence, which makes it very useful for high-speed situations where rounding errors can be significant.

Let's use the electron from the previous example and Evaluate to verify the energy-momentum relation. Its energy squared, courtesy of Eq. (SR.15), equals the right-hand side of Eq. (SR.18).

$$E^2 = \left((10.785)(0.511\,\mathrm{MeV})\right)^2 = 30.373\,\mathrm{MeV}^2$$

$$c^2p^2 + m^2c^4 = c^2\left(5.4874\frac{\mathrm{MeV}}{c}\right)^2 + \left(0.511\frac{\mathrm{MeV}}{c^2}\right)^2 c^4 = 30.373\,\mathrm{MeV}^2$$

What about Newton? All our relativistic expressions must reduce to the appropriate Newtonian result when the speed is small compared with the speed of light. Let's do a 4-term Power Series on the relativistic momentum where we Expand in Powers of u.

$$p = \frac{mu}{\sqrt{1-\dfrac{u^2}{c^2}}} = mu + \frac{1}{2c^2}mu^3 + O\left(u^5\right)$$

The first term in the expansion is the nonrelativistic momentum mu. Let's do the same for the relativistic kinetic energy.

$$K = \frac{mc^2}{\sqrt{1-\dfrac{u^2}{c^2}}} - mc^2 = \tfrac{1}{2}mu^2 + \frac{3}{8c^2}mu^4 + O\left(u^6\right)$$

The first term in this expansion is the nonrelativistic kinetic energy $\frac{1}{2}mu^2$. In each case, the leading term in the expansion is the nonrelativistic Newtonian expression and the first correction is of order u^2/c^2, an exceedingly small quantity for speeds much less than the speed of light. Newton wasn't wrong, he was just approximating.

Relativistic Collisions

Collisions between subatomic particles provide information about the structure of matter and the interactions between those particles. These collisions conserve the total momentum and total energy but not the total mass. They are usually high-speed collisions that involve matter-energy transformation, so we must use the relativistic forms for momentum and energy to analyze them.

Example SR.15 *Catch a few Z_0's*
A positron collides with an electron that is at rest.

$$e^+ + e^- \to Z_0$$

How much kinetic energy does the positron need to produce a Z_0 particle?
Solution. Energy conservation tells us that

$$K + 2mc^2 = E_z$$

where K is the kinetic energy of the incident positron, m is the mass of a positron or electron, and E_z is the total energy of the Z_0. Momentum conservation tells us the positron's initial momentum equals the final momentum of the Z_0. Given that the rest energy of a Z_0 is $M_z c^2 = 91.187\,\text{GeV}$, the positron will have a very large value of gamma. Use the energy-momentum relation to rewrite momentum conservation.

$$\left(K + mc^2\right)^2 - m^2c^4 = E_z^2 - M_z^2c^4$$

Tell *SNB* to assume all the variables are positive. Create expressions for energy and momentum conservation, and use Solve Exact to find K, E_z.

$$\begin{bmatrix} K + 2mc^2 = E_z \\ \left(K + mc^2\right)^2 - m^2c^4 = E_z^2 - M_z^2c^4 \end{bmatrix},$$

$$\text{Solution is: } \left[K = \tfrac{1}{2m}\left(c^2M_z^2 - 4c^2m^2\right), E_z = \tfrac{1}{2}\tfrac{c^2}{m}M_z^2\right]$$

Now use Substitute (with Evaluate) to calculate the required kinetic energy.

$$K = \left[\frac{1}{2}\frac{M_z^2c^2}{m} - 2mc^2\right]_{m=0.511\times10^{-3}\,\text{GeV}/c^2, M_z=91.187\,\text{GeV}/c^2} = 8.136\,1\times10^6\,\text{GeV}$$

This is over 8 million GeV, well beyond the reach of any current collider. Instead, let's have the positron and electron meet in a head-on, zero-total-momentum collision that creates an at-rest Z_0. Create a new expression for energy conservation and use Solve Exact to find K.

$$2K + 2mc^2 = M_zc^2, \text{ Solution is: } \tfrac{1}{2}c^2M_z - c^2m$$

Let's see how much kinetic energy this collision needs.

$$K = \left[\tfrac{1}{2}M_zc^2 - mc^2\right]_{m=0.511\times10^{-3}\,\text{GeV}/c^2, M_z=91.187\,\text{GeV}/c^2} = 45.593\,\text{GeV}$$

This is about $178,450$ times less energy, and it explains why most colliders take the trouble to align head-on collisions. ◄

The reference frame where the total momentum is zero is called the **center-of-mass frame** (CoM frame). The minimum energy for a given reaction occurs when the products are at rest in the CoM frame. If only one particle remains after the reaction, then it must be at rest in the CoM frame. The lab frame is often the frame where the target particle is at rest.

Energy and momentum obey the same transformation rules as time and space coordinates. A process with total energy E_{tot} and total momentum P_{tot} in the frame S has

$$\frac{1}{c}E_{\text{tot}}^{'} = \gamma\left(\frac{1}{c}E_{\text{tot}} - \beta\, P_{\text{tot}}\right) \qquad \text{(SR.19a)}$$

$$P_{\text{tot}}^{'} = \gamma\left(P_{\text{tot}} - \beta\,\frac{1}{c}E_{\text{tot}}\right) \qquad \text{(SR.19b)}$$

in frame S' moving at $\beta = v/c$. The total momentum in S' is zero when

$$\beta_{\text{com}} = \frac{cP_{\text{tot}}}{E_{\text{tot}}} \qquad \text{(SR.20)}$$

which makes S' the center-of-mass frame.

Example SR.16 *Just passing through*
In the lab frame S, a particle of mass m and speed $u = \frac{4}{5}c$ strikes a target particle of mass $5m$ at rest. What is the speed of the CoM frame S'? Find the speeds of the two particles in S'.
Solution. The total energy in the lab frame is the incident particle's energy plus the target particle's rest energy.

$$E_{\text{tot}} = \gamma mc^2 + 5mc^2 = \frac{mc^2}{\sqrt{1-\left(\frac{4}{5}\right)^2}} + 5mc^2 = \frac{20}{3}c^2 m$$

Only the incident particle has momentum in the lab frame.

$$P_{\text{tot}} = \gamma mu = \frac{m\left(\frac{4}{5}c\right)}{\sqrt{1-\left(\frac{4}{5}\right)^2}} = \frac{4}{3}cm$$

Let's use Eq. (SR.20) to Evaluate the velocity of the CoM frame.

$$\beta_{\text{com}} = \frac{v_{\text{com}}}{c} = \frac{c\left(\frac{4}{3}mc\right)}{\frac{20}{3}mc^2} = \frac{1}{5}$$

Substitute (with Evaluate) the values for u and v into the addition of velocities Eq. (SR.5b) to get the projectile's velocity in the CoM frame.

$$u' = \left[\frac{u-v}{1-\frac{uv}{c^2}}\right]_{u=4c/5,v=c/5} = \frac{5}{7}c$$

The target particle's velocity in the center-of-mass frame is $-\frac{1}{5}c$ since it's at rest in the lab frame. ◀

All collisions conserve energy and momentum in all inertial reference frames. Elastic collisions also conserve the total kinetic energy and the total rest energy. KE is conserved only when there is no mass-energy transformation so the total mass is constant.

Inelastic collisions transform kinetic energy into mass (or mass into KE). In a completely inelastic collision, two particles move toward each other along a straight line, collide and stick together. This is physically equivalent to the decay of one particle into two. One process is the other run in reverse.

Example SR.17 *When 1 + 1 doesn't always equal 2*
A particle with rest mass m and energy three times its rest energy collides with an identical particle at rest. The collision is completely inelastic and the particles fuse together. Find the mass and velocity of the resulting particle.
Solution. We're given the incident particle's energy $E = 3mc^2$ which tells us it has $\gamma = 3$. We can use Solve Exact and Eq. (SR.1) to find its initial velocity.

$$3 = \left(1 - \frac{u^2}{c^2}\right)^{-1/2}, \text{ Solution is: } \frac{2\sqrt{2}}{3}c$$

Energy conservation tells us that

$$E + mc^2 = \gamma_f\, Mc^2 \Longrightarrow 4mc^2 = \frac{Mc^2}{\sqrt{1 - \dfrac{V^2}{c^2}}}$$

where $\gamma_f = \left(1 - V^2/c^2\right)^{-1/2}$ refers to the final particle M. Momentum is also conserved, so the incident particle's momentum equals the fused particle's final momentum.

$$\gamma mu = \gamma_f\, MV \Longrightarrow 2\sqrt{2}mc = \frac{MV}{\sqrt{1 - \dfrac{V^2}{c^2}}}$$

Sometimes *SNB* can't handle radicals in the denominator, so let's square both equations. Tell *SNB* to assume the variables are all positive so it doesn't return the extraneous negative solutions, and then use Solve Exact to find M, V.

$$\begin{bmatrix} \left(4mc^2\right)^2 = \dfrac{\left(Mc^2\right)^2}{1 - \frac{V^2}{c^2}} \\ \left(2\sqrt{2}mc\right)^2 = \dfrac{\left(MV\right)^2}{1 - \frac{V^2}{c^2}} \end{bmatrix}, \text{ Solution is: } \left[M = 2\sqrt{2}m, V = \tfrac{1}{2}\sqrt{2}c\right]$$

The mass $M = 2\sqrt{2}m$ moves off at speed $V = c/\sqrt{2}$. Notice that mass M is larger than twice the total initial mass $2m$. Some of the incident particle's kinetic energy was converted into mass.

$$\frac{\Delta KE}{\Delta mc^2} = \frac{(\gamma_f - 1)\, Mc^2 - (\gamma - 1)\, mc^2}{Mc^2 - mc^2} = \frac{(\sqrt{2} - 1)\, 2\sqrt{2}mc^2 - 2mc^2}{2\sqrt{2}mc^2 - 2mc^2} = -1$$

See Problem 24 for a more general approach to this problem. ◂

Example SR.18 *Why aren't they called protrons?*
A proton-proton collision can create a π^0 meson if there is enough energy.

$$p + p \to p + p + \pi^0$$

What is the minimum lab-frame kinetic energy the incident proton must have to produce a meson if the target proton is at rest?

Solution. When the incident proton has the minimum kinetic energy, the collision creates particles that are at rest in the center-of-mass frame so they must have the same speed (and gamma) in the lab frame. Energy conservation tells us that

$$K + 2Mc^2 = \gamma\,(2M + m)\,c^2$$

where M is the proton mass, m is the meson mass, and K is the kinetic energy of the incident proton in the lab frame. Momentum conservation tells us that

$$P = \gamma\,(2M + m)\,u$$

where u is the speed of the final particles in the lab frame. It's also the speed of the center-of-mass frame relative to the lab. Let's use the energy-momentum relation to rewrite momentum conservation.

$$\left(K + Mc^2\right)^2 - M^2c^4 = c^2\gamma^2\,(2M + m)^2\,u^2$$

Tell *SNB* to assume all the variables are positive and use Solve Exact (and a little editing by-hand) to find exact expressions for K, γ, u.

$$\begin{bmatrix} K + 2Mc^2 = \gamma\,(2M+m)\,c^2 \\ \left(K + Mc^2\right)^2 - M^2c^4 = c^2\gamma^2\,(2M+m)^2\,u^2 \\ \gamma^2 = \left(1 - \frac{u^2}{c^2}\right)^{-1} \end{bmatrix}, \text{ Solution is:}$$

$$\left[K = \left(2 + \frac{m}{2M}\right)mc^2, u = \frac{\sqrt{m\,(4M+m)}}{2M+m}c, \gamma = 1 + \frac{m}{2M}\right]$$

Instead of using Substitute to get numerical values, let's use Solve Exact again with the appropriate values for the rest energy of the proton $\left(Mc^2 = 938.3\,\mathrm{MeV}\right)$ and meson $\left(mc^2 = 135.0\,\mathrm{MeV}\right)$. We don't need gamma, so let's recast the conservation equations in terms of $\beta = u/c$. One click of the Solve Exact button gives us numerical answers.

$$\begin{bmatrix} K + 2\,(938.3) = \dfrac{1}{\sqrt{1-\beta^2}}\,(2\,(938.3) + 135.0) \\ (K + 938.3)^2 - 938.3^2 = (2\,(938.3) + 135.0)^2\,\dfrac{\beta^2}{1-\beta^2} \end{bmatrix},$$

Solution is: $[K = 279.71, \beta = 0.360\,16]$

In the lab frame, the proton must have at least $K = 279.71\,\mathrm{MeV}$ to create a π^0 meson, and when it does the meson will be moving at $u = 0.360\,16\,c$. In the center of mass frame, both protons are initially moving and all particles are formed at rest so $\beta' = 0$.

$2K' + 2\,(938.3) = 2\,(938.3) + 135.0$, Solution is: 67.5

The kinetic energy of each proton in the CoM frame is half the meson's rest energy. ◀

Relativistic Dynamics

There is a place for Newton's 2nd Law and accelerated motion in Special Relativity. The SR descriptions of events made by observers in inertial reference frames are valid even if those events are accelerating. The accelerating object is at rest in its own reference frame, which you can consider a series of co-moving inertial frames.

Suppose an accelerating object moving at velocity $\overrightarrow{u}$ is seen by observers in inertial frame S. Newton's 2nd Law is valid in the form

$$\overrightarrow{F} = \frac{d\overrightarrow{p}}{dt} \tag{SR.21}$$

when $\overrightarrow{p}$ is the relativistic momentum of Eq. (SR.14). This leads us to the **relativistic Newton's 2nd Law** which says

$$\overrightarrow{F} = \gamma_u m \overrightarrow{a} + \gamma_u^3 m \left(\frac{\overrightarrow{a} \cdot \overrightarrow{u}}{c^2} \right) \overrightarrow{u} \tag{SR.22}$$

where $\overrightarrow{a} = d\overrightarrow{u}/dt$ is the object's acceleration in frame S. The solution of this equation gives the accelerating object's velocity as a function of time $\overrightarrow{u}(t)$ as seen in S.

Newton's simple relationship between net force and acceleration is only valid for small velocities. The relativistic force depends on both the acceleration and velocity, and the force and acceleration are not always in the same direction. The relativistic 2nd Law isn't quite so complicated when the acceleration is parallel or perpendicular to the velocity.

$$\overrightarrow{F} = \begin{cases} \gamma_u^3 m \overrightarrow{a} & \overrightarrow{a} \parallel \overrightarrow{u} \\ \gamma_u m \overrightarrow{a} & \overrightarrow{a} \perp \overrightarrow{u} \end{cases} \tag{SR.23}$$

Let's test the relativistic 2nd Law with a look at motion with a constant acceleration that's parallel to the velocity. The constant acceleration $\alpha \equiv F/m$ is measured in the accelerating object's non-inertial reference frame. The relativistic velocity as a function of time for a constant acceleration is

$$u(t) = \frac{\gamma_0 u_0 + \alpha t}{\sqrt{1 + \left(\frac{\gamma_0 u_0 + \alpha t}{c} \right)^2}} \tag{SR.24}$$

where u_0 is the object's initial speed and γ_0 is the corresponding initial gamma factor. Visit the Relativistic Interlude in Appendix B for the gory details of this calculation.

This result has a very cool limit. Tell *SNB* to assume the speed of light and the acceleration are positive.

$\text{assume}(c > 0) = (0, \infty)$ $\text{assume}(\alpha > 0) = (0, \infty)$

Then Simplify the following expression.

$$\lim_{t\to\infty} \left(\frac{\gamma_0 u_0 + \alpha t}{\sqrt{1 + \left(\frac{\gamma_0 u_0 + \alpha t}{c} \right)^2}} \right) = c$$

It takes an infinite amount of time to reach the speed of light with a constant acceleration. This is consistent with the cosmic speed limit that's a consequence of the 2nd Postulate.

Figure SR.6 shows the velocity as a function of time from Eq. (SR.24) for three different initial speeds

$$u_0 = 0, \pm\tfrac{1}{2}c$$

when the constant acceleration is

$$\alpha = +1\,c/\,\mathrm{y}.$$

In each case, the velocity approaches but never reaches the speed of light.

Our new relativistic 2nd Law is consistent with the 2nd Postulate!

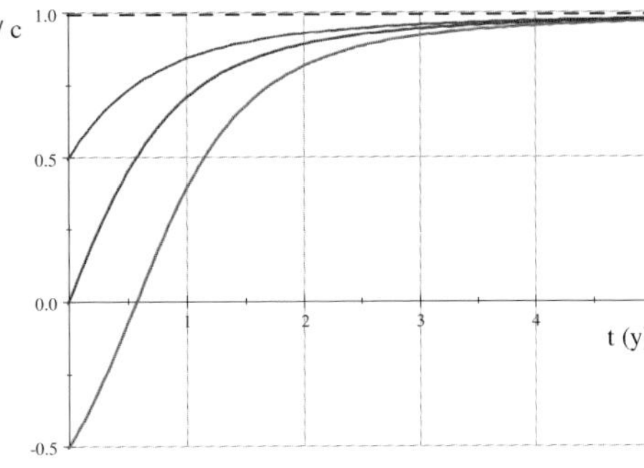

Figure SR.6 SR $u(t)$

Example SR.19 *Beyond full impulse*

A starship moving at $\frac{1}{2}c$ relative to frame S accelerates at a constant $1c/\,\mathrm{y}$. How long does the starship need to reach a speed of $0.95c$? How fast is it going after 1 year?

Solution. First, let's Evaluate the starship's initial gamma factor.

$$\gamma_0 = \left(1 - \left(\tfrac{1}{2}\right)^2\right)^{-1/2} = \tfrac{2}{3}\sqrt{3}$$

Use Eq. (SR.24) with the appropriate numerical values to create and Evaluate an expression for the starship's velocity as a function of time (in years).

$$\frac{u}{c} = \left[\frac{\dfrac{\gamma_0 u_0 + \alpha t}{c}}{\sqrt{1+\left(\dfrac{\gamma_0 u_0 + \alpha t}{c}\right)^2}}\right]_{\alpha=1c,u_0=+c/2,\gamma_0=2/\sqrt{3}} = \frac{\frac{1}{3}\sqrt{3}c + ct}{c\sqrt{\frac{1}{c^2}\left(\frac{1}{3}\sqrt{3}c + ct\right)^2 + 1}}$$

Create an expression (with a little editing by-hand) setting u/c equal to 0.95, leave the insertion point anywhere in it and click the Solve Exact button.

$$0.95 = \frac{\frac{1}{\sqrt{3}} + t}{\sqrt{\left(\frac{1}{\sqrt{3}} + t\right)^2 + 1}}, \text{ Solution is: } 2.465\,1$$

As seen from frame S, it takes the starship $2.4651\,\mathrm{y}$ to reach a speed of $0.95c$. Use Substitute with Evaluate Numerically to calculate the speed after 1 year.

$$u = \left[\frac{\frac{1}{\sqrt{3}} + t}{\sqrt{\left(\frac{1}{\sqrt{3}} + t\right)^2 + 1}}c\right]_{t=1} = 0.844\,57c$$

According to Newton, the speed after one year would be $u_0 + \alpha t = \frac{3}{2}c$, which is more than the speed of light. ◄

This next example may cause you to reconsider a career in Star Fleet.

Example SR.20 *Full stop? Now?*

A starship approaches Star Base 47 moving at a constant speed of $\frac{1}{2}c$. How much time does it take the starship to stop with a constant acceleration of $10c/\mathrm{y}$? What acceleration is needed to stop it in 1 day?

Solution. We need to calculate the time it takes a relativistic starship to go from $u_0 = \frac{1}{2}c$ to $u = 0$ when the constant acceleration is $\alpha = -10c/\mathrm{y}$. Create the following expression setting the speed as a function of time equal to zero and use Solve Exact to find the time.

$$0 = \left[\frac{\gamma_0 u_0 + \alpha t}{\sqrt{1 + \left(\frac{\gamma_0 u_0 + \alpha t}{c}\right)^2}}\right]_{\alpha=-10.0c/\mathrm{y},u_0=+c/2,\gamma_0=2/\sqrt{3}},$$

Solution is: $1.8219 \times 10^6\ \mathrm{s}$

Solve Exact returned the answer in seconds. This is not a particularly illuminating result, so let's use the Standard Method to convert this time into days.

$$1.8219 \times 10^6\ \mathrm{s} \times \frac{1\ \mathrm{d}}{86400\ \mathrm{s}} = 21.087\ \mathrm{d}$$

As seen from the star base, it takes the starship more than 21 days to stop, even with an acceleration of almost $10g$.

To find the acceleration needed for a 1-day stop, create the following expression setting the speed equal to zero and using $1\ \mathrm{d} = 1/365.24\ \mathrm{y}$ for the time. Then use Solve Exact to find the acceleration α.

$$0 = \left[\frac{\gamma_0 u_0 + \alpha t}{\sqrt{1 + \left(\frac{\gamma_0 u_0 + \alpha}{c}\right)^2}}\right]_{t=1y/365.24,u_0=+c/2,\gamma_0=2/\sqrt{3}}, \text{ Solution is: } -210.87\frac{c}{y}$$

Let's use Evaluate Numerically to convert this acceleration into units of g.

$$\alpha = -210.87 \times \frac{3 \times 10^8 \frac{\mathrm{m}}{\mathrm{s}}}{1\ \mathrm{y}} \times \frac{g}{9.8067 \frac{\mathrm{m}}{\mathrm{s}^2}} = -204.42g$$

This is a huge acceleration! Inertial dampers notwithstanding, it's not likely the ship or crew could survive this approximately $200g$ acceleration for a day. ◀

Relativity or not, once we know the velocity as a function of time we can use the results of Chapter 2 to calculate the position. The relativistic position as a function of time for constant-acceleration motion is

$$x(t) = x_0 + \frac{c^2}{\alpha}\left(\sqrt{1+\left(\frac{\gamma_0 u_0 + \alpha t}{c}\right)^2} - \gamma_0\right) \tag{SR.25}$$

where x_0 is the initial position.

Figure SR.7 shows the position as a function of time with $x_0 = 0$ for the same acceleration and 3 initial speeds used in Figure SR.6.

For large values of time, all three position-versus-time graphs are linear with the same slope.

What do you predict for the value of that slope? Make a prediction, then take a look at Problem 39.

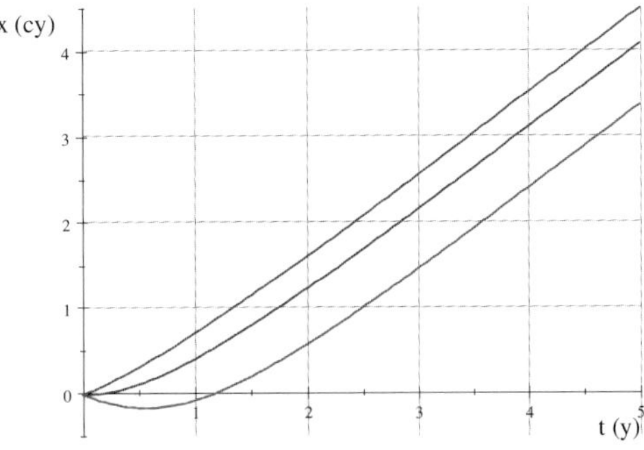

Figure SR.7 SR $x(t)$

Example SR.21 *Beyond full impulse 2*

A starship moving at $\frac{1}{2}c$ relative to frame S experiences a constant acceleration of $1c/\text{y}$. How far does the starship travel to reach a speed of $0.95c$? How much time does it take the starship to travel 1 light-year?

Solution. Use Eq. (SR.25) with the appropriate numerical values to create and Simplify an expression for the starship's position as a function of time (in years).

$$x = \left[\frac{c^2}{\alpha}\left(\sqrt{1+\left(\frac{\gamma_0 u_0 + \alpha t}{c}\right)^2} - \gamma_0\right)\right]_{\alpha=1c, u_0=+c/2, \gamma_0=2/\sqrt{3}}$$

$$= c\sqrt{t^2 + \tfrac{2}{3}\sqrt{3}t + \tfrac{4}{3}} - \tfrac{2}{3}\sqrt{3}c$$

Use Evaluate Numerically and the result of Example SR.19 to calculate the distance.

$$x = \left[\sqrt{t^2 + \tfrac{2}{3}\sqrt{3}t + \tfrac{4}{3}} - \tfrac{2}{3}\sqrt{3}\right]_{t=2.4651} = 2.0479$$

As seen from frame S, after 2.4651 years the ship has travelled 2.0479 light-years.

Create an expression setting the position equal to 1.0 light-year, leave the insertion point anywhere in it, and use Solve Exact to find the time t.

$$1.0 = \left(\sqrt{t^2 + \tfrac{2}{3}\sqrt{3}t + \tfrac{4}{3}} - \tfrac{2}{3}\sqrt{3}\right), \text{ Solution is: } 1.3312$$

It takes our starship 1.3312 y to travel 1 light-year. ◀

The relativistic velocity as a function of position for constant-acceleration motion is

$$u(x) = c\sqrt{1 - \left(\gamma_0 + \frac{\alpha}{c^2}(x - x_0)\right)^{-2}} \tag{SR.26}$$

where x_0 is the initial position and γ_0 contains all the initial-speed dependence.

Here is another cool limit. Tell *SNB* to assume the speed of light and the acceleration are positive and then Simplify the following expression.

$$\lim_{x\to\infty}\left(c\sqrt{1 - \left(\gamma_0 + \frac{\alpha}{c^2}(x - x_0)\right)^{-2}}\right) = c$$

It takes an infinite amount of space to reach the speed of light with a constant acceleration. No matter how far or for how long an object is pushed, its speed will never equal or exceed the cosmic speed limit!

In the accelerating frame, a constant force F is applied to an object with rest mass m and produces the constant acceleration $\alpha = F/m$. The object has mass $\gamma_u m$ in frame S, where its increasing mass means it has more inertia and a decreasing acceleration. The acceleration is *not* constant in S. Einstein called his paper that introduced $E = mc^2$ "Does the Inertia of a Body Depend on its Energy Content?" for good reason. The faster an object moves, the more inertia it has.

Example SR.22 *Full stop? Here?*
How far does our starship travel as it comes to rest from an initial speed of $\frac{1}{2}c$ with an acceleration of magnitude $10c/\,\mathrm{y}$?
Solution. We need to calculate the change in position for a relativistic starship as it goes from $u_0 = \frac{1}{2}c$ to $u = 0$ when the acceleration is $\alpha = -10c/\,\mathrm{y}$. Create and Evaluate the following expression for the speed as a function of position.

$$\frac{u}{c} = \left[\sqrt{1 - \left(\gamma_0 + \frac{\alpha x}{c^2}\right)^{-2}}\right]_{\alpha=-10c/\,\mathrm{y},u_0=+c/2,\gamma_0=2/\sqrt{3}} = \sqrt{1 - \frac{1}{\left(\frac{2}{3}\sqrt{3} - 10\frac{x}{c\,\mathrm{y}}\right)^2}}$$

Take this expression without the $c\,\mathrm{y}$ so we can use Solve Numeric to find the distance. The $c\,\mathrm{y}$ just sets the units so we don't need it as long as we work in light-years and years.

$$0 = \sqrt{1 - \frac{1}{\left(\frac{2}{\sqrt{3}} - 10x\right)^2}}, \text{ Solution is: } \{[x = 0.015\,47]\}$$

In the star base frame, the ship travels $0.015\,47c\,\mathrm{y}$ while stopping. Let's use Evaluate to convert that into astronomical units.

$$x = 0.015\,47c\,\mathrm{y} \times \frac{63239\,\mathrm{AU}}{1c\,\mathrm{y}} = 978.31\,\mathrm{AU}$$

This is approximately 12 times the diameter of Pluto's orbit around the Sun. Our starship can't stop on a dime. ◀

Four-Vectors

Special Relativity tells us space and time are parts of a unified 4-dimensional space-time. Space-time diagrams provide a geometric representation of the relativistic effects of space-time. Four-vectors provide a useful and elegant mathematical tool to describe and calculate those effects in 4 dimensions.

Each **4-vector** has four components, one for each dimension of space-time, so a generic 4-vector $\boldsymbol{b}$ is

$$\boldsymbol{b} = (b_t, b_x, b_y, b_z) \tag{SR.27}$$

where b_t is the time component and b_x, b_y, b_z are the three space components. The square of the 4-vector's magnitude is

$$|\boldsymbol{b}|^2 = \boldsymbol{b}^T \boldsymbol{\eta}\, \boldsymbol{b} \tag{SR.28}$$

where $\boldsymbol{b}^T$ is the vector's transpose.

In relativistic lingo, $\boldsymbol{\eta}$ is called the metric tensor (or **metric** for short). The metric describes the geometry of space-time, giving the recipe for calculating the separation between points in space-time. Special Relativity's metric is a diagonal 4×4 matrix.

$$\boldsymbol{\eta} = \begin{bmatrix} +1 & 0 & 0 & 0 \\ 0 & -1 & 0 & 0 \\ 0 & 0 & -1 & 0 \\ 0 & 0 & 0 & -1 \end{bmatrix} \tag{SR.29}$$

All the elements of the SR metric in Cartesian coordinates are constant, so space-time is flat in Special Relativity. No gravity!

In *SNB* we write a 4-vector as a 1-column, 4-row matrix. The order of the components ensures the metric, 4-vectors, and the Lorentz Transformation matrix are all consistent. The time component is always first, followed by the three spatial components in the usual x-y-z order.

$$\boldsymbol{b} = \begin{bmatrix} b_t \\ b_x \\ b_y \\ b_z \end{bmatrix}$$

To calculate a vector's Transpose, put an uppercase T in a Superscript after the vector and Evaluate the expression.

$$\boldsymbol{b}^T = \begin{bmatrix} b_t \\ b_x \\ b_y \\ b_z \end{bmatrix}^T = \begin{bmatrix} b_t & b_x & b_y & b_z \end{bmatrix}$$

Our generic 4-vector's transpose is a 4-column, 1-row matrix that we can use to Evaluate its magnitude.

$$|\boldsymbol{b}|^2 = \begin{bmatrix} b_t & b_x & b_y & b_z \end{bmatrix} \begin{bmatrix} +1 & 0 & 0 & 0 \\ 0 & -1 & 0 & 0 \\ 0 & 0 & -1 & 0 \\ 0 & 0 & 0 & -1 \end{bmatrix} \begin{bmatrix} b_t \\ b_x \\ b_y \\ b_z \end{bmatrix} = b_t^2 - b_x^2 - b_y^2 - b_z^2$$

The quantity $|\boldsymbol{b}|^2$ can be positive, negative or zero.

Three important 4-vectors are the 4-position $\boldsymbol{s}$, 4-velocity $\boldsymbol{u}$, and 4-momentum $\boldsymbol{p}$.

$$\boldsymbol{s} = (ct, x, y, z) \tag{SR.30a}$$

$$\boldsymbol{u} = \gamma_u (c, u_x, u_y, u_z) \tag{SR.30b}$$

$$\boldsymbol{p} = \left(\frac{E}{c}, \gamma_u m u_x, \gamma_u m u_y, \gamma_u m u_z\right) \tag{SR.30c}$$

$$= \gamma_u (mc, mu_x, mu_y, mu_z) \tag{SR.30d}$$

An object's 4-momentum $\boldsymbol{p} = m\boldsymbol{u}$ is its rest mass times its 4-velocity.

The magnitude of the 4-position is the space-time interval between the event at those coordinates and the origin.

$$|\boldsymbol{s}|^2 = \begin{bmatrix} ct & x & y & z \end{bmatrix} \begin{bmatrix} +1 & 0 & 0 & 0 \\ 0 & -1 & 0 & 0 \\ 0 & 0 & -1 & 0 \\ 0 & 0 & 0 & -1 \end{bmatrix} \begin{bmatrix} ct \\ x \\ y \\ z \end{bmatrix} = c^2t^2 - x^2 - y^2 - z^2$$

As we've already seen, this interval is the same in all inertial reference frames. This is a general result: the magnitude of any 4-vector is absolute and does not depend on the frame of reference.

Example SR.23 *We still call them 4-vectors*
Calculate the magnitude of an object's 4-velocity and 4-momentum it its rest frame and in a frame where it's moving along the x-axis with speed u.
Solution. The object is moving on the x-axis, so we only need one spatial dimension and the 2×2 versions of the metric and 4-vectors.

$$\boldsymbol{\eta} = \begin{bmatrix} +1 & 0 \\ 0 & -1 \end{bmatrix} \qquad \boldsymbol{b} = \begin{bmatrix} b_t \\ b_x \end{bmatrix}$$

In the object's rest frame S', its velocity is zero so its 4-velocity is $\boldsymbol{u}' = \begin{bmatrix} c \\ 0 \end{bmatrix}$.
Use Eq. (SR.26) to create and Simplify an expression for the magnitude squared.

$$|\boldsymbol{u}'|^2 = \begin{bmatrix} c & 0 \end{bmatrix} \begin{bmatrix} +1 & 0 \\ 0 & -1 \end{bmatrix} \begin{bmatrix} c \\ 0 \end{bmatrix} = c^2$$

Repeat the process in frame S where the 4-velocity is $\boldsymbol{u} = \begin{bmatrix} \gamma_u c \\ \gamma_u u \end{bmatrix}$.

$$|\boldsymbol{u}|^2 = \begin{bmatrix} \dfrac{c}{\sqrt{1 - \dfrac{u^2}{c^2}}} & \dfrac{u}{\sqrt{1 - \dfrac{u^2}{c^2}}} \end{bmatrix} \begin{bmatrix} +1 & 0 \\ 0 & -1 \end{bmatrix} \begin{bmatrix} \dfrac{c}{\sqrt{1 - \dfrac{u^2}{c^2}}} \\ \dfrac{u}{\sqrt{1 - \dfrac{u^2}{c^2}}} \end{bmatrix} = c^2$$

The magnitude of the 4-velocity is always the speed of light! You can think of all objects as moving through space-time at the speed of light. The 4-momentum $\boldsymbol{p}$ is the rest mass times the 4-velocity, so its magnitude is always mc. ◀

The Lorentz Transformation gives us another chance to use *SNB*'s powerful matrix capability. The 4×4 matrix for the $S \rightarrow S'$ transformation of Eqs. (SR.6) is

$$\boldsymbol{\mathcal{L}} = \begin{bmatrix} \gamma & -\gamma\frac{v}{c} & 0 & 0 \\ -\gamma\frac{v}{c} & \gamma & 0 & 0 \\ 0 & 0 & 1 & 0 \\ 0 & 0 & 0 & 1 \end{bmatrix} \tag{SR.31}$$

where v is the relative velocity along the common x-x' axes between the two frames. The $S \rightarrow S'$ transformation equations for a 4-vector's components take a simple form.

$$\boldsymbol{b}' = \boldsymbol{\mathcal{L}}\,\boldsymbol{b} \tag{SR.32}$$

We can use the symmetry rule to generate the $S' \rightarrow S$ LT matrix.

$$\boldsymbol{\mathcal{L}}' = \begin{bmatrix} \gamma & +\gamma\frac{v}{c} & 0 & 0 \\ +\gamma\frac{v}{c} & \gamma & 0 & 0 \\ 0 & 0 & 1 & 0 \\ 0 & 0 & 0 & 1 \end{bmatrix} \tag{SR.33}$$

The $S' \rightarrow S$ transformation equations also take a simple form.

$$\boldsymbol{b} = \boldsymbol{\mathcal{L}}'\,\boldsymbol{b}' \tag{SR.34}$$

Let's look at the $S \rightarrow S'$ transformation in more detail with Evaluate, and then Factor and Expand applied in-place to the result.

$$\begin{bmatrix} b'_t \\ b'_x \\ b'_y \\ b'_z \end{bmatrix} = \begin{bmatrix} \gamma & -\gamma\frac{v}{c} & 0 & 0 \\ -\gamma\frac{v}{c} & \gamma & 0 & 0 \\ 0 & 0 & 1 & 0 \\ 0 & 0 & 0 & 1 \end{bmatrix} \begin{bmatrix} b_t \\ b_x \\ b_y \\ b_z \end{bmatrix} = \begin{bmatrix} \gamma\left(b_t - \dfrac{v}{c}b_x\right) \\ \gamma\left(b_x - \dfrac{v}{c}b_t\right) \\ b_y \\ b_z \end{bmatrix}$$

It's not surprising that the components of a 4-vector and space-time coordinates transform exactly the same way. In fact, 4-vectors are defined so their components transform per the Lorentz Transformation so their magnitudes are the same in all IRFs.

The Lorentz Transformation, which leaves the magnitude of 4-vectors unchanged, is mathematically similar to the rotation matrix, which leaves the magnitude of 3-vectors unchanged. The similarity is evident when we write the LT as

$$\boldsymbol{\mathcal{L}} = \begin{bmatrix} \cosh\theta & -\sinh\theta & 0 & 0 \\ -\sinh\theta & \cosh\theta & 0 & 0 \\ 0 & 0 & 1 & 0 \\ 0 & 0 & 0 & 1 \end{bmatrix} \tag{SR.35}$$

where $\gamma \equiv \cosh\theta$. The parameter θ is called the rapidity.

$$\boldsymbol{s}' = \boldsymbol{\mathcal{L}}\,\boldsymbol{s} = \begin{bmatrix} \cosh\theta & -\sinh\theta & 0 & 0 \\ -\sinh\theta & \cosh\theta & 0 & 0 \\ 0 & 0 & 1 & 0 \\ 0 & 0 & 0 & 1 \end{bmatrix} \begin{bmatrix} ct \\ x \\ y \\ z \end{bmatrix} = \begin{bmatrix} ct\cosh\theta - x\sinh\theta \\ x\cosh\theta - ct\sinh\theta \\ y \\ z \end{bmatrix}$$

The LT is a coordinate rotation in 4-d space-time.

Example SR.24 *After you, 2*

Use the Lorentz Transformation matrices to verify the results of Example SR.8.

Solution. This problem uses only one spatial dimension so all we need are the 2×2 versions of the LT matrices.

$$\boldsymbol{\mathcal{L}} = \begin{bmatrix} \gamma & -\gamma\frac{v}{c} \\ -\gamma\frac{v}{c} & \gamma \end{bmatrix}$$

$$\boldsymbol{\mathcal{L}}' = \begin{bmatrix} \gamma & +\gamma\frac{v}{c} \\ +\gamma\frac{v}{c} & \gamma \end{bmatrix}$$

In the Kirk-Spock frame S, the 4-displacement between the events is $\Delta\boldsymbol{s} = \begin{bmatrix} +6\,\text{m} \\ +9\,\text{m} \end{bmatrix}$.

We need to Evaluate an expression for γ when $v = \frac{2}{3}c$.

$$\gamma = \frac{1}{\sqrt{1 - \left(\frac{2}{3}\right)^2}} = \tfrac{3}{5}\sqrt{5}$$

Create and Evaluate an expression for the $S \rightarrow S'$ transformation of $\Delta\boldsymbol{s}$.

$$\Delta\boldsymbol{s}' = \boldsymbol{\mathcal{L}}\,\Delta\boldsymbol{s} = \begin{bmatrix} \dfrac{3}{\sqrt{5}} & -\left(\dfrac{3}{\sqrt{5}}\right)\dfrac{2}{3} \\ -\left(\dfrac{3}{\sqrt{5}}\right)\dfrac{2}{3} & \dfrac{3}{\sqrt{5}} \end{bmatrix} \begin{bmatrix} +6\,\text{m} \\ +9\,\text{m} \end{bmatrix} = \begin{bmatrix} 0 \\ 3\sqrt{5}\,\text{m} \end{bmatrix}$$

The intervals in S' are $c\Delta t' = 0$ and $\Delta x = +9\,\text{m}$.

Repeat the process for the $S' \rightarrow S$ transformation of $\Delta\boldsymbol{s}'$.

$$\Delta\boldsymbol{s} = \boldsymbol{\mathcal{L}}'\,\Delta\boldsymbol{s}' = \begin{bmatrix} \dfrac{3}{\sqrt{5}} & +\left(\dfrac{3}{\sqrt{5}}\right)\dfrac{2}{3} \\ +\left(\dfrac{3}{\sqrt{5}}\right)\dfrac{2}{3} & \dfrac{3}{\sqrt{5}} \end{bmatrix} \begin{bmatrix} 0 \\ 3\sqrt{5}\,\text{m} \end{bmatrix} = \begin{bmatrix} 6\,\text{m} \\ 9\,\text{m} \end{bmatrix}$$

We recover the original intervals in S. Let's calculate the magnitude of these two vectors.

$$|\Delta\boldsymbol{s}'|^2 = \begin{bmatrix} 0 & 3\sqrt{5}\,\text{m} \end{bmatrix} \begin{bmatrix} +1 & 0 \\ 0 & -1 \end{bmatrix} \begin{bmatrix} 0 \\ 3\sqrt{5}\,\text{m} \end{bmatrix} = -45\,\text{m}^2$$

$$|\Delta\boldsymbol{s}|^2 = \begin{bmatrix} 6\,\text{m} & 9\,\text{m} \end{bmatrix} \begin{bmatrix} +1 & 0 \\ 0 & -1 \end{bmatrix} \begin{bmatrix} 6\,\text{m} \\ 9\,\text{m} \end{bmatrix} = -45\,\text{m}^2$$

Four-vectors have different components in different frames, but their magnitudes are the same in all IRFs. This invariance has physical implications. The 4-velocity's invariant magnitude means the speed of light is the same in all IRFs. The 4-momentum's invariant magnitude means a particle's rest mass is the same in all IRFs. ◀

Example SR.25 *Ride the meson*
Transform the 4-momentum of the π^0 meson created in Example SR.18 to its rest frame, and use the result to determine the meson's momentum and energy in this frame.
Solution. The meson's rest frame S' moves relative to the lab frame S at the speed u we calculated in Example SR.18. Per the Lorentz Transformation, the meson's 4-momentum in S' is

$$\begin{aligned} \boldsymbol{p}'_\pi &= \begin{bmatrix} E'_\pi/c \\ p'_\pi \end{bmatrix} \\ &= \boldsymbol{\mathcal{L}}\,\boldsymbol{p}_\pi \\ &= \begin{bmatrix} \gamma & -\gamma\frac{u}{c} \\ -\gamma\frac{u}{c} & \gamma \end{bmatrix} \begin{bmatrix} E_\pi/c \\ p_\pi \end{bmatrix} \end{aligned}$$

where $E_\pi = \gamma mc^2$ and $p_\pi = \gamma mu$.
Tell *SNB* to assume the two masses are positive.

$$\text{assume}\,(M > 0) = (0, \infty)$$

$$\text{assume}\,(m > 0) = (0, \infty)$$

Let's use our results for γ and u from Example SR.18 to Simplify the off-diagonal terms in the LT matrix.

$$\gamma\frac{u}{c} = \left(1 + \frac{m}{2M}\right)\frac{\sqrt{m\,(4M+m)}}{2M+m} = \frac{1}{2M}\sqrt{m}\sqrt{4M+m}$$

Let's Simplify this expression for the meson's 4-momentum in S'.

$$\boldsymbol{p}'_\pi = \begin{bmatrix} 1+\frac{m}{2M} & -\frac{\sqrt{m\,(4M+m)}}{2M} \\ -\frac{\sqrt{m\,(4M+m)}}{2M} & 1+\frac{m}{2M} \end{bmatrix} \begin{bmatrix} \left(1+\frac{m}{2M}\right)mc \\ \frac{\sqrt{m\,(4M+m)}}{2M}mc \end{bmatrix} = \begin{bmatrix} cm \\ 0 \end{bmatrix}$$

The π^0 meson is at rest in S' so its energy is its rest energy so $E'_\pi = mc^2$ and its has zero momentum so $p'_\pi = 0$.
Let's use Simplify to check the magnitude of the meson's 4-momentum in S.

$$\begin{aligned} |\boldsymbol{p}_\pi|^2 &= \begin{bmatrix} \left(1+\frac{m}{2M}\right)mc & \frac{\sqrt{m\,(4M+m)}}{2M}mc \end{bmatrix} \begin{bmatrix} +1 & 0 \\ 0 & -1 \end{bmatrix} \begin{bmatrix} \left(1+\frac{m}{2M}\right)mc \\ \frac{\sqrt{m\,(4M+m)}}{2M}mc \end{bmatrix} \\ &= c^2m^2 \end{aligned}$$

The magnitude is the same in both frames. ◄

Problems

1. Convert the speed of light in a vacuum into
   a. miles per second,
   b. miles per hour,
   c. meters per nanosecond,
   d. astronomical units per year, and
   e. light-years per year.

2. How far is one light-hour?

3. What is the value for gamma when an object is moving at half the speed of light?

4. What is the value for gamma when an object is moving at 99.9% the speed of light?

5. What is the value for gamma when an object is moving at 99.99% the speed of light?

6. For what speed does $\gamma = 10$?

7. For what speed does $\gamma = 100$?

8. Reproduce the graph of γ versus v shown in Figure SR.1.

9. Verify that the current human speed record of $25000\,\mathrm{mi/h}$ is about $c/27000$.

10. Use *SNB*'s Power Series... command to calculate the lowest-order velocity-dependent term in the expansion of γ. Verify numerically that for humans this is less than 1 part-per-billion.

11. Show that for small speeds the relativistic gamma factor is approximately
$$\gamma \approx 1 + \tfrac{1}{2}\frac{v^2}{c^2}\,.$$
What is the next term in the expansion?

12. Prove that $\gamma^2 v^2 = c^2\left(\gamma^2 - 1\right)$.

13. Verify that $c\gamma = \sqrt{c^2 + \gamma^2 v^2}$.

14. Consider the three events $s_1 = (2,1)$, $s_2 = (1,5)$, and $s_3 = (5,4)$ in frame S.
   a. Plot the three events on a space-time diagram.
   b. Calculate the three space-time intervals.
   c. Label each interval as space-like, time-like or light-like on your diagram.

15. What are the coordinates of the three events from the previous problem in a frame moving at $v = \frac{1}{5}c$? Verify by direct calculation that the three intervals are absolute.

16. Make the space-time diagram for the event and frames in Example SR.7.

17. Let's use Example SR.8 to explore my claim that the smallest space interval between two events is in the frame where the events are simultaneous.

    a. Show that Eqs. (SR.8a) and (SR.8b) take the form

$$c\Delta t' = \frac{1}{\sqrt{1-\beta^2}}(6-9\beta)$$

$$\Delta x' = \frac{1}{\sqrt{1-\beta^2}}(9-6\beta)$$

    where $\beta \equiv v/c$.

    b. Use basic calculus to show $\Delta x'$ has its minimum at $\beta = \frac{2}{3}$.

    c. Show that $c\Delta t' = 0$ when $\Delta x'$ has its minimum value.

    d. Plot the expressions for $c\Delta t'$ and $\Delta x'$ with $0 < \beta < 1$ on the same graph. Comment on where $c\Delta t'$ changes sign and $\Delta x'$ has its minimum.

18. Show that the angle between the x and x' axes on a space-time diagram obeys the relation $\tan\theta = v/c$.

19. How fast must an object be moving so it has momentum $p = mc$?

20. Calculate the speed and momentum of a proton whose kinetic energy is $5\,\mathrm{MeV}$. Compare your answers with those in Example SR.13.

21. How much kinetic energy must a proton have so it has the same speed as the electron in Example SR.14?

22. Verify by explicit calculation that the total momentum in the CoM frame of Example SR.16 is zero. What is the total energy in this frame?

23. Use the addition of velocities to verify the target particle in Example SR.16 moves at velocity $u_5' = -\frac{1}{5}c$ in the center of mass frame.

24. Let's generalize the inelastic collision of Example SR.17. A particle with rest mass m and energy N times its rest energy collides with an identical particle at rest. The collision is completely inelastic and they fuse together.

    a. Find the mass and velocity of the resulting particle in terms of N.

    b. Show that $N = 1$ gives the minimum energy for the incident particle.

    c. What is the minimum mass of the resulting particle?

25. What is the speed and momentum of the incident proton in Example SR.18?

26. A proton-proton collision can create a proton-antiproton pair if there is sufficient energy.

$$p + p \rightarrow p + p + p + \overline{p}$$

    What is the minimum kinetic energy the incident proton must have to produce the pair if the target proton is at rest?

27. Consider a reaction where mass m_1 collides with mass m_2 which is initially at rest. The collision is completely inelastic and the two masses stick together to form an object with mass M. Calculate the mass and speed of new object in terms of the initial masses and kinetic energy.

28. A particle of mass M at rest decays into two unequal masses m_1 and m_2. Show that the square of the momentum of each final particle is

$$p^2 = \frac{c^2}{4M^2}\left(\left(M^2 - (m_1 + m_2)^2\right)\left(M^2 - (m_1 - m_2)^2\right)\right)$$

29. Consider electron-positron pair production in the vicinity of an electron.

$$\gamma + e^- \to e^+ + e^- + e^-$$

Assume the process occurs at threshold so the incident photon has the minimum energy needed to create the pair and the three final particles all have the same speed.

a. Obtain an expression for the final speed of the particles in terms of the photon energy E_γ.

b. Prove that the threshold photon energy is $4mc^2$.

Note: photons are massless particles that are subject to Eq. (SR.18).

30. Convert $a = 1\,c/\,\mathrm{y}$ into the usual metric unit of acceleration.

31. Show that the acceleration due to gravity near the Earth's surface is $g = 1.032\,3\,c/\,\mathrm{y}$.

32. Prove that $1\,c/\,\mathrm{y}$ is equal to one light-year per year-squared. Give a physical interpretation of this in words.

33. The acceleration α in Eq. (SR.23), measured in the rest frame of the accelerating object, is constant. The acceleration a, measured in inertial reference frame S, is not. Show that

$$a\,(t) = \frac{\alpha}{\left(1 + \dfrac{\gamma_0 u_0 + \alpha t}{c}\right)^{3/2}}$$

and plot $a\,(t)$ for the same initial conditions as in Figure SR.6.

34. A starship moving at $-\frac{1}{2}c$ experiences a constant acceleration of $+1c/\,\mathrm{y}$. How long does it take the starship to reach a speed of $0.95c$? How fast is it going after 1 year? ($3.619\,8\,\mathrm{y}$, $0.389\,31c$)

35. A starship initially at rest experiences a constant acceleration of $+1c/\,\mathrm{y}$. How long does it take the starship to reach a speed of $0.95c$? How fast is it going after 1 year? ($3.042\,4\,\mathrm{y}$, $0.707\,11c$)

36. Show that for *small* values of time, the speed as a function of time for a relativistic object moving with a constant acceleration is approximately $u\,(t) \approx u_0 + \alpha t$.

37. Show that for *large* values of time, the speed as a function of time for a relativistic object moving with a constant acceleration is approximately $u\,(t) \approx c$.

38. Show that for *small* values of time, the position as a function of time for a relativistic object moving with a constant acceleration is approximately $x(t) \approx x_0 + \frac{1}{2}\alpha t^2$ when it starts from rest.

39. Show that for *large* values of time, the position as a function of time for a relativistic object moving with a constant acceleration is approximately $x(t) \approx x_0 + ct$ when it starts from rest.

40. Show that for *small* values of displacement, the speed as a function of position for a relativistic object moving with a constant acceleration reduces to $u^2(x) = u_0^2 + 2\alpha(x - x_0)$.

41. Show that for *large* values of displacement, the speed as a function of position for a relativistic object moving with a constant acceleration is approximately $u(x) \approx c$.

42. Verify the following limits for relativistic motion under a constant acceleration.

    a. $\lim_{x \to 0} \left(u(x)\right) = u_0$ b. $\lim_{x \to \infty} \left(u(x)\right) = c$

43. Make a graph of velocity versus position for a relativistic object moving with a constant acceleration of $1\, c/\,\mathrm{y}$ for three different initial velocities $\left(u_0 = 0, \pm\frac{1}{2}c\right)$. Take $x_0 = 0$.

44. Let's use the relativistic position as a function of time to verify the result of Example SR.22. Show that the slowing starship's position is given by

$$x(t) = \frac{1}{10}\left(\frac{2}{\sqrt{3}} - \sqrt{\left(\frac{1}{\sqrt{3}} - 10t\right)^2 + 1}\right)$$

and use this result to calculate the distance the starship travels while it is slowing down. You may want to take a look at Example SR.20.

45. Show that the square of a 4-vector's magnitude can be written as $|\boldsymbol{b}|^2 = b_t^2 - \vec{b} \cdot \vec{b}$ using the standard 3-d dot product.

46. Verify that the magnitude of a generic 4-momentum equals mc.

47. Verify that the product of the Lorentz Transformation matrix $\boldsymbol{\mathcal{L}}$ times the inverse $\boldsymbol{\mathcal{L}}'$ is the identity matrix. Give a physical interpretation of this.

48. Suppose an electron and a positron travel in opposite directions, collide head-on, and briefly cling together as a single particle called a clingon. The electron has total energy $E_e = 25\,\mathrm{GeV}$, the positron has $E_p = 50\,\mathrm{GeV}$, and they both have rest energy $mc^2 = 0.511\,\mathrm{MeV}$.

    a. Write down the lab-frame four-momenta for the two incident particles in terms of E_e, E_p, and m.

    b. Show that the electron-positron rest energy is negligible here.

    c. Use the clingon's 4-momentum in the lab frame to calculate its rest energy Mc^2.

49. Calculate the speed of the clingon in the previous problem. Use the Lorentz Transformation to transform the clingon's 4-momentum into the frame where it is at rest. What are the clingon's final momentum and energy in this frame?

50. Use the Lorentz Transformation matrix and the 4-momentum to verify my claim that energy and momentum obey the same transformation rules as time and space coordinates.

51. Show that if $\gamma \equiv \cosh\theta$ then $\gamma\beta = \sinh\theta$.

52. The Borg arrive at the star Wolf 359 (7.8 light-years away) just as the starship Enterprise moves past Earth at $\frac{12}{13}c$. The Earth and the ship are at the origin in their own frames and the ship passed the Earth at $t = t' = 0$.

    a. Write the position 4-vector for the event of the Borg's arrival in the Earth frame.

    b. What is the position 4-vector for the same event in the ship's frame?

53. An electron moving at $\frac{4}{5}c$ passes through an experimental tubular doohickey. To the electron, the doohickey has length L_e. Write the 4-displacement for the space and time intervals between the electron entering and leaving the doohickey in the electron's frame. Transform the 4-displacement to the doohickey's frame.

54. An electron moving at $\frac{4}{5}c$ passes through an experimental tubular doohickey that has rest length L_0. Write the 4-displacement for the space and time intervals between the electron entering and leaving the doohickey in the doohickey's frame. Transform the 4-displacement into the electron's frame.

55. Suppose a particle has this 4-momentum in the lab frame.

$$\boldsymbol{p} = \begin{bmatrix} E/c \\ p \end{bmatrix}$$

    a. What is its 4-momentum in a frame moving at speed u relative to the lab?

    b. Show that the velocity of the particle's rest frame is $u = c^2 p/E$. Verify that in its rest frame the particle has energy mc^2.

56. Consider a reaction where mass m with energy $E = \gamma mc^2$ collides with mass m which is initially at rest. The collision is completely inelastic and the two masses stick together to form an object with mass M.

    a. Show that the total 4-momentum before the collision in the lab frame is

$$\boldsymbol{p}_0 = \begin{bmatrix} (\gamma + 1)\, mc \\ \sqrt{\gamma^2 - 1}\, mc \end{bmatrix}$$

    b. Show that the speed of the center of mass frame is $\beta_{\text{com}} = \sqrt{\dfrac{\gamma - 1}{\gamma + 1}}$.

    c. What is the final 4-momentum in the center of mass frame?

    d. Transform $\boldsymbol{p}_0$ into the center of mass frame and show that $M = \sqrt{2(\gamma + 1)}\, m$.

A Topics in Classical Physics

In this appendix, we'll use *SNB* to explore some interesting problems in classical physics. These problems are not usually found in textbooks, so they are not typically part of the physics curriculum, but they are interesting and fun. Some would make worthwhile undergraduate research projects. They are worthy of our time and attention. In other words, I like them!

Newton's Nose-Cone Problem

In Book II of the *Principia*, Newton considered the shape of a solid of rotation that experiences the least air resistance as it moves at constant speed along the direction of its axis. [4, 22] It is a testament to his genius that Newton was formulating and solving such a problem in the 1680s, spawning a research field that is still active. [20]

Figure NC.1 illustrates the physical situation.

The nose cone has height H, base radius R, and the function $z(x)$ describes its shape.

The 3-dimensional nose cone is created by rotating this function about the z-axis, so it has radial (cylindrical) symmetry and a circular cross section.

The term "nose cone" refers to the generic shape and not necessarily an actual "cone", although that is a shape we will consider.

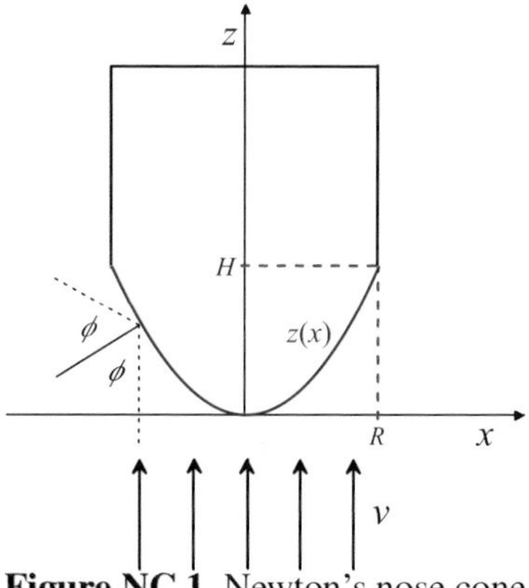

Figure NC.1 Newton's nose cone

Each air molecule collides elastically with the nose cone once, and the total drag force they exert is

$$F_D = 2\rho v^2 \int \cos^3 \phi \, dA \qquad \text{(NC.1)}$$

where v is the relative speed, ρ is the air density, and the integral is over the nose cone's surface. The **drag coefficient** contains all the geometric information about the effect of the nose cone's shape on air resistance, giving its effective cross-sectional area.

$$C_D = \int \cos^3 \phi \, dA \qquad \text{(NC.2)}$$

When the nose cone is a flat circle of radius R, the deflection angle is $\phi = 0°$ and the drag coefficient is the area of the circle (πR^2).

Doing Physics with Scientific Notebook: A Problem-solving Approach, First Edition. Joseph Gallant.

We want the nose-cone shape that minimizes the air resistance, so we need to relate the drag coefficient to the shape.

Figure NC.2 shows the geometry that relates the nose-cone shape z to the deflection angle ϕ.

The area of the strip defined by rotating the arc ds about the z-axis is

$$dA = 2\pi x\, ds$$

Trigonometry tells us $dx = \cos\phi\, ds$

and the Pythagorean Theorem says

$$\cos^2\phi = \frac{1}{1+\left(\dfrac{dz}{dx}\right)^2}$$

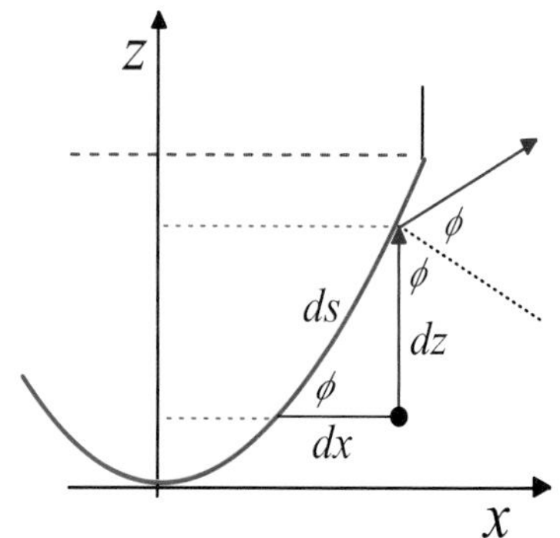

Figure NC.2 Solid of rotation

We can combine these results with Eq. (NC.2) to get an expression for the drag coefficient. It's often easier to use the dimensionless **reduced drag coefficient** (RDC), which is defined as $C \equiv C_D/\left(\pi R^2\right)$.

$$C = \frac{1}{R^2}\int_0^R \frac{2x}{1+\left(\dfrac{dz}{dx}\right)^2}\, dx \tag{NC.3}$$

This is the effective fractional cross-sectional area of the nose cone. When the nose cone is a flat circle of radius R, the RDC equals one. We can use this to test our results: in the zero-height $H \to 0$ limit, all our RDC results should reduce to one.

When the shape is a function of x/R, we can simplify the integral by letting $u \equiv x/R$.

$$C = \int_0^1 \frac{2u}{1+\dfrac{1}{R^2}\left(\dfrac{dz}{du}\right)^2}\, du \tag{NC.4}$$

This expression is particularly useful when we must approximate the RDC numerically.

Simple Shapes

Let's start with simple shapes using a general power-law form.

$$z(x) = H\left(\frac{x}{R}\right)^n \tag{NC.5a}$$

$$z'(x) = n\frac{H}{R}\left(\frac{x}{R}\right)^{n-1} \tag{NC.5b}$$

The flat nose cone has $n = 0$ or $n = \infty$. The RDC for a simply-shaped nose cone is

$$C = \int_0^1 \frac{2u}{1+\eta^2\, u^{2n-2}}\, du \tag{NC.6}$$

where $\eta \equiv nH/R$ and $u \equiv x/R$.

For a given exponent, the RDC depends on the ratio H/R. This scaling is a general result: for a given simple shape, you can change the size of the nose cone but the RDC stays the same if the ratio H/R remains constant.

Cone ($n = 1$): For a conic nose cone, the shape is linear and its derivative is constant.

$$\begin{aligned} z(x) &= H\frac{x}{R} \\ z'(x) &= \frac{H}{R} \end{aligned}$$

We can use Evaluate and Eq. (NC.6) to calculate the RDC.

$$C_{\text{cone}} = \frac{1}{1+\dfrac{H^2}{R^2}} \int_0^1 2u\,du = \frac{1}{\dfrac{H^2}{R^2}+1}$$

This is a general result for the RDC of a cone-shaped nose cone.

$$C_{\text{cone}} = \frac{1}{1+\dfrac{H^2}{R^2}} \tag{NC.7}$$

As the ratio gets larger, the cone's RDC gets smaller, so for a given radius long thin cones produce less air resistance. When $H = R$, the RDC is $C_{\text{cone}} = \frac{1}{2}$, so this nose-cone cone has half as much air resistance as the flat surface with the same radius.

Parabola ($n = 2$): For a parabolic nose cone, the shape is quadratic and its derivative is linear.

$$\begin{aligned} z(x) &= H\left(\frac{x}{R}\right)^2 \\ z'(x) &= 2\frac{H}{R}\frac{x}{R} \end{aligned}$$

Let's use Eq. (NC.6), the definite integral fudge (with Simplify) and Combine + Logs in-place.

$$C_{\text{para}} = \left[\int \frac{2u}{1+\frac{4H^2}{R^2}u^2}\,du\right]_{u=0}^{u=1} = \frac{1}{4H^2}R^2\left(\ln\frac{1}{R^2}\left(4H^2+R^2\right)\right)$$

A little editing by-hand, using Expand in-place, produces the following general result.

$$C_{\text{para}} = \frac{R^2}{4H^2}\ln\left(1+\frac{4H^2}{R^2}\right) \tag{NC.8}$$

For the sake of comparison, let's Evaluate this at $H = R$.

$$C_{\text{para}} = \left[\frac{R^2}{4H^2}\ln\left(1+\frac{4H^2}{R^2}\right)\right]_{H=R} = \tfrac{1}{4}\ln 5$$

You can apply Evaluate or Evaluate Numerically to Eq. (NC.6) and verify this.

$$C_{\text{para}} = \int_0^1 \frac{2u}{1+4u^2}\,du = \tfrac{1}{4}\ln 5 = 0.402\,36$$

When the height and radius are equal, the parabolic nose cone has almost 20% less air resistance than the conic nose cone.

Ellipse: Newton included a hemispherical nose cone in his analysis. Let's generalize that with an off-center ellipse that reduces to Newton's hemisphere when $R = H$.

$$z(x) = H\left(1 - \sqrt{1 - \frac{x^2}{R^2}}\right)$$

We'll need its derivative so let's Evaluate this expression.

$$\frac{dz}{du} = \frac{d}{du}\left(H\left(1 - \sqrt{1-u^2}\right)\right) = H\frac{u}{\sqrt{1-u^2}}$$

We can use Simplify and Eq. (NC.4) to calculate the general RDC.

$$C_{\text{ell}} = \left[\int \frac{2u}{1 + \left(\frac{H}{R}\frac{u}{\sqrt{1-u^2}}\right)^2} du\right]_{u=0}^{u=1}$$
$$= \frac{R^2}{\left(H^2 - R^2\right)^2}\left(H^2 \ln \frac{H^2}{H^2 - R^2} - H^2 \ln \frac{R^2}{H^2 - R^2} - H^2 + R^2\right)$$

A little editing by-hand, using Combine + Logs and Expand in-place, produces the following general result.

$$C_{\text{ell}} = \frac{1 - \frac{H^2}{R^2}\left(1 - \ln \frac{H^2}{R^2}\right)}{\left(1 - \frac{H^2}{R^2}\right)^2} \tag{NC.9}$$

As the ratio gets larger, the ellipse's RDC gets smaller, so for a given radius long thin ellipses produce less air resistance.

Let's see if we agree with Newton's result that the hemispherical nose cone has an RDC of $C_{\text{hemi}} = \frac{1}{2}$. The denominator in Eq. (NC.9) equals zero when $H = R$, so we cannot Evaluate this expression. Instead we can take the limit.

$$C_{\text{hemi}} = \lim_{H \to R}\left(\frac{1 - \frac{H^2}{R^2}\left(1 - \ln \frac{H^2}{R^2}\right)}{\left(1 - \frac{H^2}{R^2}\right)^2}\right) = \frac{1}{2}$$

You can also apply Evaluate to Eq. (NC.4) and verify this result.

$$C_{\text{hemi}} = \int_0^1 \frac{2u}{1 + \left(\frac{u}{\sqrt{1-u^2}}\right)^2} du = \frac{1}{2}$$

When the height and radius are equal, the hemispherical nose cone offers the same air resistance as the conic nose cone. The hemisphere has a flatter tip which increases the air resistance, but it's steeper along the edge which reduces the air resistance.

Figure NC.3 shows what these three simple shapes look like.

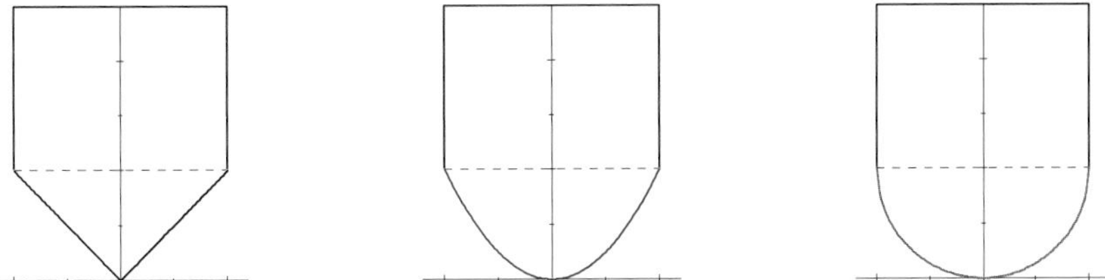

Figure NC.3 Cone, parabola, and hemisphere for $H = R$

The following table summarizes our results for the air resistance of nose cones with simple shapes.

Shape	RDC: $C\left(\frac{H}{R}\right)$	$C(H = R)$
Flat $(n = 0, \infty)$	1	1
Cone $(n = 1)$	$\dfrac{1}{1 + \frac{H^2}{R^2}}$	$\frac{1}{2} = 0.5$
Parabola $(n = 2)$	$\frac{R^2}{4H^2} \ln\left(1 + \frac{4H^2}{R^2}\right)$	$\frac{1}{4} \ln 5 = 0.402\,36$
Ellipse	$\dfrac{1 - \frac{H^2}{R^2}\left(1 - \ln \frac{H^2}{R^2}\right)}{\left(1 - \frac{H^2}{R^2}\right)^2}$	$\frac{1}{2} = 0.5$

Table NC.1

Minimizing Exponent ($H = R$)

The results for the flat, conic, and parabolic nose cones suggest that for a given value of the ratio H/R there is an exponent that minimizes the RDC of our simple shapes. Once we pick a value for the ratio, it's easy to find the exponent that minimizes the air resistance with *SNB*.

Using Eq. (NC.6) we can create a graph of the reduced drag coefficient versus exponent for the simple shapes.

Figure NC.4 shows this graph when the height and radius are equal.

The graph shows the minimum RDC occurs near $n = 2.2$ when $H = R$.

Let's find the minimizing exponent with *SNB*'s data-fitting capability.

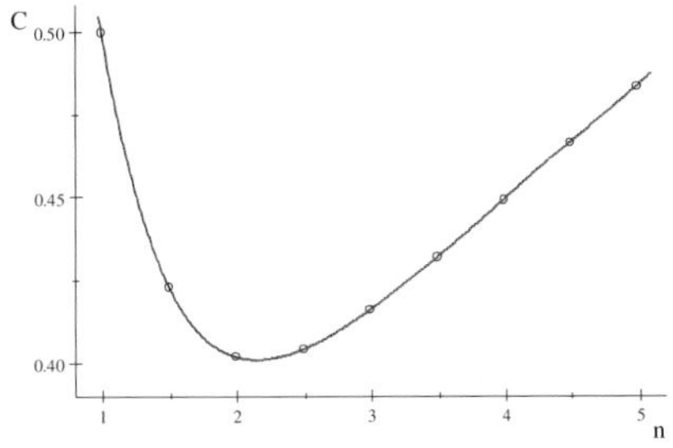

Figure NC.4 RDC vs. Exponent

Near a maximum or a minimum, you can always approximate a function with a quadratic fit. The closer you are to the maximum/minimum the more accurate the resulting fit. We'll need a 2-column matrix for our fit. Use Substitute with Evaluate Numerically and Eq. (NC.6) to calculate the RDCs so you can easily change the value of the exponent.

$$C = \left[\int_0^1 \frac{2u}{1 + n^2 u^{2n-2}} du \right]_{n=2.15} = 0.401\,42$$

Here are the resulting values and the degree-2 polynomial fit.

$$\begin{bmatrix} n & C \\ 2.13 & 0.401\,45 \\ 2.14 & 0.401\,43 \\ 2.15 & 0.401\,42 \\ 2.16 & 0.401\,41 \\ 2.17 & 0.401\,41 \\ 2.18 & 0.401\,42 \\ 2.19 & 0.401\,44 \\ 2.20 & 0.401\,46 \end{bmatrix}, \text{Polynomial fit: } C = 3.690\,5 \times 10^{-2} n^2 - 0.159\,65n + 0.574\,08$$

To find the exponent that minimizes the RDC, set the derivative of the polynomial fit to zero and use Solve Exact.

$$0 = \frac{d}{dn}\left(3.690\,5 \times 10^{-2} n^2 - 0.159\,65n + 0.574\,08\right), \text{ Solution is: } 2.163\,0$$

Use Evaluate Numerically and the minimizing exponent $n = 2.163\,0$ to find the corresponding minimum value for the RDC.

$$C = \left[\int_0^1 \frac{2u}{1 + n^2 u^{2n-2}} du \right]_{n=2.163\,0} = 0.401\,41$$

This is slightly smaller than the RDC for $n = 2.16$.

$$\left[\int_0^1 \frac{2u}{1 + n^2 u^{2n-2}} du \right]_{n=2.1630} - \left[\int_0^1 \frac{2u}{1 + n^2 u^{2n-2}} du \right]_{n=2.16} = -1.970\,8 \times 10^{-7}$$

Of all simple-shaped nose cones, this one feels the least air resistance when $H = R$.

$$z(x) = x^{2.1630}$$

The corresponding minimum RDC is

$$C = 0.401\,41$$

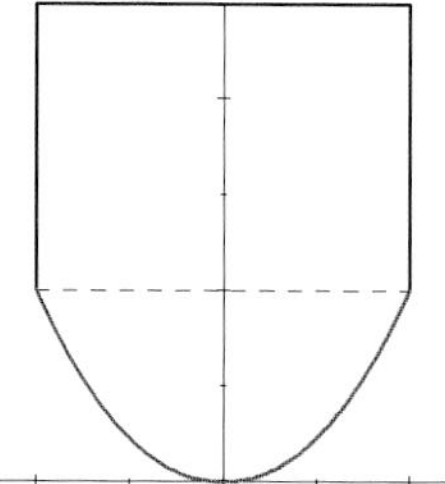

Figure NC.5 The minimum, so far...

Frusta and Fudges

Toward his goal of finding the solid of least resistance, Newton next considered the air resistance felt by a frustum.

A frustum is a cone with the apex lopped off so that the tip is parallel to the base.

Straight lines connect the tip and base.

The circular tip has radius a, and the nose cone still has height H and base radius R.

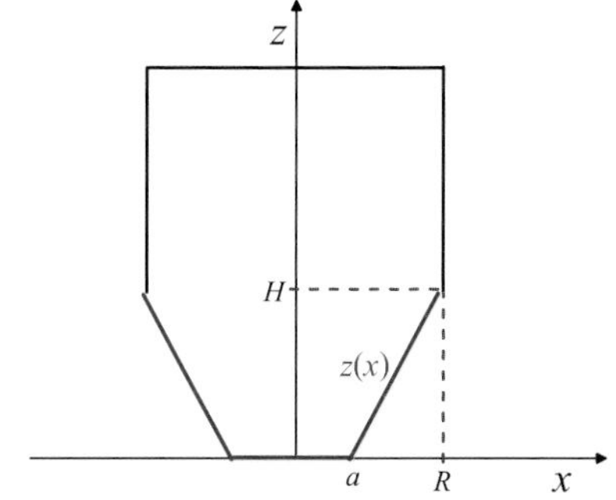

Figure NC.6 Frustum

The reduced drag coefficient of a frustum is just the sum of the contributions from the flat tip and the rest.

$$C_{\text{frus}} = \frac{a^2}{R^2} + \frac{1}{R^2}\int_a^R \frac{2x}{1+\left(\dfrac{dz}{dx}\right)^2}\,dx \tag{NC.10}$$

The scaling here depends on two ratios. The frustum's RDC stays the same as long as both H/R and a/R remain constant.

Newton's Frustum

Here are the mathematical expressions for a frustum's shape and its derivative.

$$z(x) = \begin{cases} 0 & \text{if} \quad x < a \\ H\left(\dfrac{x-a}{R-a}\right) & \text{if} \quad x \geq a \end{cases} \tag{NC.11a}$$

$$z'(x) = \begin{cases} 0 & \text{if} \quad x < a \\ \dfrac{H}{R-a} & \text{if} \quad x \geq a \end{cases} \tag{NC.11b}$$

The derivative is constant, so let's use Eqs. (NC.10) and (NC.11b) to create the following expression and Expand the expression in the parentheses in-place.

$$\frac{1}{1+\left(\dfrac{H}{R-a}\right)^2}\left(\frac{1}{R^2}\int_a^R 2x dx\right) = \frac{1}{1+\left(\dfrac{H}{R-a}\right)^2}\left(1-\frac{1}{R^2}a^2\right)$$

A little editing by-hand produces the following result.

$$C_{\text{frus}} = \frac{a^2}{R^2} + \frac{1-\dfrac{a^2}{R^2}}{1+\left(\dfrac{H}{R-a}\right)^2} \tag{NC.12}$$

This is an exact expression for the RDC of a frustum.

To find the tip-width that minimizes the RDC for given values of R and H, take the derivative of Eq. (NC.12) with respect to a and use Solve Exact to find a.

$$0=\frac{d}{da}\left(\frac{a^2}{R^2}+\frac{1-\dfrac{a^2}{R^2}}{1+\left(\dfrac{H}{R-a}\right)^2}\right), \text{ Solution is: } \frac{1}{R}\left(-\tfrac{1}{2}H\sqrt{H^2+4R^2}+\tfrac{1}{2}H^2+R^2\right)$$

Now take this exact expression for the minimizing tip-width and Substitute the appropriate values.

$$a_{\text{min}}=\left[\frac{1}{R}\left(-\tfrac{1}{2}H\sqrt{H^2+4R^2}+\tfrac{1}{2}H^2+R^2\right)\right]_{H=1,R=1}=\tfrac{3}{2}-\tfrac{1}{2}\sqrt{5}=0.381\,97$$

If all you want is the numerical answer, you could do this in one step.

$$0=\frac{d}{da}\left(\left[\frac{a^2}{R^2}+\frac{1-\dfrac{a^2}{R^2}}{1+\left(\dfrac{H}{R-a}\right)^2}\right]_{H=1,R=1}\right), \text{ Solution is: } \tfrac{3}{2}-\tfrac{1}{2}\sqrt{5}$$

We can use Substitute to find the minimum RDC for a frustum with $R=H$.

$$C_{\text{frus}}=\left[\frac{a^2}{R^2}+\frac{1-\dfrac{a^2}{R^2}}{1+\left(\dfrac{H}{R-a}\right)^2}\right]_{H=1,R=1,a=0.381\,97}=0.381\,97$$

Newton's flat-tipped frustum offers less air resistance than all of our simple shapes! There are two competing effects: the flat tip has large air resistance but it allows the nose cone to have steeper sides, which reduces the air resistance.

Adjusted Frustum

Let's see if we can do better. We'll keep the flat tip but adjust the curve connecting it to the base. Here are the mathematical expressions for this shape and its derivative.

$$z(x)=\begin{cases}0 & x<a\\ H\left(\dfrac{x-a}{R-a}\right)^n & x\geq a\end{cases}\tag{NC.13a}$$

$$z'(x)=\begin{cases}0 & x<a\\ \dfrac{nH}{R-a}\left(\dfrac{x-a}{R-a}\right)^{n-1} & x\geq a\end{cases}\tag{NC.13b}$$

This shape has two adjustable parameters, the exponent n and the tip-width a.

There is a tip-width that minimizes the RDC for each value of the exponent. We seek the minimum minimum: the tip-width and exponent that give the overall minimum RDC for a given height and base radius. Then we'll see if it's less than Newton's frustum.

With the variable change $u = \frac{x-a}{R-a}$ $\left(\text{so } du = \frac{1}{R-a}\,dx\right)$ and Eq. (NC.10), we can express the RDC in terms of a dimensionless integral.

$$C_{\text{adj}} = \frac{a^2}{R^2} + \left(1 - \frac{a}{R}\right)^2 \int_0^1 \frac{2\left(\frac{a}{R-a} + u\right)}{1 + \left(\frac{nH}{R-a}\right)^2 u^{2n-2}}\,du \tag{NC.14}$$

In most cases we'll have to approximate this integral numerically.

Once you've picked values for R and H, use the following numerical procedure for minimizing the two-parameter RDC.

1. Define functions for the RDC and its derivatives.
2. Use the derivatives to find the minimum RDC for several sets of parameters.
3. Make a 3-dimensional plot of the data $\begin{bmatrix} a & n & C \end{bmatrix}$ and approximately locate the overall minimum RDC.
4. Make a quadratic fit near the overall minimum and find the minimizing parameters a_{min} and n_{min}.
5. Evaluate the RDC function $C_{\text{min}} = C\left(a_{\text{min}}, n_{\text{min}}\right)$.

Let's find the values for the two parameters that minimize the RDC when $H = R = 1$.

Use New Definition and Eq. (NC.14) to define a function for the RDC.

$$C(a,n) = a^2 + (1-a)^2 \int_0^1 \frac{2\left(\frac{a}{1-a} + u\right)}{1 + \left(\frac{n}{1-a}\right)^2 u^{2n-2}}\,du$$

Let's check this definition by using it to Evaluate the result for Newton's frustum.

$$C(0.381\,97, 1) = 0.381\,97$$

The explicit expressions for the derivatives are cumbersome, but we don't need them since we're solving this problem numerically. Let *SNB* do the work and use New Definition to define this function for the derivative with respect to the tip width.

$$D(a,n) = 2a + 2\left[\int_0^1 \frac{\partial}{\partial q}\left(\frac{(1-q)^2\left(\frac{q}{1-q} + u\right)}{1 + \left(\frac{1}{1-q}\right)^2 n^2 u^{2n-2}}\right) du\right]_{q=a}$$

Notice I used Substitute and a dummy differentiation variable. This forces *SNB* to calculate the derivative at the appropriate value before it does the numerical integration.

You can also define a function for the derivative with respect to the exponent.

$$N(a,n) = 2(1-a)^2 \left[\int_0^1 \frac{\partial}{\partial q} \left(\frac{\left(\frac{a}{1-a}+u\right)}{1+\left(\frac{1}{1-a}\right)^2 q^2 u^{2q-2}} \right) du \right]_{q=n}$$

Let's use these definitions and create one of the data points in our 3-d matrix of the minimum RDC. The minimum RDC for $n = 1.3$ is near $a = 0.33$ (a plot is helpful here). Use Evaluate Numerically to create data for the derivative versus tip-width near the minimum, and then do a Multiple Regression fit.

$$\begin{bmatrix} a & d \\ 0.315 & D(0.315, 1.3) \\ 0.320 & D(0.320, 1.3) \\ 0.325 & D(0.325, 1.3) \\ 0.330 & D(0.330, 1.3) \\ 0.335 & D(0.335, 1.3) \\ 0.340 & D(0.340, 1.3) \end{bmatrix} = \begin{bmatrix} a & d \\ 0.315 & -2.4831 \times 10^{-2} \\ 0.32 & -1.6522 \times 10^{-2} \\ 0.325 & -8.0930 \times 10^{-3} \\ 0.33 & 4.5789 \times 10^{-4} \\ 0.335 & 9.1311 \times 10^{-3} \\ 0.34 & 1.7928 \times 10^{-2} \end{bmatrix},$$

Regression is: $d = 1.7103a - 0.56378$

The derivative is zero at the minimum, so use Solve Exact to find the minimizing tip-width.

$0 = 1.7103a - 0.56378$, Solution is: 0.32964

Now that we know the minimum RDC for $n = 1.3$ happens when $a = 0.32964$, we can use Evaluate Numerically and our defined function to calculate it.

$C(0.32964, 1.3) = 0.37611$

We just calculated one of the data points we'll need: $\begin{bmatrix} a & n & C \\ 0.32964 & 1.3 & 0.37611 \end{bmatrix}$

Figure NC.7 shows a 3-d graph of fifteen such points for the minimum RDCs as a function of tip-width and exponent.

Here are the ranges for the exponent n and tip width a.

$$1 < n < 1.5$$

$$0.28 < a < 0.38$$

The overall minimum RDC is near $n = 1.2$ and $a = 0.35$

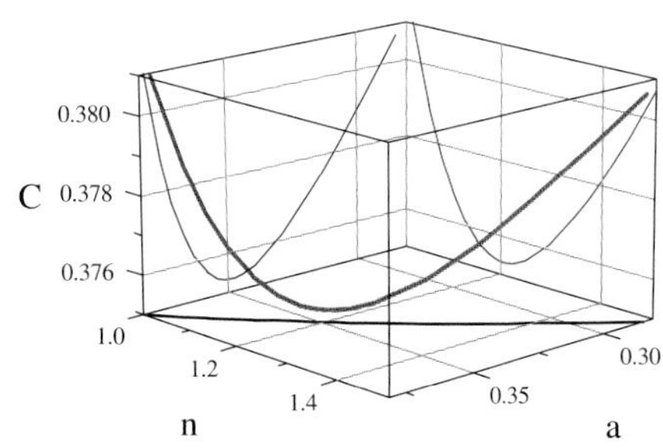

Figure NC.7 Adjustable adjustments

Here are five data points near the overall minimum shown in Figure NC.7. Select the two columns for the exponent and RDC, and fit a parabola.

$$\begin{bmatrix} a & n & c \\ 0.352\,96 & 1.18 & C(0.352\,96, 1.18) \\ 0.351\,16 & 1.19 & C(0.351\,16, 1.19) \\ 0.349\,32 & 1.20 & C(0.349\,32, 1.20) \\ 0.347\,48 & 1.21 & C(0.347\,48, 1.21) \\ 0.345\,61 & 1.22 & C(0.345\,61, 1.22) \end{bmatrix} = \begin{bmatrix} a & n & c \\ 0.352\,96 & 1.18 & 0.375\,36 \\ 0.351\,16 & 1.19 & 0.375\,32 \\ 0.349\,32 & 1.2 & 0.375\,30 \\ 0.347\,48 & 1.21 & 0.375\,30 \\ 0.345\,61 & 1.22 & 0.375\,32 \end{bmatrix},$$

Polynomial fit: $c = 0.100\,00n^2 - 0.241\,n + 0.520\,5$

Use Calculus + Find Extrema to find the minimizing exponent.

$0.100\,00n^2 - 0.241\,n + 0.520\,5$

Candidate(s) for extrema: $\{0.375\,30\}$, at $\{[n = 1.205]\}$

All we need now is the minimizing tip-width. Let's use the same five points and the RDC's derivative with respect to the exponent and fit a straight line near the minimum.

$$\begin{bmatrix} a & d \\ 0.352\,96 & N(0.352\,96, 1.18) \\ 0.351\,16 & N(0.351\,16, 1.19) \\ 0.349\,32 & N(0.349\,32, 1.20) \\ 0.347\,48 & N(0.347\,48, 1.21) \\ 0.345\,61 & N(0.345\,61, 1.22) \end{bmatrix} = \begin{bmatrix} a & d \\ 0.352\,96 & -5.601\,9\times10^{-3} \\ 0.351\,16 & -3.182\,1\times10^{-3} \\ 0.349\,32 & -9.074\,4\times10^{-4} \\ 0.347\,48 & 1.242\,2\times10^{-3} \\ 0.345\,61 & 3.261\,9\times10^{-3} \end{bmatrix},$$

Regression is: $d = 0.419\,84 - 1.204\,9a$

Set the linear fit for the derivative to zero, and use Solve Exact to find minimizing tip-width:

$0 = 0.419\,84 - 1.204\,9a$, Solution is: $0.348\,44$

The overall minimum value for the RDC happens when $n_{\min} = 1.205$ and $a_{\min} = 0.348\,44$.

$C(0.348\,44, 1.205) = 0.375\,30$

Let's compare the two frusta results.

$$100\frac{0.375\,30 - 0.381\,97}{0.381\,97} = -1.746\,2$$

Our adjusted frustum experiences about 1.75% less air resistance than Newton's frustum!

Joe's Fudge

Newton argued that the solid of least resistance has a flat tip connected to the base with a non-linear curve. Constructing the curve geometrically as a series of thin frusta, Newton showed that his minimizer has initial conditions $z(a) = 0$ and $z'(a) = 1$ such that $z(R) = H$. Let's construct a shape that satisfies these conditions and see if it reduces the RDC.

$$z(x) = \begin{cases} 0 & x < a \\ (x-a) + (H - R + a)\left(\dfrac{x-a}{R-a}\right)^n & x \geq a \end{cases} \tag{NC.15}$$

This shape is similar to the adjusted frustum and has the same two adjustable parameters (the exponent n and the tip-width a). We'll need its derivative to minimize the reduced drag coefficient.

$$\frac{dz}{dx} = 1 + n\frac{H - R + a}{R - a}\left(\frac{x-a}{R-a}\right)^{n-1} \tag{NC.16}$$

The procedure here for minimizing the RDC is identical to the one we used on the adjusted frustum. When $H = R = 1$, the minimum RDC happens for $a_{\text{min}} = 0.351\,97$ and $n_{\text{min}} = 1.717\,1$.

Once we have the minimizing parameters, we can use Evaluate Numerically on the following expression (see Problem 16) to find the minimum RDC.

$$C_{\text{min}} = \left[a^2 + (1-a)\int_0^1 \frac{2\,(a + (1-a)\,u)}{1 + \left(1 + \frac{an}{1-a}u^{n-1}\right)^2}\,du\right]_{a=0.351\,97, n=1.717\,1} = 0.374\,83$$

Let's compare my fudge to Newton's frustum.

$$pd = 100\frac{0.374\,83 - 0.381\,97}{0.381\,97} = -1.869\,3$$

The fudge experiences almost 2% less air resistance than Newton's frustum! Of course, my fudge relies entirely on his results.

Figure NC.8 shows Newton's frustum, the adjusted frustum, and my Newton-based fudge with the minimum air resistance for $H = R$.

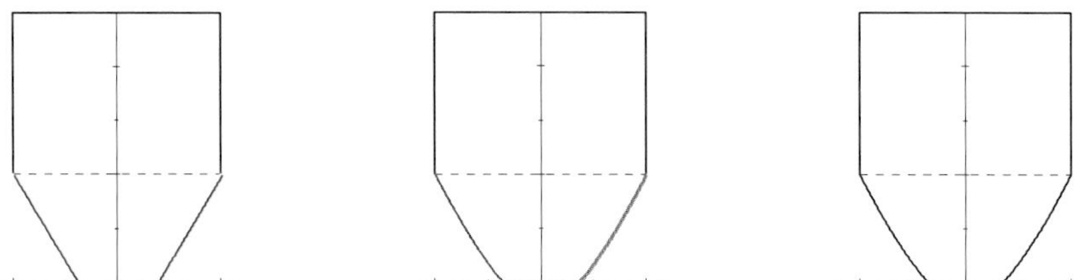

Figure NC.8 Frusta, adjusted and fudged, for $H = R$

Newton's Minimizer

In modern terms, Newton's minimizer satisfies the Euler-Lagrange equation

$$z'' = \frac{(z')^3 + z'}{3x(z')^2 - x} \tag{NC.17}$$

subject to the initial conditions $z(a) = 0$ and $z'(a) = 1$. The solution produces the flat-tipped nose cone with the smallest RDC for tip-width a and height $H = z(R)$.

This differential equation has no interesting exact solution, so we'll solve it numerically. We'll use the numerical ODE solution to plot the shape and its derivative to do the integration in Eq. (NC.10). *SNB* won't accept the variable name z' in an integral, so we'll use the dummy variable $q = z'$.

It's easy to calculate Newton's minimizer for a given tip-width and radius. Just set up a matrix with Eq. (NC.17), the dummy definition, and the two initial conditions, and use Solve ODE + Numeric. Here's an example for $a = 0.35$ and $R = 1$.

$$\begin{bmatrix} q' = \dfrac{q^3 + q}{3xq^2 - x} \\ z' = q \\ z(0.35) = 0 \\ q(0.35) = 1 \end{bmatrix}$$, Functions defined: q, z

Once you have the numeric ODE solutions, apply Evaluate Numerically to the numerical integration per Eq. (NC.10) using the dummy variable.

$$C = \frac{0.35^2}{1^2} + \frac{1}{1^2}\int_{0.35}^{1} \frac{2x}{1 + q(x)^2}\,dx = 0.373\,95$$

Evaluate the solution at $R = 1$ to find the nose-cone's height.

$$H = z(1) = 1.002\,5$$

Newton's flat-tipped minimizer with tip-width $a = 0.35$ and radius $R = 1$ has height $H = 1.002\,5$.

It's a little more complicated to find Newton's minimizer for a given radius and height because the initial conditions don't depend on height (or the radius for that matter) and we don't know the value of the slope at $x = H$.

We can use the above procedure to generate height-versus-tip-width data near the desired height. Then we'll fit a polynomial to that data and use it to find the minimizing tip-width for that height. For the sake of comparison with our earlier results, let's find Newton's minimizer for $R = H = 1$.

Here are 6 data points near $H = 1$. There's no reason to expect a linear relationship between the height and tip-width, so let's try a quadratic fit.

$$\begin{bmatrix} a & H \\ 0.348 & 1.0077 \\ 0.349 & 1.0051 \\ 0.350 & 1.0025 \\ 0.351 & 0.99985 \\ 0.352 & 0.99724 \\ 0.353 & 0.99464 \end{bmatrix}, \text{Polynomial fit: } H = -0.71429a^2 - 2.1144a + 1.83$$

Set the fit equal to the desired height ($H = 1$ in this case) and use Solve Exact to find the corresponding tip-width.

$$1 = -0.71429a^2 - 2.1144a + 1.83, \text{ Solution is: } 0.35094, -3.3111$$

Take the positive solution and use it in the initial conditions to calculate Newton's minimizing shape.

$$\begin{bmatrix} q' = \dfrac{q^3 + q}{3xq^2 - x} \\ z' = q \\ z(0.35094) = 0 \\ q(0.35094) = 1 \end{bmatrix}, \text{Functions defined: } q, z$$

Let's check that our nose cone has the right height.

$$H = z(1) = 1.0$$

Use Evaluate Numerically and Eq. (NC.10) to find the minimum RDC.

$$C_{\text{min}} = \frac{0.35094^2}{1^2} + \frac{1}{1^2} \int_{0.35094}^{1} \frac{2x}{1 + q(x)^2} dx = 0.37481$$

Newton won! The RDC for his nose-cone is smaller than all our previous results.

Here's Newton's minimizer for $H = R = 1$.

Its tip-width is $a_{\text{min}} = 0.35094$ and the RDC is $C_{\text{min}} = 0.37481$.

This is the flat-tipped nose cone that feels the least air resistance.

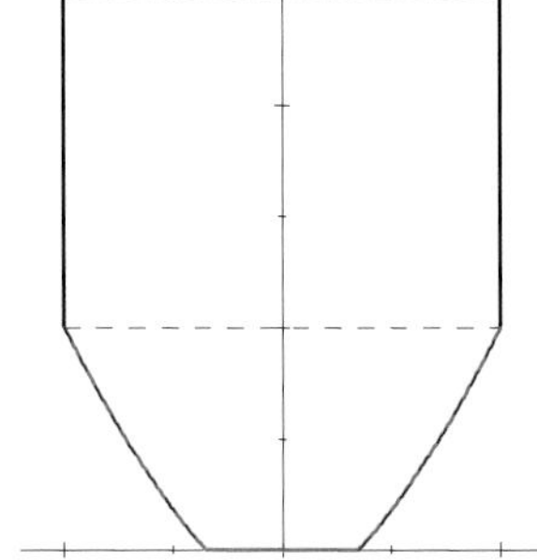

Figure NC.9 Newton's minimizer

Indented Tips and *the* Minimizer

For over 300 years, Newton's solution stood as *the* minimizer but it's only the radially-symmetric flat-tipped minimizer. Radially-symmetric nose cones with indented tips can have smaller RDCs than Newton's. An indented tip has a smaller contribution to the RDC, which allows for a wider tip that makes the outer part steeper.

There are now two varying parts of our nose cone. The indented tip shape z_{tip} is determined by the single-impact condition: no incident air molecules can bounce from one side of the tip to the other. The outer part $z(x)$ is still a solution to Eq. (NC.17) with the initial conditions $z(a) = 0$ and $z'(a) = d_0$. We want the values of a and d_0 that minimize the RDC such that $z(R) = H$.

The indentation reduces the tip's contribution to the RDC from a^2/R^2 to $\kappa a^2/R^2$ so that its effective area is $\kappa\pi a^2$. The tip reduction factor is then $\kappa \equiv C_{\text{tip}} R^2/a^2$ and we can adapt Eq. (NC.3) to calculate C_{tip}.

The problem boils down to finding the value of the initial slope d_0 that produces the smallest RDC. Remarkably, there's a nearly-analytical solution that relates the minimizing initial slope to the reduced RDC due to the indented tip and yields the minimum RDC (see Ref. [20]).

Here's the procedure to calculate the radially-symmetric minimizer for a given H/R and tip shape.

1. Use the tip shape z_{tip} to calculate the tip reduction factor κ.

$$\kappa = \int_0^1 \frac{2u}{1 + \frac{1}{a^2}\left(\frac{dz_{\text{tip}}}{du}\right)^2}\, du \qquad \text{(NC.18a)}$$

2. Calculate the minimizing initial slope d_0 of the outer part.

$$d_0 = \sqrt{\frac{3 - 2\kappa + \sqrt{9 - 8\kappa}}{2\kappa}} \qquad \text{(NC.18b)}$$

3. Find a numerical solution for the final slope d_f.

$$\frac{H}{R} = \frac{d_f}{4\left(1 + d_f^2\right)^2}\left(3\left(d_f^4 - d_0^4\right) + 4\left(d_f^2 - d_0^2\right) - 4\ln\frac{d_f}{d_0}\right) \qquad \text{(NC.18c)}$$

4. Calculate the minimizing tip-width.

$$a_{\text{min}} = R\,\frac{d_f}{d_0}\left(\frac{1 + d_0^2}{1 + d_f^2}\right)^2 \qquad \text{(NC.18d)}$$

5. Calculate the minimum RDC.

$$C_{\text{min}} = \frac{1}{1 + d_f^2} + \left(d_f - \frac{H}{R}\right)\frac{2d_f}{\left(1 + d_f^2\right)^2} \qquad \text{(NC.18e)}$$

This method works for any radially-symmetric tip (including Newton's flat-tipped minimizer with $\kappa = 1$) that satisfies the single-impact condition.

Suppose our $H = R = 1$ nose cone has a symmetric parabola for its indented tip.

$$z_{\text{tip}} = \frac{a}{2}\left(1 - \frac{x^2}{a^2}\right)$$

1. This tip produces about 69% of the air resistance of a flat tip with the same radius.

$$\kappa = \int_0^1 \frac{2u}{1 + \left(\frac{1}{2}\frac{d}{du}\left(1 - u^2\right)\right)^2}\,du = \ln 2 = 0.693\,15$$

2. We need the initial value of the slope for the outer part.

$$d_0 = \left[\sqrt{\frac{3 - 2\kappa + \sqrt{9 - 8\kappa}}{2\kappa}}\right]_{\kappa=0.693\,15} = 1.582\,7$$

3. We can use Solve Numeric to find the final value of the slope.

$$1 = \frac{d_f}{4\left(1 + d_f^2\right)^2}\left(3\left(d_f^4 - 1.582\,7^4\right) + 4\left(d_f^2 - 1.582\,7^2\right) - 4\ln\frac{d_f}{1.582\,7}\right),$$

Solution is: $\{[d_f = 2.231\,5]\}$

4. That's all we need to calculate the minimizing tip radius.

$$a_{\min} = \left[R\frac{d_f}{d_0}\left(\frac{1 + d_0^2}{1 + d_f^2}\right)^2\right]_{R=1, d_0=1.582\,7, d_f=2.231\,5} = 0.484\,41$$

5. Here's the minimum RDC for this nose cone with a symmetric parabolic tip.

$$C_{\min} = \left[\frac{1}{1 + d_f^2} + \left(d_f - \frac{H}{R}\right)\frac{2d_f}{\left(1 + d_f^2\right)^2}\right]_{H=R, d_f=2.231\,5} = 0.320\,95$$

This nose cone has a wider tip, steeper sides, and feels significantly less air resistance than Newton's flat-tipped minimizer with the same height and radius.

The tip-shape that minimizes the air resistance reflects *all* incident particles through the points $x = \pm a$. This optimum single-impact condition relates the tip's shape to its slope.

$$\frac{a + x}{z_{\text{tip}}(x)} = \frac{2\, z'_{\text{tip}}(x)}{z'_{\text{tip}}(x)^2 - 1} \tag{NC.19}$$

A particle incident at $x = \pm a$ deflects along the x-axis so $z'_{\text{tip}}(a) = -1$. This condition (plus continuity at the tip radius) completely determines the depth, width, and shape of the minimizing indented tip.

The minimizing tip-shape is an off-centered parabola!

$$z_{\text{tip}}(x) = \frac{a}{4}\left(4 - \left(1 + \frac{x}{a}\right)^2\right) \tag{NC.20}$$

This tip produces about 59% of the air resistance of a flat tip with the same radius.

$$\kappa = \int_0^1 \frac{2u}{1 + \left(\frac{1}{4}\frac{d}{du}\left(4 - (1+u)^2\right)\right)^2}\, du = \pi + 12\ln 2 - 4\ln 5 - 4\arctan 2$$
$$= 0.593\,01$$

This is the best we can do for a radially symmetric nose cone. This minimizer has $d_0 = 1.8080$ and $d_f = 2.398\,6$, with a tip radius $a_{\text{min}} = 0.530\,10$ and an RDC $C_{\text{min}} = 0.295\,19$ that's 27% smaller than Newton's flat-tipped minimizer.

Let's calculate the outer part of the nose cone so we can plot it.

$$\begin{bmatrix} q' = \dfrac{q^3 + q}{3xq^2 - x} \\ z' = q \\ z\,(0.530\,10) = 0 \\ q\,(0.530\,10) = 1.8080 \end{bmatrix}, \text{ Functions defined: } q, z$$

Here's a plot of this minimizer (and Newton's) for $H = R = 1$.

Its tip-width is $a_{\text{min}} = 0.530\,10$ and the RDC is $C_{\text{min}} = 0.295\,19$.

Notice its wider tip and steeper sides as a result of the indented tip.

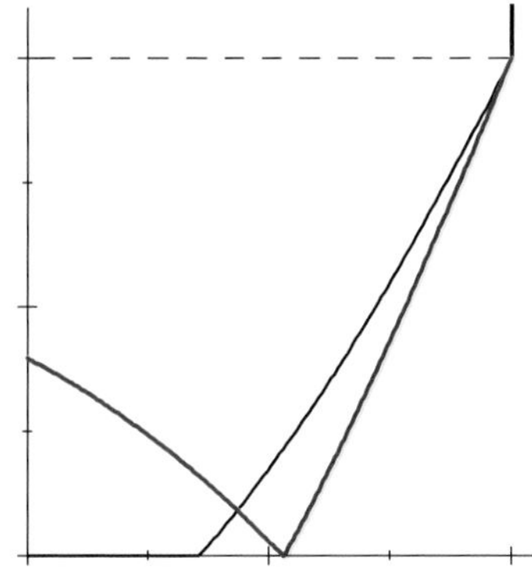

Figure NC.10 Less than Newton

This nose cone is the radially-symmetric minimizer. It feels less air resistance than any other radially-symmetric nose cone with the same height and radius.

So what does *the* minimizer look like? No one really knows. It's been proven that *the* minimizer is not radially symmetric, which complicates things since the integral in Eq. (NC.1) no longer reduces to 1-dimension. There's a critical value for the ratio H/R below which there are an infinite number of minimizers. The search continues, and so must we.

The Shape of the Eiffel Tower

The Eiffel Tower is one of the most recognizable structures in the world. Various mathematical models have been proposed to explain the Tower's distinctive shape. [35] Here we explore the possibility that its shape is based on simple physics: the maximum torque created by the wind is balanced by the torque due to the Tower's weight. [23]

We want to find the mathematical function $f(z)$ that describes the Tower's shape, giving the half-width as a function of height.

Here's the physical situation:

A horizontal wind pushes on the right edge of the Tower (of height H) creating a counter-clockwise torque while the Tower's weight exerts a clockwise torque.

The point of contact between the ground and the left edge of the Tower is the pivot point.

The positive z direction is up and positive $f(z)$ is to the left along the $+x$-axis.

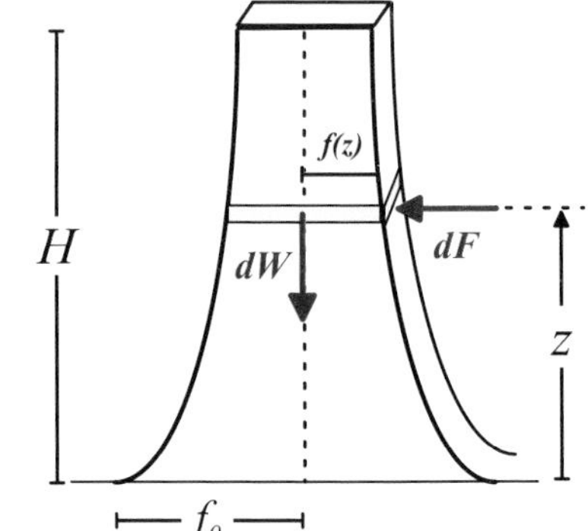

Figure ET.1 Tower torque

Let's treat the Tower as an idealized rigid structure with uniform density. The weight of the slice at height z with thickness dz is proportional to its volume

$$dW = \rho\, dV = 4\rho g\, f^2(z)\, dz$$

where ρ is the density of the Tower and g is the acceleration due to gravity. The lever arm for the weight is f_0, the half-width at ground level. The force exerted by the wind on each slice is proportional to the surface area

$$dF = P\, dA = 2P\, f(z)\, dz$$

where P is the maximum wind pressure the Tower can withstand at height z without toppling. The lever arm for the wind force is the height at which the force acts.

Balancing the torques yields $\frac{1}{2}\, f^2\,(z) = z\, w(z)\, f\,(z)$ where

$$w(z) \equiv \frac{P\,(z)}{4\rho g\, f_0} \tag{ET.1}$$

is the dimensionless, possibly height-dependent maximum wind pressure the Tower can withstand without toppling. This balance may determine the Tower's shape but there's something wrong here: this relation fails at the ground where $z = 0$.

There are two approaches we can take. Since the wind exerts no torque on the bottom of the Tower because the lever arm there is zero, we can add a constant counter-torque that is ultimately due to forces on the Tower from the ground.

$$\tfrac{1}{2}\, f^2\,(z) = z\, w(z)\, f(z) + \tfrac{1}{2}\, f_0^2 \tag{ET.2}$$

With this expression and Solve Exact we can find our solution.

$\frac{1}{2}\, f^2 = z\, w\, f - \frac{1}{2}\, f_0^2$, Solution is: $\sqrt{w^2 z^2 - f_0^2} + wz, wz - \sqrt{w^2 z^2 - f_0^2}$

The other possibility is to take the derivative of the torque-balance condition and solve the resulting differential equation for the shape function. Use New Definition on $w(z)$ and $f(z)$ without specifying them, and Evaluate the derivatives.

$$\frac{d}{dz}\left(z\,w(z)\,f(z)\right) = f(z)\,w(z) + z f(z)\frac{\partial w(z)}{\partial z} + z w(z)\frac{\partial f(z)}{\partial z}$$

$$\frac{d}{dz}\left(\tfrac{1}{2}f^2(z)\right) = f(z)\frac{\partial f(z)}{\partial z}$$

A little editing by-hand produces an ordinary differential equation for $f(z)$.

$$f(z)\frac{df}{dz} = z f(z)\frac{dw}{dz} + w(z)\left(f(z) + z\frac{df}{dz}\right) \qquad \text{(ET.3)}$$

This equation does have a general solution, but *SNB* can only solve it for particular $w(z)$ functions (see Problem 35). The general solution depends only on the width of the base ($f_0 = 62.5\,\mathrm{m}$ for the real Tower) and the maximum wind pressure.

$$f(z) = z\,w(z) \pm \sqrt{f_0^2 + z^2 w^2(z)} \qquad \text{(ET.4)}$$

The Tower is widest at the bottom so we want the negative solution for our function.

$$f(z) = z\,w(z) - \sqrt{f_0^2 + z^2 w^2(z)} \qquad \text{(ET.5)}$$

This gives the right (negative) edge of the Tower and the left (positive) edge is $-f(z)$.

All that remains is to find a physically reasonable maximum wind pressure that reproduces the Tower's shape. The simplest form of Eq. (ET.5) has a constant $w(z) = w_0$. Let's look at two points on the Tower, one high (at $z = 218\,\mathrm{m}$, $f = -6.65\,\mathrm{m}$) and one low (at $z = 14.6\,\mathrm{m}$, $f = -53.2\,\mathrm{m}$), and use Solve Exact to see if a constant maximum wind pressure works.

$-53.2 = 14.6\,w_0 - \sqrt{62.5^2 + 14.6^2 w_0^2}$, Solution is: $0.692\,66$

$-6.65 = 218\,w_0 - \sqrt{62.5^2 + 218^2 w_0^2}$, Solution is: 1.332

No! It turns out that a constant wind pressure reproduces the Tower's general shape but not the quantitative details (see Problem 34).

Figure ET.2 shows the results (plotted as points) if we repeat this process for the whole Tower. This suggests a maximum wind pressure that is constant and small at low heights but increases through the intermediate heights to approach a larger constant value near the top.

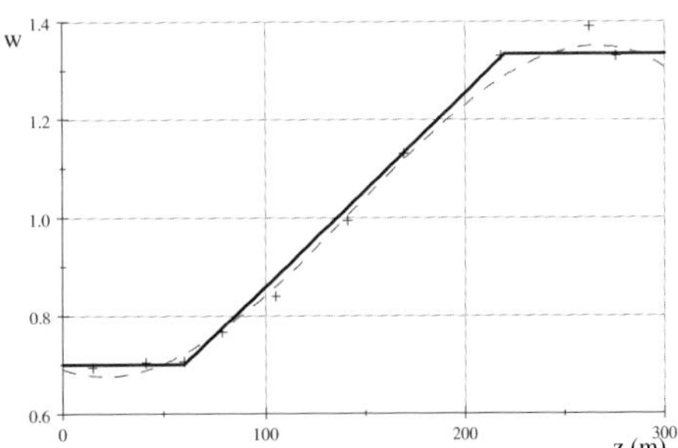

Figure ET.2 The wind

The force (and torque) from the wind increases with height, but the Tower's cross-sectional area decreases.

We can parameterize such a function using *SNB*'s piecewise-defined functions.

$$w(z) = \begin{cases} \frac{7}{10} & \text{if } z < 60 \\ \frac{19}{4800}z + \frac{37}{80} & \text{if } 60 \le z \le 220 \\ \frac{4}{3} & \text{if } 220 < z \end{cases}$$

Figure ET.2 shows this function as solid black lines and the dashed line is a cubic polynomial fit to the wind-pressure data. The expression for the intermediate linear increase comes from a simple linear fit connecting two points.

$$\begin{bmatrix} z & w \\ 60 & 7/10 \\ 220 & 4/3 \end{bmatrix}, \text{Regression is: } w = \frac{19}{4800}z + \frac{37}{80}$$

You can treat piecewise-defined functions like any other function, so you can plot, integrate, differentiate, define, and include them in the definition of other functions. So let's use New Definition on the wind pressure and this expression for the shape function with the numerical value for the base width.

$$f(z) = z\,w(z) - \sqrt{62.5^2 + z^2 w^2(z)}$$

Once we've made the Definitions for $f(z)$ and $w(z)$, we can use Plot 2D on expressions of the form $(\pm f(z), z)$, checking the Equal Scaling Along Each Axis box on the Axes tab of the Plot Properties dialog box.

There are three parts to Figure ET.3:

The circles are the data points for the Tower's width, which you'll find in the problems at the end of this chapter.

The dark black lines are our shape function with the piecewise-defined varying wind pressure.

They are superimposed over a figure of the Tower.

Pas trop mal! But...

There isn't much evidence Eiffel based his design on balancing the torques.

But such a strategy, with a physically reasonable wind profile, accurately reproduces the Tower's shape.

Figure ET.3 *La Tour Eiffel*

An Interesting Classical Orbit

A figure in a popular classical mechanics textbook ([10], page 92) shows a highly oscillatory classical orbit (reproduced here in Figure IO.1).

Although the text's caption describes an

"orbit for motion in a central force deviating slightly from a circular orbit"

the figure shows far more significant deviations which suggests a question:

What central force will produce such an orbit for arbitrarily large deviations?

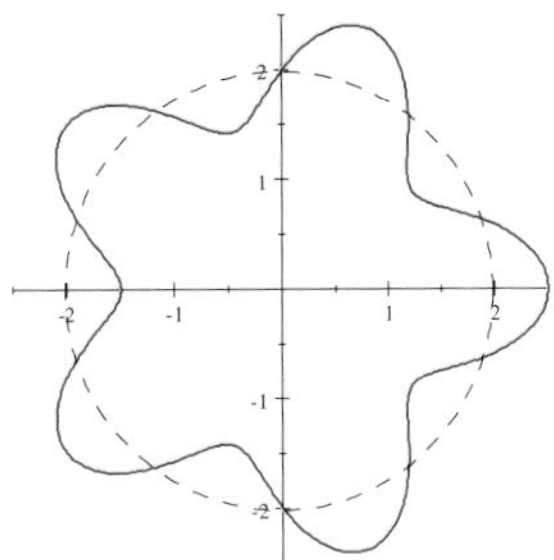

Figure IO.1 Slightly deviant?

The polar equation for this oscillatory orbit is

$$r\,(\phi) = R + a\cos n\phi \tag{IO.1}$$

where R is the radius of the circle about which the orbit oscillates, n is the number of cycles completed during one orbit and a is the amplitude of the oscillations. The initial position is $R_0 = R + a$ and the initial value of its derivative is zero. This orbit's shape does not depend on the angular momentum, but its orbital period does. The orbit in Figure IO.1 has $R = 2$, $a = \frac{1}{2}$ and $n = 5$.

Classical orbits obey Newton's 2nd Law, which for a 2-d system under a central force is

$$\frac{d^2r}{dt^2} = -\,f(r) + \frac{\ell^2}{r^3} \tag{IO.2}$$

where $f(r)$ is the central force (per unit mass) and $\ell^2/\,r^3$ is the "centripetal force" which derives from the acceleration and angular momentum conservation. The minus sign means $f\,(r)$ is positive for bound orbits so the force is attractive.

To find the central force for which a particular orbit is a solution to the 2nd Law, it's convenient to use the inverse variable $u(\phi) = r^{-1}(\phi)$. The equation of motion is then

$$f(u) = \ell^2 u^2 \left(\frac{d^2u}{d\phi^2} + u\right) \tag{IO.3}$$

where ℓ is the conserved angular momentum.

Let's use New Definition to define the inverse variable u in terms of this particular orbit.

$$u = \frac{1}{R + a\cos n\phi}$$

We can use Eq. (IO.3) and Substitute $\cos n\phi = (r - R)\,/a$ (with Expand) to eliminate all the explicit ϕ-dependence and replace u with r.

$$f(r) = \left[\ell^2 u^2 \left(\frac{d^2u}{d\phi^2} + u\right)\right]_{\cos n\phi=(r-R)/a}$$

$$= \frac{1}{r^3}\ell^2 - \frac{n^2}{r^3}\ell^2 - 2R^2\frac{n^2}{r^5}\ell^2 + 2a^2\frac{n^2}{r^5}\ell^2 + 3R\frac{n^2}{r^4}\ell^2$$

A little editing by-hand gives the central force for the oscillating orbit.

$$f(r) = \ell^2 \left(\frac{1-n^2}{r^3} + \frac{3n^2 R}{r^4} - \frac{2\,n^2\left(R^2-a^2\right)}{r^5} \right) \tag{IO.4}$$

The force consists of competing attractive and repulsive terms whose net effect is to "turn off" the centripetal force and "push and pull" to create the oscillating orbit.

There are at least three well-known limits to this force and orbit.

1. Unstable circle: $r(\phi) = R$
Our orbit becomes a circle of radius R centered at the origin when we turn off the oscillations by setting their amplitude to zero.

$$r(\phi) = [R + a\cos n\phi]_{a=0} = R$$

With these values, the force is the limiting case for Cotes' spirals.

$$f(r) = \left[\ell^2 \left(\frac{1-n^2}{r^3} + \frac{3n^2 R}{r^4} - \frac{2\,n^2\left(R^2-a^2\right)}{r^5} \right) \right]_{n=0,a=0} = \frac{1}{r^3}\ell^2$$

As we saw in Chapter 9, the circular orbit for this force is unstable.

2. Cardioid: $r(\phi) = a\;(1+\cos\phi)$
A cardioid's polar equation tells us it's the $n = 1$, $a = R$ limit of our oscillating orbit. We can verify this using Substitute (with Factor).

$$r(\phi) = [R + a\cos n\phi]_{n=1,R=a} = a\,(\cos\phi + 1)$$

With these values, the force has a simple r^{-4} dependence.

$$f(r) = \left[\ell^2 \left(\frac{1-n^2}{r^3} + \frac{3n^2 R}{r^4} - \frac{2\,n^2\left(R^2-a^2\right)}{r^5} \right) \right]_{n=1,R=a} = 3\frac{a}{r^4}\ell^2$$

When the force is $f(r) = 3\ell^2 R/r^4$ the resulting orbit is a cardioid ([13], page 144).

3. Circle through the origin: $r(\phi) = a\cos\phi$
That's the polar equation for a circle of diameter a that passes through the origin, and it's the $n = 1$, $R = 0$ limit of our oscillating orbit.

$$r(\phi) = [R + a\cos n\phi]_{n=1,R=0} = a\cos\phi$$

With these values, the force has a simple r^{-5} dependence.

$$f(r) = \left[\ell^2 \left(\frac{1-n^2}{r^3} + \frac{3n^2 R}{r^4} - \frac{2\,n^2\left(R^2-a^2\right)}{r^5} \right) \right]_{n=1,R=0} = 2\frac{a^2}{r^5}\ell^2$$

The resulting orbit is a circle that passes through the origin ([6], page 621).

These three orbits are shown in Figure IO.2. The unstable circle with the origin at its center has $R = 2$ and the circle through the origin has $a = 2$. The cardioid is the geometric shape that resembles a heart tipped on its side. This cardioid has $a = 1$.

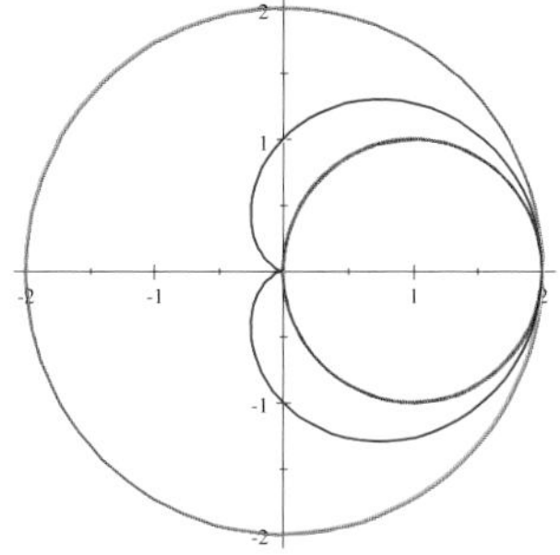

Figure IO.2 Three orbits...

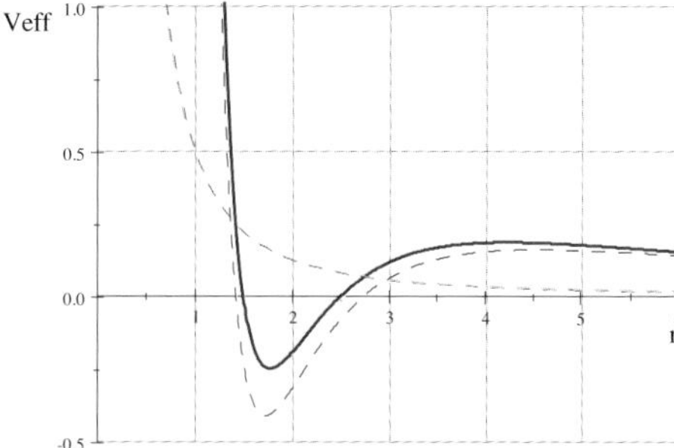

Figure IO.3 ...and an EP

We can calculate the effective potential with the force and the centripetal term.

$$V_{\text{eff}}(r) = \int f(r)dr + \frac{1}{2}\frac{\ell^2}{r^2} \tag{IO.5}$$

Use the force from Eq. (IO.4) to create and Expand an expression for $V_{\text{eff}}(r)$.

$$V_{\text{eff}}(r) = \ell^2 \int \left(\frac{1-n^2}{r^3} + \frac{3n^2 R}{r^4} - \frac{2n^2\left(R^2-a^2\right)}{r^5} \right) dr + \frac{1}{2}\frac{\ell^2}{r^2}$$

$$= \frac{1}{2}\frac{n^2}{r^2}\ell^2 + \frac{1}{2}R^2\frac{n^2}{r^4}\ell^2 - \frac{1}{2}a^2\frac{n^2}{r^4}\ell^2 - R\frac{n^2}{r^3}\ell^2$$

A little editing by-hand gives the effective potential for the oscillating orbit.

$$V_{\text{eff}}(r) = \ell^2 n^2 \left(\frac{1}{2r^2} - \frac{R}{r^3} + \frac{R^2-a^2}{2r^4} \right) \tag{IO.6}$$

Again we see the centripetal term is gone, and the overall $\ell^2 n^2$ factor means the circular orbits and orbital stability will not depend on either n or ℓ. Figure IO.3 shows the effective potential for the orbit of Figure IO.1 along with centripetal (the dashed curve that's always positive) and physical (the other dashed curve) terms.

As with all non-circular bound orbits under a central force, the equation $V_{\text{eff}}(r) = E$ has two roots which give the radial extrema. The energy of our oscillatory orbit is zero.

$$E = V_{\text{eff}}(R_0) = \left[\ell^2 n^2 \left(\frac{1}{2r^2} - \frac{R}{r^3} + \frac{R^2-a^2}{2r^4} \right) \right]_{r=R+a} = 0$$

Let's use Solve Exact to find the two radial extrema.

$$0 = \frac{1}{2r^2} - \frac{R}{r^3} + \frac{R^2-a^2}{2r^4}, \text{ Solution is: } R-a, R+a$$

The maximum and minimum distances are $r = R \pm a$, as expected.

We can find any circular orbits by minimizing the effective potential and using Solve Exact to find r.

$$0 = \frac{d}{dr}\left(\frac{1}{2r^2} - \frac{R}{r^3} + \frac{R^2 - a^2}{2r^4}\right), \text{ Solution is: } \tfrac{3}{2}R - \tfrac{1}{2}\sqrt{R^2 + 8a^2}, \tfrac{3}{2}R + \tfrac{1}{2}\sqrt{R^2 + 8a^2}$$

Unlike a simple power-law potential, the oscillating-orbit effective potential allows two circular orbits.

$$R_{\text{cir}} = \frac{R}{2}\left(3 \pm \sqrt{1 + \frac{8a^2}{R^2}}\right) \tag{IO.7}$$

The minus sign refers to the stable circular orbit between the radial extrema, and the plus sign refers to the radius of the unstable circular orbit beyond the oscillatory orbit's maximum radius. Here's the stable circular orbit for the effective potential in Figure IO.3, courtesy of Evaluate and Evaluate Numerically.

$$R_{\text{cir}} = \left[\tfrac{3}{2}R - \tfrac{1}{2}\sqrt{R^2 + 8a^2}\right]_{R=2,a=1/2} = 3 - \tfrac{1}{2}\sqrt{6} = 1.7753$$

Classical orbits are stable when they occur near a local minimum of the effective potential. The stability condition requires a positive curvature at the minimum, so let's Evaluate the effective potential's second derivative at the stable circular orbit.

$$0 < \left[\frac{d^2}{dr^2}\left(\frac{1}{2\,r^2} - \frac{R}{r^3} + \frac{R^2 - a^2}{2\,r^4}\right)\right]_{r=\left(3R-\sqrt{R^2+8a^2}\right)/2}$$

$$= -\frac{64}{\left(\sqrt{R^2 + 8a^2} - 3R\right)^6}\left(\tfrac{1}{2}R^2 - \tfrac{3}{2}R\sqrt{R^2 + 8a^2} + 4a^2\right)$$

Before we find the maximum allowed amplitude for our oscillating orbit, tell *SNB* to assume the variables are real and R is positive.

$$\text{assume}\,(\text{real}) = \mathbb{R}$$

$$\text{additionally}\,(R > 0) = (0, \infty)$$

Now use Solve Exact to find the maximum value of a.

$$0 = \tfrac{1}{2}R^2 - \tfrac{3}{2}R\sqrt{R^2 + 8a^2} + 4a^2, \text{ Solution is: } R, -R$$

The orbit is stable when $|\,a\,| < R$ so the amplitude of the oscillations must be less than the radius of the circle about which the orbit oscillates.

Like the force law that produces a circle whose force center is anywhere in the circle ([30]), deriving this force law and investigating it for other orbits or limits would make an interesting project for undergraduate physics students. The solution to this problem requires only the use of elementary differential calculus and algebra, but the math is complicated enough that this problem would be an excellent introduction to doing physics with *SNB*.

Fisher's Crystal

This topic comes from a friend who is a physics teacher, author, and former student. [36] It's my response to a question he asked me regarding a science fiction story he wrote. It doesn't have much to do with physics, but it's an example of the fun and interesting problems physicists often encounter.

In his story, there is an eight-sided crystal with an embedded 4-armed double-X, one arm for each space-time dimension. Each arm connects the center of one face through the origin to the center of the opposite face. Can we construct the double-X so that the arms of each X are perpendicular to each other and the crystal faces?

Figure FC.1 shows Fisher's 8-sided crystal. It consists of two square pyramids sharing the same base which lies in the x-y plane.

The base has four corners that are on the x- or y-axes and four sides that have length s.

Both pyramids have height h so the slope of each edge is $\pm m$ where $m \equiv \sqrt{2}\ h/s$.

If you're having trouble visualizing the angles, take a quick peek at Figure FC.7.

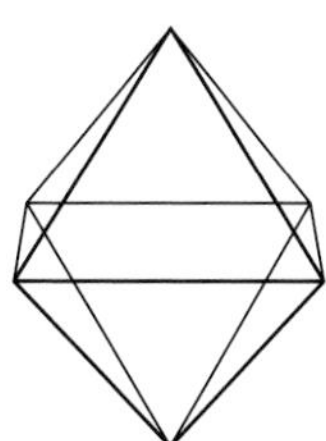

Figure FC.1 Fisher's Crystal

Our goal is to calculate various angles, so this looks like a perfect problem for vectors! First we'll locate the center of one side, then we'll use symmetry to know the centers of all the sides. We'll use vectors to calculate the angle between two arms of the X and the angle between a face and one arm of the X. After that, we'll just connect the dots to create the arms of the X. Vectors are your friends!

The Face Center $\overrightarrow{C}$

Figure FC.2 shows the side in the octant above the first quadrant from two perspectives. Edge 1 connects the apex at $(0, 0, h)$ to the vertex at $\left(s/\sqrt{2}, 0, 0\right)$ on the x-axis. Edge 2 connects the same apex to the vertex at $\left(0, s/\sqrt{2}, 0\right)$ on the y-axis.

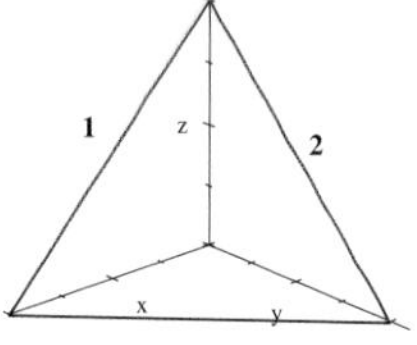

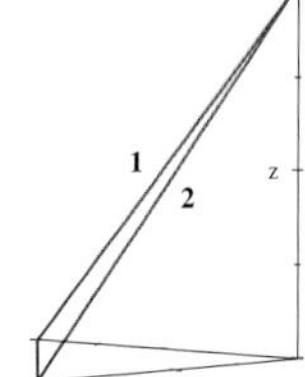

Figure FC.2 Above the... ...alpha quadrant

We can construct the vector connecting two points with the difference between the tip and tail coordinates. The vector describing edge 1 (↙) is in the x-z plane.

$$\overrightarrow{V}_1 = \left(\tfrac{1}{\sqrt{2}}s, 0, 0\right) - (0, 0, h) = \left[\tfrac{1}{2}\sqrt{2}s \quad 0 \quad -h\right]$$

The midpoint of edge 1 is at $\overrightarrow{M}_1 = \frac{1}{2}\overrightarrow{V}_1$.

The vector describing edge 2 (↘) is in the y-z plane.

$$\overrightarrow{V}_2 = \left(0, \tfrac{1}{\sqrt{2}}s, 0\right) - (0,0,h) = \begin{bmatrix} 0 & \tfrac{1}{2}\sqrt{2}s & -h \end{bmatrix}$$

The midpoint of edge 2 is at $\overrightarrow{M}_2 = \frac{1}{2}\overrightarrow{V}_2$.

Each face is a triangle so the **face center** $\overrightarrow{C} \equiv (x_c, y_c, z_c)$ is located at the intersection of the perpendicular bisectors of each edge. There is no simple 3-d analog to the linear equation $y = mx + b$ so we'll have to find the center parametrically.

$$\overrightarrow{C} = \overrightarrow{M}_1 + \widehat{n}_1\, t = \overrightarrow{M}_2 + \widehat{n}_2\, t \tag{FC.1}$$

The parameter t is a dummy variable (and not time). We'll have to construct the unit vectors ($\widehat{n}_1$ and $\widehat{n}_2$) perpendicular to the edges at the midpoints.

There are three conditions on these unit vectors.
1. They are perpendicular to the edge:

$$\widehat{n}_1 \cdot \overrightarrow{V}_1 = 0 \tag{FC.2a}$$

$$\widehat{n}_2 \cdot \overrightarrow{V}_2 = 0 \tag{FC.2b}$$

2. They are on the face:

$$\widehat{n}_1 \cdot \left(\overrightarrow{V}_1 \times \overrightarrow{V}_2\right) = 0 \tag{FC.3a}$$

$$\widehat{n}_2 \cdot \left(\overrightarrow{V}_1 \times \overrightarrow{V}_2\right) = 0 \tag{FC.3b}$$

3. They are normalized: $\|\widehat{n}_1\| = \|\widehat{n}_2\| = 1$

Those 3 conditions are all we need since our vectors have 3 components, so let's construct $\widehat{n}_1$. *SNB* does not assume vector components are real unless told otherwise.

$$\text{assume}\,(\text{real}) = \mathbb{R}$$

Let's Evaluate the dot product in condition 1.

$$0 = \begin{bmatrix} n_x & n_y & n_z \end{bmatrix} \cdot \begin{bmatrix} \tfrac{1}{\sqrt{2}}s & 0 & -h \end{bmatrix} = \tfrac{1}{2}\sqrt{2}sn_x - hn_z$$

We can relate two components by using Solve Exact for n_x in terms of n_z.

$$0 = \tfrac{1}{\sqrt{2}}n_x s - n_z h, \text{ Solution is: } \sqrt{2}\frac{h}{s}n_z$$

The x-component is $n_x = mn_z$ where $m \equiv \sqrt{2}\, h/s$ is the edge slope. We'll need to Evaluate the cross product to impose condition 2.

$$\overrightarrow{V}_1 \times \overrightarrow{V}_2 = \begin{bmatrix} \tfrac{1}{\sqrt{2}}s & 0 & -h \end{bmatrix} \times \begin{bmatrix} 0 & \tfrac{1}{\sqrt{2}}s & -h \end{bmatrix} = \begin{bmatrix} \tfrac{1}{2}\sqrt{2}hs & \tfrac{1}{2}\sqrt{2}hs & \tfrac{1}{2}s^2 \end{bmatrix}$$

Let's Evaluate the dot product in condition 2.

$$0 = \begin{bmatrix} n_x & n_y & n_z \end{bmatrix} \cdot \begin{bmatrix} \tfrac{1}{\sqrt{2}}hs & \tfrac{1}{\sqrt{2}}hs & \tfrac{1}{2}s^2 \end{bmatrix} = \tfrac{1}{2}s^2 n_z + \tfrac{1}{2}\sqrt{2}hsn_x + \tfrac{1}{2}\sqrt{2}hsn_y$$

Substitute for n_x and h in terms of m, and use Solve Exact (and Factor in-place) to find n_y in terms of n_z.

$$0 = \left[\tfrac{1}{2}s^2 n_z + \tfrac{1}{\sqrt{2}}hsn_x + \tfrac{1}{\sqrt{2}}hsn_y\right]_{n_x = mn_z, h = sm/\sqrt{2}}, \text{ Solution is: } -n_z\frac{m^2+1}{m}$$

Use the results for n_x and n_y to create an expression for the magnitude of $\widehat{n}_1$, and use Solve Exact to find n_z. You can find *SNB*'s double-lined Norm symbol by clicking the Brackets button on the Math Objects toolbar.

$$1 = \left\| \begin{bmatrix} mn_z & -\frac{m^2+1}{m}n_z & n_z \end{bmatrix} \right\|, \text{ Solution is: } \frac{m}{\sqrt{2m^4+3m^2+1}}$$

We now have the components for both $\widehat{n}_1$ and $\widehat{n}_2$ because the two vectors are related by the symmetry swap $x \leftrightarrow y$ $\left(\text{take a look at } \overrightarrow{V}_1 \text{ and } \overrightarrow{V}_2\right)$.

$$\widehat{n}_1 = \sqrt{\frac{1}{(2m^2+1)(m^2+1)}} \begin{bmatrix} -m^2 & m^2+1 & -m \end{bmatrix} \quad \text{(FC.4a)}$$

$$\widehat{n}_2 = \sqrt{\frac{1}{(2m^2+1)(m^2+1)}} \begin{bmatrix} m^2+1 & -m^2 & -m \end{bmatrix} \quad \text{(FC.4b)}$$

We can verify $\widehat{n}_1$ is a unit vector with Simplify.

$$\|\widehat{n}_1\| = \left\| \sqrt{\frac{1}{(2m^2+1)(m^2+1)}} \begin{bmatrix} -m^2 & m^2+1 & -m \end{bmatrix} \right\| = 1$$

The expressions for the unit vectors are complicated, but we don't need the overall normalization constant to find the face center. We can use non-unit vectors $\overrightarrow{n}_1$ and $\overrightarrow{n}_2$ that point in the same directions.

Let's Evaluate the expression from Eq. (FC.1) relating edge 1 to the face center.

$$\overrightarrow{C}_1 = \overrightarrow{M}_1 + \overrightarrow{n}_1 t = \tfrac{1}{2\sqrt{2}}\begin{bmatrix} 1 & 0 & m \end{bmatrix} s + \begin{bmatrix} -m^2 & m^2+1 & -m \end{bmatrix} t$$
$$= \begin{bmatrix} \tfrac{1}{4}\sqrt{2}s - m^2 t & t(m^2+1) & \tfrac{1}{4}\sqrt{2}ms - mt \end{bmatrix}$$

Let's do the same for edge 2.

$$\overrightarrow{C}_2 = \overrightarrow{M}_2 + \overrightarrow{n}_2 t = \tfrac{1}{2\sqrt{2}}\begin{bmatrix} 0 & 1 & m \end{bmatrix} s + \begin{bmatrix} m^2+1 & -m^2 & -m \end{bmatrix} t$$
$$= \begin{bmatrix} t(m^2+1) & \tfrac{1}{4}\sqrt{2}s - m^2 t & \tfrac{1}{4}\sqrt{2}ms - mt \end{bmatrix}$$

Notice the symmetry in the x-y coordinates of the two center vectors. Let's set the two x-components equal and find t with Solve Exact.

$$\frac{\sqrt{2}}{4}s - m^2 t = t(m^2+1), \text{ Solution is: } \sqrt{2}\frac{s}{8m^2+4}$$

Substitute (with Simplify) the value for t into expressions for the center's coordinates.

$$x_c = \left[t\left(m^2+1\right)\right]_{t=\sqrt{2}s/(8m^2+4)} = \sqrt{2}s\frac{m^2+1}{8m^2+4}$$

$$z_c = \left[\frac{\sqrt{2}}{4}ms - mt\right]_{t=\sqrt{2}s/(8m^2+4)} = \sqrt{2}m^3\frac{s}{4m^2+2}$$

A little editing by-hand gives us the coordinates for the center of the side in the octant above the first quadrant.

$$x_c = \frac{1}{2\sqrt{2}}\frac{m^2+1}{2m^2+1}s \qquad \text{(FC.5a)}$$

$$y_c = x_c \qquad \text{(FC.5b)}$$

$$z_c = \frac{1}{2\sqrt{2}}\frac{2m^3}{2m^2+1}s \qquad \text{(FC.5c)}$$

The other seven centers are located symmetrically at $(\pm x_c, \pm y_c, \pm z_c)$. The vector form gives us the position vector of the center.

$$\vec{C} = \frac{1}{2\sqrt{2}}s\left(\frac{m^2+1}{2m^2+1}, \frac{m^2+1}{2m^2+1}, \frac{2m^3}{2m^2+1}\right) \qquad \text{(FC.6)}$$

Let's use the vector form and Substitute the values $s = \sqrt{2}$ and $m = 1$ (so $h = 1$) I used to create the previous figures.

$$\vec{C} = \left[\frac{1}{2\sqrt{2}}s\left(\frac{m^2+1}{2m^2+1}, \frac{m^2+1}{2m^2+1}, \frac{2m^3}{2m^2+1}\right)\right]_{m=1, s=\sqrt{2}} = \begin{bmatrix}\frac{1}{3} & \frac{1}{3} & \frac{1}{3}\end{bmatrix}$$

Figure FC.3 shows the same side as Figure FC.2 with the face-center indicated.

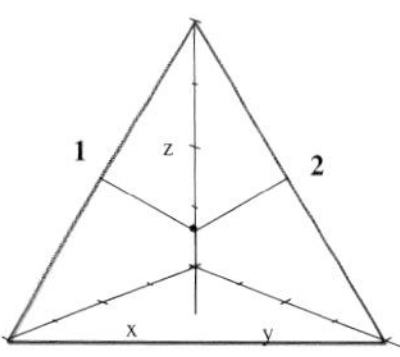

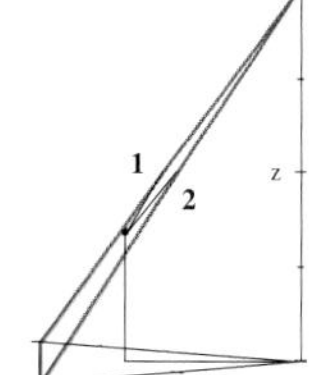

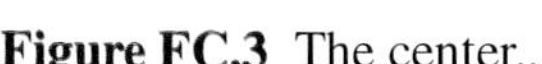

Figure FC.3 The center... ...no chewy nougat

Arm Angle θ_X and Face Angle θ_F

The **arm angle** θ_X is formed by the position vectors for the centers at (x_c, y_c, z_c) and $(-x_c, -y_c, z_c)$. The calculation is easier if we use symmetry and Eq. (FC.6), since θ_X is twice the angle between one arm and the z-axis.

$$\cos\tfrac{1}{2}\theta_X = \frac{\vec{C}\cdot\hat{z}}{\left\|\vec{C}\right\|\,\|\hat{z}\|} \qquad \text{(FC.7)}$$

The arm starts above the 1st quadrant and bounces off the x-y plane at the origin into the 3rd quadrant.

Let's Simplify the complicated vector stuff.

$$\frac{\frac{1}{2\sqrt{2}}s\left(\frac{m^2+1}{2m^2+1},\frac{m^2+1}{2m^2+1},\frac{2m^3}{2m^2+1}\right)\cdot(0,0,1)}{\left\|\frac{1}{2\sqrt{2}}s\left(\frac{m^2+1}{2m^2+1},\frac{m^2+1}{2m^2+1},\frac{2m^3}{2m^2+1}\right)\right\|\|(0,0,1)\|}=\sqrt{2}\frac{m^3}{\sqrt{m^4+1}\sqrt{2m^2+1}}$$

The angle between the arms of the X depends only on the edge-slope m.

$$\theta_{\mathrm{X}}=2\cos^{-1}\frac{\sqrt{2}m^3}{\sqrt{(m^4+1)(2m^2+1)}} \tag{FC.8}$$

Let's use Solve Numeric to find the crystal with an arm angle of $90°$.

$$\frac{\pi}{2}=2\cos^{-1}\frac{\sqrt{2}m^3}{\sqrt{(m^4+1)(2m^2+1)}},\ \text{Solution is: }\{[m=1.1990]\}$$

Figure FC.4 shows an $m=1.1990$ crystal that has one embedded X with $\theta_{\mathrm{X}}=90°$.

Remember that θ_{X} is the angle between two arms that are symmetric with the z-axis.

The "F" and "X" labels show you the location of the face (F) and arm (X) angles.

The face angle connects the origin, the face center, and the center of the side of the base.

It turns out the face angle for this crystal is $\theta_{\mathrm{F}}=75.529°$ and that's our next topic.

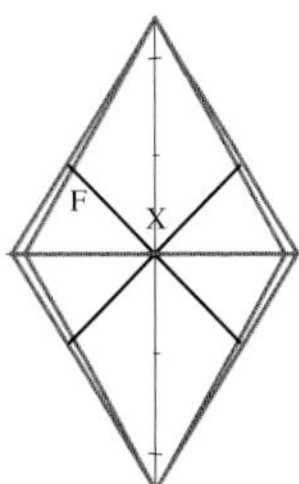

Figure FC.4 $\theta_{\mathrm{X}}=90°$

The **face angle** θ_{F} is the angle between the surface of the face and the position vector of the face center. We need a vector to describe the crystal face. The cross product $\overrightarrow{V}_1\times\overrightarrow{V}_2$ would suffice, but it turns out the calculation is easier if we use a different vector.

$$\cos\theta_{\mathrm{F}}=\frac{\overrightarrow{C}\cdot\overrightarrow{S}}{\left\|\overrightarrow{C}\right\|\left\|\overrightarrow{S}\right\|} \tag{FC.9}$$

The vector $\overrightarrow{S}$ points from the apex to the center of a side of the base.

$$\overrightarrow{S}=\left[(0,0,h)-\left(\tfrac{1}{2\sqrt{2}}s,\tfrac{1}{2\sqrt{2}}s,0\right)\right]_{h=ms/\sqrt{2}}=\left[-\tfrac{1}{4}\sqrt{2}s\quad-\tfrac{1}{4}\sqrt{2}s\quad\tfrac{1}{2}\sqrt{2}ms\right]$$

Let's Simplify this complicated vector stuff.

$$\frac{\frac{1}{2\sqrt{2}}s\left(\frac{m^2+1}{2m^2+1},\frac{m^2+1}{2m^2+1},\frac{2m^3}{2m^2+1}\right)\cdot\left[-1\quad-1\quad2m\right]\frac{\sqrt{2}}{4}s}{\frac{1}{2\sqrt{2}}s\left\|\left(\frac{m^2+1}{2m^2+1},\frac{m^2+1}{2m^2+1},\frac{2m^3}{2m^2+1}\right)\right\|\left\|\left[-1\quad-1\quad2m\right]\right\|\frac{\sqrt{2}}{4}s}=\frac{m^2-1}{\sqrt{m^4+1}}$$

The face angle also depends only on the edge-slope m.

$$\theta_{\mathrm{F}} = \cos^{-1}\frac{m^2-1}{\sqrt{m^4+1}} \tag{FC.10}$$

Let's use Solve Numeric to find the crystal with an face angle of $45°$.

$$\left[\begin{array}{c} 45 = \frac{180}{\pi}\cos^{-1}\left(\frac{m^2-1}{\sqrt{m^4+1}}\right) \\ m \in (0,2) \end{array}\right], \text{ Solution is: } \{[m = 1.9319]\}$$

Eq. (FC.8) and Evaluate Numerically will tell us the arm angle for this crystal.

$$\theta_{\mathrm{X}} = \left[\frac{360}{\pi}\cos^{-1}\frac{\sqrt{2}m^3}{\sqrt{(m^4+1)(2m^2+1)}}\right]_{m=1.9319} = 49.791$$

Figure FC.5 shows three crystals with difference face angles.

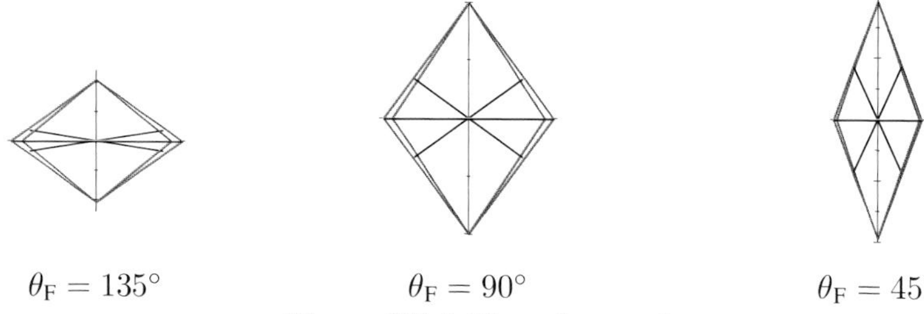

Figure FC.5 Three face angles

Here's a table of the numerical values for the four crystals in the previous two figures.

Figure	FC.5	FC.5	FC.4	FC.5
$m = \sqrt{2}\,h/s$	0.51764	1	1.1990	1.9319
θ_{F}	135°	90°	75.529°	45°
θ_{X}	162.41°	109.47°	90°	49.791°

Table FC.1

Figure FC.6 shows plots of the arm and face angles as functions of the edge-slope m. Both angles are decreasing functions of m, and they equal $90°$ for different values of m.

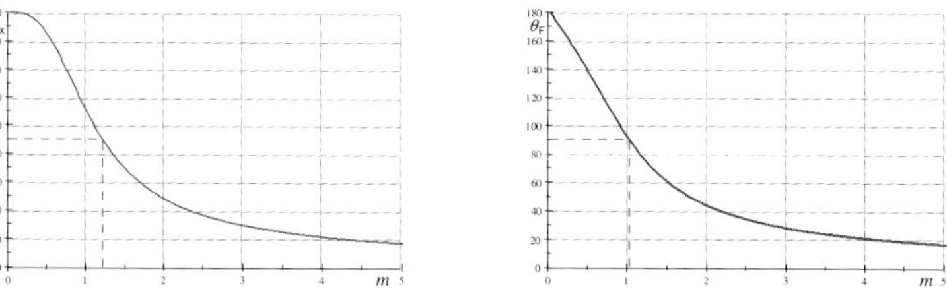

Figure FC.6a θ_{X} vs. m **Figure FC.6b** θ_{F} vs. m

Both angles equal $180°$ at $m = 0$, the zero-height limit in the x-y plane.

The two graphs look very similar when $m > 3$ and this suggests an interesting limit.

$$\frac{\theta_F}{\theta_X} \to \lim_{m\to\infty}\left(\frac{\cos^{-1}\dfrac{m^2-1}{\sqrt{m^4+1}}}{2\cos^{-1}\dfrac{\sqrt{2}m^3}{\sqrt{m^4+1}\sqrt{2m^2+1}}}\right) = 1$$

The angles are approximately the same for tall, thin crystals. Let's use Evaluate Numerically to compare the two angles at $m = 5$.

$$\frac{\theta_F}{\theta_X} = \left[\frac{\cos^{-1}\dfrac{m^2-1}{\sqrt{m^4+1}}}{2\cos^{-1}\dfrac{\sqrt{2}m^3}{\sqrt{m^4+1}\sqrt{2m^2+1}}}\right]_{m=5} = 0.981\,03$$

They differ by less than 2%.

Here is an $m = 1$ crystal with its embedded 4-armed double-X.

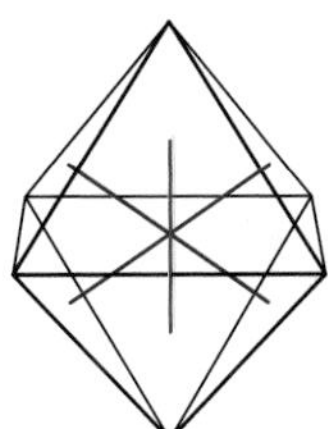

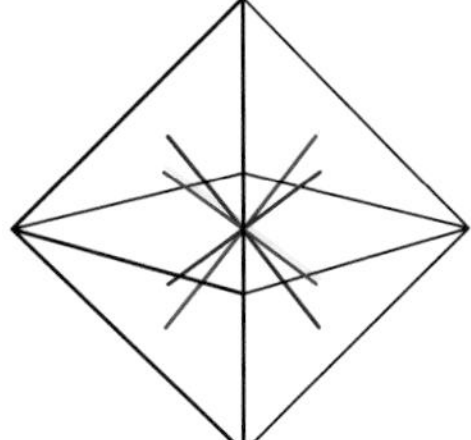

Figure FC.7 Two views of Fisher's Crystal

The easiest way to plot a crystal is using 3-column matrices to draw lines that connect the vertices. Here are the general forms of the matrices you'll need.

One base: $\frac{1}{\sqrt{2}}\begin{bmatrix} s & 0 & 0 \\ 0 & s & 0 \\ -s & 0 & 0 \\ 0 & -s & 0 \\ s & 0 & 0 \end{bmatrix}$ Four edges: $\begin{bmatrix} \frac{1}{\sqrt{2}}s & 0 & 0 \\ 0 & 0 & \pm h \\ -\frac{1}{\sqrt{2}}s & 0 & 0 \end{bmatrix}, \begin{bmatrix} 0 & \frac{1}{\sqrt{2}}s & 0 \\ 0 & 0 & \pm h \\ 0 & -\frac{1}{\sqrt{2}}s & 0 \end{bmatrix}$

Four X-arms: $\begin{bmatrix} x_c & y_c & \pm z_c \\ 0 & 0 & 0 \\ -x_c & -y_c & \pm z_c \end{bmatrix}, \begin{bmatrix} x_c & -y_c & \pm z_c \\ 0 & 0 & 0 \\ -x_c & y_c & \pm z_c \end{bmatrix}$

All my plots used $s = \sqrt{2}$ which means $h = m$.

It is not possible to construct a Fisher's Crystal with equal arm and face angles. Perhaps a story could involve a crystal that oscillates between $\theta_F = 90°$ and $\theta_X = 90°$ using a multiphase quantum-resonance bypass that reverses the polarity of the flux capacitor. Perhaps I should leave the writing of science fiction stories to the experts.

Problems

Newton's Nose-Cone Problem

1. Compare the reduced drag coefficients for a conic and elliptical nose cone when $H = 2R$ and $H = \frac{1}{2}R$.

2. Calculate and plot the ratio of $C_{\text{cone}}/C_{\text{ell}}$ in terms of the ratio $\eta \equiv \frac{H}{R}$. Is the graph consistent with the results of the previous problem?

3. Show that the reduced drag coefficient of a parabolic nose cone can be written as
$$C = \frac{\ln\left(1+\eta^2\right)}{\eta^2}$$
where $\eta \equiv 2\frac{H}{R}$.

4. Show that a nose cone with shape $z(x) = H\left(\frac{x}{R}\right)^3$ has a reduced drag coefficient of
$$C = \frac{\arctan\eta}{\eta}$$
where $\eta \equiv 3\frac{H}{R}$.

5. Show that the reduced drag coefficient of the nose cone $z(x) = H\left(\frac{x}{R}\right)^{1/2}$ is
$$C = 1 - 2\eta^2 + 2\eta^4 \ln\left(1 + \frac{1}{\eta^2}\right)$$
where $\eta \equiv \frac{1}{2}\frac{H}{R}$.

6. Show that the reduced drag coefficient of the nose cone $z(x) = H\left(\frac{x}{R}\right)^{3/2}$ is
$$C = \frac{2}{\eta^2}\left(1 - \frac{\ln\left(1+\eta^2\right)}{\eta^2}\right)$$
where $\eta \equiv \frac{3}{2}\frac{H}{R}$.

7. Use Evaluate Numerically and Eq. (NC.4) to verify the results from the previous problems when $\frac{H}{R} = 1$, 2, and 3.

8. Verify that the RDCs from the previous problems reduce to the expected value in the zero-height limit.

9. Find the minimizing exponent and the minimum RDC for the simple-shaped nose cone $z(x) = H\left(\frac{x}{R}\right)^n$ when
   a. $\frac{H}{R} = \frac{1}{2}$,
   b. $\frac{H}{R} = 2$, and
   c. $\frac{H}{R} = 3$.

10. Show that Eq. (NC.14) produces the exact expression for a frustum's RDC.

11. Verify that the general expression for the reduced drag coefficient of a frustum reduces to the expected result when

    a. the height is zero, or

    b. the tip-width is zero.

12. Show that the tip-width that minimizes the RDC of a frustum is

$$a_{\text{min}} = R\left(1 + \frac{1}{2}\frac{H^2}{R^2}\left(1 - \sqrt{1 + 4\frac{R^2}{H^2}}\right)\right).$$

13. Show the adjusted frustum satisfies the initial condition $z'(a) = 0$ as long as $n > 1$.

14. Verify that the shape of Eq. (NC.15) satisfies Newton's minimizing conditions.

15. a. Verify that the shape of Eq. (NC.15) has this derivative.

$$\frac{dz}{dx} = 1 + n\frac{H-R+a}{R-a}\left(\frac{x-a}{R-a}\right)^{n-1}$$

    b. Verify that the derivative satisfies these conditions.

$$\frac{dz}{dx} = \begin{cases} 1 & x = a \\ 1 - n\left(1 - \frac{H}{R-a}\right) & x = R \end{cases}$$

16. Show that the RDC for Joe's fudge can be written as

$$C = \frac{a^2}{R^2} + \left(1 - \frac{a}{R}\right)^2 \int_0^1 \frac{2\left(\frac{a}{R-a} + u\right)}{1 + \left(1 + \frac{H-R+a}{R-a}\, n\, u^{n-1}\right)^2} du$$

    with the same variable change used for the adjusted frustum.

17. Find the minimizing exponent and tip-width, and the minimum RDC for Joe's fudge when

    a. $\frac{H}{R} = \frac{1}{2}$,

    b. $\frac{H}{R} = 1$, and

    c. $\frac{H}{R} = 2$.

18. Joe's fudge and Newton's minimizer have the same initial value for the nose-cone's slope. Compare their *final* values of the slope for several different tip-widths and height-to-radius ratios.

19. Find Newton's minimizing tip-width when

    a. $\frac{H}{R} = \frac{1}{2}$,

    b. $\frac{H}{R} = 2$, and

    c. $\frac{H}{R} = 3$.

    Note: It's OK to take a look at the graph in the next problem.

20. Reproduce the following graph of the minimizing tip-width versus nose-cone height for Newton's minimizer. Fit an appropriate polynomial and use it to make a prediction for $H = \frac{3}{4}R$.

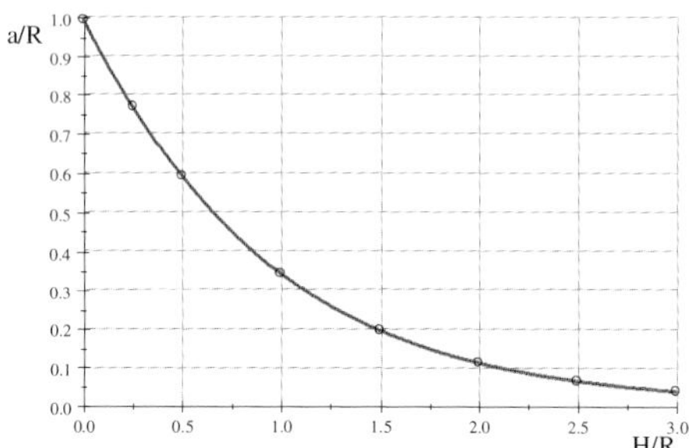

As the height gets larger, the minimizing tip-width decreases but never vanishes.

21. There's another way to solve for Newton's minimizer, using the differential form of Eq. (NC.3)

$$\frac{dC}{dx} = \frac{1}{R^2}\frac{2x}{1+(z')^2}$$

with the additional initial condition $C(a) = a^2/R^2$. The RDC is $C(R)$.

a. Set up a matrix like this one with appropriate numerical values and solve the differential equations for the shape $z(x)$ and the RDC function $C(x)$ simultaneously.

$$\begin{bmatrix} z'' = \frac{(z')^3 + z'}{3x(z')^2 - x} \\ C' = \frac{2x}{1+(z')^2} \\ z(a) = 0 \\ z'(a) = 1 \\ C(a) = a^2/R^2 \end{bmatrix}$$

b. What is the RDC of this shape?

c. Plot the shape of this nose cone.

22. Show that the contribution to the RDC from any tip is

$$C_{\text{tip}} = \tfrac{1}{R^2}\int_0^a \frac{2x}{1+\left(\frac{dz_{\text{tip}}}{dx}\right)^2}\,dx$$

where a is the tip's radius.

23. Use the result of previous problem and the substitution $u = x/a$ to derive Eq. (NC.18a).

24. Verify that the method of finding the radially-symmetric minimizer produces Newton's result for a flat tip.

25. Let's explore the single-impact condition at the origin.

    a. Look at Figure NC.2 and convince yourself that $\tan\phi = \dfrac{dz}{dx}$.

    b. Prove the deflection angle for a particle incident at the origin reflecting through $x = \pm a$ satisfies $\tan 2\phi = a/z_{\text{tip}}(0)$.

    c. Use these results and the appropriate double-angle formula and derive Eq. (NC.19) for $x = 0$.

26. Let's explore the single-impact condition in general.

    a. Reassure yourself that $\tan\phi = \dfrac{dz}{dx}$ is still valid.

    b. Prove the deflection angle for a particle reflecting through $x = \pm a$ satisfies
$$\tan 2\phi = \frac{a+x}{z_{\text{tip}}(0)} .$$

    c. Use these results and the appropriate double-angle formula and derive Eq. (NC.19).

27. Let's explore the single-impact condition at $x = a$.

    a. Convince yourself that the deflection angle is $\phi = 45°$.

    b. Use part a. of the previous problem and argue that $z'_{\text{tip}}(a) = -1$.

28. Verify that the minimizing tip of Eq. (NC.20) satisfies the single-impact condition at $x = 0$, $x = a$, and for unspecified x.

29. Show that the minimum RDC for a nose cone with an indented tip can be written as
$$C_{\text{min}} = \frac{\kappa a^2}{R^2} + \frac{1}{R^2}\int_a^R \frac{2x}{1+z'(x)^2}\,dx .$$

30. Consider the generic centered polynomial $z_{\text{tip}} = z_0\left(1 - \dfrac{x^n}{a^n}\right)$ where z_0 is the depth of the indented tip.

    a. Apply the single-impact condition at the origin and prove that $z_0 = a/\sqrt{3}$ minimizes the tip's air resistance when $n = 1$.

    b. Apply the single-impact condition at $x = a$ and prove that $z_0 = a/n$ minimizes the tip's air resistance when $n \neq 1$.

31. Show that the inverted-cone tip $z_{\text{tip}} = \frac{1}{\sqrt{3}}(a - x)$ produces 75% of the air resistance of a flat tip with the same radius. This is the deepest such tip that satisfies the single-impact criterion.

32. Show that the centered-polynomial tip $z_{\text{tip}} = \dfrac{a}{n}\left(1 - \dfrac{x^n}{a^n}\right)$ produces about 61% of the air resistance of a flat tip with the same radius when $n = \frac{3}{2}$.

The Shape of the Eiffel Tower

Here are my data (in meters) for the shape of the Eiffel Tower:

$$\begin{bmatrix} f & 5.30 & 5.32 & 6.65 & 10.0 & 13.3 & 19.6 & 26.6 & 33.2 & 39.9 & 53.2 \\ z & 275 & 262 & 218 & 169 & 141 & 105 & 78.5 & 60.0 & 41.2 & 14.6 \end{bmatrix}$$

33. Verify by direct calculation that the solutions presented in Eq. (ET.4) satisfy both Eqs. (ET.2) and (ET.3).

34. Let's verify my claims about constant wind pressures and the Tower's shape.

    a. Transpose the data matrix to the form $\begin{bmatrix} f(z) & z \end{bmatrix}$ and plot the data as points.

    b. Plot the shape function for constant wind pressures $w_0 = 0.7, 1,$ and 1.33.

    c. Use your plot to analyze where the functions fit the data and where they don't.

35. Use Solve ODE + Exact on Eq. (ET.3) with the following wind pressures.

    a. Constant: $w(z) = c$

    b. Linear: $w(z) = bz + c$

    c. Quadratic: $w(z) = az^2 + bz + c$

    Here's what the matrix looks like for the linear wind pressure:

$$\begin{bmatrix} f\dfrac{df}{dz} = bz\,f + (bz + c)\left(f + z\dfrac{df}{dz}\right) \\ f(0) = f_0 \end{bmatrix}$$

36. Show that all three solutions from the previous problem are of the form predicted by Eq. (ET.4). You may find *SNB*'s Factor in-place useful.

    Here are my data for the wind pressure (shown in Figure ET.2).

$$\begin{bmatrix} z & 275 & 262 & 218 & 169 & 141 & 105 & 78.5 & 59.9 & 41.2 & 14.6 \\ w & 1.33 & 1.39 & 1.33 & 1.13 & 0.994 & 0.84 & 0.766 & 0.705 & 0.704 & 0.693 \end{bmatrix}$$

37. Fit a linear, quadratic, and cubic polynomials to these data and use them to plot the Tower's shape function. Which fit fits best?

38. We treated the Tower as an idealized rigid solid structure that can withstand a maximum wind pressure of approximately $4\rho g f_0 w_{\max} \approx 33.3\ \mathrm{kN/m^2}$ without toppling. What wind speed produces such a pressure?

39. Because of its open-lattice and flexible design, the real Eiffel Tower offers only about 12% as much wind resistance as our idealized structure. Show that if this is the case, then the maximum physical wind pressure is $P(z) = 2970\,\frac{\mathrm{N}}{\mathrm{m}^2}\,w(z)$.

40. Compare the mass of the air in a box just large enough to enclose the Tower ($125^2\ \mathrm{m}^2 \times 300\ \mathrm{m}$) with the Tower's mass $\left(7.30 \times 10^6\ \mathrm{kg}\right)$.

41. Calculate the arc-length travelled by Semolina Pilchard when she climbed up the Eiffel Tower. Yes I know, but it sounds like a woman's name to me. Goo goo g'joob.

42. Let's see if the Eiffel Tower is a witch.

    a. Use the expression $V = \int_0^H (2f(z))^2\,dz$ to calculate the Tower's volume.

    b. Calculate the Tower's density (its mass is significantly larger than a typical duck's).

    c. If the Tower were a uniform solid structure, would it float in water?

    It is good to be wise in the ways of science.

43. The scale model of the Eiffel Tower in Las Vegas is an approximately half-scale reproduction of the original. Calculate an appropriate maximum wind pressure and use it to plot the model's shape function.

44. Repeat the previous problem for a 1:3 scale reproduction of the Eiffel Tower.

45. Exponentials can also describe the Tower's shape (see [35]). Plot my data for both edges of the Tower. On the $+z$ side, plot $\left(2.6958e^{0.01117(283-z)}, z\right)$ for its lower half, $\left(4.7439e^{0.00721(283-z)}, z\right)$ for the upper half, and $\left(4.2547e^{0.00892(283-z)}, z\right)$ for the whole Tower. On the $-z$ side, add my result for $w(z) = 0.7$ (lower), $w(z) = 1.33$ (upper), and varying wind (whole). Compare the results of the two approaches.

An Interesting Classical Orbit

46. Find the force that produces the oscillating orbit $r(\phi) = R + a\sin n\phi$.

47. Take the same three well-known limits to the force and orbit of the previous problem discussed in the text. Plot the three orbits for the same numerical values shown in Figure IO.2. Compare the two plots.

48. What is the angle between consecutive maxima of the generic oscillating orbit given in Eq. (IO.1)? Verify your answer for Figure IO.1.

49. Use the radius of the stable circular orbit from Eq. (IO.7) to confirm the stability condition of the oscillating orbit.

50. Let's explore my claim that the oscillating-orbit's shape does not depend on the angular momentum but its period does. Use Solve ODE + Numeric on Eq. (IO.2) for the orbital parameters in Figure IO.1 with $\ell = \frac{1}{2}$, 1, and 2. Plot $r(t)$ each orbit and use the graph to approximate the orbital periods. What is the relationship between the angular momentum and the orbital period?

51. Plot the effective potential for the oscillating orbit with $R = 2$ and $n = 5$ for $a = 2$ (so that $a = R$) and $a = 3$ (so that $a > R$). Are there any stable orbits?

52. Plot the effective potential for the three orbits shown in Figure IO.2.

53. Plot the effective potential for the three orbits you calculated in Problem 46.

54. Show that when $a = 0$, the effective potential for the oscillating orbit has a local minimum at $r = R$ and a local maximum at $r = 2R$.

Fisher's Crystal

55. Show that the z-component of the center of the side in the octant above the first quadrant can be written

$$z_c = \frac{m^2}{2m^2 + 1}\, h$$

where $m \equiv \sqrt{2}\, h/s$.

56. Show that the vector I used for the midpoint of edge 1 can be written

$$\vec{M}_1 = \tfrac{1}{2\sqrt{2}}\, s\, \begin{bmatrix} 1 & 0 & m \end{bmatrix}$$

where $m \equiv \sqrt{2}\, h/s$.

57. Using the definitions given in the text, calculate $\vec{S} \cdot \left(\vec{V}_1 \times \vec{V}_2\right)$. What does it mean?

58. Show that the vector describing the side of the base ($\longleftarrow$) in the first quadrant of the x-y plane is

$$\vec{V}_3 = \left(\tfrac{1}{\sqrt{2}}s, -\tfrac{1}{\sqrt{2}}s, 0\right)$$

and verify that $\vec{V}_3 \times \vec{V}_1 = \vec{V}_1 \times \vec{V}_2$. What does that mean?

59. Plot the arm angle θ_X and the face angle θ_F on the same graph. At what value of m are the angles within 5% of each other?

60. Find the height of Fisher's Crystal that has

    a. $\theta_F = 60°$ and   b. $\theta_X = 60°$.

61. Find the height of Fisher's Crystal that has $\theta_X = 45°$. What is the face angle for this crystal?

62. The Great Pyramid at Giza is 480.7 ft high and has a square base with sides that are 756 ft long.

    a. Calculate the arm and face angles of this pyramid.

    b. Plot the Great Crystal this pyramid makes.

    c. Just for kicks, calculate the ratio $2\, s/h$ and compare it with any number you think is relevant.

63. Repeat the previous problem for Khafre's Pyramid at Giza. Khafre's Pyramid is 448 ft high and has a square base with sides 706 ft long.

64. Repeat the repeat of the previous-previous problem for the pyramid in front of the Louvre. The Louvre Pyramid is 20.6 m high and has a square base with sides 35 m long. Extraterrestrials didn't build this pyramid either.

65. Write a short science fiction story where an ancient real-life pyramid is actually the top of a half-buried Fisher's Crystal.

B Topics in Modern Physics

In this appendix, we'll use *SNB* to explore some interesting problems in modern physics. Modern Physics includes Special Relativity (SR), General Relativity (GR), and Quantum Mechanics (QM). These problems are often mentioned in physics textbooks, but usually not in great detail. *SNB* helps us dig a little deeper and fill in some of the details.

The Tale of the Traveling Triplets

The study of Special Relativity involves the analysis of physical events from different inertial reference frames (IRF). To emphasize the counter-intuitive nature of SR, twins often occupy those IRFs. Here we add a third observer to a standard SR problem.

Our story involves three intrepid explorers, triplets and members of Starfleet. Homer stays on his home-planet Earth. Constance and Axel leave Earth at the same time but travel in different starships to the star Vega, which is 25 light-years away. Constance makes the trip at a constant speed on Starship-C, while on Starship-A Axel uses a constant acceleration to mimic gravity. At the start of the trips, all three are at the origin of their respective reference frames and they are 20 y old.

Trip 1: Constance goes to Vega

The two IRFs for this trip are the Earth-Vega frame (at rest relative to each other) and the Starship-C frame where observers move at a constant speed of $v = 0.928\,47\,c$. The gamma factor for this relative speed tells us relativistic effects are significant.

$$\gamma = \left(1 - \frac{v^2}{c^2}\right)^{-1/2} = \frac{1}{\sqrt{1 - 0.928\,47^2}} = 2.692\,5$$

Earth-Vega Frame (S: Homer)
The Earth-Vega distance in this frame is the rest length of the space interval between them $\Delta x = 25\,c\,\text{y}$. The two events (Constance leaving Earth, arriving at Vega) occur at different places in S, so the elapsed time Δt is the dilated time, not the proper time.

$$\Delta t = \frac{\Delta x}{v} = \frac{25c\,\text{y}}{0.928\,47\,c} = 26.926\,\text{y}$$

As measured by Homer, it takes Starship-C 26.926 y to complete the $25\,c\,\text{y}$ trip to Vega.

Doing Physics with Scientific Notebook: A Problem-solving Approach, First Edition. Joseph Gallant.

Starship-C Frame (S': **Constance**)

As seen from Starship-C, Earth and Vega are moving at speed v so the space interval between them is contracted.

$$\Delta x' = \frac{\Delta x}{\gamma} = \frac{25c\,\text{y}}{2.692\,5} = 9.285\,1c\,\text{y}$$

The time interval is the proper time since in this frame the two events happen at the same place (the origin of S').

$$\Delta t' = \frac{\Delta x'}{v} = \frac{9.285\,1\,c\,\text{y}}{0.928\,47\,c} = 10.0\,\text{y}$$

As measured by Constance, it takes her starship $10.0\ \text{y}$ to complete the $9.285\,1\,c\,\text{y}$ trip to Vega. We also could use Time Dilation to calculate the elapsed time on the ship.

$$\Delta t' = \frac{\Delta t}{\gamma} = \frac{26.926\ \text{y}}{2.692\,5} = 10.0\,\text{y}$$

Immediately upon reaching Vega, Constance embarks on a return trip at the same speed When she arrives back on Earth, Constance's age is $20\,\text{y} + 2\,(10.0\ \text{y}) = 40.0\,\text{y}$ and Homer's age is $20\,\text{y} + 2\,(26.926\ \text{y}) = 73.852\,\text{y}$. She is almost 34 years younger than her triplet! Here's a fun way to think about this: a baby born on the day Constance left would be almost $54\,\text{y}$ old when she returned. The baby is 14 years older than her!

This scenario is sometimes called the Twin Paradox and it has been used as an argu ment against Special Relativity. Here's the apparent paradox: since both Homer and Constance see the other moving, they both see the other's slow-running clocks, so when they reunite they both expect to find the other is younger. They obviously *both* can't be younger, so SR must be wrong.

All inertial reference frames are equivalent per the 1st Postulate, so is the paradox valid?

No! Its resolution can best be seen on a space-time diagram.

Homer stayed in one IRF at $x = 0$, but Constance changed frames when she turned around and headed home.

Changing reference frames is a physically detectable act that establishes Constance was moving through space.

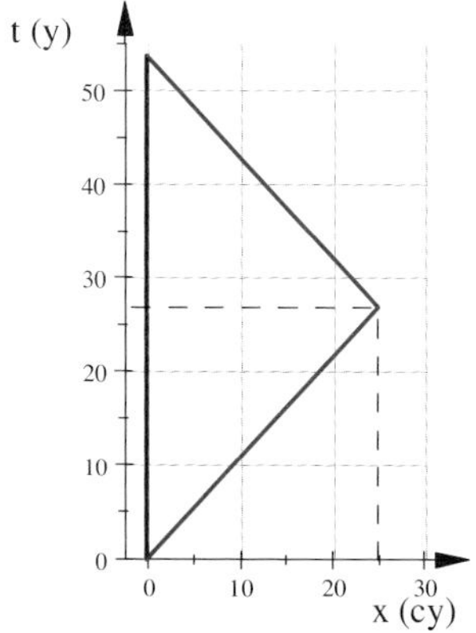

Figure TT.1 STD #1

Remember the one-sentence version of Special Relativity: the faster you move through space, the slower you move through time!

Relativistic Interlude: Constant Acceleration

Before we continue our analysis of the Traveling Triplets, we need to take a small detour to consider straight-line motion with a constant acceleration. Common misconceptions aside, SR can deal with accelerated motion. It's not always simple (the expressions are complicated) or useful (neither force nor acceleration are invariant) but we can analyze accelerated frames in SR.

Suppose a starship moves with a constant acceleration. The ship's instantaneous rest frame S'' is a non-inertial frame that moves at varying speed with respect to observers at rest in the inertial reference frame S.

A starship passenger could easily measure its acceleration a'' in S'' with simple free-fall experiments. A constant acceleration as seen in S would eventually give the starship a speed greater than the speed of light, violating the 2nd Postulate. The ship's velocity and acceleration are along the same axis, so its acceleration is $a = a''/\gamma^3$ in S.

Velocity as a function of Time

Since we used v for a constant speed, let's use u for the ship's varying speed so that $a = du/dt$. Let's also use $a'' \equiv g$ to remind us the acceleration mimics gravity.

$$a = \frac{a''}{\gamma^3} \Longrightarrow \frac{du}{dt} = \left(1 - \frac{u^2}{c^2}\right)^{3/2} g \qquad \text{(TT.1)}$$

The solution of this equation gives the ship's velocity as a function of time $u(t)$ as seen in frame S.

$$g(t - t_0) = \int \frac{du}{\left(1 - \frac{u^2}{c^2}\right)^{3/2}} \qquad \text{(TT.2)}$$

We'll look at general motion where the ship's initial velocity is u_0 and its initial position is x_0. Let's change to the dimensionless variable $\beta = u/c$ and Expand the right-hand side of Eq. (TT.2) using the definite-integral fudge.

$$c\left[\int \frac{d\beta}{\left(1-\beta^2\right)^{3/2}}\right]_{u_0/c}^{u/c} = \frac{u}{\sqrt{1 - \frac{1}{c^2}u^2}} - \frac{u_0}{\sqrt{1 - \frac{1}{c^2}u_0^2}}$$

Let's define γ_0 as the gamma factor for the initial velocity, restore the left-hand side of Eq. (TT.2) with $t_0 = 0$ and use Solve Exact to find u.

$$gt = \frac{u}{\sqrt{1 - \frac{u^2}{c^2}}} - \gamma_0 u_0, \text{ Solution is: } c\frac{\gamma_0 u_0 + gt}{\sqrt{c^2 + g^2t^2 + 2gt\gamma_0 u_0 + \gamma_0^2 u_0^2}}$$

Rearrange the term inside the square root by-hand, and then use Expand and Factor in-place on the result (except for the "+1").

$$\frac{c^2 + g^2t^2 + 2gt\gamma_0 u_0 + \gamma_0^2 u_0^2}{c^2} = \frac{g^2}{c^2}t^2 + \frac{1}{c^2}\gamma_0^2 u_0^2 + \frac{2}{c^2}gt\gamma_0 u_0 + 1 = \frac{\left(\gamma_0 u_0 + gt\right)^2}{c^2} + 1$$

This gives us the relativistic velocity as a function of time for a constant acceleration as measured in the non-accelerated frame S.

$$u(t) = \frac{\gamma_0 u_0 + gt}{\sqrt{1 + \left(\frac{\gamma_0 u_0 + gt}{c}\right)^2}} \tag{TT.3}$$

Position as a function of Time

Relativity or not, once we know the velocity as a function of time, we can use the results of Chapter 2 to calculate the position.

$$x(t) = x_0 + \int_0^t u(t)\, dt$$

This one works best with Simplify, assume (positive), and the definite-integral fudge.

$$\left[\int \left(\frac{\gamma_0 u_0 + gt}{\sqrt{1 + \left(\frac{\gamma_0 u_0 + gt}{c}\right)^2}}\right) dt\right]_{t=0}^{t=t}$$
$$= \frac{c}{g}\left(\sqrt{c^2 + g^2t^2 + 2gt\gamma_0 u_0 + \gamma_0^2 u_0^2} - \sqrt{c^2 + \gamma_0^2 u_0^2}\right)$$

The first term inside the parentheses is the same one we encountered calculating the velocity, and the second term is just $c\gamma_0$.

Here is the SR position as a function of time in S.

$$x(t) = x_0 + \frac{c^2}{g}\left(\sqrt{1 + \left(\frac{\gamma_0 u_0 + gt}{c}\right)^2} - \gamma_0\right) \tag{TT.4}$$

Velocity as a function of Position

This one is a lot easier than it looks. Rearrange the position-versus-time result from Eq. (TT.4) to get an equation relating the time stuff to the space stuff.

$$\sqrt{1 + \left(\frac{\gamma_0 u_0 + gt}{c}\right)^2} = \gamma_0 + \frac{g}{c^2}(x - x_0)$$
$$\gamma_0 u_0 + gt = c\sqrt{\left(\gamma_0 + \frac{g}{c^2}(x - x_0)\right)^2 - 1}$$

Edit Eq. (TT.3) by-hand, replace the time stuff with the space stuff, and put it all under the radical to get the relativistic expression for velocity as a function of position.

$$u(x) = c\sqrt{1 - \left(\gamma_0 + \frac{g}{c^2}(x - x_0)\right)^{-2}} \tag{TT.5}$$

Remember g is the constant acceleration as measured on the ship!

On the ship?

The time, position, and velocity in the above results are all measured in the non-accelerated frame S. We certainly want to know how much time elapses on the starship. The usual Time Dilation result is only valid for constant-speed motion.

$$\Delta t'' = \frac{\Delta t}{\gamma} = \sqrt{1 - \frac{v^2}{c^2}}\,\Delta t$$

Let's generalize it to account for the starship's varying speed. The time interval between two events in the starship's frame is

$$\Delta t'' = \int_{t_1}^{t_2} \sqrt{1 - \frac{u^2(t)}{c^2}}\,dt \tag{TT.6}$$

where $\Delta t = t_2 - t_1$ is the time interval as measured in S.

The velocity in Eq. (TT.3) looks simpler in terms of a dimensionless variable.

$$\xi \equiv \frac{\gamma_0 u_0 + gt}{c} \Longrightarrow \frac{u}{c} = \frac{\xi}{\sqrt{1+\xi^2}}$$

We need to Simplify the expression inside the radical in Eq. (TT.6).

$$1 - \left(\frac{u}{c}\right)^2 = 1 - \left(\frac{\xi}{\sqrt{1+\xi^2}}\right)^2 = \frac{1}{\xi^2+1}$$

Now we can Expand the integral $\left(\text{with } dt = \frac{c}{g}\,d\xi\right)$.

$$\Delta t'' = \frac{c}{g}\int_{\xi_1}^{\xi_2} \frac{d\xi}{\sqrt{1+\xi^2}} = \frac{c}{g}\operatorname{arcsinh}\xi_2 - \frac{c}{g}\operatorname{arcsinh}\xi_1$$

Restoring the explicit time-dependence gives us an expression for the proper time interval as measured on the accelerating ship.

$$\Delta t'' = \frac{c}{g}\operatorname{arcsinh}\left(\frac{\gamma_1 u_1 + gt_2}{c}\right) - \frac{c}{g}\operatorname{arcsinh}\left(\frac{\gamma_1 u_1 + gt_1}{c}\right) \tag{TT.7a}$$

The change to $\gamma_1 u_1$ reminds you that the initial speed and corresponding gamma factor refer to the start of the time interval at time t_1.

Let's set $t_1 = t_1'' = 0$ so the clocks are synchronized and use Eq. (TT.7a) and Solve Exact to find t in terms of t''.

$$t'' = \frac{c}{g}\left(\operatorname{arcsinh}\frac{\gamma_0 u_0 + gt}{c} - \operatorname{arcsinh}\frac{\gamma_0 u_0}{c}\right),$$

$$\text{Solution is: } -\frac{1}{g}\left(\gamma_0 u_0 - c\sinh\left(\operatorname{arcsinh}\frac{1}{c}\gamma_0 u_0 + \frac{1}{c}gt''\right)\right)$$

This gives us an expression for the time interval as measured in the Earth-Vega frame.

$$\Delta t = \frac{c}{g}\sinh\left(\frac{gt_2''}{c} + \operatorname{arcsinh}\frac{\gamma_1 u_1}{c}\right) - \frac{c}{g}\sinh\left(\frac{gt_1''}{c} + \operatorname{arcsinh}\frac{\gamma_1 u_1}{c}\right) \tag{TT.7b}$$

We can use clocks in either frame to measure time intervals.

We also want to know how far the starship travels in S''. The distance between any two events in S'' is not constant. As the ship moves faster, the distance between two events gets shorter, so the spatial interval is time-dependent. The usual Length Contraction result is only valid for constant-speed motion.

$$\Delta x'' = \frac{\Delta x}{\gamma} = \sqrt{1 - \frac{v^2}{c^2}}\ \Delta x$$

Let's generalize it to account for the starship's varying speed. The space interval between two events in the starship's frame is

$$\Delta x'' = \int_{x_1}^{x_2} \sqrt{1 - \frac{u^2(x)}{c^2}}\ dx \qquad \text{(TT.8)}$$

where $\Delta x = x_2 - x_1$ is the space interval as measured in S.

Let's use Eq. (TT.5) and Evaluate to get an expression for the integrand.

$$\left[1 - \frac{u^2}{c^2}\right]_{u=c\sqrt{1-(\gamma_1+g(x-x_1)/c^2)^{-2}}} = \frac{1}{\left(\gamma_1 + \frac{1}{c^2} g\,(x - x_1)\right)^2}$$

Now we can Evaluate the integral in Eq. (TT.8).

$$\Delta x'' = \left[\int \frac{dx}{\gamma_1 + \frac{g}{c^2}(x - x_1)}\right]_{x=x_1}^{x=x_2} = \frac{c^2}{g} \ln \frac{1}{g}\left(\gamma_1 c^2 - g x_1 + g x_2\right) - \frac{c^2}{g} \ln \frac{c^2}{g}\gamma_1$$

A little editing by-hand gives us an expression for the space interval as measured on the accelerating ship.

$$\Delta x'' = \frac{c^2}{g} \ln\left(1 + \frac{g}{\gamma_1 c^2}\,\Delta x\right) \qquad \text{(TT.9a)}$$

The γ_1 again reminds you that "initial" refers to the start of the space interval at position x_1. Here is the expression for the space interval as measured in the Earth-Vega frame.

$$\Delta x = \frac{\gamma_1 c^2}{g}\left(e^{g\Delta x''/c^2} - 1\right) \qquad \text{(TT.9b)}$$

We can use meter sticks in either frame to measure space intervals.

The relationship between the proper time t'' and IRF time t is much simpler when the accelerating ship starts from rest (so $u_0 = 0$ and $\gamma_0 = 1$).

$$\frac{gt}{c} = \sinh \frac{gt''}{c}$$

This leads to two easily-derived results. The first uses Eq. (TT.4) to give the ship's position in S as a function of proper time.

$$x\left(t''\right) = x_0 + \frac{c^2}{g}\left(\cosh\left(\frac{gt''}{c}\right) - 1\right) \qquad \text{(TT.10)}$$

The second uses Eq. (TT.3) to give the ship's velocity as a function of proper time.

$$u\left(t''\right) = c \tanh \frac{gt''}{c} \qquad \text{(TT.11)}$$

Observers in both frames agree on the instantaneous relative velocity between frames.

Trip 2: Axel goes to Vega

The two reference frames for this trip are Homer's Earth-Vega frame S and Axel's Starship-A frame S'' where observers travel to Vega with a constant acceleration. For the first half of the trip, the ship is speeding up and acceleration is $g = +1\,c/\,\mathrm{y}$ and for the second it's slowing down with $g = -1\,c/\,\mathrm{y}$. To keep the expressions relatively simple, we'll sometimes set $c = 1$. All times are in years and all distances are in light-years.

Earth-Vega Frame (S: Homer)

We can use the Interlude results to create expressions for the starship's position and velocity in S. Here are the ship's position and velocity for the first half of the trip.

$$x_1(t) = \left[\frac{c^2}{g}\left(\sqrt{1+\left(\frac{\gamma_0 u_0 + gt}{c}\right)^2} - \gamma_0\right)\right]_{g=+1,c=1,u_0=0,\gamma_0=1} = \sqrt{t^2+1} - 1$$

$$u_1(t) = \left[\frac{\gamma_0 u_0 + gt}{\sqrt{1+\left(\frac{\gamma_0 u_0 + gt}{c}\right)^2}}\right]_{g=+1,c=1,u_0=0,\gamma_0=1} = \frac{t}{\sqrt{t^2+1}}$$

Use the position and Solve Exact to get the time of the ship's arrival at the midpoint.

$12.5 = \sqrt{t^2+1} - 1$, Solution is: 13.463

The ship reaches the midpoint at $t_m = 13.463\,\mathrm{y}$. The trip is symmetric, so total time interval is $\Delta t = 2t_m = 26.926\,\mathrm{y}$. As measured by Homer, it takes Starship-A 26.926 y to complete the $25\,c\,\mathrm{y}$ trip to Vega.

This explains my choice for Constance's speed. For the sake of comparison, the two trips have the same time and space intervals (and average velocity) as measured in the Earth-Vega frame.

$$u_{\text{ave}} = \frac{\Delta x}{\Delta t} = \frac{25\,c\,\mathrm{y}}{26.926\ \mathrm{y}} = 0.928\,47c$$

The second half of the trip starts at the midpoint so we'll need the ship's speed there (using Eq. (TT.5)) and the corresponding gamma factor.

$$\frac{u_m}{c} = \sqrt{1-\left(1+\tfrac{1c/\mathrm{y}}{c^2}\,(12.5c\mathrm{y})\right)^{-2}} = 0.997\,25$$

$$\gamma_m = \frac{1}{\sqrt{1-\frac{u_m^2}{c^2}}} = \frac{1}{\sqrt{1-0.997\,25^2}} = 13.493$$

So $\gamma_m u_m = (13.493)\,(0.997\,25\,c) = 13.456c$.

Here's the ship's velocity as a function of time for the second half of the trip.

$$u_2(t) = \left[\frac{\gamma_m u_m + g(t - t_m)}{\sqrt{1 + \left(\dfrac{\gamma_m u_m + g(t - t_m)}{c}\right)^2}}\right]_{g=-1,c=1,u_m=0.99725,\gamma_m=13.493,t_m=13.463}$$
$$= \frac{26.919 - t}{\sqrt{(t - 26.919)^2 + 1}}$$

This velocity equals zero at $t = 26.919\,\text{y}$ which gives a slightly different arrival time than our above result. This is due mostly to rounding errors in the gamma factor, which is a rapidly changing function at large velocities.

Starship-A Frame (S'': Axel)

Now let's look at the time and space intervals as measured on Starship-A using the Interlude results relating the intervals in S to those in S''. Remember, the starship is at rest in S'' and the total intervals are between Earth leaving and Vega arriving.

Let's start with the intervals for the first half of the trip when the ship is speeding up at $g = +1c/\,\text{y}$. Eqs. (TT.7a) and (TT.9a) give us the time and space intervals.

$$\Delta t_1'' = \frac{c}{g}\,\text{arcsinh}\,(t_m) = \frac{c}{1c/\,\text{y}}\,\text{arcsinh}\,(13.463) = 3.294\,5\,\text{y}$$
$$\Delta x_1'' = \frac{c^2}{g}\ln\left(1 + \frac{g}{c^2}\frac{\Delta x_1}{\gamma_0}\right) = \frac{c^2}{1c/\,\text{y}}\ln\left(1 + \frac{12.5}{1}\right) = 2.602\,7c\,\text{y}$$

Here are the intervals for the second half of the outbound trip to Vega when the ship is slowing down at $g = -1c/\,\text{y}$.

$$\Delta x_2'' = \frac{c^2}{g}\ln\left(1 + \frac{g}{c^2}\frac{\Delta x_2}{\gamma_m}\right) = \frac{c^2}{-1c/\,\text{y}}\ln\left(1 - \frac{12.5}{13.493}\right) = 2.609\,2c\,\text{y}$$
$$\Delta t_2'' = \frac{c}{g}\,\text{arcsinh}\left(\frac{\gamma_m u_m + g t_2}{c}\right) - \frac{c}{g}\,\text{arcsinh}\left(\frac{\gamma_m u_m + g t_m}{c}\right)$$
$$= \frac{c}{-1c/\,\text{y}}\Big(\text{arcsinh}\,(13.456 - 26.926) - \text{arcsinh}\,(13.456 - 13.463)\Big) = 3.288\,0\,\text{y}$$

These values are not exactly equal to the first-half results, again due to rounding errors. Nonetheless, we'll use them to calculate the total intervals.

$$\Delta t'' = \Delta t_1'' + \Delta t_2'' = 3.294\,5\,\text{y} + 3.288\,0\,\text{y} = 6.582\,5\,\text{y}$$
$$\Delta x'' = \Delta x_1'' + \Delta x_2'' = 2.602\,7c\,\text{y} + 2.609\,2c\,\text{y} = 5.211\,9c\,\text{y}$$

As measured by Axel on Starship-A, it takes the ship $6.582\,5\,\text{y}$ to travel $5.211\,9\,c\,\text{y}$ from Earth to Vega.

Immediately upon reaching Vega, Axel embarks on a return trip with the same acceleration. When he arrives back on Earth, Axel's age is $20\,\mathrm{y} + 2\,(6.582\,5\,\mathrm{y}) = 33.165\,\mathrm{y}$ and Homer's age is $20\,\mathrm{y} + 2\,(26.926\ \mathrm{y}) = 73.852\,\mathrm{y}$. Axel is 40 years younger than Homer!

Here is the space-time diagram showing the world lines of Homer and Axel in frame S.

Axel's world line consists of 4 curves of the form $(x\,(t)\,,t)$ using Eq. (TT.4) with the appropriate values for all 4 parts (to and from Vega, $a'' = \pm\,g$) of his trip.

The red dashed line shows Constance's world line for comparison.

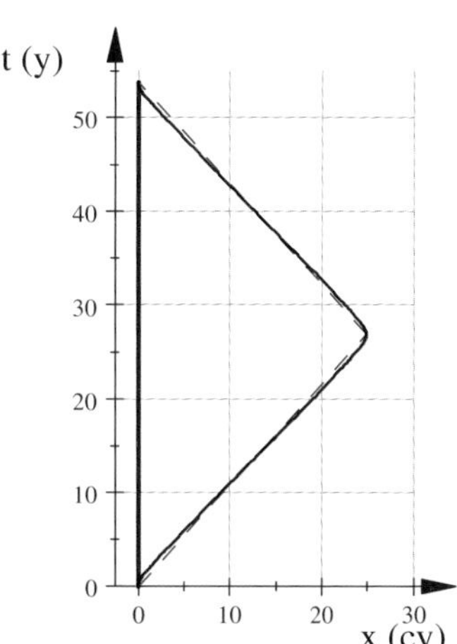

Figure TT.2 STD #2

There's not much difference between the world lines of Constance and Axel, but that's not surprising. From Homer's perspective, the two trips take the same amount of time, travel the same distance, and have the same average speed. Their trajectories through space-time are similar.

We can use the first-half intervals to calculate Earth-Vega's average speed in S'' for the first half of the trip.

$$u''_{\text{ave}} = \frac{\Delta x''_1}{\Delta t''_1} = \frac{2.602\,7\,c\,\mathrm{y}}{3.294\,5\,\mathrm{y}} = 0.790\,01c$$

We could also use Eq. (TT.11) to calculate the average speed.

$$u''_{\text{ave}} = \frac{1}{\Delta t''_1}\int_0^{\Delta t''_1} u\,(t'')\;dt'' = \frac{1}{3.294\,5}\,c\int_0^{3.294\,5} \tanh t''\;dt'' = 0.790\,02c$$

This is *not* the same average velocity Homer measures in S. The accelerating frame S'' is not an inertial reference frame, so there is no average-velocity symmetry between the two frames. The two observers do not agree on the average velocity.

What happens on the way to Vega...

If you've endured our story this far, you might well be asking yourself "huh?" or some other 3-letter interrogatory. As seen by Homer, Constance and Axel leave Earth, arrive at Vega, and return home all at the same time. Their trips cover the same distance in the same elapsed time in frame S. Yet when they all reunite on Earth, the three triplets are three different ages! Constance is 40, Axel is 33, and Homer is 74 years old.

Why is Axel younger than Constance? Let's take a look at the velocity as a function of time for Axel and Constance as measured in frame S. Figure TT.3 shows Axel's velocity $u(t)$ per Eq. (TT.3) for his trip from Earth to Vega.

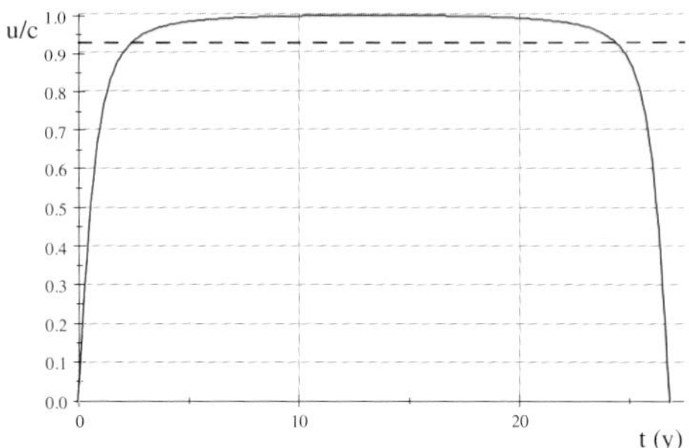

Figure TT.3 Axel's $u(t)$

The dashed line is Constance's constant speed of $v = 0.928\,47\,c$. Let's find the time (in frame S) it takes Axel to reach her speed.

$$0.928\,47 = \frac{t}{\sqrt{t^2+1}}, \text{ Solution is: } 2.499\,9$$

For most of the 26.926 year trip, Axel is moving significantly faster than Constance. Since he moves through space faster than Constance, he moves through time slower.

What really happens during these trips? It's instructive to look at the difference between Axel's position $x(t)$ per Eq. (TT.4) and Constance's position vt as seen in the Earth-Vega frame S. Figure TT.4 shows a plot of $x(t) - vt$ for the 26.926-year outbound trip.

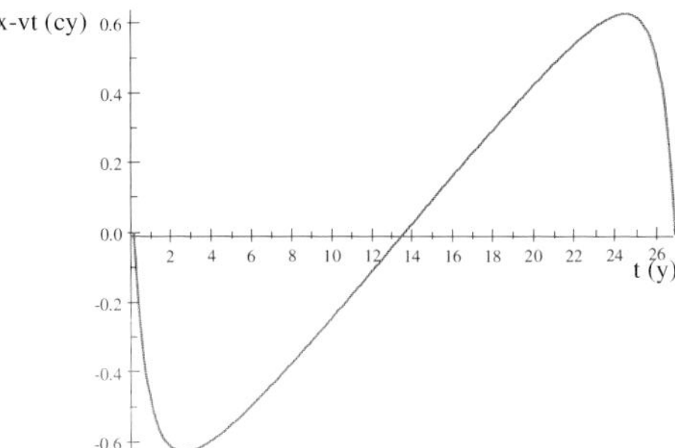

Figure TT.4 What's the difference?

Constance starts her trip already moving but Axel starts from rest, so initially she is ahead of him and the difference is negative. The space interval between them increases until Axel's speed equals her speed at $t = 2.499\,9$ y. After that, Axel is moving faster and he starts to catch up until they arrive at the midpoint at the same time. At that point, he moves ahead of her but he also starts slowing down. The space interval between them again increases until his speed equals hers. Then she starts to catch up and they arrive at Vega at the same time.

Orbits in General Relativity

In Chapter 9 we examined Newtonian gravity and planetary orbits. Now we look at the same problem in General Relativity (GR) and consider orbits around a spherically-symmetric, non-rotating black hole. Far from a black hole the GR orbits aren't all that different from Newton's ellipses, but piloting a starship near a black hole presents new and interesting challenges.

Mathematically, the GR equations of motion are identical to Eqs. (9.5)–(9.7) with this central interaction.

$$f(r) = \frac{GM}{r^2} + \frac{3GM\ell^2}{c^2 r^4} \tag{GR.1}$$

This is Newton's inverse-square gravity plus an r^{-4} term with the same sign that depends on the angular momentum (ℓ). The equation for the radial position $r(\tau)$ as a function of time is

$$\frac{d^2 r}{d\tau^2} = -\frac{GM}{r^2} - \frac{3GM\ell^2}{c^2 r^4} + \frac{\ell^2}{r^3} \tag{GR.2}$$

where τ is the local time measured by the orbiter. The equations for the orbit $r(\phi)$ are

$$\frac{d^2 r}{d\phi^2} - \frac{2}{r}\left(\frac{dr}{d\phi}\right)^2 - r = -\frac{GM}{c^2}\left(3 + \frac{c^2}{\ell^2} r^2\right) \tag{GR.3a}$$

$$\frac{d^2 u}{d\phi^2} + u = \frac{GM}{c^2}\left(3u^2 + \frac{c^2}{\ell^2}\right) \tag{GR.3b}$$

where $u(\phi) = 1/r(\phi)$.

Physically, the two theoretical explanations of gravity are completely different. The equations of motion may be similar, but Newton and Einstein offer vastly different interpretations. According to Newton, gravity is an inverse-square force that acts at a distance; like any force, gravity changes an object's motion in accordance with Newton's 2nd Law. According to Big Al, gravity is a geometric effect; the presence of matter curves space-time and that curvature affects the motion of anything (including light) moving through that space-time.

GR predicts the existence of black holes. You can think of a black hole as a spherical region isolated from the rest of the universe. Nothing, not even light, can escape from inside a black hole. That's why they're black. The boundary between inside and outside a black hole is called the event horizon. The event horizon is not a physical surface, but it marks the point of no return.

The event horizon's size is given by the Schwarzschild radius

$$R_s = \frac{2GM}{c^2} \tag{GR.4}$$

where M is the back hole's mass. For a 1-solar-mass black hole, the Schwarzschild radius is only about 3 kilometers.

$$R_s = \frac{2\left(1.3271 \times 10^{20}\,\mathrm{m^3/s^2}\right)}{\left(2.9979 \times 10^{8}\,\mathrm{m/s}\right)^2} = 2953.2\,\mathrm{m}$$

That's a lot of mass in a small volume.

The relativistic analog to Eq. (9.32) is

$$\frac{1}{2}\left(\frac{dr}{d\tau}\right)^2 - \frac{GM}{r} - \frac{GM\ell^2}{c^2r^3} + \frac{1}{2}\frac{\ell^2}{r^2} = \frac{1}{2}\left(\frac{E^2}{c^2} - c^2\right) \tag{GR.5}$$

where E is the orbiting object's energy and τ is the local time. This is equivalent to a particle moving in the one-dimensional effective potential

$$V_{\text{eff}}(r) = -\frac{GM}{r} - \frac{GM\ell^2}{c^2r^3} + \frac{1}{2}\frac{\ell^2}{r^2} \tag{GR.6}$$

with energy per-unit-mass $\frac{1}{2}\left(\frac{E^2}{c^2} - c^2\right)$.

In General Relativity, it's often useful to give r in units of GM/c^2, ℓ in GM/c, and t in GM/c^3, particularly for numerical calculations. In a geometrized unit system where $G = c^2 = 1$, all these quantities are in terms of M, which has units of length. Here are the GR interaction and effective potential in these units.

$$f(r) = \frac{M}{r^2} + \frac{3M\ell^2}{r^4} \tag{GR.7a}$$

$$V_{\text{eff}}(r) = -\frac{M}{r} - \frac{M\ell^2}{r^3} + \frac{1}{2}\frac{\ell^2}{r^2} \tag{GR.7b}$$

Figure GR.1 shows a typical effective potential in general relativity with $\ell = 4.5$ and $M = 1$ (we'll take $M = 1$ for all our numerical results).

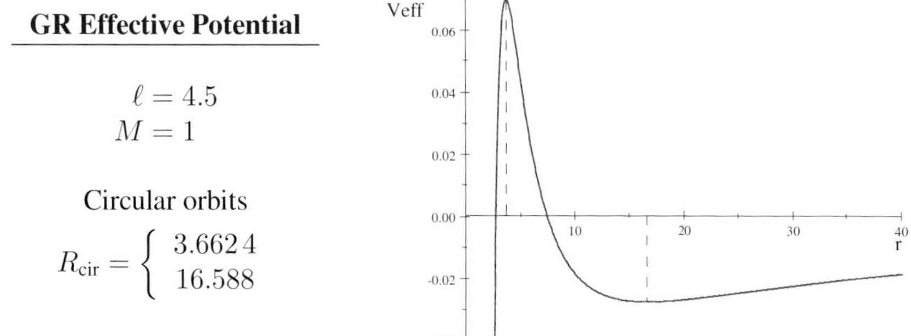

Figure GR.1 GR effective potential

The GR effective potential is obviously different than its Newtonian counterpart. There are two circular orbits, a stable circular orbit at the local minimum and an unstable circular orbit at the local maximum. Here's the effective potential in Figure GR.1.

$$V_{\text{eff}} = \left[-\frac{M}{r} - \frac{M\ell^2}{r^3} + \frac{1}{2}\frac{\ell^2}{r^2}\right]_{M=1,\ell=4.5} = \frac{10.125}{r^2} - \frac{1}{r} - \frac{20.25}{r^3}$$

A little calculus and Solve Exact finds both circular orbits.

$$0 = \frac{d}{dr}\left(-\frac{1}{r} + \frac{10.125}{r^2} - \frac{20.25}{r^3}\right), \text{ Solution is: } 3.662\,4, 16.588$$

When $\ell = 4.5$ the inner unstable circular orbit is at $r = 3.662\,4$ and the outer stable circle is at $r = 16.588$.

Angular Momentum

There are no free parameters in General Relativity. The gravitational interaction is completely specified, so our orbital analysis consists of understanding the allowed initial conditions and their consequences. Keeping with our custom, we'll start our orbits at radial extremes which leaves us to vary the initial radial position R_0 and the angular momentum ℓ.

There are four particular values of the angular momentum that will help our analysis of objects in bound stable orbits.

1. Minimum Angular Momentum: ℓ_{min}

Newton's Law of Universal Gravitation allows stable bound orbits for any non-zero value of angular momentum. This is not the case in General Relativity.

Let's start by finding general expressions for the radii of the two circular orbits. We'll need the derivative of the effective potential.

$$\frac{dV_{\text{eff}}}{dr} = \frac{d}{dr}\left(-\frac{M}{r} + \tfrac{1}{2}\frac{\ell^2}{r^2} - \frac{M\ell^2}{r^3}\right) = \frac{1}{r^4}\left(Mr^2 - r\ell^2 + 3M\ell^2\right)$$

Set the numerator equal to zero and use Solve Exact to solve for r.

$$0 = Mr^2 - r\ell^2 + 3M\ell^2,$$

$$\text{Solution is: } \frac{1}{2M}\ell\left(\ell + \sqrt{\ell^2 - 12M^2}\right), \frac{1}{2M}\left(\ell^2 - \ell\sqrt{\ell^2 - 12M^2}\right)$$

The two circular orbits are at

$$R_{\text{out}} = \frac{\ell^2}{2M}\left(1 + \sqrt{1 - \frac{12M^2}{\ell^2}}\right) \qquad \text{(GR.8a)}$$

$$R_{\text{in}} = \frac{\ell^2}{2M}\left(1 - \sqrt{1 - \frac{12M^2}{\ell^2}}\right) \qquad \text{(GR.8b)}$$

where "out" refers to the outer stable circle and "in" to the inner unstable one. These physical radii are real, so the term inside the radical must always be positive. This puts a lower limit on the angular momentum for stable bound orbits.

$$\ell_{\text{min}} = \sqrt{12}M \qquad \text{(GR.9)}$$

This minimum angular momentum does not depend on the initial position R_0. A bound particle starting at any radial position with $\ell < \ell_{\text{min}}$ will eventually cross the event horizon.

When the angular momentum equals ℓ_{min} the radius of both circular orbits is $6M$.

$$R_{\text{in}} \to \left[\frac{\ell^2}{2M}\left(1 - \sqrt{1 - \frac{12M^2}{\ell^2}}\right)\right]_{\ell=\sqrt{12}M} = 6M$$

$$R_{\text{out}} \to \left[\frac{\ell^2}{2M}\left(1 + \sqrt{1 - \frac{12M^2}{\ell^2}}\right)\right]_{\ell=\sqrt{12}M} = 6M$$

As the angular momentum increases, the radius of the unstable circular orbit decreases, but it's always larger than $3M$.

$$R_{\text{in}} \to \lim_{\ell\to\infty} \left(\frac{\ell^2}{2M}\left(1-\sqrt{1-\frac{12M^2}{\ell^2}}\right)\right) = 3M$$

The inner unstable circle is always found in the range $3M \le R_{\text{in}} \le 6M$. As the angular momentum increases, the radius of the stable circular orbit increases, so it's always greater than $6M$.

Figure GR.2 shows the relativistic effective potential (with $M = 1$) for five different values of angular momentum. The lowest curve has $\ell = 3$, which is less than $\ell_{\text{min}} \approx 3.464\,1$, so it doesn't allow any stable bound orbits. The next higher curve has $\ell = \ell_{\text{min}}$ and allows one unstable circular orbit (the little "plateau") at $r = 6$.

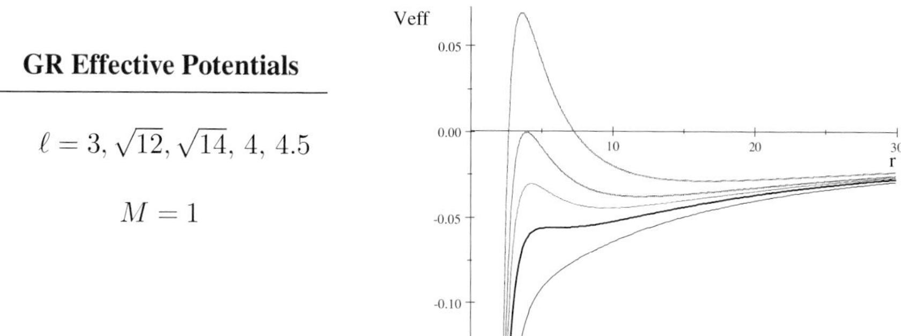

Figure GR.2 More GR effective potentials

The next three curves all have $\ell > \ell_{\text{min}}$ and display the distinctive shape we discussed earlier. These larger values of the angular momentum allow a variety of bound stable orbits.

2. Stable Circular Orbits: ℓ_{cir}

When $\ell > \ell_{\text{max}}$ there is always a stable circular orbit at the local minimum in the effective potential. We've already calculated the derivative of the effective potential, so let's take that expression and use Solve Exact to solve for ℓ.

$$0 = Mr^2 - r\ell^2 + 3M\ell^2,$$

$$\text{Solution is: } -\frac{r}{r-3M}\sqrt{M(r-3M)},\ \frac{r}{r-3M}\sqrt{M(r-3M)}$$

A little editing by-hand produces an expression for the angular momentum needed for a stable circular orbit.

$$\ell_{\text{cir}} = \sqrt{\frac{MR_0}{1-\dfrac{3M}{R_0}}} \tag{GR.10}$$

A particle starting at $R_0 > 6M$ with $\ell = \ell_{\text{cir}}$ will move in a stable circular orbit.

3. Maximum Angular Momentum: ℓ_{max}

The relativistic effective potential always approaches zero from below as r goes to infinity. This puts an upper limit on the angular momentum for bound orbits. Let's apply Solve Exact to the condition $V_{\text{eff}} = 0$ and solve for ℓ.

$$0 = -\frac{M}{r} - \frac{M\ell^2}{r^3} + \frac{1}{2}\frac{\ell^2}{r^2},$$

Solution is: $-\sqrt{2}\dfrac{r}{r-2M}\sqrt{M(r-2M)}, \sqrt{2}\dfrac{r}{r-2M}\sqrt{M(r-2M)}$

There are two solutions, one each for clockwise or counterclockwise motion. A little editing by-hand produces an expression for the maximum angular momentum.

$$\ell_{\text{max}} = \sqrt{\frac{2MR_0}{1-\dfrac{2M}{R_0}}} \tag{GR.11}$$

A particle starting at R_0 with $\ell \geq \ell_{\text{max}}$ is not bound and will move off to infinity. Solving this equation for R_0 gives the closest initial position a stable bound orbit can have for a given value of angular momentum.

$$R_0 \geq \frac{\ell^2}{4M}\left(1 + \sqrt{1 - \frac{16M^2}{\ell^2}}\right) \tag{GR.12}$$

Take a look at the term inside the radical: these expressions are only valid for $\ell \geq 4M$. Let's Substitute the smallest value $\ell = 4M$ into the radii of the circular orbits.

$$R_{\text{in}} \to \left[\frac{\ell^2}{2M}\left(1 - \sqrt{1 - \frac{12M^2}{\ell^2}}\right)\right]_{\ell=4M} = 4M$$

$$R_{\text{out}} \to \left[\frac{\ell^2}{2M}\left(1 + \sqrt{1 - \frac{12M^2}{\ell^2}}\right)\right]_{\ell=4M} = 12M$$

The closest approach for all orbits with $\ell = 4M$ is between $4M$ and $12M$. Let's Substitute $\ell = 4M$ and $R_{\text{in}} = 4M$ into the effective potential.

$$V_{\text{eff}}(R_{\text{in}}) \to \left[-\frac{M}{r} - \frac{M\ell^2}{r^3} + \frac{1}{2}\frac{\ell^2}{r^2}\right]_{r=4M,\ell=4M} = 0$$

The local maximum of the effective potential is zero when $\ell = 4M$, positive for all orbits with $\ell > 4M$, and negative for $\ell < 4M$.

4. Asymptotic Circles: ℓ_{asym}

An object starting from an initial position outside the stable circular orbit, with just the right angular momentum, can asymptotically approach the inner unstable circular orbit. This condition for this orbit is $V_{\text{eff}}(R_0) = V_{\text{eff}}(R_{\text{in}})$ where R_{in} is the radius of the inner unstable circle. Figure GR.3a shows the GR effective potential under this condition with $R_0 = 10$ and $R_{\text{in}} = 5$.

We can find that just-right angular momentum using Substitute (with Solve Exact) and Eq. (GR.8b) to find ℓ.

$$\left[-\frac{M}{r}-\frac{M\ell^2}{r^3}+\frac{1}{2}\frac{\ell^2}{r^2}\right]_{r=R_0}=\left[-\frac{M}{r}-\frac{M\ell^2}{r^3}+\frac{1}{2}\frac{\ell^2}{r^2}\right]_{r=\frac{\ell^2}{2M}\left(1-\sqrt{1-\frac{12M^2}{\ell^2}}\right)},$$

Solution is: $\pm 4MR_0\sqrt{\dfrac{1}{(R_0-2M)(6M+R_0)}}, \pm R_0\sqrt{\dfrac{M}{R_0-3M}}$

The first two solutions are our asymptotic circles and the second two are stable circles. (To save space the "$\pm$" combines four cumbersome solutions into two expressions.)

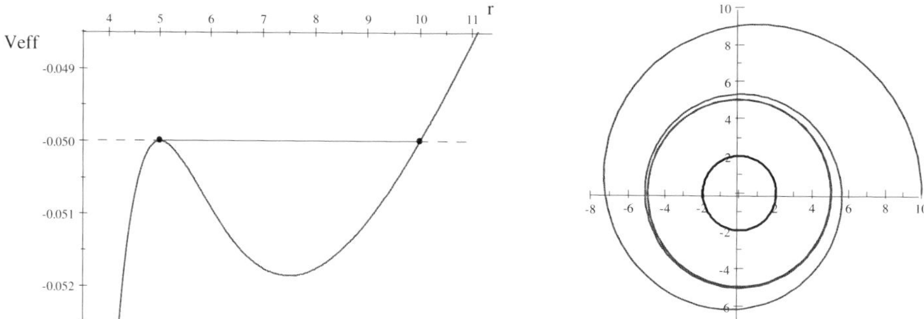

Figure GR.3a Asymptotic... **Figure GR.3b** ...Circles

A little editing by-hand produces an expression for the angular momentum for the asymptotic circle.

$$\ell_{\text{asym}}=\frac{4M}{\sqrt{\left(1+\dfrac{6M}{R_0}\right)\left(1-\dfrac{2M}{R_0}\right)}} \tag{GR.13}$$

Now that we have ℓ_{asym} we can use Solve Exact to solve for r to find the final radius in terms of R_0.

$$\left[-\frac{M}{R_0}+\frac{1}{2}\frac{\ell^2}{R_0^2}-\frac{M\ell^2}{R_0^3}=-\frac{M}{r}+\frac{1}{2}\frac{\ell^2}{r^2}-\frac{M\ell^2}{r^3}\right]_{\ell=\frac{4MR_0}{\sqrt{(R_0-2M)(6M+R_0)}}},$$

Solution is: $R_0, 4M\dfrac{R_0}{R_0-2M}$

The two solutions are the initial and final values of r. A little more editing by-hand produces an expression for the radius of the asymptotic circle.

$$R_{\text{asym}}=\frac{4M}{1-\dfrac{2M}{R_0}} \tag{GR.14}$$

A particle starting at $R_0 \geq 6M$ with $\ell=\ell_{\text{asym}}$ will spiral into an unstable circular orbit with radius R_{asym}. Figure GR.3b shows four rotations of the asymptotic circle with $R_0=10$, $\ell_{\text{asym}}=5/\sqrt{2}$ and $R_{\text{asym}}=5$. The circle at $r=2$ is the event horizon.

Precessing Ellipses and Periodic Orbits

When two objects interact gravitationally, Newton says all bound orbits are ellipses. As the asymptotic circle suggests, GR orbits are more interesting and diverse. We'll start far away from the event horizon and look at Earth-like orbits. Then we'll move closer to the black hole to see GR's effects on orbits.

Nearly-Elliptical Orbits

For Earth-like orbits, the predictions of Newton and Einstein aren't much different. We can explain this by comparing the two terms in Eq. (GR.1) in a ratio.

$$\frac{\frac{3GM\ell^2}{c^2 r^4}}{\frac{GM}{r^2}} = \frac{3\ell^2}{c^2 r^2} \approx \frac{3\left(2\pi\,\mathrm{AU}^2\,/\,\mathrm{y}\right)^2}{(63239\,\mathrm{AU}\,/\,\mathrm{y})^2\,(1\,\mathrm{AU})^2} = 2.961\,5\times 10^{-8}$$

GR's contribution is 3 hundred-million times smaller than Newton's gravity. The Earth is too far from the Sun to see GR's effects on an *SNB* plot of its orbit (see Problem 25).

During its motion around the Sun, a planet moves radially in-and-out. Its closest point to the Sun is called its perihelion. Newtonian orbits are fixed ellipses (with apsidal angle $\Phi_A = \pi$) so the perihelion doesn't move. The elliptical orbits we calculated in Chapter 9 look the same after 1 orbit or 47.

For Earth-like orbits, GR predicts a small but non-zero perihelion shift that we can calculate. We'll calculate the rates at which the angular and radial motions change and use them to predict the perihelion shift for one orbit.

Let's use the definition of angular momentum (and Simplify) to calculate the orbital angular velocity of GR's stable circular orbit of radius R and angular momentum ℓ_{cir}.

$$\omega_{\text{ang}}^2 = \left[\frac{\ell^2}{r^4}\right]_{r=R,\ell=\sqrt{GMR/\left(1-\frac{3R_s}{2R}\right)}} = -2G\frac{M}{R^2\,(3R_s - 2R)}$$

A little editing by-hand produces this expression for the orbital angular velocity.

$$\omega_{\text{ang}}^2 = \frac{GM}{R^3}\frac{1}{1-\frac{3}{2}\frac{R_s}{R}} \tag{GR.15}$$

We can use the effective potential (and Simplify) to calculate the angular frequency of the small radial oscillations about the local minimum (that same stable circle) of the effective potential.

$$\omega_{\text{rad}}^2 = \left[\frac{d^2}{dr^2}\left(-\frac{GM}{r} - \frac{1}{2}\frac{R_s\ell^2}{r^3} + \frac{1}{2}\frac{\ell^2}{r^2}\right)\right]_{r=R,\ell=\sqrt{GMR/\left(1-\frac{3R_s}{2R}\right)}}$$

$$= \frac{1}{R^3\,(3R_s - 2R)}\,(6GMR_s - 2GMR)$$

A little more editing by-hand produces an expression for the radial frequency.

$$\omega_{\text{rad}}^2 = \frac{GM}{R^3}\frac{1-3\dfrac{R_s}{R}}{1-\frac{3}{2}\dfrac{R_s}{R}} = \omega_{\text{ang}}^2\left(1-3\frac{R_s}{R}\right) \tag{GR.16}$$

Both ω_{ang}^2 and ω_{rad}^2 equal their nonrelativistic value $\left(GM/R^3\right)$ times a small relativistic correction. Notice the radial frequency is slightly smaller than the angular velocity.

The general expression for the apsidal angle is $\Phi_A = \frac{\omega_{\text{ang}}}{\omega_{\text{rad}}}\pi$ and periodic orbits occur when $\omega_{\text{ang}}/\omega_{\text{rad}}$ equals a ratio of integers. The angle between one perihelion and the next is twice the apsidal angle, so the perihelion shift for one orbit is $2\Phi_A - 2\pi$.

$$\Delta\phi = 2\pi\left(\frac{\omega_{\text{ang}}}{\omega_{\text{rad}}} - 1\right) \tag{GR.17}$$

For Newtonian ellipses the ratio equals one, so the precession shift is zero. The in-and-out radial motion completes exactly one cycle when the angular motion completes one orbit. The perihelion doesn't move!

The ratio R_s/R is very small for Earth-like orbits, so we can use Eq. (GR.16) to do a Power Series expansion. Tell *SNB* to Expand in Powers of R_s (which is really expanding in powers of R_s/R) to 2 terms.

$$\frac{\omega_{\text{ang}}}{\omega_{\text{rad}}} = \frac{1}{\sqrt{1-3\dfrac{R_s}{R}}} = 1 + \frac{3}{2R}R_s + O\left(R_s^2\right)$$

Now Substitute this into the expression for the perihelion shift.

$$\Delta\phi \approx 2\pi\left[\frac{\omega_{\text{ang}}}{\omega_{\text{rad}}} - 1\right]_{\omega_{\text{ang}}=\left(1+\frac{3}{2}\frac{R_s}{R}\right)\omega_{\text{rad}}} = 3\frac{\pi}{R}R_s$$

Here is the GR prediction for the perihelion shift of Earth-like orbits.

$$\Delta\phi = 3\pi\frac{R_s}{R} \tag{GR.18}$$

This is a positive number, so the perihelion *advances* by $\Delta\phi$ with each orbit. Since the radial frequency is smaller than the angular velocity, the radial oscillation isn't quite finished after one orbit.

In our solar system, Mercury is the closest planet to our Sun so it experiences the largest GR effects. According to observation, gravity causes Mercury's perihelion to shift by about $574''$ per century. Most (all but about $43''$) comes from the small Newtonian effects of the other planets. Let's see what GR says about the rest.

Let's calculate the Sun's Schwarzschild radius in AUs.

$$R_s = \frac{2\left(4\pi^2\,\text{AU}^3/\text{y}^2\right)}{\left(63239\,\text{AU}/\text{y}\right)^2} = 1.974\,3\times10^{-8}\,\text{AU}$$

We need to look up Mercury's average distance from the Sun, and use it to calculate the perihelion advance.

$$\Delta\phi = 3\pi \frac{1.974\,3 \times 10^{-8}\,\text{AU}}{0.38710\,\text{AU}} = 4.806\,9 \times 10^{-7}$$

That's in radians per orbit, so let's use Mercury's orbital period and convert the perihelion advance into degrees per century.

$$\Delta\phi = 4.806\,9 \times 10^{-7} \frac{\text{rad}}{\text{orb}} \times \frac{\text{orb}}{0.24085\,\text{y}} \times 100\,\text{y} \times \frac{360\,^\circ}{2\pi\,\text{rad}} = 1.143\,5 \times 10^{-2\,\circ}$$

The predictions of Newton and Einstein differ by only about one hundredth of a degree *per century*! Let's convert this to seconds-of-arc per century.

$$\Delta\phi = 0.011\,435\,^\circ \times \frac{3600\,''}{1\,^\circ} = 41.166\,''$$

Our simple first-order calculation is very close to the famous $43\,''$ discrepancy!

Precessing Elliptical Orbits

Relativistic effects on orbits are much more significant near a black hole, so we'll have to get closer than 1 AU if we want to see them on an orbital plot. Let's try starting at $R_0 = 100\,GM/c^2$.

$$R_0 = 100 \frac{1.327\,1 \times 10^{20}\,\text{m}^3/\,\text{s}^2}{(2.9979 \times 10^8\,\text{m}/\,\text{s})^2} \times \frac{1\,\text{km}}{1000\,\text{m}} = 147.66\,\text{km}$$

These orbits start a mere $148\,\text{km}$ $(92\,\text{mi})$ from the black hole, where (using $\ell \approx \ell_{\text{cir}}$) the GR term in Eq. (GR.1) is about 3% of Newton's gravity.

$$\frac{3\ell^2}{c^2 r^2} \approx \frac{3\left(\dfrac{10GM}{c}\right)^2}{c^2\left(\dfrac{100GM}{c^2}\right)^2} = \frac{3}{100}$$

Let's numerically solve Eq. (GR.3a) for the GR orbit with $R_0 = 100$ and $\ell = 6.650\,2$ so we can look at the orbit.

$$\begin{bmatrix} r'' - \dfrac{2}{r}(r')^2 - r = -\left(3 + \dfrac{r^2}{6.650\,2^2}\right) \\ r(0) = 100 \\ r'(0) = 0 \end{bmatrix}, \text{ Functions defined: } r$$

Figure GR.4a shows the result: a precessing elliptical orbit with angle $\Delta\phi = \frac{1}{6}\pi$ between consecutive perihelia that is significantly different than its Newtonian counterpart.

The perihelion of the GR orbit has precessed by 90° after only $3\frac{1}{4}$ orbits. The dashed curve is the Newtonian ellipse with the same initial conditions and the black hole's event horizon is shown to scale.

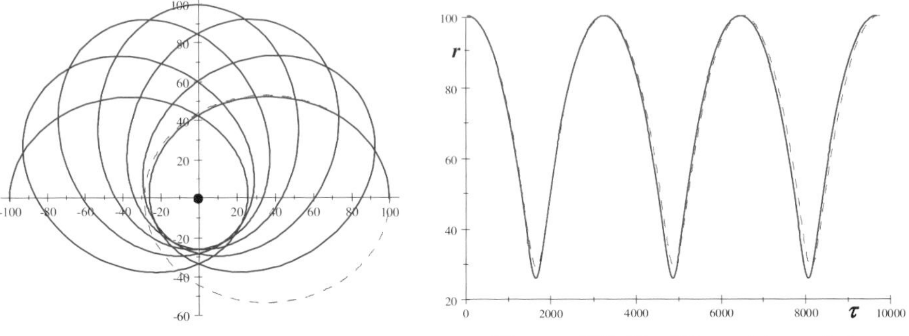

Figure GR.4a GR orbit $r\left(\phi\right)$ **Figure GR.4b** GR position $r\left(\tau\right)$

The radial position after one orbit is not the initial position.

$$r\left(2\pi\right) = 85.171$$

Let's numerically solve Eq. (GR.2) (with the same R_0 and ℓ) so we can look at the GR $r\left(\tau\right)$ radial position as a function of local time.

$$\begin{bmatrix} r'' = -\dfrac{1}{r^2} - \dfrac{3\left(6.650\,2^2\right)}{r^4} + \dfrac{6.650\,2^2}{r^3} \\ r(0) = 100 \\ r'(0) = 0 \end{bmatrix}, \text{Functions defined: } r$$

Figure GR.4b shows three cycles of $r\left(\tau\right)$ along with the time dependence for the Newtonian ellipse (the dashed curve). Even this close to the black hole the two curves for the radial position as a function of local time are not all that different.

We can use Evaluate Numerically (with a little help from the graph) to estimate the time it takes for one complete revolution and the time for the orbit to precess once.

$$r\left(2496.9\right) = 85.172$$

$$r\left(3206.4\right) = 100.000$$

Of course, these times are in units of GR/c^3 so let's convert them into seconds. Here's the local time for one complete orbit.

$$\tau_1 = 2496.9 \times \frac{1.327\,1 \times 10^{20}\,\mathrm{m^3/\,s^2}}{\left(2.9979 \times 10^8\,\mathrm{m\,s^{-1}}\right)^3} = 1.229\,9 \times 10^{-2}\,\mathrm{s}$$

That's about 12.3 *milliseconds*! The $6\frac{1}{2}$ orbits shown in Figure GR.4a take only $79.9\,\mathrm{ms}$ as measured aboard an orbiting starship. Inertial dampers on full!

Periodic Orbits

With few exceptions, periodic orbits are highly dependent on the initial conditions. Most GR orbits are not periodic. Only particular values of the initial position and angular momentum produce periodic orbits, but many such orbits are possible.

So far we've seen the two extremes for orbits around black holes, which range from slowly precessing ellipses to asymptotic circles. The closer the orbit is to the black hole, the less it looks like an ellipse and the more it resembles an asymptotic circle. Asymptotic circles have an infinite apsidal angle (and period) so they provide the lower limit on the angular momentum of GR periodic orbits.

There are no non-circular period-1 periodic orbits and every orbit has an apsidal angle larger than π. Orbits around a non-rotating black hole *all* precess in the direction of the orbital motion because the additional term in Eq. (GR.1) has the same sign as Newton's inverse-square gravity. For a given periodicity P, the periodic orbits range from $\Phi_A = P\pi$ to $\Phi_A = \pi P/(P-1)$, including only the ratios of relatively prime integers.

Figure GR.5a shows a plot of angular momentum versus apsidal angle for periodic orbits with $R_0 = 10$ (the lowest curve), 20, and $100\,GM/c^2$ (the highest). Each curve is a family of periodic orbits with the same initial position, showing orbits with apsidal angles ranging from $\frac{6}{5}\pi$ to 3π.

As the angular momentum gets very large, the apsidal angle tends toward π. These are the large-radius slowly-precessing near-elliptical orbits. The apsidal angle increases as the angular momentum decreases towards the asymptotic-circle orbits which have an infinite apsidal angle.

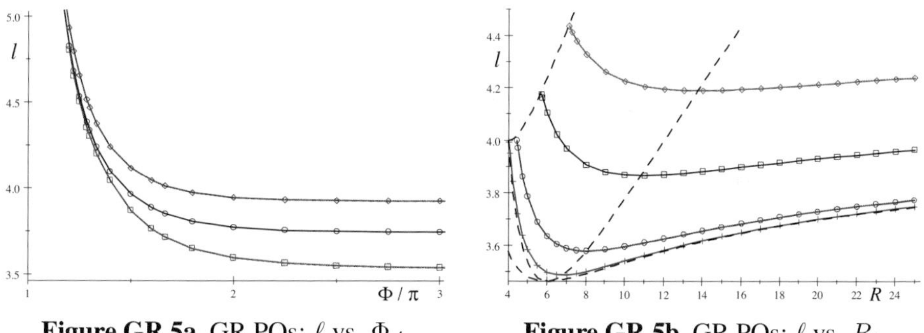

Figure GR.5a GR POs: ℓ vs. Φ_A **Figure GR.5b** GR POs: ℓ vs. R_0

Figure GR.5b shows a plot of angular momentum versus initial radius for periodic orbits with $\Phi_A = 3\pi$ (the lowest curve), 2π, $\frac{3}{2}\pi$, and $\frac{4}{3}\pi$ (the highest). Each curve is a family of periodic orbits with the same apsidal angle. A line of constant angular momentum across any family gives the maximum and minimum radii for the orbit with that ℓ and Φ_A. Each family includes a stable circular orbit at its minimum radius. Every orbit with $\Phi_A = P\pi$ has angular momentum $\ell < 4$ for all values of R_0.

All four special values of the angular momentum appear on this graph. The ℓ-axis starts at ℓ_{min} and the three dashed curves are ℓ_{cir} from Eq. (GR.10), ℓ_{max} from Eq. (GR.11), and ℓ_{asym} from Eq. (GR.13). Notice that two of those curves include the point $(R_0 = 4, \ell = 4)$.

We can use Eq. (GR.3b) to calculate the periodic orbit with $R_0 = 10$ and $\ell = 3.538\,67$.

$$\begin{bmatrix} u'' + u = \dfrac{1}{3.538\,67^2} + 3u^2 \\ u(0) = \frac{1}{10} \\ u'(0) = 0 \end{bmatrix}, \text{ Functions defined: } u$$

To see the orbit, plot $\dfrac{1}{u}$ with Plot 2D + Polar.

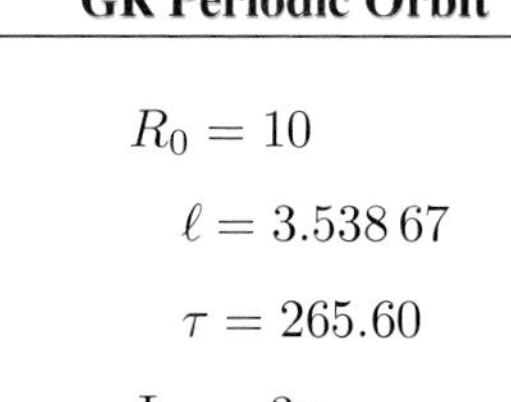
GR Periodic Orbit

$R_0 = 10$

$\ell = 3.538\,67$

$\tau = 265.60$

$\Phi_A = 3\pi$

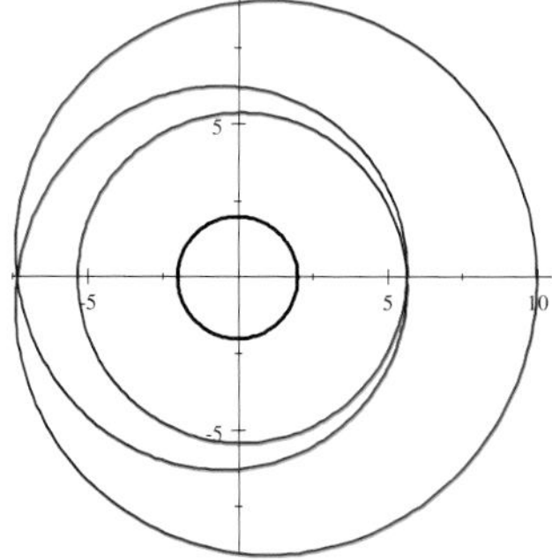

Figure GR.6 GR period-3 orbit

This period-3 orbit spirals in for 3π radians to reach its minimum radius, then spirals out. The angular momentum for this orbit is less than 0.1% different from the asymptotic-circle value.

$$100\frac{3.538\,67 - 5/\sqrt{2}}{5/\sqrt{2}} = 0.08870\,2$$

Let's see how close the minimum comes to the event horizon (at $r = 2$) and the asymptotic circle for this initial radius (at $r = 5$).

$$r_{\min} = \frac{1}{u(3\pi)} = 5.316\,5$$

Given the dire consequences of missing the unstable asymptotic circular orbit and crossing the event horizon, this seems close enough.

Periodic orbits have many applications. They provide a direct connection between classical and quantum mechanics. Periodic orbits are also aesthetically pleasing.

Be the Ball: Embedding Diagrams

The one-sentence version of General Relativity is "matter tells space-time how to curve and space-time tells matter how to move". It's useful to form a mental picture of any physical situation you're trying to understand, and there's a better way to visualize the geometry of curved space-time than 2-dimensional orbital plots. Since these GR orbits are 2-d, we can use the third dimension to visualize the curvature. Embedding diagrams show a slice of curved 2-d space-time embedded in flat 3-dimensional space. [34]

The Schwarzschild metric gives the appropriate description of the space-time geometry around a non-rotating object. At a given moment in time ($dt = 0$) the metric in the equatorial plane ($\theta = \pi/2$) is

$$ds^2 = \frac{1}{1 - \dfrac{2\,m\,(r)}{r}}\,dr^2 + r^2\,d\phi^2 \tag{GR.19}$$

where $m\,(r)$ is mass within a sphere of radius r. Let's compare this with the metric for 3-dimensional space in cylindrical coordinates.

$$ds^2 = dz^2 + dr^2 + r^2\,d\phi^2 \tag{GR.20}$$

Setting these expressions equal produces an equation for the embedding function.

$$\left(\frac{dz}{dr}\right)^2 = \frac{1}{1 - \dfrac{2\,m\,(r)}{r}} - 1 \tag{GR.21}$$

Let me emphasize that in spite of its letter-name, z has no physical significance. It is an artificial variable designed to mimic the correct space-time geometry and does *not* represent a space-time dimension.

For a star of uniform density, the mass as a function of radial position is

$$m\,(r) = \begin{cases} M\dfrac{r^3}{R^3} & r < R \\ M & r > R \end{cases} \tag{GR.22}$$

where R is the radius and M is the total mass of the star.

Before we integrate Eq. (GR.21) let's Simplify the right-hand side for inside the star.

$$\frac{1}{1 - \dfrac{2}{r}M\dfrac{r^3}{R^3}} - 1 = 2M\frac{r^2}{R^3 - 2Mr^2}$$

Now we can integrate using the definite integral fudge.

$$z_{\text{in}} = \sqrt{2M}\left[\int \frac{r}{\sqrt{R^3 - 2Mr^2}}\,dr\right]_{r=R}^{r=r}$$
$$= \sqrt{2M}\left(\frac{1}{2M}\left(\sqrt{R^2\,(R - 2M)} - \sqrt{R^3 - 2Mr^2}\right)\right)$$

Repeat these two steps for outside the star.

$$\frac{1}{1 - \dfrac{2M}{r}} - 1 = 2\frac{M}{r - 2M}$$

$$z_{\text{out}} = \sqrt{2M}\left[\int \frac{1}{\sqrt{r - 2M}}\,dr\right]_{r=R}^{r=r} = \sqrt{2M}\left(2\sqrt{r - 2M} - 2\sqrt{R - 2M}\right)$$

A little editing by-hand produces an expression for the embedding function that depends on both the star's mass and size.

$$z(r) = \begin{cases} \sqrt{\dfrac{R^3}{2M} - R^2} - \sqrt{\dfrac{R^3}{2M} - r^2} & r < R \\ 2\sqrt{2M}\left(\sqrt{r - 2M} - \sqrt{R - 2M}\right) & r > R \end{cases} \tag{GR.23}$$

This defines an embedding surface with the same geometry as the Schwarzschild metric in the equatorial plane. Notice that $z(r)$ is continuous at the star's surface, ensuring the geometry is too.

Let's use this expression and New Definition to look at the space-time around a small, dense star with $R = 3M$ and $M = 1$.

$$z(\rho) = \begin{cases} \sqrt{\dfrac{3^3}{2} - 3^2} - \sqrt{\dfrac{3^3}{2} - \rho^2} & \rho < 3 \\ 2\sqrt{2}\left(\sqrt{\rho - 2} - \sqrt{3 - 2}\right) & \rho > 3 \end{cases}$$

Create the expression $(\rho, \theta, z(\rho))$ and apply Plot 3D + Cylindrical. Using ρ and θ avoids any definition conflicts when we calculate and plot orbits numerically in terms of r and ϕ. Figure GR. 7 shows the resulting plot from three perspectives (with a Tilt of 60, 0, and 90, respectively). All three have the following Properties: $0 \leq \theta \leq 2\pi$, $0 \leq \rho \leq 13$, Axes Type: None, Equal Scaling Along Each Axis checked, and Surface Style: Wire Frame.

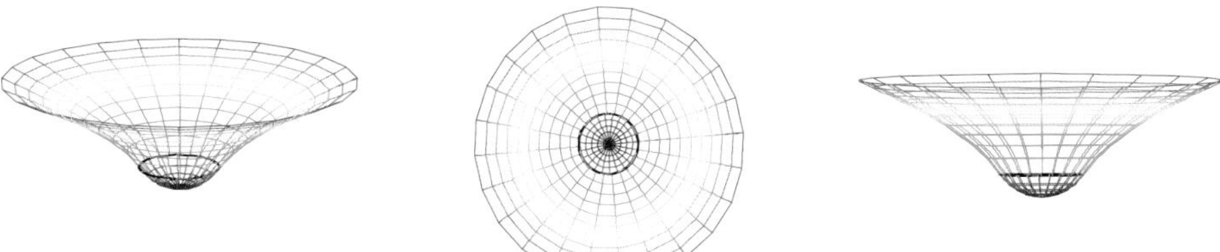

Figure GR.7 Be the ball...

This 2-dimensional surface plot is the key to the embedding diagram. Only points on the embedded surface have any physical meaning, so objects near the star move only on the surface. In a popular analogy, the curvature of space is modeled by placing a heavy ball on the surface of a rubber sheet. The ball represents the star and the rubber sheet is our embedded surface.

Where does the actual "visualizing" happen? Look at the middle view of Figure GR.7. The circles represent r-coordinates at constant intervals ($r = 9$, $r = 10$ and so on). In flat space-time, the radial distance between consecutive circles is also constant. The view on the right shows this is not the case. The circles near the star are further apart than those far away. The Schwarzschild metric tells us the distance between two points along a radius (where $d\phi = 0$) in curved space-time.

$$\Delta s = \int ds = \int_{r_1}^{r_2} \sqrt{\frac{1}{1 - \dfrac{2M}{r}}}\, dr \tag{GR.24}$$

SNB can do this integral but the result is cumbersome so let's do it numerically.

Here's the radial distance between coordinates $r = 103$ and $r = 104$ (using Evaluate Numerically).

$$\Delta s_{103\to 104} = \int_{103}^{104} \sqrt{\frac{1}{1-\frac{2}{r}}}\, dr = 1.009\,8$$

It's only about 1% larger than the flat-space result $\Delta s = 1$. Here's the radial distance between coordinates $r = 3$ and $r = 4$.

$$\Delta s_{3\to 4} = \int_{3}^{4} \sqrt{\frac{1}{1-\frac{2}{r}}}\, dr = 1.542\,2$$

It's about 54% larger than the flat-space result! The closer you get to the star, the more space is stretched! Far from the star, space isn't stretched very much at all.

We can include GR orbits on an embedding diagram to show the effects of curved space-time near a black hole. GR's 4-d description of space-time ends at the event horizon, so there is no "in" for a black hole. Let's use Eq. (GR.23) and New Definition for a black hole with $R = 2M$ and $M = 1$.

$$z(\rho) = \begin{cases} 0 & \rho < 2 \\ \sqrt{8(\rho - 2)} & \rho > 2 \end{cases}$$

To include GR orbits on the embedding diagram, we need both $r(\tau)$ and $\phi(\tau)$ so let's solve Eqs. (9.5) numerically for the same orbit in Figure GR.6.

$$\begin{bmatrix} r'' = -\frac{1}{r^2} - \frac{3(3.53867)^2}{r^4} + \frac{3.53867^2}{r^3} \\ \phi' = \frac{3.53867}{r^2} \\ r(0) = 10 \\ r'(0) = 0 \\ \phi(0) = 0 \end{bmatrix}, \text{ Functions defined: } r, \phi$$

To include the orbit, add the expression $\left(r(\tau), \phi(\tau), z(r(\tau))\right)$ to the embedding diagram. Figure GR.8 shows the result from the same perspectives (with $2 \le \rho \le 11$).

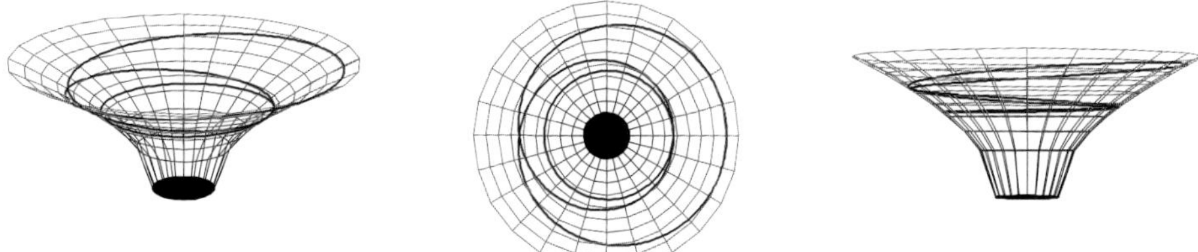

Figure GR.8 ...na na na na na

Embedding diagrams do not show the actual appearance of a black hole. They are just a cool way to visualize the space-time geometry around a black hole.

Classical Lifetime of a Hydrogen Atom

You may reasonably wonder why a topic called the "classical" anything is in the Modern Physics section. Rutherford's planetary model of the atom explained his experimental scattering results, but it failed to explain the stability of atoms. That failure was a seminal moment in the development of quantum mechanics.

In Maxwell's classical theory, electromagnetic radiation is generated by accelerating electrical charges. In Rutherford's atomic model, electrons orbit the nucleus in circular orbits. Since circular motion requires an acceleration, classical physics predicts the electron will emit radiation, lose energy, and quickly spiral into the nucleus.

We'll make one simplifying assumption: during its spiral plunge toward the nucleus, the electron is always in a nearly circular orbit. We're approximating the electron's path as a series of slightly smaller circles, so its velocity is tangential and its acceleration is radial. The electron's spiral plunge is more spiral than plunge.

Let's also introduce the classical electron radius.

$$r_e = \frac{kq^2}{mc^2} = \frac{\left(8.9876 \times 10^9\,\mathrm{N\,m^2/\,C^2}\right)\left(1.6022 \times 10^{-19}\,\mathrm{C}\right)^2}{\left(9.1094 \times 10^{-31}\,\mathrm{kg}\right)\left(2.9979 \times 10^8\,\mathrm{m/\,s}\right)^2} = 2.818\,1 \times 10^{-15}\,\mathrm{m}$$

Physically, the classical radius is a charged particle's approximate size so its rest energy is completely due to its electric potential energy, taking no quantum mechanical effects into account. A charge q has electric potential energy kq^2/r equal to its rest energy mc^2 if it is confined to a spherical volume of radius r_e. Mathematically, the classical radius makes the equations simpler.

Missed It By *That* Much

We need four bits of physics to solve this problem. The first is an old friend, Newton's 2nd Law for circular motion. The second is Coulomb's Law, which tells us the electric force between two charged particles. The electron's equation of motion is

$$ma = m\frac{v^2}{r} = \frac{k\,q^2}{r^2} \tag{CL.1}$$

where $k = 8.987\,6 \times 10^9\,\mathrm{N\,m^2/\,C^2}$ is what *SNB* calls the Coulomb force constant and q is the charge on a proton. (*SNB* reserves the symbol e for the base of natural logarithms.)

The third is the Larmor formula, which tells us the rate at which radiated energy is lost by an accelerating point charge. The total radiated power is

$$\frac{dE}{dt} = -\tfrac{2}{3}\frac{k\,q^2}{c^3}\,a^2 \tag{CL.2}$$

where a is the acceleration and c is the speed of light.

The fourth bit of physics we need is the electron's total energy.

$$E = \tfrac{1}{2}mv^2 - \frac{kq^2}{r} \tag{CL.3}$$

Our plan is to express everything in terms of the radius, solve the resulting equation for $r\,(t)$, and then calculate the time it takes the electron to reach the origin.

Let's Substitute (with Simplify) the acceleration from Eq. (CL.1) into the Larmor formula.

$$\frac{dE}{dt} = \left[-\frac{2}{3}\frac{kq^2}{c^3}a^2\right]_{a=kq^2/(mr^2)} = -\frac{2}{3c^3}\frac{k^3}{m^2}\frac{q^6}{r^4}$$

Let's do the same thing with the velocity and the total energy.

$$E = \left[\frac{1}{2}mv^2 - \frac{kq^2}{r}\right]_{v=\sqrt{kq^2/(mr)}} = -\frac{1}{2}k\frac{q^2}{r}$$

The derivative of the energy with respect to radius will be useful.

$$\frac{dE}{dr} = \frac{d}{dr}\left(-\frac{1}{2}k\frac{q^2}{r}\right) = \frac{1}{2}k\frac{q^2}{r^2}$$

The chain rule gives us another expression for the rate of change of energy.

$$\frac{dE}{dt} = \frac{1}{2}\frac{kq^2}{r^2}\frac{dr}{dt} \tag{CL.4}$$

Setting the expressions for the power equal to each other gives an equation for $r(t)$.

$$\frac{dr}{dt} = -\frac{4}{3}\frac{r_e^2}{r^2}c \tag{CL.5}$$

This is an equation we can easily separate and integrate.

$$\int_{r_0}^{r} r^2 dr = \frac{1}{3}r^3 - \frac{1}{3}r_0^3$$

A little editing by-hand produces this result for the radial position as a function of time.

$$r^3(t) = r_0^3 - 4r_e^2 ct \tag{CL.6}$$

All that remains is to set this equal to zero and use Solve Exact to find the lifetime.

$$0 = r_0^3 - 4r_e^2 ct, \text{ Solution is: } \frac{1}{4c}\frac{r_0^3}{r_e^2}$$

When the atom is in its ground state, the electron's most probable position is the Bohr radius a_0 so let's use that for the initial position.

$$T_0 = \frac{1}{4\left(2.9979\times10^8\,\frac{\text{m}}{\text{s}}\right)}\frac{\left(0.52918\times10^{-10}\,\text{m}\right)^3}{\left(2.8181\times10^{-15}\,\text{m}\right)^2} = 1.556\times10^{-11}\,\text{s}$$

In the ground state, the atom's lifetime is about 15.6 trillionths of a second! The result for the lifetime is very sensitive to the initial position r_0. Suppose we use the expectation value for the $n = 2$, $\ell = 1$ state.

$$\langle r\rangle = \left[\frac{1}{2}a_0\left(3n^2 - \ell(\ell+1)\right)\right]_{n=2,\ell=1} = 5a_0$$

This initial position gives a different lifetime.

$$T_2 = \frac{1}{4\left(2.9979 \times 10^8 \,\frac{\mathrm{m}}{\mathrm{s}}\right)} \frac{\left(5 \times 0.52918 \times 10^{-10}\,\mathrm{m}\right)^3}{\left(2.818\,1 \times 10^{-15}\,\mathrm{m}\right)^2} = 1.945 \times 10^{-9}\,\mathrm{s}$$

In this excited state, the atom's lifetime is about 2 billionths of a second! No matter the electron's initial position, classical theory predicts an extremely short lifetime. Experimentally, hydrogen atoms appear to have infinite lifetimes. This absurdly large discrepancy between classical theory and experiment suggests that Newton's 2nd Law and Maxwell's E&M theory do not apply to individual atoms.

Can Special Relativity Save the Day?

Let's see if Special Relativity can explain the stability of a hydrogen atom. Our simplifying assumption of nearly circular orbits means the electron's velocity and acceleration are perpendicular, so the SR version of Eq. (CL.1) is

$$\gamma m a = \gamma m \frac{v^2}{r} = \frac{k\,q^2}{r^2} \tag{CL.7}$$

where m is the electron's rest mass.

Since the acceleration and velocity are perpendicular, the Lorentz Transformation tells us the radial acceleration in the electron's reference frame is $a'' = \gamma^2\,a$. The relativistic energy loss per unit time is then

$$\frac{dE}{dt} = -\tfrac{2}{3}\frac{k\,q^2}{c^3}\,\gamma^4 a^2 \tag{CL.8}$$

where γ is the electron's relativistic gamma factor. We'll also need the electron's relativistic total energy.

$$E = \gamma m c^2 - \frac{kq^2}{r} \tag{CL.9}$$

Let's Substitute (with Simplify) the acceleration from Eq. (CL.7) into the relativistic Larmor formula.

$$\frac{dE}{dt} = \left[-\tfrac{2}{3}\frac{k\,q^2}{c^3}\gamma^4 a^2\right]_{a=kq^2/(\gamma m r^2)} = -\frac{2}{3c^3}\frac{k^3}{m^2}\frac{q^6}{r^4}\gamma^2$$

We can use Eq. (CL.9) and the chain rule to get another expression for the rate of change of energy.

$$\frac{dE}{dt} = mc^2\frac{d\gamma}{dt} + \frac{kq^2}{r^2}\frac{dr}{dt} \tag{CL.10a}$$

$$= mc^2\left(\frac{d\gamma}{dr} + \frac{r_e}{r^2}\right)\frac{dr}{dt} \tag{CL.10b}$$

Setting the two relativistic expressions for the power equal to each other gives us the SR equation for $r\,(t)$.

$$\left(r_e\frac{d\gamma}{dr} + \frac{r_e^2}{r^2}\right)\frac{dr}{dt} = -\tfrac{2}{3}\frac{r_e^4}{r^4}\gamma^2 c \tag{CL.11}$$

Now we need γ in terms of r, so let's recast Eq. (CL.7).

$$\gamma \frac{v^2}{c^2} = \frac{k\,q^2}{mc^2 r} = \frac{r_e}{r} \tag{CL.12}$$

Tell *SNB* to assume $(\gamma > 1)$ and use Solve Exact to find v in terms of γ.

$\gamma = \left(1 - v^2/c^2\right)^{-1/2}$, Solution is: $-\frac{c}{\gamma}\sqrt{\gamma^2 - 1}, \frac{c}{\gamma}\sqrt{\gamma^2 - 1}$

Now we can use Eq. (CL.12) and Solve Exact to find γ as a function of r.

$\frac{1}{\gamma}\left(\gamma^2 - 1\right) = \frac{r_e}{r}$, Solution is: $\frac{1}{2r}\left(r_e + \sqrt{r_e^2 + 4r^2}\right)$

As is often the case in relativity, the expressions are getting complicated so let's write them in terms of a dimensionless variable $\rho = r/r_e$.

$$\gamma^2 = \left[\left(\frac{r_e}{2r}\left(1 + \sqrt{1 + 4\frac{r^2}{r_e^2}}\right)\right)\right]^2_{r=\rho r_e} = \frac{1}{4\rho^2}\left(\sqrt{4\rho^2 + 1} + 1\right)^2$$

$$r_e\frac{d\gamma}{dr} + \frac{r_e^2}{r^2} = \left[r_e\frac{d}{dr}\left(\frac{r_e}{2r}\left(1 + \sqrt{1 + 4\frac{r^2}{r_e^2}}\right)\right) + \frac{r_e^2}{r^2}\right]_{r=\rho r_e} = \frac{1}{2\rho^2} - \frac{1}{2\rho^2\sqrt{4\rho^2 + 1}}$$

With a little editing by-hand and judicious use of Simplify and Expand (both in-place), we get the full SR equation for $r(t)$ in terms of ρ.

$$\frac{d\rho}{dt} = -\tfrac{4}{3}\frac{1}{\rho^2}\frac{c}{r_e} \times \frac{\sqrt{4\rho^2 + 1}}{4\rho^2}\frac{\left(\sqrt{4\rho^2 + 1} + 1\right)^2}{\sqrt{4\rho^2 + 1} - 1} \tag{CL.13a}$$

This equation is much too complicated to solve for $r(t)$, but it's still useful. It relates the electron's positional rate of change to its nonrelativistic counterpart times some complicated relativity stuff (CRS) that's always positive and greater than 1.

$$\frac{dr}{dt} = \left(\frac{dr}{dt}\right)_{\text{NR}} \times CRS \tag{CL.13b}$$

The rate is *faster* so the atom's lifetime is *smaller* when we include the effects of SR. It looks like SR can't resolve the issue of atomic stability.

Even though we can't solve Eq. (CL.13a) exactly, we can use it to find the approximate lifetime numerically.

$$T = \tfrac{3}{4}\frac{r_e}{c}\int_0^{\rho_0} \frac{4\rho^4}{\sqrt{4\rho^2 + 1}}\frac{\sqrt{4\rho^2 + 1} - 1}{\left(\sqrt{4\rho^2 + 1} + 1\right)^2}\,d\rho \tag{CL.14}$$

We'll need $\rho_0 \equiv r_0/r_e$ for the ground state where $r_0 = a_0$.

$$\rho_0 = \frac{a_0}{r_e} = \frac{0.52918 \times 10^{-10}\,\text{m}}{2.818\,1 \times 10^{-15}\,\text{m}} = 18778.$$

We can use Evaluate Numerically to find the SR lifetime of the ground state.

$$T_0 = \frac{3}{4}\frac{2.8181\times10^{-15}\,\mathrm{m}}{2.9979\times10^{8}\,\mathrm{m/s}}\int_0^{18778}\frac{4\rho^4}{\sqrt{4\rho^2+1}}\frac{\sqrt{4\rho^2+1}-1}{\left(\sqrt{4\rho^2+1}+1\right)^2}\,d\rho = 1.5559\times10^{-11}\,\mathrm{s}$$

The atom's ground-state lifetime is still about 15.6 trillionths of a second! Let's compare the NR lifetime with the SR result.

$$\frac{\mathrm{NR}-\mathrm{SR}}{\mathrm{SR}} = \frac{\frac{1}{4\left(2.9979\times10^{8}\frac{\mathrm{m}}{\mathrm{s}}\right)}\frac{\left(0.52918\times10^{-10}\,\mathrm{m}\right)^3}{(2.8181\times10^{-15}\,\mathrm{m})^2}}{\frac{3}{4}\frac{2.8181\times10^{-15}\,\mathrm{m}}{2.9979\times10^{8}\,\mathrm{m/s}}\int_0^{18778}\frac{4\rho^4}{\sqrt{4\rho^2+1}}\frac{\sqrt{4\rho^2+1}-1}{\left(\sqrt{4\rho^2+1}+1\right)^2}\,d\rho} - 1 = 1.0385\times10^{-4}$$

The fractional difference is only about 10^{-4} and it's even smaller for higher initial states. Let's do a 3-term Power Series expansion in powers of r_e of the complicated relativity stuff in Eq. (CL.13a).

$$CRS = \left[\frac{\sqrt{4\rho^2+1}}{4\rho^2}\frac{\left(\sqrt{4\rho^2+1}+1\right)^2}{\sqrt{4\rho^2+1}-1}\right]_{\rho=r/r_e} = 1+\frac{3}{2r}r_e+\frac{5}{4r^2}r_e^2+O\left(r_e^3\right)$$

The first-order SR correction is of order r_e/r and we can Evaluate it using $r\approx\frac{3}{4}a_0$ as an average value (see Problem 39).

$$\frac{3}{2}\frac{r_e}{r}\approx\frac{3}{2}\frac{2.8181\times10^{-15}\,\mathrm{m}}{\frac{3}{4}0.52918\times10^{-10}\,\mathrm{m}} = 1.0651\times10^{-4}$$

Again the effects of SR are about 10^{-4}. This is consistent with $\gamma^2-1=\gamma\,r_e/r$ which says $\gamma\approx1$ unless r is very small and the electron is moving very fast. For all but the very end of the electron's spiral plunge, the relativistic effects are exceedingly small (see Problem 40).

In the nonrelativistic limit $\rho\gg1$, the complicated relativity stuff goes to 1 and Eqs. (CL.13) reduce to the NR result of Eq. (CL.5).

$$CRS = \frac{\sqrt{4\rho^2+1}}{4\rho^2}\frac{\left(\sqrt{4\rho^2+1}+1\right)^2}{\sqrt{4\rho^2+1}-1}\to\frac{\sqrt{4\rho^2}}{4\rho^2}\frac{\left(\sqrt{4\rho^2}\right)^2}{\sqrt{4\rho^2}} = 1$$

Special Relativity *cannot* save the day. The effects of SR actually *reduce* the atom's lifetime with a small correction. There is no way to explain the stability of the hydrogen atom without the laws of Quantum Mechanics.

Quantum Mechanical Bound States

Classical Mechanics (CM) cannot explain the stability of the hydrogen atom and much of anything in the atomic world. Newton's Laws and Maxwell's Equations are macroscopic laws that make wrong and sometimes absurd predictions about individual atoms.

Microscopic systems obey the laws of Quantum Mechanics (QM), a theory based on the Uncertainty Principle and the matter waves of Wave-Particle Duality. QM is the most successful scientific theory ever. It's been tested by experiment and so far its predictions are undefeated. It explains an amazing range of phenomena, from the behavior of atoms to the structure of white dwarfs and neutron stars.

The **Uncertainty Principle** relates the uncertainties in simultaneous measurements of an object's position and momentum to Plank's small constant.

$$\Delta x \, \Delta p \geq \tfrac{1}{2} \, \hbar \tag{QM.1}$$

The more we know about where an object is, the less we know about where it's going. The more we know about where an object is going, the less we know about where it is. There is a fundamental limit to what we can know and when we can know it.

The 1-dimensional **Schrödinger Equation** for an object of mass m is

$$-\frac{\hbar^2}{2m}\frac{\partial^2 \Psi}{\partial x^2} + V(x)\,\Psi = i\hbar\,\frac{\partial \Psi}{\partial t} \tag{QM.2}$$

where $V(x)$ is the potential energy. The Schrödinger Equation is to QM what Newton's 2nd Law is to CM. It's ***the*** fundamental equation of non-relativistic dynamics.

The constant $\hbar$ ("h-bar") is **Planck's constant** divided by 2π.

$$\hbar \equiv \frac{h}{2\pi} = 1.0546 \times 10^{-34}\,\mathrm{J\,s} = 6.5821 \times 10^{-16}\,\mathrm{eV\,s} \tag{QM.3}$$

In the language of QM, you dot your o's and cross your h's.

The general plan has two parts. The mathematical part is to solve the Schrödinger Equation with a given potential energy $V(x)$ for the wave function $\Psi(x,t)$. The physics part is to use $\Psi(x,t)$ to predict observable results.

The **wave function** gives a complete description of the state of the system. You can think of the wave function as the amplitude of the matter wave. It tells us "how much" wave there is at a particular place and time. The basic connection between the behavior of the object and the properties of its wave function is the **probability density** $|\Psi|^2$.

A solution of the Schrödinger Equation in the form

$$\Psi(x,t) = \psi(x)\; e^{-i(E/\hbar)t} \tag{QM.4}$$

is called a **stationary state**. Such a state is significant because it has definite energy with no uncertainty and its probability density does not vary in time.

$$|\Psi|^2 \equiv \Psi^*(x,t)\,\Psi(x,t) = |\psi(x)|^2 \tag{QM.5}$$

This provides the QM explanation for the stability of atoms. In QM, an electron in an atom isn't an accelerating charge. It's in a stationary state with a probability density that is constant in time. This represents a static charge distribution that does not radiate. The atom is stable!

For stationary states, we use the 1-d, time-independent Schrödinger Equation (TISE)

$$-\frac{\hbar^2}{2m}\frac{d^2\psi}{dx^2} + V(x)\,\psi = E\,\psi \qquad \text{(QM.6)}$$

where E is the object's total energy.

Not every mathematical solution to the Schrödinger Equation is a physically acceptable wave function. Wave functions are mathematical descriptions of the particle's *physical* state. Since they tell us information about the particle's physical state, they must be well-behaved.

The wave function of a physically observable system must satisfy these conditions:

1. It's a solution of the Schrödinger Equation.
2. It's normalizable.
3. It's a continuous function.
4. Its derivative is also a continuous function.

The second condition requires the wave function to approach zero as $x \to \pm\infty$ so the area under the curve is finite. The wave function is **normalized** by integrating the probability density over all space.

$$\int_{\text{all space}} |\psi(x)|^2\, dx = 1 \qquad \text{(QM.7)}$$

The total probability always adds up to 1.

Conditions 3 and 4 are boundary conditions on the wave function. The wave function of a confined object interacts with the boundary and the boundary conditions impose restrictions on the wave function. This results in discrete bound states with quantized energy values that come directly from the boundary conditions.

Once we have a general solution to the Schrödinger Equation, we'll apply the appropriate boundary conditions to it, normalize it, and find the allowed quantized energy values. Then it will be a wave function, and we'll use it to calculate physical observables.

In QM, the **expectation value** is the predicted average value of the results of many measurements. In general, the expectation values for position and momentum can depend on time, but for stationary states they do not.

$$\langle x\rangle = \int \psi^*(x)\, x\,\psi(x)\, dx \qquad \text{(QM.8a)}$$

$$\langle p\rangle = \int \psi^*(x)\,\frac{\hbar}{i}\frac{\partial\psi}{\partial x}\, dx \qquad \text{(QM.8b)}$$

These expectation values have the same role in QM that their classical counterparts $x(t)$ and $p(t)$ have in CM. The probability that a particle in a stationary state is in a certain space interval is also time-independent.

$$P(x_1 < x < x_2) = \int_{x_1}^{x_2} |\psi(x)|^2\, dx \qquad \text{(QM.9)}$$

The one-sentence version of Quantum Mechanics is "if an object can't be somewhere, then it probably isn't". If Classical Mechanics says it is impossible for an object to be somewhere, then QM says the probability of finding it there is probably small.

Infinite Square Well ("Particle in a Box")

The **infinite square well** (ISW) is a simple problem with an exact solution that's a good introduction to the goofy world of Quantum Mechanics. My old professor and advisor called it "the easiest Quantum Mechanics problem there is" so it's a good place to start. This problem is also known as "the particle in a box" for obvious reasons.

We'll orient our infinite well symmetrically about the origin (see Figure QM.1).

Infinite SW Potential

$$V(x) = \begin{cases} 0 & |x| < \frac{1}{2}a \\ \infty & |x| > \frac{1}{2}a \end{cases}$$

Width: a

Depth: ∞

Shaded grey areas show the classically forbidden regions.

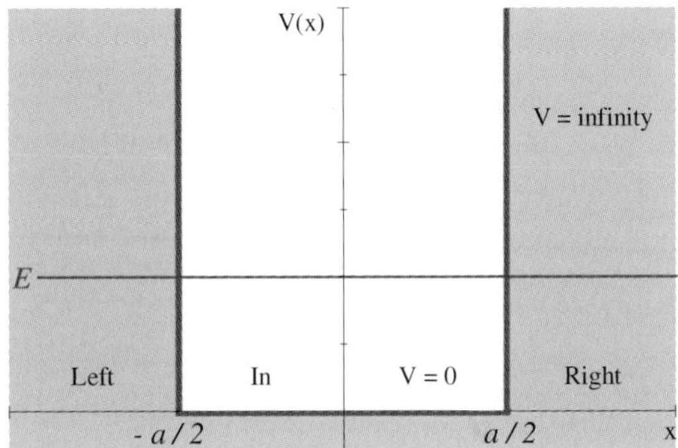

Figure QM.1 The ∞ box

Classically, the particle is free to move in the 1-d well between $x = \pm a/2$ but it never leaves the well no matter how much energy it has.

Quantumly, we're looking for bound stationary-state wave functions. Outside the well (in the Left and Right regions) the wave function must be zero. Otherwise the left-hand side of the TISE would be infinite but not the right-hand side. That means there is zero probability of finding the particle outside the well. Even in the goofy world of QM, an infinite barrier completely holds the particle in. Of course, there aren't many infinite barriers in the real world.

Inside the well, the Schrödinger Equation takes the form

$$\frac{d^2\psi_{\rm I}}{dx^2} + k^2\,\psi_{\rm I} = 0 \qquad \text{(QM.10)}$$

where $k^2 \equiv 2mE/\hbar^2$ is a positive constant. You should know the general solution to this equation, but if you don't, *SNB* (perhaps with a little trial and error) will help.

$$\begin{bmatrix} \frac{d^2\psi_{\rm I}}{dx^2} + k^2\psi_{\rm I} = 0 \\ \psi_{\rm I}(0) = A \\ \psi_{\rm I}'(0) = kB \end{bmatrix}, \text{ Exact solution is: } \{A\cos kx + B\sin kx\}$$

At this point, the typical strategy is to apply the boundary conditions $\psi_{\rm I}(\pm a/2) = 0$ to the general solution and notice the results have symmetry (see for example [2], page 236). Instead, let's invoke symmetry first. Because our square well potential is symmetric about the origin, the resulting probability density must also be symmetric so the wave functions have either odd or even parity.

Even-parity solution: $\psi_{\mathrm{I}}(x) = \psi_{\mathrm{I}}(-x)$
This wave function does not pass through the origin, but its slope there is zero.

$$\begin{bmatrix} \dfrac{d^2\psi_{\mathrm{I}}}{dx^2} + k^2\psi_{\mathrm{I}} = 0 \\ \psi_{\mathrm{I}}(0) = A \\ \psi_{\mathrm{I}}'(0) = 0 \end{bmatrix}, \text{ Exact solution is: } \{A\cos kx\}$$

The boundary conditions require $\cos\frac{ka}{2} = 0$ which means $ka = n\pi$ where $n = 1, 3, 5\ldots$ is an odd integer.

Odd-parity solution: $\psi_{\mathrm{I}}(x) = -\psi_{\mathrm{I}}(-x)$
This wave function passes through the origin, and has a non-zero slope there.

$$\begin{bmatrix} \dfrac{d^2\psi_{\mathrm{I}}}{dx^2} + k^2\psi_{\mathrm{I}} = 0 \\ \psi_{\mathrm{I}}(0) = 0 \\ \psi_{\mathrm{I}}'(0) = kB \end{bmatrix}, \text{ Exact solution is: } \{B\sin kx\}$$

The boundary conditions require $\sin\frac{ka}{2} = 0$ which means $ka = n\pi$ where $n = 2, 4, 6\ldots$ is an even integer.

In both cases, the allowed values of k are restricted to

$$k = \frac{n\pi}{a} \tag{QM.11}$$

where the positive integer n is called the quantum number. Here we see the boundary conditions on the wave function inside the box producing discrete states.

We need to normalize these wave functions. Tell *SNB* to assume $(n > 0)$ and assume $(a > 0)$, and also use additionally $(n, \text{integer})$. Then Evaluate the appropriate integrals. Integrating over all space here means between $\pm a/2$ since the wave function vanishes everywhere else.

$$1 = B^2 \int_{-a/2}^{+a/2} \sin^2 \frac{n\pi x}{a}\, dx = \tfrac{1}{2}B^2 a$$

$$1 = A^2 \int_{-a/2}^{+a/2} \cos^2 \frac{n\pi x}{a}\, dx = \tfrac{1}{2}A^2 a$$

In both cases, the normalization constant $A = B = \sqrt{2/a}$ does not depend on the quantum number. The normalized wave functions for the infinite square well are

$$\psi_n(x) = \begin{cases} \sqrt{\dfrac{2}{a}}\cos\dfrac{n\pi x}{a} & n = 1, 3, 5\ldots \quad \textbf{even parity} \\ \sqrt{\dfrac{2}{a}}\sin\dfrac{n\pi x}{a} & n = 2, 4, 6\ldots \quad \textbf{odd parity} \end{cases} \tag{QM.12}$$

inside the well and zero everywhere outside. Notice that the even-parity wave functions have odd quantum numbers and the odd-parity wave functions have even quantum numbers. Quantum goofiness abounds.

We can use Eq. (QM.11), the definition of k^2 in terms of the energy, and Solve Exact to find the energy.

$$\frac{2mE}{\hbar^2} = \frac{n^2\pi^2}{a^2}, \text{ Solution is: } \tfrac{1}{2}\frac{\pi^2}{a^2 m}n^2\hbar^2$$

In CM our boxed-in particle can have any energy, but in QM only these quantized values are allowed.

$$E_n = n^2 \frac{\pi^2\hbar^2}{2ma^2} \qquad n = 1, 2, 3 \ldots \qquad \text{(QM.13)}$$

The ground-state energy E_1 is the particle's minimum energy, and we can Evaluate it for an electron in a 1 nm-wide well.

$$E_1 = \frac{\hbar^2 c^2}{2mc^2}\frac{\pi^2}{a^2} = \frac{(197.33\,\text{eV nm})^2}{2\,(511000\,\text{eV})}\frac{\pi^2}{(1\,\text{nm})^2} = 0.376\,04\,\text{eV}$$

Figure QM.2 shows the first 4 wave functions and energy values for a particle in an infinite square well.

The wave functions have the right symmetry and the allowed energy levels E_n are not evenly spaced.

The nth energy level is related to the ground-state energy.

$$E_n = n^2\ E_1$$

The particle in a box can never have zero energy.

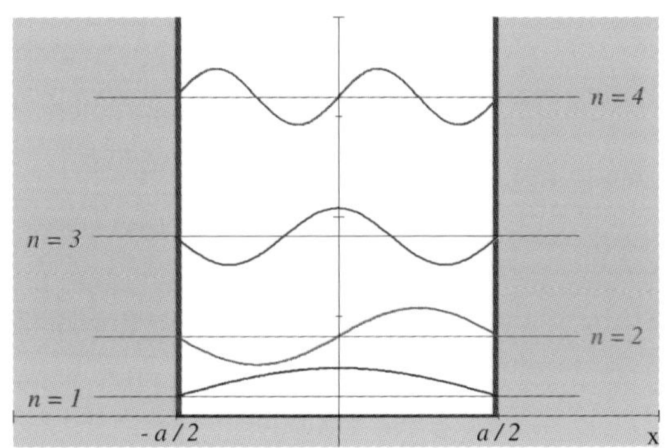

Figure QM.2 Boxed-in $\psi_n(x)$

The probability of finding the particle to the right of the origin is 50% for all states.

$$\frac{2}{a}\int_0^{a/2} \sin^2 \frac{n\pi x}{a}\,dx = \frac{1}{2}$$

The probability of finding the particle in the $n = 2$ state between $x = a/8$ and $x = 3a/8$ is almost 41%.

$$\frac{2}{a}\int_{a/8}^{3a/8} \sin^2 \frac{2\pi x}{a}\,dx = \frac{1}{2\pi} + \frac{1}{4} = 0.409\,15$$

When the particle occupies an even-parity state, the expectation value for its position does not depend on the state.

$$\langle x \rangle = \frac{2}{a}\int_{-a/2}^{+a/2} x \cos^2 \frac{n\pi x}{a}\,dx = 0$$

On average, measurements find the particle at the origin of our symmetric well.

Finite Square Well

The **finite square well** is the logical extension of the infinitely deep well. The particle is still confined in a 1-d box, but the box has finite walls so the problem is more realistic. The physical difference between the two systems is the particle's non-zero probability of being outside the finite well, even though CM says it doesn't have enough energy to scale the potential barrier. The mathematical difference is the lack of a completely analytical solution but we'll use *SNB* to solve the equations numerically.

We'll orient our finite well symmetrically about the origin (see Figure QM.3).

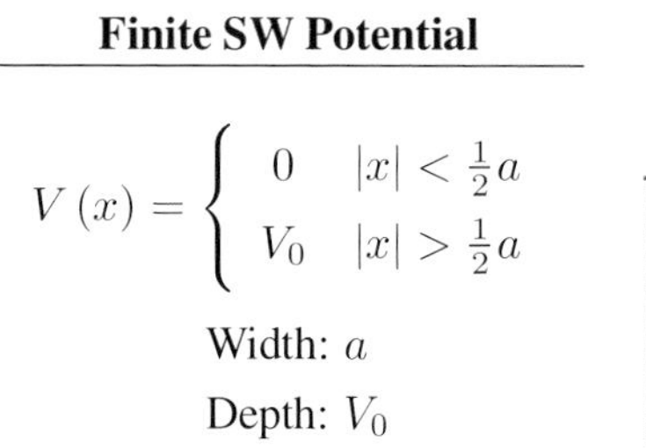

Shaded grey areas show the classically forbidden regions.

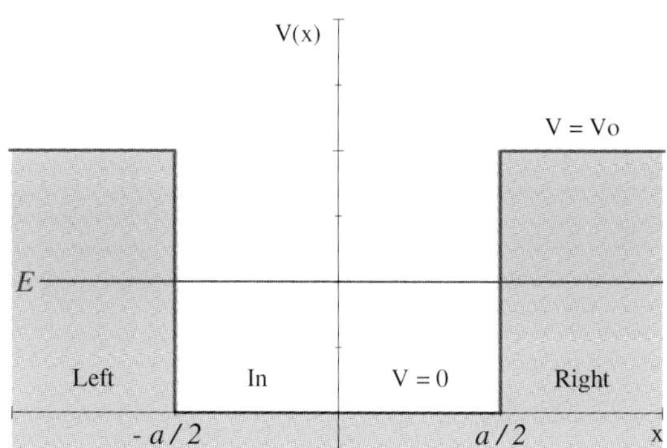

Figure QM.3 A finite box

Classically, the object oscillates between $x = \pm a/2$ when its energy is less than V_0 and its motion is unconstrained when its energy is larger than V_0.

We're looking for stationary bound states with energy $E < V_0$. Inside the well, the Schrödinger Equation takes the same form

$$\frac{d^2\psi_I}{dx^2} + k^2\,\psi_I = 0 \tag{QM.14}$$

with the same positive constant $k^2 \equiv 2mE/\hbar^2$ as for the infinite well. The solutions are the same

$$\psi_I(x) = \begin{cases} A\cos kx & \textbf{even parity} \\ A\sin kx & \textbf{odd parity} \end{cases}$$

but the boundary conditions are different. The potential energy outside is not infinite, so the wave function does not vanish at $x = \pm a/2$.

Outside the well, the Schrödinger Equation takes the form

$$\frac{d^2\psi_O}{dx^2} - \kappa^2\,\psi_O = 0 \tag{QM.15}$$

where $\kappa^2 \equiv 2m(V_0 - E)/\hbar^2$ is a positive constant for bound states ($E < V_0$). You should know the general solution to this equation too, but again *SNB* can help you.

$$\begin{bmatrix} \dfrac{d^2\psi_O}{dx^2} - \kappa^2\,\psi_O = 0 \\ \psi_O(0) = C + D \\ \psi_O'(0) = \kappa(C - D) \end{bmatrix}, \text{ Exact solution is: } \{Ce^{x\kappa} + De^{-x\kappa}\}$$

Outside the well the wave function is exponential.

When we apply the condition that the wave function must approach zero as $x \to \pm\infty$ to the general solution, we get the wave function in the Left and Right regions.

$$\begin{aligned}\psi_R(x) &= D\,e^{-\kappa x}\\ \psi_L(x) &= C\,e^{+\kappa x}\end{aligned}$$

We can construct the even-parity states by setting $C = +D$.

$$\psi_{\text{even}}(x) = \begin{cases} \text{Left} & \text{In} & \text{Right} \\ D\,e^{+\kappa x} & A\cos kx & D\,e^{-\kappa x} \end{cases}$$

The boundary condition that requires the wave function be continuous at $x = +a/2$ yields a relation between the two constants.

$$\psi_I(a/2) = \psi_R(a/2): \quad A\cos\tfrac{ka}{2} = De^{-\kappa a/2} \Longrightarrow D = A\cos\tfrac{ka}{2}e^{\kappa a/2}$$

Here's the even-parity wave function so far.

$$\psi_{\text{even}}(x) = \begin{cases} \text{Left} & \text{In} & \text{Right} \\ A\cos\frac{ka}{2}e^{\kappa a/2}\,e^{+\kappa x} & A\cos kx & A\cos\frac{ka}{2}e^{\kappa a/2}\,e^{-\kappa x} \end{cases}$$

Let's use Simplify and then Expand to normalize the wave function.

$$\frac{1}{A^2} = 2\left[\int \cos^2 kx\,dx\right]_{x=0}^{x=a/2} + 2\left[\int \left(e^{\kappa a/2}\cos\tfrac{ka}{2}\,e^{-\kappa x}\right)^2 dx\right]_{x=a/2}^{x=+\infty}$$

$$= \tfrac{1}{2}a + \frac{1}{2\kappa}\cos ak + \frac{1}{2\kappa} + \frac{1}{2k}\sin ak$$

The even-state normalization constant is $A = \left(\frac{1}{2}\,a + \dfrac{1}{2\kappa}\,(1+\cos ka) + \dfrac{1}{2k}\sin ka\right)^{-1/2}$.

We can construct the odd-parity states by setting $D = -C$.

$$\psi_{\text{odd}}(x) = \begin{cases} \text{Left} & \text{In} & \text{Right} \\ C\,e^{+\kappa x} & B\sin kx & -C\,e^{-\kappa x} \end{cases}$$

The continuity condition at $x = +a/2$ yields a relation between the two constants.

$$\psi_I(-a/2) = \psi_L(-a/2): \quad -B\sin\tfrac{ka}{2} = C\,e^{-\kappa a/2} \Longrightarrow C = -B\sin\tfrac{ka}{2}\,e^{+\kappa a/2}$$

Here's the odd-parity wave function so far.

$$\psi_{\text{odd}}(x) = \begin{cases} \text{Left} & \text{In} & \text{Right} \\ -B\sin\frac{ka}{2}e^{+\kappa a/2}\,e^{+\kappa x} & B\sin kx & B\sin\frac{ka}{2}e^{+\kappa a/2}\,e^{-\kappa x} \end{cases}$$

Let's use Expand to normalize the wave function.

$$\frac{1}{B^2} = 2\left[\int \sin^2 kx\,dx\right]_{x=0}^{x=a/2} + 2\left[\int \left(e^{\kappa a/2}\sin\tfrac{ka}{2}\,e^{-\kappa x}\right)^2 dx\right]_{x=a/2}^{x=+\infty}$$

$$= \tfrac{1}{2}a - \frac{1}{2\kappa}\cos ak + \frac{1}{2\kappa} - \frac{1}{2k}\sin ak$$

The odd-state normalization constant is $B = \left(\frac{1}{2}\,a + \dfrac{1}{2\kappa}\,(1-\cos ka) - \dfrac{1}{2k}\sin ka\right)^{-1/2}$.

The **even-parity** finite-well wave functions are

$$\psi_{\text{even}}(x) = A \begin{cases} \cos\frac{ka}{2}\, e^{\kappa a/2}\, e^{+\kappa x} & x < -\frac{a}{2} \\ \cos kx & -\frac{a}{2} < x < +\frac{a}{2} \\ \cos\frac{ka}{2}\, e^{\kappa a/2}\, e^{-\kappa x} & x > +\frac{a}{2} \end{cases} \tag{QM.16}$$

where $A = \left(\frac{1}{2}\,a + \dfrac{1}{2\kappa}\,(1+\cos ka) + \dfrac{1}{2k}\,\sin ka\right)^{-1/2}$.

The **odd-parity** finite-well wave functions are

$$\psi_{\text{odd}}(x) = B \begin{cases} -\sin\frac{ka}{2}\, e^{+\kappa a/2}\, e^{+\kappa x} & x < -\frac{a}{2} \\ \sin kx & -\frac{a}{2} < x < +\frac{a}{2} \\ +\sin\frac{ka}{2}\, e^{+\kappa a/2}\, e^{-\kappa x} & x > +\frac{a}{2} \end{cases} \tag{QM.17}$$

where $B = \left(\frac{1}{2}\,a + \dfrac{1}{2\kappa}\,(1-\cos ka) - \dfrac{1}{2k}\,\sin ka\right)^{-1/2}$.

These are exact expressions for the wave functions, but we need to determine the values $k = k_n$ that give the allowed energy values. Those energy values come from the boundary conditions on the wave function and its derivative.

$$\frac{\psi_{\text{I}}'(a/2)}{\psi_{\text{I}}(a/2)} = \frac{\psi_{\text{R}}'(a/2)}{\psi_{\text{R}}(a/2)} \tag{QM.18}$$

Here's the energy condition for the even-parity wave functions.

$$\left[\frac{1}{A\cos kx}\frac{d}{dx}A\cos kx = \frac{1}{De^{-\kappa x}}\frac{d}{dx}De^{-\kappa x}\right]_{x=a/2} = -k\frac{\sin\frac{1}{2}ak}{\cos\frac{1}{2}ak} = -\kappa$$

Here's the energy condition for the odd-parity wave functions.

$$\left[\frac{1}{B\sin kx}\frac{d}{dx}B\sin kx = \frac{1}{Ce^{-\kappa x}}\frac{d}{dx}Ce^{-\kappa x}\right]_{x=a/2} = k\frac{\cos\frac{1}{2}ak}{\sin\frac{1}{2}ak} = -\kappa$$

A little editing by-hand produces the energy quantization conditions for a particle bound in a finite square well.

$$k\,\tan\frac{ka}{2} = \kappa \qquad \textbf{(Even)}$$

$$-k\,\cot\frac{ka}{2} = \kappa \qquad \textbf{(Odd)}$$

This is analogous to Eq. (QM.11) for the infinite well, but these equations have no algebraic solution. We'll have to solve them numerically or graphically, but that's a problem *SNB* can handle easily.

We'll start by noticing that we can write the constant κ in terms of k.

$$\kappa \equiv \sqrt{\frac{2m\,(V_0 - E)}{\hbar^2}} = \sqrt{\frac{2mV_0}{\hbar^2} - k^2} \tag{QM.19}$$

The graph of κ is a circle with "radius" $k_0 \equiv \sqrt{2mV_0/\hbar^2}$.

Let's place an electron in a well that's $V_0 = 4\,\text{eV}$ deep and $a = 1\,\text{nm}$ wide, find its energy, and construct its wave function when it's in the ground state.

$$k_0^2 = \frac{2mV_0}{\hbar^2} = \frac{2\,(511000\,\text{eV})\,(4\,\text{eV})}{(197.33\,\text{eV nm})^2} = \frac{104.98}{\text{nm}^2}$$

For these well values, the "radius" is $k_0 = 10.246\,\text{nm}^{-1}$.

Figure QM.4 shows a plot of κ plus the even and odd boundary conditions for the $4\,\text{eV}$-deep, $1\,\text{nm}$-wide well.

The quarter-circle κ intersects the boundary-condition curves 4 times, so this well supports 4 bound states (2 even-parity and 2 odd-parity).

Again we see quantum bound states are discrete as a direct consequence of the boundary conditions acting on the wave function.

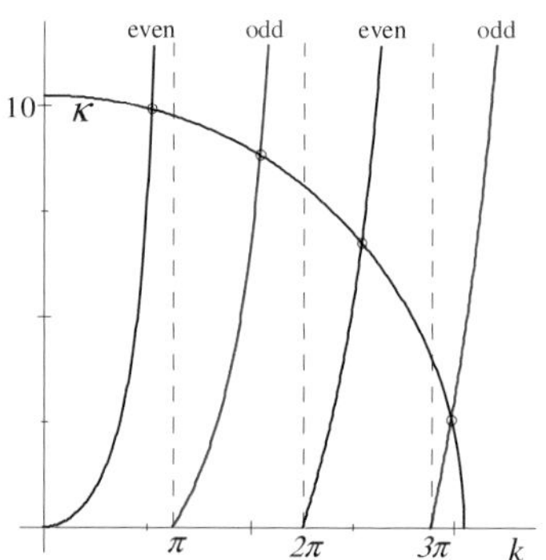

Figure QM.4 *SNB* transcends a problem

The first intersection corresponds to the ground-state energy and occurs between $k = 0$ and $k = \pi$. We can use Solve Numeric to find the value of k_1.

$$\left[\begin{array}{c}\sqrt{104.98 - k_1^2} = k_1 \tan\dfrac{k_1}{2} \\ k_1 \in (0, \pi)\end{array}\right], \text{ Solution is: } \{[k_1 = 2.623\,7]\}$$

That's all we need to Evaluate the ground-state energy.

$$E_1 = \frac{\hbar^2}{2m}k_1^2 = \frac{(197.33\,\text{eV nm})^2}{2\,(511000\,\text{eV})}\left(2.623\,7\,\text{nm}^{-1}\right)^2 = 0.262\,28\,\text{eV}$$

This is significantly less than the infinite-well ground-state energy. Even though the electron is confined in a well with the same width, the wave function has a longer wavelength and less energy than its infinite-well counterpart.

We need κ for the wave function outside the well.

$$\kappa_1 = \left[\sqrt{104.98 - k_1^2}\right]_{k_1=2.623\,7} = 9.904\,4$$

We need the even-parity normalization constant and the outside-the-well coefficients.

$$A_1 = \left[\left(\frac{a}{2} + \frac{1+\cos k_1 a}{2\kappa_1} + \frac{\sin k_1 a}{2k_1}\right)^{-1/2}\right]_{a=1,k_1=2.623\,7,\kappa_1=9.904\,4} = 1.290\,0$$

$$D_1 = 1.290\,0\left[e^{\kappa_1 a/2}\cos\frac{k_1 a}{2}\right]_{a=1,k_1=2.623\,7,\kappa_1=9.904\,4} = 46.736$$

The expression for the complete wave function is cumbersome, so let's use New Definition and *SNB*'s piecewise-defined functions to define the ground-state wave function.

$$\psi_1(x) = \begin{cases} 46.736\, e^{+9.904\,4\,x} & x < -\frac{1}{2} \\ 1.290\,0 \cos 2.623\,7\,x & -\frac{1}{2} < x < +\frac{1}{2} \\ 46.736\, e^{-9.904\,4\,x} & x > +\frac{1}{2} \end{cases}$$

Let's check the normalization. Make sure you use Evaluate or Simplify to calculate a definite integral of a piecewise function. Evaluate Numerically will not work.

$$\int_{-\infty}^{\infty} \psi_1^2(x)\, dx = 1.000\,1$$

The normalization is accurate to one part in ten thousand. We can confidently calculate the probability that the electron is outside the well $\left(\text{either } x > +\frac{1}{2} \text{ or } x < -\frac{1}{2}\right)$.

$$2\int_{1/2}^{\infty} \psi_1^2(x)\, dx = 1.101\,7 \times 10^{-2}$$

The probability that the electron is outside the well is about 1.1%. This is the physics of **quantum tunneling**. The wave function in the classically forbidden region is not zero, so neither is the probability of finding the electron there. If it can't be there, then it probably isn't! The electron is very likely to be between $x = \pm 1/4$.

$$\int_{-1/4}^{+1/4} \psi_1^2(x)\, dx = 0.722\,58$$

Let's see if the ground state satisfies the Uncertainty Principle. Due to the well's symmetry, the momentum and position expectation values both vanish.

$$\langle x \rangle = \int_{-\infty}^{\infty} x\psi_1^2(x)\, dx = 0.0 \qquad \langle p \rangle = \frac{\hbar}{i}\int_{-\infty}^{\infty} \psi_1(x)\frac{\partial \psi_1(x)}{\partial x} dx = 0.0$$

The expectation values of their squares do not.

$$\langle x^2 \rangle = \int_{-\infty}^{\infty} x^2\psi_1^2(x)\, dx = 4.803\,9 \times 10^{-2}$$

$$\langle p^2 \rangle = \left(\frac{\hbar}{i}\right)^2 \int_{-\infty}^{\infty} \psi_1(x)\frac{\partial^2 \psi_1(x)}{\partial x^2} dx = 5.727\,7\hbar^2$$

Tell *SNB* to assume $(\hbar > 0)$ and Simplify the product of the uncertainties.

$$\Delta x \Delta p = \sqrt{\langle x^2 \rangle - \langle x \rangle^2}\sqrt{\langle p^2 \rangle - \langle p \rangle^2} = \sqrt{(0.0480\,39)\,(5.727\,7\hbar^2)} = 0.524\,55\hbar$$

The finite-well ground state satisfies the Uncertainty Principle!

The k-value for the first excited state is between $k = \pi$ and $k = 2\pi$.

$$\begin{bmatrix} \sqrt{104.98 - k_2^2} = -k_2 \cot \frac{k_2}{2} \\ k_2 \in (\pi, 2\pi) \end{bmatrix}, \text{ Solution is: } \{[k_2 = 5.215\,2]\}$$

That's all we need to Evaluate the first excited-state energy.

$$E_2 = \frac{\hbar^2}{2m} k_2^2 = \frac{(197.33\,\mathrm{eV\,nm})^2}{2\,(511000\,\mathrm{eV})} \left(5.215\,2\,\mathrm{nm}^{-1}\right)^2 = 1.036\,3\,\mathrm{eV}$$

The ratio of these energies is slightly less than the n_2^2/n_1^2 value of the infinite well.

$$\frac{E_2}{E_1} = \frac{1.036\,3\,\mathrm{eV}}{0.262\,28\,\mathrm{eV}} = 3.951\,1$$

The energies for the finite-well states are not only smaller than those for the infinite well, they're closer together.

Figure QM.5 shows the available wave functions and energy values for an electron in a $4\,\mathrm{eV}$-deep, $1\,\mathrm{nm}$-wide finite square well.

States with more energy have a higher probability of being in the classically forbidden regions.

Like the infinite well states, the nth state has $n - 1$ nodes so each state has one more node than the state just below it.

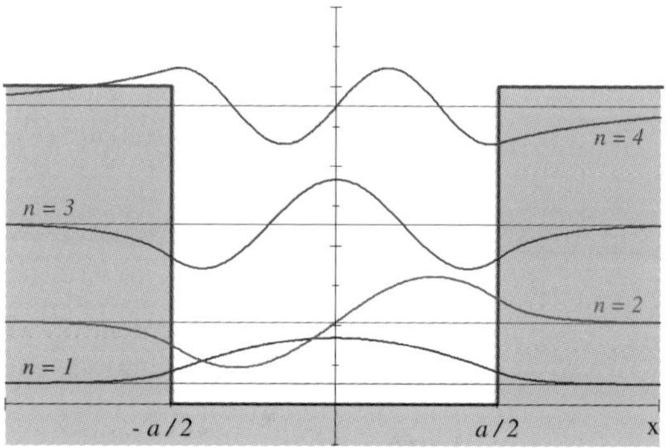

Figure QM.5 Not-so-boxed-in $\psi_n(x)$

It's a general property that bound-state wave functions for states with more energy have more nodes (all in the classically allowed region), more curvature, and oscillate more. More energy means more wiggle in the wave function!

Note well the following well notes:

1. There's bound to be a bound state.

No matter how shallow or narrow the 1-d finite well, there is always at least one bound state. Even small values of κ intersect the even-parity quantization condition at least once. The number of bound states is the largest integer that's less than $1 + k_0 a/\pi$, which we can calculate with *SNB*'s Floor function and Evaluate.

$$N_{\text{bound}} = \left\lfloor 1 + \frac{(10.246)\,(1)}{\pi} \right\rfloor = 4$$

There are 4 bound states for our $4\,\mathrm{eV}$-deep, $1\,\mathrm{nm}$-wide finite well.

2. Very deep is nearly infinite.

Infinite wells are very deep, so as the depth of the finite well becomes large we should recover the ISW results. *SNB* can find the first state analytically.

$$ka = 2\tan^{-1}(\infty) = \pi$$

For an electron in a very deep $1\,\mathrm{GeV}$ well that's $1\,\mathrm{nm}$ wide, Solve Numeric returns $k \approx 2\pi$ for the first odd-parity state.

$$\left[\begin{array}{c}\sqrt{\frac{10^9}{0.038109} - k^2} = -k\cot\frac{k}{2} \\ k \in (5,7)\end{array}\right], \text{ Solution is: } \{[k = 6.2831]\}$$

As the well deepens, κ gets larger and the intersection points on Figure QM.4 all approach the $ka = n\pi$ result for a $1\,\mathrm{nm}$-wide infinite square well. The larger κ also suppresses the exponential wave function outside the well.

3. You can narrow it down.

Suppose we deepen and narrow our finite well so $V_0 \to \infty$ and $a \to 0$ but $aV_0 \equiv \beta_0$ stays fixed and finite. Let's use the even-parity quantization condition in the form $ka\tan(ka/2) = \kappa a$. Since the width is small, we can expand the left-hand side in a Power Series to two terms in powers of a.

$$ka\tan\frac{ka}{2} = \tfrac{1}{2}a^2k^2 + O\left(a^4\right)$$

Write the quantization condition with κ in terms of k and β_0, replace k with $2mE/\hbar^2$ and use Solve Exact to find the energy.

$$\tfrac{1}{2}a^2\left(\tfrac{2mE}{\hbar^2}\right) = \sqrt{\tfrac{2ma\beta_0}{\hbar^2} - a^2\left(\tfrac{2mE}{\hbar^2}\right)}, \text{ Solution is: } \left\{-\tfrac{1}{a^2m}\left(\hbar^2 - \hbar\sqrt{\hbar^2 + 2am\beta_0}\right)\right\}$$

Edit the result by-hand, and expand again to two terms in powers of a.

$$E = \frac{\hbar^2}{a^2m}\left(\sqrt{1 + \frac{2am\beta_0}{\hbar^2}} - 1\right) = \frac{1}{a}\beta_0 - \tfrac{1}{2}m\frac{\beta_0^2}{\hbar^2} + O(a)$$

Restore the explicit dependence on the well's depth and width.

$$E = \left[\frac{\beta_0}{a} - \frac{m\beta_0^2}{2\hbar^2}\right]_{\beta_0 = aV_0} = V_0 - \tfrac{1}{2}a^2\frac{m}{\hbar^2}V_0^2$$

The result is the bound state energy for the delta-function potential $V(x) = -aV_0\,\delta(x)$. The delta-function potential is just a very narrow, extremely deep well. Keeping aV_0 fixed and finite ensures the area inside stays constant as the finite well shrinks into a delta function.

V-shaped Linear Well

The constant force problem in CM is taught in most introductory physics courses. It applies to the gravitational force on an object near the Earth's surface and the force on a charged particle in a constant electric field. The corresponding linear potential problem in QM, which approximates potentials found in some transistors in semiconductor physics, is often overlooked because its solution involves special functions. This is a complication *SNB* can help you overcome.

We can use the linear potential energy $V(x) = F\,|x|$ to define a V-shaped well, which we'll orient symmetrically about the origin (see Figure QM.6).

V-shaped Potential

$$V(x) = \begin{cases} +F\,x & x > 0 \\ -F\,x & x < 0 \end{cases}$$

Slope: $\pm F$

Shaded grey areas show the classically forbidden regions.

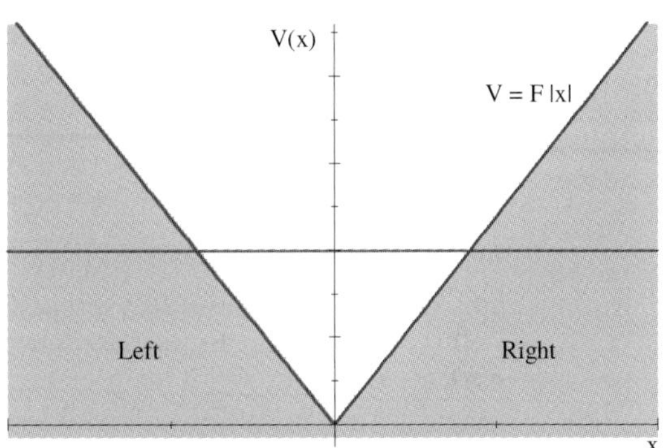

Figure. QM.6 V is for...

Classically, the object moves under the influence of a constant restoring force that always pushes toward the origin.

Again we're looking for stationary states. To the right of the origin, the Schrödinger Equation takes the form

$$\frac{d^2\psi_{\mathrm{R}}}{dx^2} + \left(k^2 - b^3 x\right)\psi_{\mathrm{R}} = 0 \tag{QM.20}$$

where $k^2 \equiv 2mE/\hbar^2$ and $b^3 \equiv 2mF/\hbar^2$ are both positive constants. To the left of the origin, the TISE takes the form

$$\frac{d^2\psi_{\mathrm{L}}}{dx^2} + \left(k^2 + b^3 x\right)\psi_{\mathrm{L}} = 0 \tag{QM.21}$$

with the same two positive constants. Note that here "right" means the entire $x > 0$ region and "left" means all $x < 0$.

You probably don't know the general solution to these equations, so here's where *SNB* (and Solve ODE + Exact) can really be a big help. Here's the general solution to the right of the origin.

$\dfrac{d^2\psi_{\mathrm{R}}}{dx^2} + \left(k^2 - b^3 x\right)\psi_{\mathrm{R}} = 0$, Exact solution is:

$$\left\{C_3 \operatorname{AiryAi}\left(bx - \frac{1}{b^2}k^2, 0\right) + C_4 \operatorname{AiryBi}\left(bx - \frac{1}{b^2}k^2, 0\right)\right\}$$

Here's the general solution to the left of the origin.

$$\frac{d^2\psi_L}{dx^2} + \left(k^2 + b^3 x\right)\psi_L = 0, \text{ Exact solution is:}$$

$$\left\{C_7 \operatorname{AiryAi}\left(-\frac{1}{b^2}k^2 - bx, 0\right) + C_8 \operatorname{AiryBi}\left(-\frac{1}{b^2}k^2 - bx, 0\right)\right\}$$

Both general solutions are linear combinations of Airy functions.

Airy functions are examples of "special functions" that are slightly more complicated than the standard trigonometric and exponential functions. *SNB* calls them AiryAi and AiryBi, but the names Ai and Bi are more common. To *SNB*, $\operatorname{AiryAi}(y, 0)$ or $\operatorname{AiryAi}(y)$ denotes the Airy function of the first kind and $\operatorname{AiryAi}(y, m)$ is the mth derivative of $\operatorname{AiryAi}(y)$. As you might guess, $\operatorname{AiryBi}(y, 0)$ or $\operatorname{AiryBi}(y)$ denotes the Airy function of the second kind and $\operatorname{AiryBi}(y, m)$ is the mth derivative of $\operatorname{AiryBi}(y)$.

Figure QM.7 shows a graph of the two kinds of Airy functions. For negative arguments, they both oscillate about zero with increasing frequency and decreasing amplitude. For positive arguments, AiryAi is positive and decreases exponentially to zero, while AiryBi is positive and increases exponentially to infinity.

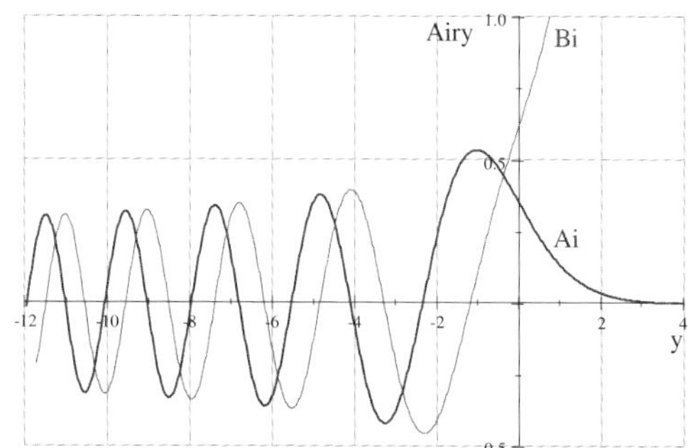

Figure QM.7 A very Airy situation

For finite potentials, the AiryBi function is not acceptable for a wave function because of its off-to-infinity behavior. The AiryAi function is an ideal candidate for a wave function when the oscillatory part is in the classically allowed region and the exponential part is in the classically forbidden region. The argument $\pm bx - k^2/b^2$ has an energy-dependent offset that slides the solution along the x-axis, so only certain energy values produce acceptable wave functions. Boundary conditions in action!

The general solution now has this form.

$$\psi_R(x) \sim \operatorname{AiryAi}\left(+bx - \frac{k^2}{b^2}\right) \tag{QM.22a}$$

$$\psi_L(x) \sim \operatorname{AiryAi}\left(-bx - \frac{k^2}{b^2}\right) \tag{QM.22b}$$

We'll have to be careful at the origin where the wave function $\psi(x = 0)$ is proportional to the Airy function of the offset $\operatorname{AiryAi}\left(-k^2/b^2\right)$. The offset is not zero!

Our V-shaped potential is symmetric about the origin, so the wave functions have either odd or even parity. The AiryAi function itself doesn't have the correct parity, but we can use the general solution to construct wave functions that do by replacing the $\pm bx$ term (which is always positive) with $b\,|x|$.

The absolute value gives the AiryAi function even-parity so the wave function is

$$\psi_{\text{even}}(x) = N_{+}\,\text{AiryAi}\left(b\,|x| - \frac{k^2}{b^2}\right) \tag{QM.23}$$

where N_{+} is the normalization constant. This wave function is not zero at the origin, and its slope there is zero when $\text{AiryAi}\left(-k^2/b^2, 1\right) = 0$. We can express this as $d_n = -k^2/b^2$ where d_n is the nth zero of AiryAi's derivative. This means $k^2 = -d_n b^2$ gives us the allowed energy values for the even states.

Adding *SNB*'s signum function changes the parity so the odd-parity wave function is

$$\psi_{\text{odd}}(x) = N_{-}\,\text{signum}(x)\,\text{AiryAi}\left(b\,|x| - \frac{k^2}{b^2}\right) \tag{QM.24}$$

where N_{-} is the normalization constant. This wave function has a non-zero slope at the origin, and vanishes there when $\text{AiryAi}\left(-k^2/b^2\right) = 0$. We can express this as $a_n = -k^2/b^2$ where a_n is the nth zero of AiryAi. This means $k^2 = -a_n b^2$ gives us the allowed energy values for the odd states.

We now have two separate expressions for the allowed energy values, one for the even states that depends on AiryAi's zeros and one for the odd states that depends on the zeros of its derivative. We want a single expression analogous to Eq. (QM.11) for *all* the allowed energy values.

The product of AiryAi times its derivative has all the zeros we need, so let's use New Definition and define a new function of a dimensionless variable. Then Evaluate the expression and see what *SNB* has to say.

$$Q(z) = \text{AiryAi}(z)\,\frac{d\,\text{AiryAi}(z)}{dz} = \text{AiryAi}(z, 0)\,\text{AiryAi}(z, 1)$$

SNB uses its own internal function for the derivative.

Figure QM.8 shows a graph of $Q(z)$. The allowed energy values depend on the zeros of this function, which are not evenly spaced like those of the sine and cosine functions. There is no simple expression to identify those zeros, so we'll have to use Solve Numeric to find them. The potential energy has no effect on the function $Q(z)$ so we only have to find the zeros once.

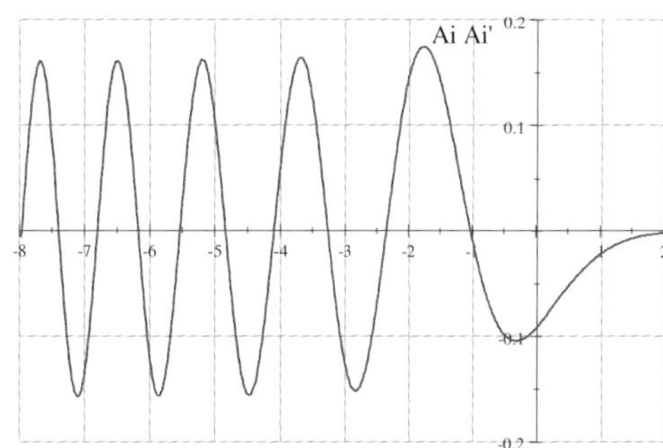

Figure QM.8 Zeros and zeros

The boundary condition on all wave functions is now $k^2 = -q_n b^2$ where q_n is the nth zero of $Q(z)$. Those zeros are either from the AiryAi function or its derivative, so this boundary condition maintains the appropriate parity in the wave functions.

We can use the definitions of k^2 and b^3 to restore the constants and finish the result for the quantized energy values of a particle in a V-shaped well.

$$E_n = -q_n \left(\frac{F^2\hbar^2}{2m}\right)^{1/3} \qquad n = 1, 2, 3\ldots \tag{QM.25}$$

This is an exact result, but to find the nth zero q_n we need to solve the transcendental equation $Q(q_n) = 0$ numerically.

The first zero q_1 gives us the ground-state energy. A look at Figure QM.8 tells us q_1 is near -1 so let's tell Solve Numeric to look between -1.5 and -0.5.

$$\begin{bmatrix} Q(q_1) = 0 \\ q_1 \in (-1.5, -0.5) \end{bmatrix}, \text{ Solution is: } \{[q_1 = -1.0188]\}$$

When you use Solve Numeric over a specified range, always list the smaller value first. The second zero q_2, which gives us the energy of the first excited state, is between -3 and -2.

$$\begin{bmatrix} Q(q_2) = 0 \\ q_2 \in (-3, -2) \end{bmatrix}, \text{ Solution is: } \{[q_2 = -2.3381]\}$$

Figure QM.8 shows the location of the first 10 zeros of $Q(z)$.

The derivation of the normalization constants requires analytic integration of AiryAi. There's not much *SNB* can do to help us with that, so I'll just tell you the answers and leave their derivation as the last problem. The normalization constants are

$$N_+ = \sqrt{\frac{b}{2}} \frac{1}{\sqrt{-q_n}\ \text{AiryAi}(q_n)} \tag{QM.26a}$$

$$N_- = \sqrt{\frac{b}{2}} \frac{1}{\text{AiryAi}(q_n, 1)} \tag{QM.26b}$$

where $\text{AiryAi}(q_n, 1)$ is the first derivative of AiryAi evaluated at q_n.

The **even-parity** wave functions for the V-shaped well are

$$\psi_n(x) = \sqrt{\frac{b}{2}} \frac{1}{\sqrt{-q_n}\ \text{AiryAi}(q_n)} \text{AiryAi}(b|x| + q_n) \qquad n = 1, 3, 5\ldots \tag{QM.27}$$

where q_n is the nth root of $Q(z)$.

Here are the **odd-parity** wave functions for the V-shaped well.

$$\psi_n(x) = \sqrt{\frac{b}{2}} \frac{\text{signum}(x)}{\text{AiryAi}(q_n, 1)} \text{AiryAi}(b|x| + q_n) \qquad n = 2, 4, 6\ldots \tag{QM.28}$$

Once again the even-parity wave functions have odd quantum numbers and the odd-parity wave functions have even quantum numbers. You get used to it.

Let's continue our custom of placing an electron in our well. We'll somewhat arbitrarily pick $b = 5\,\mathrm{nm}^{-1}$ to set the force and energy scales.

$$F = \frac{\hbar^2 b^3}{2m} = \frac{(197.33\,\mathrm{eV\,nm})^2\left(5\,\mathrm{nm}^{-1}\right)^3}{2\,(511000\,\mathrm{eV})} = 4.762\,6\frac{\mathrm{eV}}{\mathrm{nm}}$$

This very small force is about 0.76 nano-Newtons.

$$F = 4.762\,6\frac{\mathrm{eV}}{\mathrm{nm}} = 7.630\,5\times 10^{-10}\frac{\mathrm{m}}{\mathrm{s}^2}\,\mathrm{kg}$$

This value of b sets the energy scale at about $1\,\mathrm{eV}$.

$$E \sim \left(\frac{F^2\hbar^2}{2m}\right)^{1/3} = \left(\frac{\left(4.762\,6\frac{\mathrm{eV}}{\mathrm{nm}}\right)^2(197.33\,\mathrm{eV\,nm})^2}{2\,(511000\,\mathrm{eV})}\right)^{1/3} = 0.952\,52\,\mathrm{eV}$$

We can use New Definition to define the ground-state wave function.

$$\psi_1(x) = \sqrt{\frac{5}{2}}\,\frac{1}{\sqrt{1.018\,8}\ \mathrm{AiryAi}\,(-1.018\,8)}\,\mathrm{AiryAi}\,(5\,|x| - 1.018\,8)$$

Let's check the wave function's normalization.

$$\int_{-\infty}^{\infty} \psi_1^2(x)\,dx = 1.0$$

That checks out, so let's calculate the probability that the electron is beyond the classical turning point. First we need the ground state energy.

$$E_1 = -q_1\left(\frac{F^2\hbar^2}{2m}\right)^{1/3} = -(-1.018\,8)\,(0.952\,52\,\mathrm{eV}) = 0.970\,43\,\mathrm{eV}$$

The classical turning point is the solution to $E = V(x_1)$.

$$x_1 = \frac{E_1}{F} = \frac{0.970\,43\,\mathrm{eV}}{4.762\,6\,\mathrm{eV}/\,\mathrm{nm}} = 0.203\,76\,\mathrm{nm}$$

There are two ways to calculate the probability we want.

$$2\int_{0.203\,76}^{\infty} \psi_1^2(x)\,dx = 0.229\,16 \qquad 1 - \int_{-0.203\,76}^{0.203\,76} \psi_1^2(x)\,dx = 0.229\,16$$

There is a 22.9% chance the electron in the ground state will be detected beyond the classical turning point. If it can't be there, then it probably isn't!

Figure QM.9 shows the first four wave functions and energy values for this V-shaped potential. States with more energy have a lower probability of being in the classically forbidden regions. As in our previous calculations, the nth state has $n-1$ nodes.

Let's see if the ground state satisfies the Uncertainty Principle.

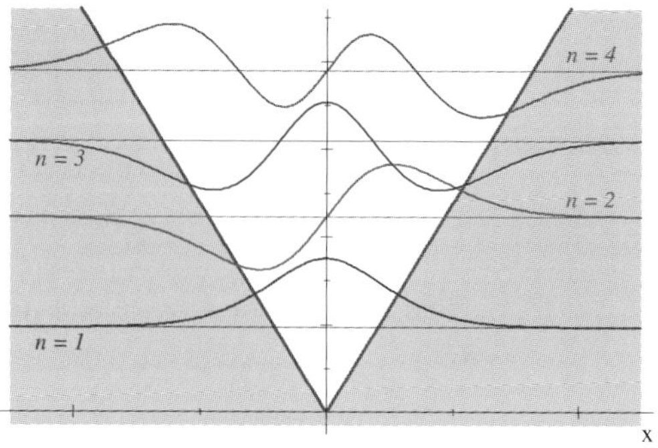

Figure QM.9 V is for $\psi_n(x)$

First we'll calculate the expectation values for the electron's position and momentum.

$$\langle x\rangle = \int_{-\infty}^{\infty} x\psi_1^2(x)\,dx = 0.0 \qquad \langle p\rangle = \frac{\hbar}{i}\int_{-\infty}^{\infty} \psi_1(x)\,\frac{d\psi_1(x)}{dx}\,dx = 0.0$$

Due to the potential's symmetry, the position and momentum expectation values both vanish. The calculation of the expectation value of x^2 is straightforward.

$$\langle x^2\rangle = \int_{-\infty}^{\infty} x^2\psi_1^2(x)\,dx = 2.9995\times10^{-2}$$

The electron's rms position is $x_{\text{rms}} = 0.17319\,\text{nm}$. The calculation of $\langle p^2\rangle$ is a little problematic. *SNB*'s brute-force calculation of the wave function's second derivative has a delta-function term at the origin due to the absolute value. We can avoid that mess by using Evaluate to quickly derive an identity.

$$x>0\text{:}\ \frac{d^2}{dx^2}\,\text{AiryAi}\,(+bx+q_n) = b^2\,(q_n+bx)\,\text{AiryAi}\,(q_n+bx,0)$$

$$x<0\text{:}\ \frac{d^2}{dx^2}\,\text{AiryAi}\,(-bx+q_n) = b^2\,(q_n-bx)\,\text{AiryAi}\,(q_n-bx,0)$$

In other words, the wave function's second derivative is exactly what the Schrödinger Equation requires for the V-shaped potential energy.

$$\frac{d^2\psi_n}{dx^2} = b^2\,(b\,|x|+q_n)\,\psi_n \tag{QM.29}$$

Now the calculation of the expectation value of p^2 is straightforward too.

$$\langle p^2\rangle = 5^2\,(\hbar/i)^2\int_{-\infty}^{\infty}\psi_1(x)\,(5\,|x|-1.0188)\,\psi_1(x)\,dx = 8.4901\hbar^2$$

Tell *SNB* to assume $(\hbar>0)$ and Simplify the product of the uncertainties.

$$\Delta x\Delta p = \sqrt{(2.9995\times10^{-2})\,(8.4901\hbar^2)} = 0.50464\hbar$$

The ground state for the V-shaped potential certainly satisfies the Uncertainty Principle!

Problems

The Tale of the Traveling Triplets

1. A spaceship leaves Earth and accelerates from rest at $1g$. How fast is it moving when it gets to Alpha Centauri, 4.37 light-years away? How much time does the trip take, as measured by clocks on Earth? As measured by clocks on the ship?

2. Verify these expression for the gamma factor of a spaceship accelerating at $1g$.

   a. $\gamma\left(x\right) = \gamma_0 + \frac{g}{c^2}\left(x - x_0\right)$ b. $\gamma\left(t\right) = \sqrt{1 + \left(\frac{\gamma_0 u_0 + gt}{c}\right)^2}$

3. Use the results of the previous problem to relate the time and space intervals in S.
$$\Delta t = \frac{c}{g}\sqrt{\left(\gamma_0 + \frac{g}{c^2}\Delta x\right)^2 - 1} - \frac{\gamma_0 u_0}{g}$$

4. Show that the acceleration of Axel's ship as seen in Homer's frame is
$$a\left(t\right) = \frac{g}{\left(1 + \left(\frac{\gamma_0 u_0 + gt}{c}\right)^2\right)^{3/2}}$$
Verify this result has the expected behavior for very short and very long times.

5. Calculate the mathematical average of Axel's velocity $u\left(t\right)$ as seen from Earth for each half of his outbound trip.

6. Plot the expression for Axel's acceleration for the first 5 years of the trip (as seen in S). What does Homer see for Axel's speed, position, and acceleration at that time?

7. Show that Axel starts to catch-up to Constance at time $t = \gamma v/g$ where t is measured in S and $\gamma = \left(1 - v^2/c^2\right)^{-1/2}$ is Constance's gamma factor.

8. Starfleet sends a starship on a round trip to Proxima Centauri, located 4.24 light-years from Earth. The ship accelerates at $+1\,c/\mathrm{y}$ for the first half of the trip, at $-1\,c/\mathrm{y}$ for the second half, and returns in the same manner. How much time elapses on the clocks at Starfleet Command? How much time elapses on the clocks aboard the starship? How far does the starship travel in its own reference frame?

9. Repeat Problem 6 for a starship sent on a round trip to the planet Vulcan (orbiting 40 Eridani A), located 16.0 light years from Earth.

10. Repeat Problem 6 for a starship sent on a round trip to Betelgeuse, located 643 light-years from Earth.

11. Repeat Problem 6 for a starship sent on a round trip to the Center of Galaxy, located $26,400$ light-years from Earth.

12. Repeat Problem 6 for a starship sent on a round trip to the Andromeda Galaxy, located 2.52 million light-years from Earth.

Orbits in General Relativity

13. How much mass would a black hole need to have a radius of 1 AU? Give your answer in kilograms and solar masses $\left(M_\odot = 1.9889 \times 10^{30}\,\text{kg}\right)$.

14. How much mass would a black hole need to have the same radius as our Sun? Give your answer in kilograms and solar masses.

15. Show that the GR effective potential can be written
$$V_{\text{eff}}(r) = -\frac{GM}{r} - \tfrac{1}{2}\frac{R_s \ell^2}{r^3} + \tfrac{1}{2}\frac{\ell^2}{r^2}$$
where R_s is the Schwarzschild radius.

16. Use the form of the effective potential from the previous problem and find the radii of the circular orbits.

17. Show that the angular momentum for a stable circular orbit of radius R is
$$\ell_{\text{cir}} = \sqrt{\frac{GMR}{1 - \dfrac{3GM}{c^2 R}}}$$
when given in conventional units.

18. Show that the radii (in conventional units) of the two circular orbits are
$$R_{\text{cir}} = \frac{\ell^2}{2GM}\left(1 \pm \sqrt{1 - \frac{12G^2M^2}{c^2\ell^2}}\right)$$

19. Show that the local maximum in the effective potential equals zero for $\ell = 4GM/c$ and $R_{\text{in}} = 4GM/c^2$ when given in conventional units.

20. Directly Substitute the expression for the radius of the unstable circular orbit into the expression for the angular momentum of asymptotic circles and reproduce the result of Eq. (GR.14) for the asymptotic radius.

21. Verify that the required angular momentum for an asymptotic circle starting from $R_0 = 10$ is $\ell = 5/\sqrt{2}$. Use these values to calculate and plot the orbit.

22. Verify by direct calculation that Newtonian orbits have equal angular velocity and radial frequency $\omega_{\text{rad}} = \omega_{\text{ang}} = \sqrt{GM/R^3}$ where R is the radius of a circular orbit.

23. Show that GR perihelion advance can be written $\Delta\phi = 6\pi\dfrac{GM}{Rc^2}$.

24. Calculate and plot the following relativistic orbits. Start at $R_0 = 25$.

   a. $\ell = 3.7455754$, $P = 3$

   b. $\ell = 3.7488329$, $P = 5$

   c. $\ell = 3.7719728$, $P = 2$

   d. $\ell = 3.8508305$, $P = 5$

   e. $\ell = 3.9651980$, $P = 3$

   f. $\ell = 4.2390436$, $P = 4$

   What are the apsidal angles for these orbits?

25. Let's verify my claim that the effects of General Relativity on the Earth's orbit are too small to be seen on an orbital plot. The eccentricity of the Earth's orbit is about 0.01671.

    a. Use Eq. (9.21) to make a 2-d polar plot of the Earth's orbit. Plot the orbit as a thick bright-red line.

    b. Calculate the angular momentum of the Earth in its orbit.

    c. Calculate the numerical approximation to the GR orbit with the same initial conditions.

    d. Add the GR orbit to your plot as a thin black line, and let it do 6 orbits. See?

26. Calculate the perihelion advance predicted by GR for Venus, Earth, Mars, and the asteroid Icarus. Look up the observed values and compare.

27. What is the apsidal angle of the precessing ellipses in Figure GR.4a?

28. The three dashed curves on Figure GR.5b are ℓ_{max}, ℓ_{cir}, and ℓ_{asym}. Which is which? State your reasons.

29. Here's another way to derive the GR orbital precession for nearly-elliptical orbits.

    a. Use the orbital equation $u'' + u = 3Mu^2 + \dfrac{M}{h^2}$ to show that

$$u = \frac{1}{6M}\left(1 - \sqrt{1 + 12\left(Mu'' - \frac{M^2}{h^2}\right)}\right)$$

    b. Expand this to 4 terms in M and argue that $(u'')^2$ is negligible so that

$$\frac{1}{\Omega^2}\frac{d^2u}{d\phi^2} + u = A$$

    where $\Omega \equiv \left(1 + \dfrac{6M^2}{h^2}\right)^{-1/2}$ and $A \equiv \dfrac{M}{h^2} + 3\dfrac{M^3}{h^4}$.

    c. Verify that the Laplace solution for $u(\phi)$ is

$$u(\phi) = A + \frac{1 - AR}{R}\cos\Omega\phi$$

    when $u_0 = 1/R$ and $u_0' = 0$.

    d. Look at the solution $u(\phi)$ and argue that $\Delta\phi = \dfrac{2\pi}{\Omega} - 2\pi$. Use Newtonian orbital results to give $\Delta\phi$ in terms R.

30. Calculate and plot the $\Phi_A = 4\pi$ GR orbit with $R_0 = 10$ and $\ell = 3.535\,72$. Estimate the orbital period and minimum radius. Compare your results with the asymptotic circle of Figure GR.3b and the periodic orbit of Figure GR.6, using the word "between" frequently.

31. Prove that the embedded surface is vertical at the event horizon. What does it mean?

32. Let's explore the perihelion shift when we combine Special Relativity and Newton's Law of Gravitation. The SR effective potential is

$$V_{\text{eff}}(r) = -\left(1+\frac{E}{c^2}\right)\frac{GM}{r} + \tfrac{1}{2}\left(\ell^2 - \frac{G^2M^2}{c^2}\right)\frac{1}{r^2}$$

where E is the orbiter's total energy per unit mass.

a. Show that the angular momentum for a circular orbit of radius R is

$$\ell_{\text{cir}} = \sqrt{\frac{G^2M^2}{c^2} + GM\left(1+\frac{E}{c^2}\right)R}\ .$$

b. Show that the orbital angular velocity is $\omega_{\text{ang}}^2 = \dfrac{GM}{R^3}\left(1+\dfrac{E}{c^2}+\dfrac{GM}{c^2R}\right)$.

c. Show that the frequency of small radial oscillations is $\omega_{\text{rad}}^2 = \dfrac{GM}{R^3}\left(1+\dfrac{E}{c^2}\right)$.

d. Show that $\Delta\phi_{\text{SR}} = \frac{1}{6}\Delta\phi_{\text{GR}}$. This and the speed-of-light limitation eliminate SR as a viable gravitational theory.

Classical Lifetime of a Hydrogen Atom

33. Show that the electron's velocity is $v \approx \sqrt{r_e/r}\,c$ in the nearly-circular-orbit approximation.

34. Show that the electron's acceleration is $a \approx c^2\,r_e/r^2$ in the nearly-circular-orbit approximation.

35. Use the results from the text and show that $\dfrac{v_r}{v_\phi} \approx 5.2\times10^{-7}$ when the electron is in the ground state.

36. Show that the nonrelativistic radial position as a function of time can be written

$$\frac{r(t)}{a_0} = \left(\frac{r_0^3}{a_0^3} - 0.064\,266\,t\right)^{1/3}$$

where t is in picoseconds. Plot $r(t)/a_0$ for the ground state.

37. Use the gamma factor as a function of r to show that the power series of the derivative $d\gamma/dr$ equals the derivative of the power series of γ.

38. Calculate the fractional difference between the NR and SR lifetimes for the $n=2$, $\ell=1$ state and show that it is approximately 10^{-6}.

39. Use Eq. (CL.6) to show the electron's time-average radial position is $\overline{r} = \frac{3}{4}\,r_0$.

40. Plot the ratio of the SR integrand for the in-fall time to the NR integrand ρ^2 (in other words, plot $1/CRS$). Let ρ range from 0 to 1000. Based on the graph, comment on the very small difference between the SR and NR lifetime results.

Quantum Mechanical Bound States

41. Convert $\hbar c$ into $\mathrm{eV\,nm}$.

42. Tell *SNB* that n is a positive integer and use Simplify to verify these identities.

$$\begin{aligned}\cos n\pi &= (-1)^n\\ \sin n\pi &= 0\end{aligned}$$

43. Show that the probability of finding the particle-in-a-box to the right of the origin is 50% for all even states.

44. Verify that the when a particle in an infinite well occupies an odd-parity state, the expectation value for its position does not depend on the state.

45. Show that the probability of finding a particle in the interval $(0, a/n)$ is $1/n$ for any state in a symmetrically-oriented infinite well but the ground state. Why isn't this valid for the ground state?

46. The root-mean-square value of the position is $x_{\text{rms}} = \sqrt{\langle x^2 \rangle}$ where

$$\langle x^2 \rangle = \int \psi^* (x)\; x^2\, \psi (x)\; dx$$

is the expectation value of x^2. Calculate x_{rms} for the first 4 particle-in-a-box states.

47. The root-mean-square value of the momentum is $p_{\text{rms}} = \sqrt{\langle p^2 \rangle}$ where

$$\langle p^2 \rangle = \left(\frac{\hbar}{i}\right)^2 \int_{-\infty}^{\infty} \psi^* (x) \frac{\partial^2 \psi (x)}{\partial x^2}\, dx$$

is the expectation value of p^2. Calculate p_{rms} for the first four particle-in-a-box states.

48. The uncertainty in some observable q is $\Delta q = \sqrt{\langle q^2 \rangle - \langle q \rangle^2}$. Verify that the first four particle-in-a-box states satisfy the uncertainty principle.

49. Verify these general results for particle-in-a-box expectation values.

$$\begin{aligned}\langle p^2 \rangle &= n^2 \pi^2 \frac{\hbar^2}{a^2}\\ \langle x^2 \rangle &= \tfrac{1}{12} a^2 \left(1 - \frac{6}{n^2 \pi^2}\right)\end{aligned}$$

50. Use the results of the previous problem and verify that all particle-in-a-box states satisfy the uncertainty principle.

51. The expectation value of the non-relativistic kinetic energy is $\dfrac{\langle p^2 \rangle}{2m}$. Show that the energy of the nth particle-in-a-box state is the expectation value of the kinetic energy.

52. Suppose an electron occupies the first excited state in a $4\,\text{eV}$-deep, $1\,\text{nm}$-wide finite square well. Show that its wave function is approximately

$$\psi_2(x) = \begin{cases} -53.447\,e^{+8.8194\,x} & x < -\frac{1}{2} \\ 1.2768\sin 5.2152\,x & -\frac{1}{2} < x < +\frac{1}{2} \\ +53.447\,e^{-8.8194\,x} & x > +\frac{1}{2} \end{cases}.$$

53. Verify that an electron in this state has about a 4.8% probability of being found outside the well.

54. What is the probability that an electron in this state will be found between $x = 1/8$ and $x = 3/8$? How does this compare with the infinite-well result?

55. Suppose an electron occupies the first even-parity excited state in a $4\,\text{eV}$-deep, $1\,\text{nm}$-wide finite square well. Show that its wave function is approximately

$$\psi_3(x) = \begin{cases} -27.180\,e^{+6.73819\,x} & x < -\frac{1}{2} \\ 1.2419\cos 7.7186\,x & -\frac{1}{2} < x < +\frac{1}{2} \\ -27.180\,e^{-6.73819\,x} & x > +\frac{1}{2} \end{cases}.$$

56. What is the probability that an electron in this state will be found inside the well?

57. What is the probability that an electron in this state will be found between $x = \pm\frac{1}{4}$?

58. Suppose an electron occupies the second odd-parity excited state in a $4\,\text{eV}$-deep, $1\,\text{nm}$-wide finite square well. Show that its wave function is approximately

$$\psi_4(x) = \begin{cases} +3.6504\,e^{+2.54058\,x} & x < -\frac{1}{2} \\ 1.0579\sin 9.9260\,x & -\frac{1}{2} < x < +\frac{1}{2} \\ -3.6504\,e^{-2.54058\,x} & x > +\frac{1}{2} \end{cases}.$$

59. What is the probability that an electron in this state will be found outside the well?

60. What is the probability that an electron in this state will be found between $x = \pm\frac{1}{4}$?

61. Calculate the energy ratios E_4/E_3 and E_3/E_2 and compare them with the corresponding ratios for the infinite well states.

62. Verify by direct calculation that the expectation values for our electron's position and momentum are zero for all three finite-well excited states.

63. Generalize the calculation in the previous problem and show that $\langle x \rangle = 0$ and $\langle p \rangle = 0$ for any particle in any state in a symmetrically-oriented finite square well.

64. Verify that our three finite-well excited states satisfy the uncertainty principle.

65. How many bound states will our $4\,\text{eV}$-deep finite well support if we double its width?

66. How many bound states will our $4\,\text{eV}$-deep finite well support if we halve its width?

67. Show that both finite-well normalization constants reduce to the expected value when $k = n\pi/a$.

68. Make a table listing the first six zeros of my function $Q(z)$.

69. Using the definitions for the V-shaped well, verify the following expression.

$$\frac{k^2}{b^2} = \frac{E}{\left(F^2\hbar^2/(2m)\right)^{1/3}}$$

70. Calculate the wave function for the first excited state of our electron in a V-shaped potential and show that it's approximately

$$\psi_2(x) = 2.2549\,\text{signum}(x)\,\text{AiryAi}(5|x| - 2.3381).$$

71. Calculate the wave function for the second excited state of our electron in a V-shaped potential and show that it's approximately

$$\psi_3(x) = -2.0937\,\text{AiryAi}(5|x| - 3.2482).$$

72. Calculate the wave function for the $n = 4$ state of our electron in a V-shaped potential and show that it's approximately

$$\psi_4(x) = -1.9688\,\text{signum}(x)\,\text{AiryAi}(5|x| - 4.0879).$$

73. Calculate the probability that the electron will be found beyond the classical turning point for the first three excited states. Compare and discuss your results.

74. Analyze $Q(z) = \text{AiryAi}(z)\dfrac{d\,\text{AiryAi}(z)}{dz}$ and verify that the even-numbered zeros (second, fourth, and so on) come from the AiryAi function, and the odd-numbered zeros (first, third, and so on) come from the derivative.

75. When $q = \beta x + \alpha$ any two Airy functions $A(q)$ and $B(q)$ satisfy this relation.

$$\int A(q)\,B(q)\,dx = \left(x + \frac{\alpha}{\beta}\right) A(q)\,B(q) - \frac{1}{\beta} A'(q)\,B'(q)$$

a. Show that $\displaystyle\int_0^\infty A(q)\,B(q)\,dx = \frac{1}{\beta} A'(\alpha)\,B'(\alpha) - \frac{\alpha}{\beta} A(\alpha)\,B(\alpha)$.

b. Let $A = B$ be from an odd wave function so that $A(\alpha) = 0$. Show that

$$\frac{1}{N_-^2} = \frac{2}{\beta}\left(A'(\alpha)\right)^2.$$

c. Let $A = B$ be from an even wave function so that $A'(\alpha) = 0$. Show that

$$\frac{1}{N_+^2} = -\alpha\frac{2}{\beta} A^2(\alpha).$$

Here are some wise words for you to ponder...

"I think it is safe to say that no one understands quantum mechanics."
Richard Feynman

"Anyone who is not shocked by quantum mechanics has not fully understood it."
Niels Bohr

Quantum goofiness. You get used to it.

References and Suggested Reading

Here are books and journal articles I consulted while writing this book. Some inspired problems that I tried to solve with *SNB*, while others answered questions that came up along the way. Most but not all of the material in these references is at an appropriate level for undergraduate physics students.

There's one non-textbook in particular I'd like to recommend. My daughter suggested I read Bill Bryson's **A Short History of Nearly Everything** and as usual she was right. It is indeed a history of nearly everything, from the big bang to us today, told by a truly gifted writer. This is the book you could give to the non-science people in your life that don't quite understand what it is you do. You just might learn a thing or two about physics and the other sciences too. I sure did!

Books

[1] R.K. Adair, **The Physics of Baseball** (2nd ed.), HarperPerennial, New York, 1994.

[2] J.J. Brehm and W.J. Mullins, **Introduction to the Structure of Matter: A Course in Modern Physics**, Wiley & Sons, 1989.

[3] Bill Bryson, **A Short History of Nearly Everything**, Broadway Books, NY, 2003.

[4] S. Chandrasekhar, **Newton's Principia for the Common Reader**, Oxford, 1995.

[5] N. de Mestre, **The Mathematics of Projectiles in Sports**, Cambridge University Press, Cambridge, UK, 1990.

[6] A.P. French, **Newtonian Mechanics**, W.W. Norton, NY, 1971.

[7] A.P. French, **Special Relativity**, W.W. Norton, NY, 1968.

[8] J. Freund, **Special Relativity for Beginners**, World Scientific, Singapore, 2008.

[9] N.J. Giordano, **Computational Physics**, Prentice Hall, NJ, 1997.

[10] H. Goldstein, **Classical Mechanics** (2nd ed.), Addison-Wesley, Reading, MA, 1980.

[11] D.J. Griffiths, **Introduction to Electrodynamics** (3rd ed.), Prentice Hall, NJ, 1999.

[12] M. Okuda and D. Okuda, **The Star Trek Encyclopedia**, Pocket Books, NY, 1997.

[13] K. Rossberg, **Analytical Mechanics**, Wiley & Sons, NY, 1983.

[14] D.V. Schroeder, **An Introduction to Thermal Physics**, Addison-Wesley, SF, CA, 2000.

[15] K.R. Symon, **Mechanics** (3rd ed.), Addison-Wesley Longman, Reading, MA, 1971.

[16] P.T. Tam, **A Physicist's Guide to *Mathematica***, Academic Press, San Diego, CA, 1997.

[17] E.F. Taylor and J.A. Wheeler, **Exploring Black Holes**, Addison Wesley Longman, San Francisco, CA, 2000.

Journal Articles

[18] P.J. Brancazio, "*Trajectory of a Fly Ball*", The Physics Teacher **23** (1), 20–23 (1985).

[19] E.H. Buyco and F.E. Davis, "*Specfic Heat of Aluminum from Zero to Its Melting Point and Beyond*", Journal of Chemical and Engineering Data **15** (4), 518–523 (1970).

[20] M. Compte and T. Lachand-Roberts, "*Newton's problem of the body of minimal resistance under a single-impact assumption*", Calc. Var. PDE **12** (2001), 173–211.

[21] A.H. Compton, "*The Variation of the Specific Heat of Solids with Temperature*", Phys. Rev. **6**, 377–389 (1915).

[22] C.H. Edwards, "*Newton's Nose-Cone Problem*", The *Mathematica* Journal **7** (1), 64–71 (1997).

[23] J. Gallant, "*The Shape of the Eiffel Tower*", Am. J. Phys. **70** (2), 160–162 (2002).

[24] C. Hirata and D. Thiessen, "*The Period of* $\overrightarrow{F} = -k\,x^n\,\widehat{x}$ *Harmonic Motion*", The Physics Teacher **33** (9), 562–564 (1995).

[25] A. Jafari, "*Experimental Test of* $\overrightarrow{F} = -k\,x^n\,\widehat{x}$", The Physics Teacher **34** (4), 196 (1996).

[26] D.A. Martinez and S.L. Queiro, "*An Elementary Solution for a Difficult Motion Problem*", The Physics Teacher **41** (9), 518–520 (2003).

[27] L.F. Moody, "*Friction factors for pipe flow*", Transactions of the ASME **66** (8), 671–684 (1944).

[28] C.E. Mungan, "*Sliding on the Surface of a Rough Sphere*", The Physics Teacher **41** (6), 326–328 (2003).

[29] N.S. Osborne, H.F. Stimson, and D.C. Ginnings, "*Thermal properties of saturated water and steam*", J. Res. Nat. Bur. Stand. **23**, 261 (1939).

[30] H. Otto Petsch, "*On Circles and Central Forces*", Am. J. Phys. **39** (8), 969 (1971).

[31] M. Sawicki, "*Angular Speed of a Compact Disk*", The Physics Teacher **44** (6), 378–380 (2006)

[32] C.E. Synolakis and H. S. Badeer, "*On combining the Bernoulli and Poiseuille equation – A plea to authors of college physics texts*", Am. J. Phys. **57** (11), 1013–1019 (1989).

[33] N.B. Tufillaro, T.A. Abbott, and D.J. Griffiths, "*Swinging Atwood's Machine*", Am. J. Phys. **63** (2), 121–126 (1984).

[34] F. Y.-H. Wang, "*Relativistic orbits with computer algebra*", Am. J. Phys. **72** (8), 1040–1044 (2004).

[35] P. Weidman and I. Pinelis, "*Model equations for the Eiffel Tower profile*", C.R. Mecanique **332**, 571–584 (2004).

Other

[36] John Fisher at The Science Cafe (http://www.freewebs.com/sciencecafe/)

[37] National Institute of Standards and Technology (http://physics.nist.gov/)

Here it is...

At the end of the 19th century, physics consisted of what we now call Classical Physics: Newton's Laws of Motion and Universal Gravitation, Maxwell's Electrodynamics, and Thermodynamics. Each was hugely successful and highly regarded individually, but conflicts and inconsistencies arose when they were used together.

Calculations using Maxwell's Electrodynamics and Thermodynamics to predict the blackbody spectrum yielded results so bad they were called the Ultraviolet Catastrophe. Calculations using Newton's and Maxwell's theories suggested observers moving at the speed of light would see light waves that were not solutions to Maxwell's Equations.

During the 20th century, most of these issues were resolved and the result is what we now call Modern Physics. Planck's quantum hypothesis resolved the Ultraviolet Catastrophe and correctly predicted the blackbody spectrum. This eventually led to the formulation of Quantum Mechanics by Schrödinger and Heisenberg. Einstein solved the speed-of-light problem and Special Relativity reconciled Newton's Laws and Maxwell's Electrodynamics. When Newton's Universal Gravitation turned out to be inconsistent with Special Relativity, Einstein devised General Relativity to include gravity.

Dirac united Special Relativity and Quantum Mechanics for spin–$\frac{1}{2}$ particles into Relativistic Quantum Mechanics. Feynman and others completed that unification with Quantum Electrodynamics, a quantum field theory that explains the interaction of matter and electromagnetic radiation.

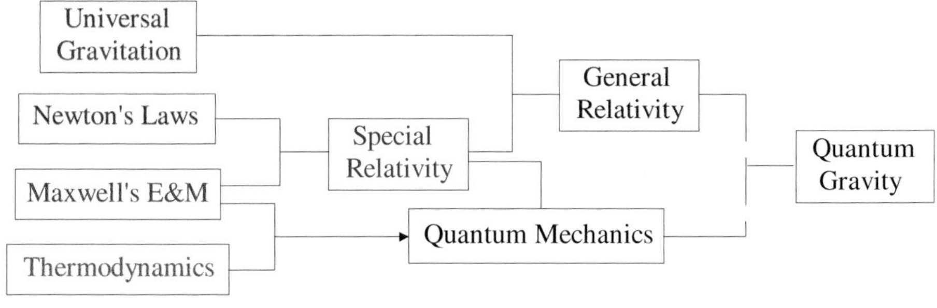

...your moment of zen?

At the start of the 21st century, physics has successful theories for all four known forces. General Relativity is the classical theory of gravity and Quantum Mechanics includes quantum field theories for electromagnetism and the two nuclear forces. Individually, GR and QM each have almost 100 years of successful, experimentally-confirmed predictions. But they produce absurd, often infinite results when they're combined.

This is one of the great unsolved problems of physics and it has implications for our understanding of black holes and the early universe. Someone will eventually solve this problem, develop a theory of Quantum Gravity, and enter the physics pantheon.

Maybe it will be you!

Index